BEYOND THE WILD BLUE

Also by Walter J. Boyne

BEYOND THE WILD BLUE

SECOND EDITION

A HISTORY OF THE UNITED STATES AIR FORCE, 1947–2007

by Walter J. Boyne
Colonel, USAF (Ret.)

Thomas Dunne Books
St. Martin's Press
New York

THOMAS DUNNE BOOKS.
An imprint of St. Martin's Press.

All photos, except where otherwise noted, are courtesy of the United States Air Force Association.

Design by Interrobang Design Studio

ISBN-13: 978-0-312-35811-2
ISBN-10: 0-312-35811-3

This book is respectfully dedicated to all the men and women—and their families—who have served the United States Air Force and its predecessor organizations.

CONTENTS

FOREWORD

by General Richard B. Myers, USAF (Ret.)
Former chairman, Joint Chiefs of Staff

Walter Boyne's second revised and updated book on the history of the U.S. Air Force is just as well written and informative as the first. And the timing of his effort parallels the precision of the Air Force as it performs its many missions around the globe. Published in commemoration of the sixtieth anniversary of the U.S. Air Force, this new addition deepens our understanding of the Air Force's history and the multiple actors and events that shaped it. This new book especially highlights how the events of 1997 to 2006 continued to shape our Air Force to meet the demands of the twenty-first century.

Debate and controversy over the potential of air power have always confronted airmen, but the events in the 1990s and early in the twenty-first century help airpower advocates immeasurably in successfully arguing their case. Boyne's book covers almost all of those airpower contributions, including the fundamentally important war in Kosovo, a war won by airpower. It's unlikely that air, land, or maritime power will achieve victory alone like this in the future. However, this one campaign underscored the prescience of our early airpower advocates.

Boyne's latest effort explains how our concept of airpower evolved after the end of the Cold War. There were many changes in Air Force doctrine, strategy, training and tactics, organization, and personnel during this era as the nation confronted the new realities of asymmetric warfare, terrorism, and conflicts fought halfway around the world. The author nicely explains the evolution from a garrison force to a mature expeditionary force that brings so much power and flexibility to the joint force commander. Along the way he points out how research and development, long an Air Force strong suit, continued to be a catalyst for success.

Most recently, from Afghanistan to Iraq, the U.S. Air Force has played an extremely critical role in fighting extremism. It adapted quickly to a new way of thinking, fighting, and supporting. This is something the Air Force has always done well throughout its history—it must be in the DNA. Who would have thought that air commanders would employ B-52s in a close air support role with precision weapons as they did in Afghanistan? The innovative use of airpower and Special Forces in major combat in Iraq will be studied for years to come. The strategic and tactical airlift over the last five years has been awe inspiring as the nation fought at vast distances from our shores. And the medevac mission, undervalued in peacetime, has done extraordinary things and saved countless lives in our battles in Afghanistan and Iraq. All this and much more are covered in rich detail, derived from authoritative sources.

Since the first edition, the Air Force has been given the executive agency for space. This critical capability continues to make vital contributions to our national security, yet remains our least understood. This book points out the fundamental but often overlooked fact that our very dependence upon our space assets renders us vulnerable and makes their defense ever more critical.

It was important to me that Boyne succeeded in capturing the powerful bond between the airmen who were privileged to operate the most technologically advanced pieces of equipment of their era and those who were equally devoted to supporting them—ensuring mission success. In this sense he evokes the spirit of today's twenty-first century Expeditionary Air Force. When the Air Force goes to war now, just as it did in the first Gulf War, the unit, not a random gathering of individuals, goes to war. The book also highlights the superb contributions of the Total Force. Employing our Active Duty, Air National Guard, and Air Force Reserve assets as one total force has raised overall U.S. Air Force mission effectiveness to a previously unimagined level.

From the perspective of a senior Air Force officer for two decades, Boyne also provides another powerful message. He reminds today's Air Force leaders and all airmen of the importance of understanding the history of their world-class institution. Indeed, his book is a case study for today's leadership on how to think creatively about strategy, operations, and tactics. He reminds us that the Air Force has a unique responsibility to constantly prepare its airmen to confront the unknown and that one of the best ways to do that is by understanding how earlier Air Force leaders wrestled with similar challenges.

I cannot overemphasize the importance of this latter contribution, in particular. As we endeavor today to adapt airpower to meet the vast challenges of the long war against terror, while maintaining unparalleled conventional capabilities, histories like this remind us that seemingly insolvable challenges *can* be resolved. They also remind us that the integrity, ingenuity, intelligence, and courage of America's airmen (and soldiers, sailors, and Marines) are the greatest assets our nation has—in peace and war.

PREFACE

Writing a history of the sixty glorious years of the United States Air Force is at once a privilege and torment. The privilege lies in reviewing the great accomplishments of the Air Force over its existence as it overcame all difficulties to become the most powerful and proficient force in the world. The torment is in being forced to leave out the tens of thousands of stories that should be told but for which space cannot be found in a single volume.

The best way to tell the story of the Air Force would be to do it in real time, recounting every adventure, every flight, every difficulty, every triumph. The poignant thing about this impossible dream is that the fascinating stories would not be confined to air-to-air combat or masterful leadership or overcoming daunting odds, but would also be found in the motor pools, administrative offices, schools, engine shops, laundries, guardhouses—everywhere. If it had been possible to do, I would have included the story of every unit of every type, past and present, in the Air Force, and with it the story of every member.

This was, of course, impossible, and I've been forced to leave out enormous masses of interesting information, confining my efforts to tracking what seemed to me to be the pivotal points of Air Force history. For every omission, I apologize.

When one contemplates the sixty-year history of the United States Air Force, many things come to mind. First among these are the predecessor services, for the roots of the USAF go back to August 1, 1907, when the U.S. Army Signal Corps established an Aeronautical Division to "take charge of all matters pertaining to military ballooning, air machines, and all kindred subjects." The branches that followed, including the Air Service, the Air Corps, and the Army Air Forces, distinguished themselves in peace and in war, with devoted personnel pushing the technical limits of the aircraft and weapons available to them.

When the United States Air Force was formed on September 18, 1947, few realized the degree of technological change that would occur over the next six decades. Three great transformations occurred during this period, as Dr. Benjamin Lambeth informs us. The first of these was the use of nuclear weapons as the primary offensive armament of bombers, fighters, and strategic missiles. The second was the post–Vietnam War introduction of new techniques such as stealth, precision-guided munitions, information superiority, and the integration of air and space assets. The third is ongoing today in achieving superiority with effective command, control, communications, and computer (C4) links and the fusion of intelligence, surveillance, and reconnaissance (ISR) information. The result has been the creation of a new portfolio of weapons, but even more important, new methods of waging war. Combined, these give the United States Air Force of today and of the future a degree of advantage never before contemplated, much less attained—one that goes far beyond the raw energy implicit in superpower status.

This new advantage was a long time in coming and represents the fruits of more than sixty years of a two-part investment in research and technology. The first part of the investment was the money the USAF has plowed into basic and applied research over the years. The second part was the dedicated people who have been brought into the R&D field and allowed to use their full talent.

The deterrent power of the Air Force in its early years depended upon a small stockpile of nuclear weapons and modernized versions of World War II bombers to deliver them. Over time, the number and variety of nuclear weapons grew, as did the types of aircraft and missiles that could carry them. The nuclear power the Strategic Air Command vested in its bombers and missiles created a Pax Americana, one that ultimately brought about victory in the Cold War as the inherent faults of the Communist system caused the Soviet Union to self-destruct.

Yet that nuclear power was unsuitable for use in limited wars, and it was necessary to redirect emphasis to conventional weapons delivered by conventional means. While these were effective, they were not decisive in either the Korean War or the Vietnam War, particularly because they had to be employed with artificial political restraints that vitiated the potential of air power.

The resources that had been poured into research and development began to bear exotic fruit toward the end of the Vietnam conflict, when the first precision-guided weapons proved their worth. In the years that followed, the Air Force saw that it was within the realm of possibility to create not only "smart weapons" but also the equipment essential to provide a new degree of battlefield awarness that would for the first time in history pierce the fog of war. Instead of just seeing "the other side of the hill," the United States Air Force could dominate the battlefield with information gathered from ground-, air-, and spaced-based systems. The preliminary degree of competence in this new concept of warfare was demonstrated in the Persian Gulf War, when the combination of such complementary elements as AWACS, J-STARS, GPS, LANTIRN, FLIR (acronyms that will be explained later), cruise missiles, and a variety of precision-guided munitions were used by highly skilled crews to provide an astounding double victory. The first victory was over the dispirited Iraqi commanders and troops, who quite literally did not understand what had happened to them. The second was over the Soviet Union, whose military leaders *did* understand what had happened, recognized that they could not compete, and elected to permit their country to dissolve. These weapons would be indispensable in Kosovo, Afghanistan, and again in Iraq in 2003, but they have not yet reached the degree of refinement necessary to exercise the same authority over the acts of radical Islamic terrorists.

The illumination of the battlefield with electronic equipment is but the first phase of the information war to be waged in the future. The equipment, strategy, and tactics of this new mode of warfare are highly classified and can only be hinted at. Against national opponents it will involve the selective destruction or manipulation of enemy command, control, communications, and intelligence, functions to secure a rapid, and relatively bloodless, victory. The Global War on Terror has revealed that human intelligence is still a vital part of warfare, one that sadly has been lacking to date in the struggle against radical Muslim terrorists.

Yet despite the revolution in warfare, the success in battle will inevitably devolve upon the same resource that it always has in the past: the people— men and women, officers, airmen, and civilians—who make up the United States Air Force. The USAF has been blessed with excellent leaders, and perhaps even more important, an intelligent and dedicated enlisted force that has always given more than could have been expected of it. In these regards, the USAF is better off than at any point in its history, with well-educated, well-motivated personnel at all levels who are devoted to their work, and who, equipped with the most advanced weapons in history, will be supremely competent to deal with the mental, moral, and physical challenges of the uncertain future.

ACKNOWLEDGMENTS

*I*t would be fair to say that millions of men and women have contributed to this history of the United States Air Force—those that served, and those who were family members—for without them, the book could not have been written.

But I've been the beneficiary of much more specific help from people I know and admire, and from strangers who, from a phone or letter request, went to great lengths to dig out the information or photos that I asked for.

First and foremost, I must express my appreciation and pay my respects to a group of leaders whose candor, insight, information, and review of my manuscript were as valuable to me as their service was to their country, which is to say, invaluable. My thanks go to General Joseph E. Ralston, USAF, Vice Chairman of the Joint Chiefs of Staff; General David C. Jones, USAF (Ret.), former Chief of Staff and Chairman of the Joint Chiefs of Staff; General Lew Allen, Jr., USAF (Ret.), former Chief of Staff; General Larry D. Welch, USAF (Ret.), former Chief of Staff; General Merrill A. McPeak, USAF (Ret.), former Chief of Staff; General Michael J. Dugan, USAF (Ret.), former Chief of Staff; General John Michael Loh, former Acting Chief of Staff and the first Commander of the Air Combat Command; General T. Michael Moseley, Chief of Staff; General John P. Jumper, former Chief of Staff; General Richard B. Myers, former Chairman of the Joint Chiefs of Staff; General Michael E. Ryan, former Chief of Staff; General Ronald R. Fogleman, former Chief of Staff; General Merrill McPeak, former Chief of Staff; General Michael Loh, former Commander of the Air Combat Command; General Richard E. Hawley, former Commander of the Air Combat Command; General Howard M. Estes III, Commander in Chief, North American Aerospace Defense Command, United States Space Command, and Air Force Space Command; General Duane Cassidy, Commander, U.S. Transportation Command; General Paul V. Hester, Commander, Pacific Air Forces; Lieutenant General David A. Deptula, Vice Commander, Pacific Air Forces; Major General Donald Shepperd, USAF (Ret.); Major General Ronald J. Bath, Director, Air Force Strategic Planning; Major General Charles D. Link, National Defense Review; Dr. Alan Gropman, Dr. Benjamin S. Lambeth; Dr. Philip S. Meilinger; Dr. Richard P. Hallion; Dr. Jacob Neufeld; Dr. George Watson; Dr. Perry Jameson; Dr. Herman S. Wolk; Dr. Daniel R. Mortensen; Colonel William J. DelGrego (and his "Skunks"); Colonel Steve "Hari" Carey; Lieutenant Colonel Paul Berg; Lieutenant Colonel Earle Thompson, Jr.; Captain Jon A. Mancuso;

Robyn S. Read; John L. Conway; Captain Kevin S. Mortensen. General Russell E. Dougherty, USAF (Ret.), former Commander in Chief of the Strategic Air Command; General Robert D. Russ, USAF (Ret.), former Commander, Tactical Air Command; General W. L. Creech, USAF (Ret.), former Commander, Tactical Air Command; General Bernard A. Schriever, USAF (Ret.), former Commander, Air Force Systems Command; General Henry Viccellio, Jr., USAF, Commander, Air Materiel Command; General Robert T. Marsh, USAF (Ret.), former Commander, Air Force Systems Command; General John A. Shaud, USAF (Ret.), former Commander in Chief, Air Training Command; General Joseph W. Ashy, USAF (Ret.), former Commander, Air Force Space Command, Commander in Chief of North American Aerospace Command and United States Space Command; Chief Master Sergeant of the Air Force, David J. Campanale, and former Chief Master Sergeant of the Air Force, James McKay. While I am sure that my manuscript, with all its limitations of space and my own limitations as a historian, did not completely satisfy all of these gentlemen, whose inside knowledge of the workings of the Air Force would fill many volumes, I do know that the book could not have been completed without their insight, information, and cooperation.

I am indebted to many others, including the members of the Office of Air Force History. There Dr. George Watson patiently tried to keep me on the correct track, while the inimitable Cargill Hall, with less patience and more invective, provided revealing comments and pointed the direction to more information. Other Air Force historians who were invaluable include Tommy R. Young, Headquarters, Air Mobility Command; Fred Johnsen, Edwards Flight Test Center; John D. Weber, Headquarters, Air Force Materiel Command, SrA. Trisha M. Morgan, Pacific Air Forces, and Thomas Manning, Air Training and Education Command.

The Public Affairs Officers and Information Officers at many levels were of inestimable value. It is remarkable to make a phone call with a specific question, and get not only the correct response, but floods of additional information as was so often the case. Many information and public affairs officers assisted, and I regret that I misplaced a whole list of them, which will surface promptly as soon as this is printed. However, I do have records of help from Lieutenant Colonel K. C. McClain, Air Force General Officer Matters Office, Captain Sam McNiel, Air Force Office of Public Affairs, Major Roger L. Overture, Air Force Institute of Technology; Master Sergeant Catherine A. Segal, Secretary of the Air Force, Public Affairs Office; Sandie Henry, Air Force Special Operations Command, and Amanda Gaylor, of the RAND Corporation. There were many more, some of whom were so efficient that they got me the information before I even had time to get their names.

At the Air Force Association, Pearlie Draughn did her usual inimitable job of ferreting out facts and photos. John Correll offered encouragement and advice, and permitted me to use the AFA Chronology. I also want to thank the Lockheed Martin Company, McDonnell Douglas, Long Beach, and the Boeing Company for assistance with photos and information. Robert F. Dorr, a distinguished author and photographer, was, as usual, very helpful. Henry Snelling did his usual superb job review of the manuscript. My daughter, Molly Boyne, not only typed out many long hours of audio tape from interviews, but compiled the bibliography, for both of which I'm grateful. Wally Meeks made his usual useful suggestions.

At St. Martin's, I've had the good fortune of having the distinguished Tom Dunne as editor, as well as his colleagues Sean Desmond, Benjamin Hart, Jeremy Katz, and Jason Rekulak. As always, I appreciate the efforts of my excellent agent and true gentleman, Jacques de Spoelberch. And I am honored to have the former chairman of the Joint Chiefs of Staff, General Richard B. Myers, USAF (Ret.), write the Foreword to this book.

1

THE MAN OF INFLUENCE

*I*n just sixty years the United States Air Force has grown from a disorganized giant, mired in the jumble of too rapid demobilization after World War II, to the most influential military service in the world today. In the process it has achieved triumphant successes that exceeded even the promise of its evocative song "The Wild Blue Yonder" while overcoming haunting failures of concept, equipment, and personnel.

Fortunately for the United States and the world, the successes have vastly outnumbered the failures in both number and degree. Until September 11, 2001, the Air Force has been a significant, if not the principal, factor in the remarkable victories of both the Gulf War and the Cold War. Every leader of the United States Air Force, from Secretary to Chief of Staff to squadron commander, would be quick to note that these triumphs were won in concert with the Army and the Navy. No matter how hotly the three services contend for roles and missions, appropriations, media attention, and public support, the serious bickering stops when it comes to battle. The concept of joint operations, so successful in World War II, was not always observed in the intervening years, but was demonstrated admirably in the Gulf War, operations in the Balkins, in Afghanistan, and in Operation Iraqi Freedom. Nonetheless, while much of what will be said applies equally to its sister services, this book will focus on the United States Air Force.

The Air Force achieved its great successes despite a number of formidable obstacles, foreign and domestic. The first and most immediate of these was the talented, focused, and effective air forces of the Soviet Union, which developed excellent equipment in massive numbers along with the strategy and tactics to use it. The USSR shared its capabilities bountifully with its satellite states, some of which were destined to become fierce opponents of the United States. The threat of the Soviet Union was real, massive, and seemingly never-ending. Soviet nuclear missile capability, exaggerated at first, soon grew to immense proportions. And while the Soviet Union is no more, its missile force, now divided among three of the survivor states, Russia, Ukraine, and Kazakhstan, not only remains but is perhaps more threatening because its control is far less certain.

There were less obvious but equally important hazards at home. The first of these was the continual requirement to cope not only with the vagaries of the Congressional budgeting process but also with the growing restrictions inherent in oversight—a kindly term for micromanagement by both Congress and the Executive Branch. The second was the telling loss of public support, almost two decades in duration, resulting from distaste for the war in Vietnam. For the first time in its history, members of the United States Air Force found themselves publicly vilified for doing what they had been ordered to do. And while the prestige of the USAF has been largely restored today, there lurks a reservoir of antimilitary sentiment still to be found in the media, in academia, and, surprisingly, in the government.

Most remarkably, even while the Air Force struggled to overcome these varied challenges, it created and maintained a unique ability to plan far into the future. The Air Force's reliance on technology was perhaps inherent in the very science of flight itself. More than the Army or the Navy, and more than the services of other nations, including the Soviet Union, the Air Force put its faith in advanced technologies. Fostered from the very start by General of the Air Force Henry H. "Hap" Arnold, and encouraged by succeeding Chiefs of Staff, the Air Force not only made the funds available for research, it granted credibility and opportunity to the military and civilian personnel who pursued technology as a career. The funding was not always constant, for wartime operational considerations invariably drained funds away from research, but the basic idea that research and development was the essential element for the continued success of the Air Force always remained.

Despite every effort to avoid the characteristics and the operating methods of a giant bureaucracy, the very size and age of the Air Force has made it one. Prescience is not normally associated with a huge organization, yet the Air Force has over the years managed to endow its leadership and its operating forces with the ability to anticipate future requirements for equipment and training. The phenomenal result has been that the Air Force, operating under the budget constraints imposed upon all the services, has managed all current crises while doing the necessary research and development to accelerate the technologies necessary for future conflicts.

For forty years the principal task of the United States Air Force was to deter offensive action by the Soviet Union. The USAF accomplished this in part by combining the experience and techniques gained in the employment of air power in World War II with an ever increasing arsenal of atomic weapons, including the intercontinental ballistic missile. The rest of the task was achieved when the Air Force, drawn reluctantly and against its instincts into the space age, responded by capitalizing on the opportunity to create an amazing array of new technologies.

At the same time, the USAF had to respond to other challenges. Some of these were of the monumental size and scope of the Korean and Vietnamese wars, while some were less threatening, like the invasions of Grenada and Panama. In addition, the USAF had to undertake disaster relief at home and abroad, as well as show the flag and project power. And all the while, it had to deal with major social issues ranging from the integration of black personnel into the service, to overcoming civilian distaste for the military during and after the Vietnam War period, to providing equal opportunity for women and minorities.

Despite the multifaceted nature of the Air Force's tasks, it was successful in almost all of them, all the while containing the Soviet Union and making the most vital contributions to winning the Cold War. It was next thrust into a peacekeeping role in the Balkans. Then, as the global war on terror became manifest, it became actively engaged in new operations in Afghanistan and Iraq.

In retrospect, the years of the Cold War have a monolithic quality, as if there had been an unchanging confrontation with the Soviet Union which the Air Force steadfastly met with unchanging means. Yet it was not so, for the nature of the threat changed almost annually, forcing a corresponding change in the Air Force's response. In the very early years, at the time of the Berlin Blockade, the Air Force's response was a hollow one, brandishing a nearly empty nuclear arsenal at a gigantic array of Soviet forces. As the years passed, the Soviet Union, through its surrogates, challenged the United States all around the world, in each instance with a minimal involvement of its own troops. Thus it fought the Korean War with North Korean and Chinese forces, supplemented by Soviet equipment, training, and limited personnel. It supported the North Vietnamese in a similar economic manner, letting another country bleed for its own purposes. The same pattern prevailed in the Middle East, in Africa, and ultimately as close as Cuba. With the Soviet Union tugging at the seams of countries all around the world, the U.S. policy of containment, begun by President Truman, was an expensive one.

Yet it was ultimately successful, despite the lack of a decision in Korea and the loss of the war in Vietnam. Over the years, the United States Air Force, both the benefactor and the beneficiary of the American system of free enterprise, was able to build air and missile forces that kept the Soviet Union within the general sphere of influence allotted to it at Yalta.

The Soviet Union was not only contained, it was strained, its military budget consuming it economically and technically. The Soviet advances in military equipment and in space exploration were obtained by investments that matched and often exceeded those of the United States, particularly as a percentage of gross national product. The tremendous expenditures were at the expense of a rational expansion of the USSR's civilian economy. The productive capacity of the Soviet Union, channeled so single-mindedly into its military efforts (for its space program was primarily for military purposes), was unable to develop an industrial base with a technology and a market structure comparable to those of its old Western enemies or of the emerging nations of Asia. The USSR's atrophied civilian industrial base made its military burden increasingly difficult to bear by 1980, and impossible to bear a decade later.

In that critical ten-year period, three separate undertakings by the United States spelled the downfall of the Soviet Union. The first was the buildup of American arms that began in 1980 and reversed the decline in strength that had occurred under the Carter administration. The Soviet economy, already almost exhausted, was strained beyond endurance by the requirement to match the American buildup.

The second undertaking was the dazzling if ultimately unfulfilled prospects of President Ronald Reagan's "Star Wars" program. The grandiose project was obviously beyond the capacity of the Soviet Union to match; the risk that the United States might succeed was too much for Soviet leaders to contemplate.

The third, and conclusive, element was the overwhelming success of our weapons in the Persian Gulf War. The invulnerability of the stealth fighter and the incredible military—and public relations—effect of precision-guided munitions completely disheartened the political and military leaders of the Soviet Union. With their economy imploding under the strain of seventy-four years of corruption and inefficiency, the Soviet leaders were finally compelled to admit that their system had failed, and to abandon—at least temporarily—their historic quest for world domination. Just as Mussolini's corrupt Fascism withered and died almost overnight, so did the Soviet Union and its single political component, the Communist Party, swiftly dissolve into a nightmare of confusion and recrimination.

The Soviet Union, suddenly exposed as a gigantic empty rust belt of industrial and political folly, simply shut down, leaving its people to its own devices, for better or for worse. Its huge military forces, overwhelming in both their conventional and nuclear might, almost instantaneously went from being a threat to the very existence of the world to embarrassing centers of poverty, unable to feed, equip, or clothe their recruits, sometimes unable even to pay their electric bills.

Yet winning the Cold War was only part of the United States Air Force's task during the first fifty years of its existence. Each decade presented a new challenge that it had to handle as a "part-time" job, subsidiary to the principal task of nuclear deterrence. Some of the challenges were internal: adapting to social change, meeting equipment deficiencies, trying each year to do more with less. Other challenges were external, from the sobering experience in Korea through the demoralizing agony of Vietnam to the exhilaration of winning the Persian Gulf War. The exertions of the global war on terror after 9/11 were compounded by additional requirements for compassionate relief efforts and the increasing interest in securing our borders from illegal immigration.

Each challenge was overcome by the men and women of the Air Force, who were simultaneously accomplishing another remarkable feat. Even as they endured the rigors and uncertainties of service life, with its frequent moves, relatively low pay, and often disagreeable jobs, the men and women of the Air Force moved into the mainstream of the American community, and indeed became the United States in microcosm. The old concept of a military base being apart from the community, a self-sufficient entity with its own standards and mores, faded away. USAF personnel increasingly broke away from the frontier outpost outlook that had characterized the military for so many years and instead became active members of their communities, owning homes, working second jobs, sending their children to school, paying taxes, and generally becoming indistinguishable from their civilian neighbors.

One of the most remarkable aspects of this transformation from a parochial group with an essentially garrison mentality into a fundamental part of American society is that it has been rarely perceived and little remarked upon, even by members of the Air Force. People both within and without the Air Force still tend to think of it as a separate social entity, as distinct from being a separate business entity. The fact is that the composition of the Air Force population is essentially identical to the composition of the American populace as a whole, and as such reflects the trends, the biases, the problems, and the potential of that populace.

One of the most interesting questions about the Air Force is how it managed to foresee its equipment and weapons needs as much one or two decades in advance. The successes obtained in World War II might be attributed to a specialized leadership, trained for twenty years with but a single goal, that of establishing air superiority with conventional weapons of the times. The postwar successes, each one perhaps as important as success in battle had been in World War II, resulted from the quick and precise execution of plans that would have been deemed grandiose if they had not succeeded. Among the most remarkable of these for the grandeur of their conception, planning, and execution are the deployment of not one but four intercontinental ballistic missile systems, the establishment of a comprehensive continental radar defense, and the systematic exploitation of the possibilities of space for war and other military purposes.

In the meantime, besides leading the way to victory in the Cold War, the Air Force has, almost off the back of its hand, fought four major and three minor wars, while leading the nation in the process of integrating minorities and women into the service. During the same interval, it has transitioned from a primarily nuclear strike force pitted against a superpower into one capable of responding to regional conflicts with conventional arms, while still maintaining a decisive nuclear capacity.

The answers to the question of the source of the Air Force's general success in operation and in anticipation will be revealed in the following chapters. In essence, the Air Force's success derived from having the right leaders at the right time at the officer level and, perhaps surprisingly but even more importantly, at the noncommissioned officer level. Obtaining those leaders derived from the Air Force's intrinsic ability to attract high-caliber personalities to serve, and from a carefully cultivated culture that allows persons of talent to reach the top. The relationship between officers, noncommissioned officers, and enlisted personnel in the USAF is unique, and stems from a tradition created in the old days when commissioned pilots realized that their lives depended upon noncommissioned crew chiefs—and treated them accordingly. The sense of mutual respect and mutual importance is pervasive in the Air Force today, and is in many ways responsible for the success of the organization.

This is not to say the Air Force has solved all the problems of democracy and is truly egalitarian, for it is not. Nevertheless, the nature of the Air Force organization has always permitted the truly talented to rise to the top, regardless of connections, schooling, or appearance. For the past thirty-five years it has been increasingly easier for truly talented persons to rise to the top regardless of race, and for the last twenty years regardless of sex. The Air Force has always led the nation, including the other services, in the trend toward true equal opportunity, and it has benefited extraordinarily from the practice.

It is said that the Israeli Air Force was born in battle, coming into being as it did in the 1947 struggle for independence. It is not stretching a point to say that the United States Air Force was born in battles, and has remained in battle of one sort or another for its entire existence. The concept of an independent air force was first articulated—prematurely—by Billy Mitchell and others during the 1920s. It was nurtured during World War II, when leaders like General Henry H. Arnold and General Carl "Tooey" Spaatz sometimes conformed operational considerations to the preparation for postwar independence. It was sustained in the demobilization collapse of our military forces after V-J day, and survived the intransigent opposition of the United States Navy. But independence merely meant a new set of wars. The simplest to deal with were actual conflicts, from Korea to Operation Iraqi Freedom and beyond, where the enemy was known and the action required was military. Much more complex were budgetary battles, public relations battles with the Navy, and internal strains as the Air Force bureaucracy grew over time.

Fortunately for the United States, the hand of Providence and good leadership prevailed, and the USAF managed to prevail in each of its battles, learning in the process, and directing its efforts ever to the future.

Many mistakes were made, some trivial, some of immense consequence. Yet the end result of the Air Force's effort was the establishment of a Pax Americana that deterred World War III and proved to be the carborundum wheel upon which the Soviet Union ground itself to extinction.

The Air Force's achievement was a compound of many elements—people, leadership, equipment, and more. Yet the very basis for its success lay in the commitment it made—and continues to make—to the necessary research and development effort to create the advanced technologies that defined it as a superpower.

Coincidentally, and fortunately, the advances in Air Force capability have characteristics that match two unique demands that have since Vietnam come to be made by the American public. The first of these demands is that we must fight our wars with a minimum number of casualties to our forces. America wants no more Vietnams where our troops are forced to fight and die in unconscionable numbers. The second of these demands is unusual in history, for it is that we must also win our wars with a minimum number of casualties inflicted on the enemy.

These requirements, noble for any nation, may be a reflexive reaction to the years of the Cold War, when the strategies of the Soviet Union and the United States had the appropriate acronym MAD, for mutual assured destruction. Under this doctrine of reciprocal deterrence, it was a given that either side had the ability to inflict unacceptable damage on the other. If the strategy failed, casualties on each side would have been in the tens or hundreds of millions, depending upon how long the madness lasted. Long before the end of the Cold War, it was recognized that the success of the mutual assured destruction concept might prevent global nuclear conflict but would have no effect upon limited wars that were not so threatening as to require the use of nuclear weapons. The war in Vietnam demonstrated this painfully to the United States, which could not find until December 1972 the will to use air power in its fullest measure against North Vietnam. In the preceding eight years of conflict, it had endured—and inflicted—heavy casualties. The public's new requirements were revealed in the victory in the Gulf, where the United Nations casualties, as minimal as they were given the scale of operations, were rightfully resented. Remarkably, there was a similar resentment of the Iraqi casualties, particularly among civilians. Other incidents—friendly-fire losses in the Persian Gulf, the tragic killings in Somalia, the long list of casualties in Iraq—all confirm the American public's attitude that wars must be fought with a minimum of bloodshed on both sides.

The demand is without precedent; no military service in history has ever had placed upon it the requirement for victory at minimum cost to both sides. Fortunately, the United States Air Force is, for the first time, in a position to meet the requirement, thanks to a program of technological refinement that extends back more than six decades, and that can be traced to a single driving personality, that of General of the Air Force Henry H. Arnold.

"HAP"

One man, Henry Harley Arnold, did the most to shape the image of the modern United States Air Force. His patient political spadework established trust with the United States Army brass and secured the essential patronage of

General George C. Marshall, the Chief of Staff. Among his colleagues, Arnold's career inspired deep friendship—and open hatred, for there were those who disagreed with his views and thought him too political and much too quick to compromise.

Yet it was Arnold's flexibility that laid the groundwork for the modern United States Air Force, and few of his hard-driven subordinates would have agreed that he was too quick to compromise. His focused energy, unrelenting, omnipresent, helped them ensure that the performance of the United States Army Air Forces, locked in the greatest war in history, exceeded all expectations. And, perhaps most important of all, Arnold was willing to depart from his own strengths and past experience to embark the new independent air force upon a course of constant technological change. More of a verbalist than a technician, he did not fully comprehend all of the change he sought, but he knew intuitively that such change was essential. He was Hap Arnold to his friends, the public, and the press, and the free world owes him much.

Arnold rose from an indifferent West Point cadet career to become the first and only five-star General of the Air Force. His roller-coaster rise to the top would see him pioneer, then abandon flying; he would become the youngest colonel in the United States Army by 1917, only to be dropped back to the rank of captain in 1920, in the helter-skelter post–World War I curtailment of the armed forces. By hard work, charm, and no little guile; he rose to the rank of major general, becoming Chief of the Air Corps on September 29, 1938. At that moment, Arnold commanded an Air Corps which numbered about 23,000 officers and men, only 2,500 of whom were rated pilots, and about 1,200 mostly obsolete combat aircraft. With thirty-five years of service behind him and retirement looming, Arnold must have felt both satisfaction in his own career and tremendous frustration at the desperately poor condition of the air force he commanded, which lagged behind all the major world powers in the numbers and quality of its aircraft.

Yet the war was imminent, and with it came changes beyond his imagination. Only seven years later, long past the time he would nominally have retired, he wore five stars as his idol General John J. Pershing had done, and commanded more than 2,400,000 troops and an armada of 70,000 planes. More important, he had completely vindicated airpower prophet Billy Mitchell's claims as he closely supervised the air victories in the European and Pacific theaters.

He invested his whole person—including his health—in this effort, and long before V-J day had taken the necessary steps to fulfill his dream of an independent air force.

Arnold was assisted by trusted colleagues, handpicked men who had sacrificed themselves, as he had, to serve in the Air Corps at a time when pay was minimal, prestige nonexistent, and danger ever present, for the fatality rate for fliers was high. The assistance of general officers like Carl "Tooey" Spaatz, Ira C. Eaker, Joseph J. McNarney, Lawrence S. Kuter, George C. Kenney, Lauris Norstad, and Hoyt Vandenberg and of civilians like Assistant Secretary for

War for Air Robert A. Lovett and his successor, Stuart Symington, was invaluable, but it was Arnold who masterminded the effort with political skill and exquisite timing.

Hap was so acutely aware of the importance of timing that early in the war he assigned Major General McNarney to head a committee whose sole purpose was to suppress the tide rising for an independent air force until after the war. Incidents like these cost Arnold in the eyes of fellow officers who wanted an independent air force immediately, but Arnold gladly bartered full alliance with the Army during the war for the Army's postwar support for independence.

His subtle preparations were to have enormous and lasting effect upon the character of the USAF, for it was Arnold—a visionary pragmatist rather than a scientist—who set the Air Force firmly upon the path of research and development. And in a more mysterious, less definable way, it was Arnold who established the culture that permitted the early identification of essential future leaders. From Arnold's era to the present, with very few exceptions, the Air Force has managed to elicit from its ranks the right leaders at the right time, anticipating the ever-changing demands they would be required to meet. Spurred on to do so by Marshall, Arnold began this selection process himself, choosing young officers early in the war for responsible positions and then, as the war drew to its victorious close, taking steps to see that these younger men were given preference over more senior officers in the postwar air force. As salubrious as this was for the nascent United States Air Force, it was the last straw for some senior officers, many of whom had never been in Arnold's camp. They, like Arnold, had endured the dreadful doldrums of the prewar era, with its low pay and lack of promotions. When war came, they were certain they had executed their new wartime responsibilities with admirable efficiency, for Arnold surely would have fired them had they not. But just as the dream of an independent air force was about to be realized, many of them were passed over, given dead-end assignments, or even politely asked to retire. This saddened Arnold, but it was a price he was prepared to pay and it was a tradition in the making.

Arnold's discreet, comprehensive preparations to ensure the emergence of an independent air force were unlike the damn-the-torpedoes working style he used in running the war effort. As such, they indicate a depth and breadth to Arnold's intellect that belie the traditional picture of the reckless airman, intent on the mission and heedless of politics. His plan was a masterful piece of work on its own; that Arnold executed it while building the world's most powerful air force and leading it to victory in World War II is little short of incredible. The smiling, silver-haired general used tact, diplomacy, and well-thought-out public relations in his effort, all talents sneered at by his detractors. Yet he won the trust, affection, and personal commitment of giants like President Franklin Delano Roosevelt and General Marshall not by these traits, but by delivering the goods. Arnold's United States Army Air Forces performed in an outstanding manner, and he personally saw to it that its loyalty to the Army and its dedication to winning the war were unwavering. He never allowed his indirect lobbying efforts for an independent air force to interfere with the resolute prosecution of the war. It was a winning strategy, for by 1943 he had created the climate within the Army, the Congress, and the executive branch that would after the war permit him to realize his dream.

The nickname "Hap" suited his public persona rather better than his supervisory style. He was extremely photogenic, and his genial, confident manner enabled him to get along well socially. A big smile and warm handshake went a long way in the promotion of airpower, and he used them continuously on the prodigious wartime travel circuit that impacted so heavily on his health.

Working for him was something else again. Arnold believed he knew what could be accomplished with air power, and he threw himself into the work of overcoming the manifold deficiencies of the Army Air Forces in the early years of World War II. Enormously energetic, he worked long hours every day, never sparing himself or anyone who worked for him. In 1940 his task seemed impossible: the Air Corps lacked everything, including planes, pilots, ground crew, and bases. There was no training base—less than 1,000 pilots per year were being trained, when the requirement was for 100,000 annually. A modern air force required thousands of skills, ranging from cooks and bakers to navigators and radar observers, but the cadre of regulars to train them simply didn't exist. The industrial base was equally inadequate, despite the infusion of help from prewar purchases of aircraft by the Allies. American aircraft designers were still building to the almost naive specifications of the prewar HIAD—Handbook of Instructions for Aircraft Designers—and blithely ignoring the lessons being taught daily in European skies about the value of armor, self-sealing tanks, and other combat necessities. He overcame each of these difficulties, sometimes by personal intervention, as when he personally saw to it that the vast civilian pilot training scheme was established, sometimes by motivation, as when he induced manufacturers to risk money without a contract in hand, and sometimes by fear, as when he would announce a forthcoming visit to a delinquent subordinate.

Hap was too impatient to be a good administrator, and far too prone to parcel out assignments to the first person—or sometimes, the first few persons—he saw, regardless of their authority, expertise, or ability. He wanted instant results and accepted no excuses for delay or failure. It was remarked of him, in rueful jest, that he was "the most even-tempered man in the Army—always angry."

Most important for the American war effort, Arnold possessed a signal quality that cost him dearly each time he exercised it. Unlike many commanders in foreign air forces, including, oddly enough, the German Luftwaffe, he did not let personal friendship interfere with his evaluation of his subordinates. The air corps in which he grew up was so very small that when World War II broke out, not only his friends and colleagues but his rivals and opponents won top positions. Yet no matter who the person was, close friend or rival, Arnold was consistent. He was sparing in praise, but if the person failed to perform to Arnold's expectations, he was relieved without mercy. When he decided that World War I ace Brigadier General Frank O'Driscoll "Monk" Hunter had failed to use his VIII Fighter Command effectively, Arnold insisted that he be replaced. Even his longtime close friend and his coauthor in prewar books on aviation Major General Ira C. Eaker was removed from his Eighth Air Force command position when Arnold felt that he was not extracting the maximum potential from his bomber force. This came as close to breaking Eaker's spirit as anything could, for he had created the Eighth and guided it to the point where it was almost ready to accomplish its task. The incident put a further strain on Arnold's own already troubled heart, adding to the burden of his twelve-hour days, and seven-day weeks. Nonetheless, he met each challenge with the fierce, professional resolution that became the model for Air Force general officer attitudes.

IMPATIENCE EARNED

Arnold was entitled to be impatient, having paid his dues and more in his long service. He had earned his wings in 1911 with Orville and Wilbur, at the Wright brothers' flying school at Simms Station in Dayton, Ohio. Besides the luster of the association with the Wrights, the school had little to offer, for while its biplanes could go higher than the interurban trolley car that daily sped past the airfield at 45 miles per hour, they could not go as fast.

Instruction began on a crude sawhorse-mounted "simulator" in which the student pilot learned how the controls moved, followed by a series of flights lasting from five to fourteen minutes, depending upon wind and weather and the instructor's nerves. After twenty-eight lessons and a total of three hours and forty-eight minutes in the air, Arnold was considered qualified as a pilot, and thus automatically able to instruct others. It was a time when flying was learned by doing. Arnold initiated the use of goggles after being hit in the eye by a bug; a colleague devised the seat belt after almost being thrown from the airplane.

Arnold knew that flying was dangerous from the start; several Wright exhibition pilots had been killed, and Orville Wright had pointed out to him a man who waited with a wagon each day outside their flying field at Simms Station—the local undertaker in Dayton. The raw statistics were grimmer even than the waiting undertaker. Between 1909 and 1913, there were a total of twenty-four qualified Army aviators; of these eleven were killed while still in training and seven were killed subsequently in crashes.

Assigned to the Signal Corps Aviation School at College Park, Maryland, Arnold, with his almost four hours of flying time, began teaching others. He liked flying. He set several records and won the first MacKay Trophy ever awarded with an unprecedented reconnaissance flight on a triangular course from College Park to Washington Barracks, D.C., then to Fort Myer, Virginia, and back to College Park. The flight exhausted him, for controlling a Wright biplane was a demanding mixture of muscle power and apprehension.

As his knowledge grew, he recognized the danger implicit not in merely being airborne, but in flying the Wright and Curtiss pusher aircraft of the period, which had the engine mounted directly behind the pilot. These aircraft, slow and prone to stall, were literal death traps in a crash, when the engine would tear loose and crush the pilot. (Pushers were banned from the U.S. Army Air Service in 1914, but not until after too many deaths.)

A number of his friends perished in a series of deadly mishaps that followed the same pattern: a stall and a fatal dive directly into the ground. Arnold was at Fort Riley, Kansas, in November 1912, earning a munificent $124 per month and conducting experiments in which he corrected artillery fire by means of a primitive one-way radio. He was flying a Wright Model C with Lieutenant A. L. P. Sands of the Field Artillery perched on the wing alongside him, taking photographs. The Model C differed only slightly in appearance from the original Wright Flyer of 1903, but it was slightly larger, with a 55-horsepower six-cylinder engine and twin chain-driven pusher propellers. The crew of two sat upright on tiny slab wood seats on the wing. Like the earlier Model B, the Model C had an inherent design flaw in the relationship of its center of gravity to the propellers' thrust line. When the aircraft was gliding at low speeds, a sudden application of power would cause the nose to pitch down—exactly the opposite of what a pilot would expect.

Crossing a line of troops at an altitude of about 400 feet, Arnold's biplane suddenly spun in a 360-degree circle and plunged toward the ground. Stall-recovery techniques were still unknown—it was widely believed that Wright's aircraft could not be recovered from a stall—and Arnold was certain that he would be killed. Somehow he managed to pull out at the last instant and land, probably having inadvertently done the correct thing—pushing forward to break the stall and gain speed before pulling back to raise the nose. He was shaken to the core, the photographer still blissfully unaware that anything was wrong.

Arnold assessed the situation and realized that the aircraft he was being asked to fly were intrinsically and unreasonably dangerous. Summoning all of his moral courage, aware that his action would be criticized, he admitted to a fear of flying and requested reassignment. It was granted, and he did not return to aviation until 1916, and then at the specific request of Major William "Billy" Mitchell, Chief of the Signal Corps' Aviation Section. Mitchell was already promoting the concept of an independent air force; in less than ten years his flamboyant demands for such a service would bring him before a court-martial board.

When Arnold returned to flying in 1916, he did so with his customary verve; the airplanes, while still dangerously primitive, were not the death traps that the pushers had been, and his love of flight returned. When war came he was promoted to the rank of colonel at the age of thirty-one, the youngest in the Army. Despite his best efforts, he did not see combat, and in the pell-mell postwar reduction in the size of the armed forces, he reverted to the rank of captain in 1920.

Arnold's hallmark of moral courage was put into play on Mitchell's behalf in 1925, when he stoutly defended the outspoken general at his court-martial, despite the fact that Mitchell was certain to be found guilty of the charges against him. Now a major, Arnold spoke out for Mitchell's concept of air power, although he knew that doing so would damage his career. He was joined in Mitchell's defense by many of the men to whom he would give important assignments during World War II. When the trial was over, Arnold was sent into exile, commanding a small detachment at the Fort Riley cavalry school, the scene of his near-crash thirteen years before.

The Mitchell trial divided the Air Corps into two groups—those who had defended Mitchell and those who had not. The former group suffered initially, but would band together to endure and then prevail. The trial matured Arnold politically. He recognized that there was not yet sufficient military, public, or Congressional support for the independent air force Mitchell had advocated, and he concentrated on winning friends at every level to acquire that support. In the process he alienated many of his contemporaries, including a man who was his chief rival for the top position in the Air Corps, Frank M. Andrews, of whom more later.

Exile at Fort Riley should have meant the end of Arnold's career, particularly because he continued to lobby covertly for Mitchell's ideas. Although discouraged about his progress—he would remain a major from 1920 to 1931—he turned down the offer of several potentially rewarding civilian jobs in aviation, not wanting to quit while under fire. Even before Mitchell's trial he had been instrumental in forming the nucleus of what became Pan American Airways, and had refused the opportunity to become president of the new line.

Mitchell had been a visionary, but he had imputed to an independent air force a strength that was far beyond the reach of contemporary technology. The aircraft of Mitchell's day lacked every necessary quality to be a war-winning weapon, not possessing the speed, range, bomb load, or defensive capability required for combat. Of equal importance was the totally inadequate defense budget of the time, which would never have permitted an air force to exist as a separate service with its own bases, personnel, and equipment. Arnold believed in air power, but Mitchell's trial had taught him to use politics and low-key diplomacy rather than confrontational methods.

Arnold pursued his career diligently, becoming a true "comeback kid" of the Air Corps. He became Commanding Officer of March Field in 1931, where he later worked both ends of the social spectrum, setting up Civilian Conservation Corps camps and at the same time drawing upon the Hollywood film community for support. In July and August 1934, he personally led a flight of ten Martin B-10 bombers in a record round-trip flight from Washington D.C. to Fairbanks, Alaska, receiving his second McKay Trophy for the achievement.

In February 1935, already identified as "bomber man," he received an unprecedented peacetime two-jump increase in grade to become a Brigadier General in Command of the 1st Wing of General Headquarters Air Force at March Field. He then moved on to become Assistant to the Chief of the Air Corps, and in March 1938 was appointed Chief of Air Corps as major general. On June 30, 1941, he became Chief of the Army Air Forces, receiving his third star in December of that year.

Arnold did not spare himself leading the Army Air Forces in World War II, working long hours and sacrificing his health; in the process, he became convinced that the future of an independent Air Force was totally bound with advanced technology.

He continued these tactics despite all the pressures of World War II, never losing sight of his goal. It should be remarked here that the impetus behind this thrust was not careerism, nor the desire for power. Arnold and his airmen recognized that the ultimate effect of air power could never be attained if air forces continued to be controlled by ground commanders. Over the next half century, this would prove to be a lesson strangely hard to learn and extremely easy to forget.

ARNOLD AND RESEARCH AND DEVELOPMENT

One could scarcely imagine two less likely friends than Hap Arnold, the tough, skillful, but not particularly technically inclined airman, and Dr. Theodor von Karman, charming, gentle, professorial, and recognized as the leading research aerodynamicist in the United States. Arnold lived for the most part the austere, almost monastic life of a military leader at war, while von Karman enjoyed academia and had a legendary love of parties where pretty women were abundant. In those days of lesser political correctness, it was said, perhaps apocryphally, that von Karman's progress through a party gathering could be marked by the jumps and surprised squeals of the women he was passing by.

Yet this unlikely combination of stern aviator and convivial theoretician was to have a profoundly beneficial effect upon the Army Air Forces in World War II, and upon the history and culture of the United States Air Force.

An almost prototypical Hungarian, von Karman studied at the Budapest Royal Polytechnic Institute, gaining honors with a degree in mechanical engineering. He went on to pursue his doctorate at Göttingen University, studying under the great Professor Ludwig Prandtl, who would become his colleague, peer, and rival. During World War I he first served as an artillery officer, and then the Austrian Air Service put his knowledge to use in aviation research. He experimented with things as diverse as helicopters (including one powered by four rotary engines driving two counterrotating propellers!), gun synchronizers, and self-sealing fuel tanks. After the war, von Karman became director of the Aachen Aeronautics Institute, where his work earned him a worldwide reputation. There he made contributions to glider design and to the design of the Zeppelin LZ-126, which was awarded to the United States as war reparations and became the ZR-3, the *Los Angeles*—the longest-lived and most successful of all U.S. Navy dirigibles. And, oddly enough, it was this experience with lighter-than-air craft that would be the key to his first interaction with Arnold.

Professor Robert A. Millikan, who headed Cal Tech in Pasadena, was an expert in recruiting eminent scientists. He courted von Karman for several years, and finally managed to persuade him to accept the role of director of the Guggenheim Aeronautical Laboratory at the California Institute of Technology (GALCIT) in Pasadena. Von Karman came to the United States in 1929, in part because of the nascent threat of the Nazis and in part because of the facilities that were made available to him both at Cal Tech and at a lighter-than-air laboratory to be built in Akron. It was an irresistible offer, and one that did well for American aviation.

Von Karman's influence at GALCIT was so great that he enhanced the entire Southern California aviation industry, attracting many graduate students who would go on to become leaders in the industry. He lifted GALCIT beyond the level of his former laboratory at Aachen; the climate he created in his American work elicited a far higher level of performance from his students and colleagues than would otherwise have been possible.

A singular irony in von Karman's work brought him his first contact with Hap Arnold. Von Karman's attention had been fixed on the effects of turbulence, in particular as it related to airship design. Sadly, before his studies were completed and published, two great airships, the U.S. Navy's *Akron* (ZRS-4) and *Macon* (ZRS-5), had crashed as a result of encounters with two very different sorts of turbulence. The *Akron* was christened in 1931 by Mrs. Herbert Hoover before a crowd estimated at 500,000. Only two years later, on April 4, 1933, after seventy-four flights and 1,700 hours in the air, she was caught off the coast of New Jersey in a violent storm that tore her apart, sending her crashing into the Atlantic with a loss of all but three of a crew of seventy-six. One of those dead was the Navy's great champion of both aircraft and airships, the Chief of the Bureau of Aeronautics, Rear Admiral William Moffett.

The *Macon* had been christened by Mrs. William A. Moffett less than a month before, on March 11, 1933, the name honoring the home district of the chairman of the House Committee on Naval Affairs, a man who was later a loyal friend of the Air Force, Congressman Carl Vinson. Like the *Akron*, an aerial aircraft carrier with four Curtiss F9C-2 lightweight fighters aboard, the *Macon* was off the California coast, approaching Point Sur, south of Monterey, and had just recovered the last of her aircraft. A sharp gust of wind suddenly struck the airship and she rolled violently, the top fin of her cruciform tail ripping away, tearing open three of the helium-filled gas bags that kept her aloft. After a gallant forty-minute fight to keep her aloft, she crashed into the sea; only two men of the crew of eighty-one were lost.

Von Karman determined the cause of these accidents through the work of Irving Krick, who explained how an atmospheric disturbance little known at the time, the collision of two powerful air masses, had overwhelmed the airship.

Krick became head of a newly formed Cal Tech meteorology department, which caught Hap Arnold's eye and cemented his relationship with von Karman. (Krick would come to prominence during the tension-filled hours before D-day, when he and his colleagues made the estimates upon which General Eisenhower decided to launch the invasion.)

Krick also proved to be the key in investigating the disaster with the *Macon,* discovering an even more unfamiliar phenomenon, the hidden turbulence that contributed to the accident.

Arnold had followed these events, attending meetings at Cal Tech on the accident discussions. (This was risky territory for him—Mitchell's court-martial had stemmed from charges he had raised over the 1925 crash of the Navy dirigible *Shenandoah.*) In the course of these meetings, von Karman found him to be sympathetic, particularly when he learned that Arnold had written some children's books (the Bill Bruce series), just as von Karman's father had done. They met again in the fall of 1938 to discuss rockets as a means of assisting heavily laden bombers to take off. Then, in 1939, Arnold showed the prescience that would do so much for the Air Force, asking von Karman to assist in the design of the precedent-breaking 20-foot, 40,000-horsepower wind tunnel at Wright Field. This led to a part-time consultant position—and the beginning of a long and productive relationship that would enter a new and decisive phase in the fall of 1944.

Von Karman relates that he was asked to meet Arnold under mysterious circumstances at La Guardia Field. He was driven to the end of the runway, where Arnold was sitting in an olive-drab staff car. Von Karman entered, and the driver left. Both men were ill. Arnold was exhausted and had already suffered several heart attacks, while von Karman was just recovering from two operations, one for intestinal cancer. Arnold told him in essence that the present war was won, but that he was concerned about the future of air power. He could not mention the atomic bomb, but it scarcely seems possible that von Karman had not had already envisaged that possibility. As the conversation developed, Arnold asked von Karman to come to the Pentagon to head up a group of fellow scientists in order to plan aeronautical research for as far as five decades into the future.

It was a heavy sacrifice, for von Karman enjoyed academic life and was not certain that he could conform to the military culture of the Pentagon. Arnold assured him that he would smooth out any differences; if alterations in style were required, he would see to it that the Pentagon conformed to von Karman.

The scientist agreed, and on November 7, 1944, Arnold formally established the group. Its prestige was heightened soon after, when von Karman was appointed director of the AAF Long Range Development and Research Program, which was ultimately codified, on December 1, 1944, as the AAF Scientific Advisory Group (SAG). The mission of the SAG was to evaluate research and development trends and prepare special studies on scientific and technical matters relevant to airpower.

Of all Arnold's many contributions to the well-being of the future United States Air Force, this was at once the most important—and the most unlikely. It was the initial mechanism that enabled the postwar leaders of the Air Force to survive a catastrophic demobilization and years of inadequate budgets. Thanks to the technology that was generated, the Air Force would be able, time after time, to pull the country's irons out of the recurrent political fires. The amazing victories of the Air Force in the Persian Gulf War were the result of a technological progression begun by von Karman and Arnold.

Von Karman's forte had been pure research, as it had been for most of the extraordinarily talented group he selected to join the SAG, and it would not have been surprising if they had resented the explicit framework for their work that Arnold laid out. They did not, because of his achievements and because of his legendary charm.

Arnold wanted to be sure that the SAG did not place too much emphasis on the results of the last war, but instead looked to the future to see what intensive research programs might develop in the fields of electronics, with emphasis on radar, aerodynamics, propulsion (both jet turbine and rocket), and, of course, the basic sciences.

Arnold kept Marshall's economy concerns in mind and asked the scientists to condition their thinking to a *de minimus* defense budget, in which the effectiveness of the relatively small numbers of personnel in the armed forces had to be multiplied by equipment.

Displaying the flexible thinking that continued to surprise his peers, Arnold asked the SAG to address also the organizational and administrative problems of military-related science, with particular emphasis on the question of what should be the optimum percentage of the budget to be invested in research and development.

Von Karman was practical enough to understand and comply with Arnold's requests. He had the bittersweet pleasure at the close of the war of taking some of his group to his own old haunts in Europe, where they traveled in uniform for safety and convenience, von Karman particularly enjoying his temporary role as a major general. The results of the trip were astounding, as everything from tons of scientific documents to a high-speed wind tunnel was scooped up and sent home.

Von Karman created a report of the trip entitled *Where We Stand*, a comprehensive presentation of the current state of aeronautical knowledge. His conclusions were both startling and far-ranging, and included the imminent arrival of supersonic aircraft, a family of missiles, including both air defense and ballistic, and immeasurably improved navigation and communication devices.

It was but a start. From the trip, and from other studies, the SAG created the multivolume epic *Toward New Horizons*, which included thirty-two reports from twenty-five authors and ranged in subject matter from aerodynamics and aircraft design to explosives and terminal ballistics.

The Scientific Advisory Group was followed by the Scientific Advisory

Board (SAB) in 1947, when the Air Research and Development Command was formed. The SAB was so successful that many other government agencies copied it, setting up similar special units of research experts. Ironically, the result was a dilution of the talent available to the SAB, a problem further compounded by the steadily increasing size and importance of the Department of Defense bureaucracy. The SAB's role has been diminished, its original mission now being carried out by "think tanks" like the RAND Corporation MITRE, and other similar civil research-and-development organizations. Yet had not Arnold not first given emphasis to research and development, and, as will be shown, had not von Karman paved the way with seminal publications, the United States Air Force would not exist as we know it today.

A tremendously important, possibly decisive, problem that neither Arnold nor von Karman could have foreseen was that over the years bureaucratic complexity would develop at a rate commensurate to or exceeding technical capacity. While Arnold could make instant decisions on both R&D and procurement, confident that he would be backed by the War Department and Congress, later Chiefs of Staff of the Air Force found that they had to convince more and more individuals and many more intermediate layers of command. As will be shown later, in time the bureaucracy of the Department of Defense and especially of Congressional staffs became a Sargasso Sea for R&D and procurement, and currently it threatens to stifle both by making them unreasonably costly and lengthy.

Arnold's contribution to research and development in the postwar United States Air Force was as great as the contribution he had made to the Army Air Forces as it fought and won its battles in World War II. In the course of the war he had suffered four heart attacks, and by war's end he was drained, as much a victim of the conflict as any of the men who had fallen in combat. He retired on June 30, 1946, to watch with pleasure as General Spaatz, who succeeded him as Commanding General of the Army Air Forces, carried on with the fight for independence. His work had been widely recognized around the world, and his list of decorations was long, but he was most pleased when on May 27, 1949, President Truman wrote to him that he had just signed a bill making Arnold the first (and to date only) permanent five-star General of the Air Force.

2

THE SORTING-OUT
PROCESS: 1943–1949

THE RIGHT SORT OF ALLIES

As impetuous and demanding as Arnold was, he also had the God-given grace to see his own faults and to seek allies who could compensate for them. During World War II, the most important of these was found in an office adjacent to his own where the patrician Robert A. Lovett, Assistant Secretary of War for Air, worked. Lovett favored Arnold and recognized both his strengths and his weaknesses, mitigating the latter when he presented Arnold's ideas to General Marshall.

Marshall had earlier been inoculated with the concept of an independent air force by his prewar friendship with Major General Frank M. Andrews, then Commander of the General Headquarters (GHQ), which constituted what little there was of Air Corps striking power. (It was said that Marshall thought so highly of Andrews that had the latter not died in the crash of his aircraft in May 1943, Marshall might have nominated him to become Supreme Allied Commander in Europe—a position ultimately occupied by General Dwight D. Eisenhower.)

Fellow officers have described Arnold as "smart" and Andrews as "intelligent"; both were wise enough to subordinate their rivalry to the good of the service. Marshall, perhaps convinced by Andrews of the importance of the power of the Air Force and by Arnold of its willing subordination to the prosecution of the war, transferred his friendship to Arnold and became an ally in the quest for postwar independence. Noted for his reserved, sometimes frigid manner, Marshall approved of Arnold to the unusual degree that, later in the war, he would use the word "affectionately" in signing his letters. Yet it is probable that without Lovett as a buffer and a facilitator, Arnold's often impulsive manner would have incurred Marshall's wrath.

And just as Lovett smoothed the way for Arnold with Marshall, so did Marshall smooth the way for Arnold with President Franklin Delano Roosevelt, who at first entertained a poor impression of him. Unfounded rumors of Arnold's drinking had almost caused Roosevelt to refuse his promotion to Chief of the Air Corps in 1938. Then, in 1940, Arnold's Congressional testimony had displeased Roosevelt, who later suggested broadly that those who did not "play ball" might well expect an assignment to Guam. Arnold's subsequent performance, particularly at the many summit conferences, eventually smoothed all this over. Marshall made Arnold a Deputy Chief of Staff for Air, remarking that he had tried to make him "as nearly as I could Chief of Staff of the Air without any restraint although he was very subordinate."

When war began, Arnold, just promoted to lieutenant general, became a member of the U.S. Joint Chiefs of Staff and the Anglo-American Combined Chiefs of Staff. This was implicit and unprecedented recognition that the Air Force was equal to and independent of sea and land power, and was done in part to match the British staffing pattern and in part to ensure that Arnold's views on air power were not filtered through an Army interpreter.

Arnold made good use of his position, which brought him into contact with the top military and political leaders of the Allied cause, beginning with his invitation to accompany President Roosevelt's party to the historic Atlantic Conference with British Prime Minister Winston Churchill in August 1941. This was the first of an endless series of trips and conferences that would consume Arnold's time—and health—in the course of the war, but would elevate him to the very top rank of military power.

Just as he made good use of his position with the world's leaders, so did he make good use of his staff in laying the administrative groundwork for the postwar air force. In March 1942, the armed forces of the United States were still reeling from the Japanese victories in the Pacific when the War Department issued Circular 59, *War Department Reorganization*. Based largely on early recommendations by Arnold and Spaatz, Circular 59 made the Army Air Forces (AAF) one of three autonomous commands, along with the Army Ground Forces (AGF) and Services of Supply, later the Army Service Forces (ASF).

As heady as the new authority was, the circular's mandate contained a stinger that would cause a flurry of activity in the last years of the war and continuing into the immediate postwar period. The new arrangements set up by Circular 59 were condemned to expire six months after the close of the war—a victim of the wording of the First War Powers Act of December 18, 1941. It thus became imperative for Arnold and his staff to take all the necessary steps to bring about a separate air force as soon as the war had ended. They knew that if the war ended and six months elapsed, their reversion to a mere component of the Army would delay independence, perhaps forever.

More than a year after Circular 59, when the war had obviously taken a turn for the better, another event occurred that enhanced the status of the Army Air Forces and boded well for postwar independence. This was the publication of

War Department Field Manual (FM) 100–20, *Command and Employment of Air Power,* July 21, 1943.

Whereas Circular 59 had been the careful expression of academic beliefs on air power and was made possible only because wartime expansion permitted the Air Force genie to escape from the Army's bureaucratic bottle, FM 100–20 was written with the knowledge—one could almost say swagger—learned in the combat over the North African desert and was not even coordinated with the War Department. FM 100–20 was described with dismay by some members of the Army as the "Army Air Forces 'Declaration of Independence'"— and so it was, for it stated unequivocally that land power and air power were coequal and that the gaining of air superiority was the first requirement for the success of any major land operation.

FM 100–20 went on to say: "The inherent flexibility of air power is its greatest asset. This flexibility makes it possible to employ the whole weight of available air power against selected areas in turn; such concentrated use of the air striking force is a battle-winning factor of first importance. Control of available air power must be centralized and command must be exercised through the Air Force commander if this inherent flexibility and ability to deliver a decisive blow are to be fully exploited. Therefore, the command of air and ground forces in a theater of operations will be vested in the superior commander charged with the actual conduct of operations in the theater, who will exercise command of air forces through the air force commander and command of ground forces through the ground force commander." This doctrine was so welcome to the Army Air Forces that a potentially troublesome element of FM 100–20 was discounted. The manual went on to describe the mission and composition of a strategic air force, a tactical air force, an air defense command, and an air service command; this separation of strategic and tactical air forces would be a bone of contention and a source of both inter- and intraservice rivalry for the next half century.

General Arnold, perhaps wishing to make evident that his apparent side-stepping of the independent air force issue had been practical politics, seized upon FM 100–20 to send a letter to each Army Air Force commander, emphasizing that "the interrelated role of air power must be constantly impressed on all airmen through the medium of command."

Although more than two years of war remained after the issuance of FM 100–20, the creation of an independent air force now seemed inevitable, despite opposition from the Navy. Drawing on wartime experience that clearly showed the efficiency of the unified command of air, land, and sea operations, Arnold now moved toward unification of the armed services as a whole. In February 1944 he increased the pressure by advocating the establishment of a single Secretary of War, with four under secretaries heading the ground forces, naval forces, air forces, and a bureau of war resources. The President would be advised by a single chief of staff and a supreme war council of the four major service commanders, a pattern not unlike that used by the British.

Implicit in this was a red flag to the Navy: the Air Force was to include "all military aviation except ship-borne units operating with the Navy and those artillery-control and 'liaison' units operating with the Army." This was directly opposite to the philosophy of the Navy, which wanted to maintain its own naval and marine air forces, and go on to establish a heavy strategic bomber force of its own.

Arnold's proposal was followed in February by the first formal blueprint for a peacetime air force, known as Initial Postwar Air Force-1 (IPWAF-1), which optimistically called for 105 air groups. The plan, which grew out of a study by Brigadier General Kuter's Post-War Division, was a striking parallel to the famous AWPD-1 of August 1941, in which Kuter had participated as a very young major. AWPD-1 had assessed what would be necessary to defeat Germany and forecast with uncanny accuracy the number of aircraft, crews, missions, and even losses that the war would require.

Kuter was a talented pilot who had flown with Claire Chennault's aerobatic demonstration team as one of the daring young men on "the Flying Trapeze," the ancestors of the famous Thunderbirds. Promoted to the rank of brigadier general at thirty-seven, Kuter was Commanding General of the 1st Bomb Wing in England and later played an important role in the Tunisian campaign. He was thus exactly the sort of leader that Arnold preferred—young, intelligent, widely experienced, and proven in combat. Arnold thought so much of him that when felled by one of his early heart attacks, he sent Kuter to represent him at the Yalta and Malta conferences.

Just as AWPD-1 had ignored costs, so did IPWAF-1—and this was a fatal flaw in the planning, for while the plan naturally emphasized spending on Air Force requirements and limiting naval and ground forces, it ran counter to General Marshall's concept of what the voters and the economy of the United States would bear in the postwar world, and to the political realities that would define "balanced power" of the U.S. military as a roughly equal division of the total defense budget between Army, Navy, and Air Force.

Marshall wished to revert to a professional peace establishment, slightly larger and better equipped than that which had existed before World War II, and capable of being reinforced quickly from the traditional civilian Army reserves. He was an advocate of Universal Military Training (UMT) to provide the necessary pool of trained personnel for the reserves. As a concept, UMT was not inconsistent with raising a large land army from reserves; however, it was of negative value to an air force, which required career specialists with years of training and experience in order to function.

General Marshall's views are somewhat surprising in retrospect; he was aware of the Manhattan Project, and he had made possible the advances in air power that were proving decisive in the war. Further, he subscribed to the idea that the next war would again be initiated by a surprise attack, and that the United States would have no major allies for at least eighteen months. He mitigated this harsh scenario somewhat by a wistful stipulation that the United

States would be aware of the possibility of war by one year, during which emergency measures would be started. Events would soon prove that this too was an incorrect assumption. Indeed, from all of his practical experience, Marshall might have been expected to conclude that while a citizens' army and navy might be practical, a professional air force of large size was essential. Instead, the memories of the totally inadequate defense budget in the years between 1918 and 1941 still weighed too heavily upon him.

His inhibitions about the expense of maintaining a large standing air force were reinforced by the extremely pessimistic forecast from the Special Planning Division of the War Department, which estimated that only $2 billion would be available annually for defense. This restriction effectively crippled the air force, which would then have had to be limited to a total of 120,000 men and sixteen groups.

The AAF reacted strongly, pitting Arnold in the unusual role of contesting with Marshall. A total force of only sixteen groups not only would be inadequate for defensive purposes, it would mean the dissolution of the base of the aviation industry, so that a subsequent buildup would be virtually impossible.

With the uncanny eye for talent that characterized him, Arnold now turned to another rising young two-star, Major General Lauris Norstad, who took over Kuter's post as Assistant Chief of Air Staff, Plans. Norstad had been a top combat air commander in England, North Africa, and Italy and had been promoted to the rank of brigadier general at thirty-six.

Norstad would be guided by one of the grand old men of the AAF, the remarkable Lieutenant General Eaker. Now Deputy Commander and Chief of Air Staff, Eaker, good soldier that he was, had swallowed his resentment at being relieved of command of "his" Eighth Air Force and gone on to do remarkable work as Commander in Chief of the Mediterranean Allied Air Forces. As he had been with the Eighth and in the Med, Eaker was once again deprived of the pleasure of being in at the kill; instead he went to work in the newly built Pentagon, its recently poured concrete walls still sweating, to control planning for the Army Air Forces interim and permanent force structures.

The Army and the Navy were equally concerned about their postwar strengths, and the Secretary of the Navy, James V. Forrestal, moved quickly, proposing legislation to establish the permanent strength of the postwar Navy. It was the first shot across the bow of an increasingly acrimonious battle that would continue long after the Air Force became an independent service.

The reason for the interservice squabble was obvious. The Navy knew that in any "vote" in a unified service, the Army and its offspring, the independent air force, would ally themselves.

Fortunately for Arnold and his staff, Forrestal's legislative gambit backfired. When President Harry S Truman succeeded Roosevelt on April 12, 1945, no one had gained his confidence more quickly or more fully than the Chief of Staff, General Marshall. Marshall cited Forrestal's request as a typical example of military parochialism and as a perfect argument for postwar unification. Truman then asked all the services to establish their postwar requirements for review.

The AAF was ready; Arnold had tasked his key personnel, Spaatz, Eaker, Norstad, and another young comer, Lieutenant General Hoyt S. Vandenberg, to set a firm goal for the postwar air force. Vandenberg, former commander of the Ninth Air Force, had, like Norstad and Kuter, combined a distinguished combat career with broader tasks, including heading an Air Mission to the Soviet Union under Ambassador Averell Harriman. The three men had something else in common: they were all highly regarded by General Spaatz, whose interpersonal relationships were perhaps less demanding than Arnold's, but who nonetheless expected extremely high levels of performance.

It is worth noting that Arnold and Spaatz might well have just accepted the potential of young officers like Kuter, Norstad, and Vandenberg as a given, without appreciating their importance to the postwar effort. Indeed, it would have been only human for either Arnold or Spaatz to have resented the rapid rise of these young officers and to have instead preferred those with whom they had spent so many long years in the trenches of the prewar Air Corps. Fortunately, both Arnold and Spaatz gladly cooperated with General Marshall's wish to have young officers brought forward. They both had the vision to recognize what would be required, the patience to accept the sometimes bitter censure from their colleagues whom they were forced to pass over, and the courage to prepare the way for their own departure from the scene. These are rare qualities in civil life, and perhaps even rarer in the military.

Thirteen days after Japan agreed to surrender, on August 28, 1945, the planning group under General Eaker established a postwar goal of a seventy-group air force with 550,000 men. (A "group" varied with the type of aircraft. Very Heavy Bombardment [VHB] groups, equipped with Boeing B-29s, had forty-five aircraft and 2,078 men on establishment. A Heavy Bombardment [HB] group of Boeing B-17s or Consolidated B-24s had seventy-two aircraft and 2,261 men. A single-engine-fighter group, with either Republic P-47s or North American P-51s, had 111 to 126 aircraft and 994 men. A troop-carrier group, with Douglas C-47s, had 80 to 110 aircraft and 883 men.) The seventy groups were to be backed up by twenty-seven Air National Guard and thirty-four Air Reserve groups. (The status of Guard and Reserve groups would be an object of controversy for years until the Air Force matured enough to provide them with adequate training, first-rate equipment, and essential missions.)

The seventy-group goal, regarded as barely sufficient by its own air force planners in peacetime, and totally inadequate for war, would remain an elusive target, swamped first by the incredible demobilization that took place immediately after the formal Japanese surrender and then by unrealistically restrictive postwar budgets.

The AAF argued that a standing force of at least seventy groups was essential for a number of reasons, not least of which were the global demands that were now thrust upon the United States as the leading world power. However, the most obvious reason was that in the next war, the United States would not be given time to recover as it had been after the debacle of Pearl Harbor. (This same factor was held to work to the disadvantage of Guard and Reserve units.) There could no longer be a question of the oceans protecting the heartland of the country while an army and an air force was mustered, trained, and equipped.

The seventy-group air force would be a goal that would be reached only temporarily before it was swept away in a series of postwar cost-cutting orgies. But even as Arnold rode out the numbers war with the planners, he was already creating a mind-set, a creative thrust, that would characterize the USAF for the next fifty years, and would, on more than one occasion, be the literal savior of the country.

FROM DEMOBILIZATION DEMENTIA TO ORDERLY INDEPENDENCE

The cold recital of administrative events leading to the unification of the services and the establishment of the independent United States Air Force was played out on the shifting sands of the wildest, most expensive and reckless demobilization in history. In the shortest time possible, given physical constraints of transport, the mighty armed services of the United States were reduced to impotence at home and abroad. The Medal of Honor recipient who had led the 44th Bomb Group, the Eight Balls, against Ploesti, General Leon W. Johnson, aptly termed the demobilization frenzy a "riot."

Incongruously, the riot occurred just as the United States was recognized for the first time as a superpower (although the term had not yet come into vogue) and its influence and interests were tacitly acknowledged as global. Political and military philosophers from Sun Tzu to Winston Churchill to Colin Powell have always agreed that political policies without military strength are hollow reeds doomed to failure, yet this was exactly the state in which demobilization had left the United States.

The political pressures for demobilization were overwhelming. "Bring the Boys Home" was the rallying cry, and "points," the credits achieved for length and place of service, were the determining factor. It was a civilian military force, and it wanted out!

At the same time, the armed forces were struggling with the problems implicit in the occupation of the territories of Germany and Japan and in withstanding any encroachment by the Soviet Union outside its designated spheres of influence.

With an incredible naiveté, the United States allowed itself to totally disarm, depending upon its atomic monopoly to maintain its world position. Ironically—and dangerously—the United States possessed almost no such capability. It had very few atomic bombs—only thirteen on June 30, 1947—and its only means of delivery were the B-29s of the famous 509th Bomb Group. The image of a powerfully armed atomic strike force was a myth, pure and simple, for not only were weapons and carriers lacking, but so were the thirty-nine-man technical teams that required two full days to do the intricate assembly and arming of those early weapons.

The situation was actually worse than described above, for a turf-conscious Atomic Energy Commission insisted (in the name of civilian control of the military) on controlling the number and type of bombs as well as all material and manufacture. The Chief of Staff of the Air Force himself had no certain knowledge of the number of bombs available to him for use by his Strategic Air Command.

Demobilization had greater effect on the Navy than the Army, and still greater on the Air Force than the Navy because of the relatively higher degree of technical skills needed. Troops were demobilized purely on the basis of the points they had earned, without regard to their skills or the demand for their services. It was democratic in the extreme, but it was disastrous to the Air Force in particular.

The quantity of equipment was not the problem—there were thousands of everything from trainers to B-29s, with tremendous pipelines of parts to support them. The problem was personnel, for there were simply not enough mechanics to repair the aircraft, or supply clerks to find the parts, or crews to man them.

Unit strengths fell as surely as if they'd been lined up to march over a cliff. The Army Air Forces had 2,253,000 military and 318,514 civilian personnel on strength on VJ-day. Four and one-half months later, AAF strength had declined to 888,769 military personnel, a figure that fell to a low of 303,600 in May 1947, when there were about 110,000 civilians still on the rolls.

The gigantic, smoothly functioning machine that had won wars in two theaters ground noisily to a halt, with equipment lost, stolen, scattered, or deteriorating because of lack of maintenance. More than 90 percent of the 350,000 aircraft mechanics departed the service. Where in wartime as many as 50 or 60 percent of the combat aircraft had been ready, the number now averaged about 18 percent, and that only because so many of the airplanes were brand-new and virtually unused. In hard numbers, this meant that of the 25,000 aircraft available, only 4,750 were fit for combat, and these were dispersed around the world.

Most telling was the drop in combat-ready groups. On August 15, 1945, there had been 218 AAF groups of all types, most of them ready to fight. By December 1946, sixteen months later, there were only fifty-two groups remaining in the AAF, and of these only *two* were combat ready.

Arnold and Spaatz had watched the tailspin of their great creation with horror, as much appalled by the lack of proficiency and readiness of the organizations that remained as they were by the lack of personnel. They watched against a backdrop of steadily deteriorating relations with the Soviet Union, whose menacing intentions were evident in the Eastern bloc countries. In March 1946, the old lion Winston Churchill had roared again, this time in Fulton, Missouri, where he used the term "iron curtain," a metaphor previously used by Kaiser Wilhelm, Vladimir Lenin, and an old enemy, Joseph Goebbels, but never so tellingly. Stating, "From Stettin in the Baltic to Trieste in the

Adriatic, an iron curtain has descended across the Continent," he went on to note that all the capitals of the ancient states of Central and Eastern Europe, Warsaw, Berlin, Prague, Vienna, Budapest, Belgrade, Bucharest, and Sofia, were subject to Moscow's increasing control. Warrior that he was, what he undoubtedly sensed at that moment in history was that with Britain exhausted, Germany devastated, France a military cipher, and the United States convulsively disarming and its nuclear strike force virtually impotent, Joseph Stalin could have raced across Europe unchallenged. Fortunately, the wounds of the Soviet Union were too recent and Stalin's misapprehensions about U.S. nuclear strength too great, and he missed his opportunity.

The magnitude of what would today be called the "downsizing" of the AAF had not surprised Arnold or Spaatz, but its rate and devastating effect had. Already a dying man, Arnold had increasingly removed himself from the picture after the victories in Europe and the Pacific. On February 15, 1946, Spaatz formally succeeded Arnold, and it fell to Spaatz to undertake the measures necessary to offset the effects of the dangerous demobilization and to prepare for the future independent air force.

Under Spaatz's leadership, the staff had in 1946 built into the old Army Air Forces the structural elements considered necessary for a new air force. The top leaders were bomber men because they had led the most powerful units of World War II. They would be for years to come—and they saw the requirement for a global capability some forty-four years before the motto "Global Reach—Global Power" became the Air Force's watchword.

The new structure, formally established on March 21, 1946, was composed of three new functional organizations, the Strategic Air Command, the Tactical Air Command, and the Air Defense Command. Four support commands included the Air Training Command, Air Materiel Command, Air Proving Ground Command, and Air Transport Command. There were also a number of overseas theater commands, relics of the unified commands, the most important of which were the United States Air Forces in Europe and the Far East Air Forces.

All planning was based on a nominal seventy-group air force, considered by almost all air force planners as the minimum number required to meet its global commitments, serve as a springboard for growth, and, not incidentally, maintain an aviation industry with the capability to build up to previous wartime production levels if necessary.

In a gesture to tradition and for ease in transfer, existing numbered air forces were reassigned into the new commands. The Strategic Air Command (SAC) had been created to fulfill the ideas of several young officers, including Generals LeMay, Vandenberg, and Norstad, to create an atomic strike force. SAC received the one-two punch of the European war, the Eighth and Fifteenth Air Forces.

The Tactical Air Command mission was more disputed; strong elements within the AAF would have preferred to maintain the ground-support force within a larger unit, the Continental Air Command. But General Eisenhower, who had supported air force independence, wanted an air force designated specifically for air-to-ground operations, and Spaatz wanted to please him. There was, besides, the specter of the Army demanding its own tactical aviation and depriving the Air Force of an important mission. The Third, Ninth, and Twelfth Air Forces were assigned to TAC.

Air Defense Command received the First, Second, Fourth, Tenth, Eleventh, and Fourteenth Air Forces—and had precious little in the way of resources for them.

Then as now, controversy swirled around the personnel given command of the new organizations. General George C. Kenney, who had mesmerized Douglas MacArthur even as he orchestrated the defeat of his Japanese opponents, was named Commander of SAC. Yet Kenney had other interests, including intensive lobbying for an independent air force. SAC was to suffer for his inattention, which might have stemmed from the letdown coming in peacetime work after his brilliant combat career in the Pacific. And it might have been a natural pique; he was senior to Spaatz, but of course had not enjoyed Spaatz's long and close relationship with Arnold. Nonetheless, Kenney would have had to have been more than human not to have felt some resentment at being passed over for the role of Commanding General of the Army Air Forces.

Major General Elwood R. Quesada was given command of TAC. The smiling, personable flier had flown with Spaatz in the 1927 flight of the *Question Mark,* and as the popular leader of the IX Tactical Air Command had been ranked by General Omar Bradley as the fourth most capable American general in the European theater. Here the personality dispute was a raw issue of promotion and turf between Vandenberg and Quesada dating back to their days in Europe. It was a battle that would ultimately leave TAC vitiated and Quesada bitter.

The Air Defense Command was headed by Lieutenant General George E. Stratemeyer, who, in turn, would run into difficulties with Vandenberg during the Korean War.

The level of importance Arnold had attached to research and development can be found in the assignment that came to Major General Curtis E. LeMay. A proven combat leader in both Europe and the Pacific, LeMay was given the job of Deputy Chief of Air Staff, Research and Development. Arnold and Spaatz didn't pick LeMay because of his scientific background— he had no more than the typical commander of the era—but because he would focus powerful energy on the task and imbue it with the importance Arnold knew it deserved.

Spaatz buttressed his new organization plan with the creation of an Air Board to review plans and policies, with Arnold's longtime foe, the obstreperous Major General Hugh J. Knerr, as its first secretary-general. Knerr, friend and adviser to Frank Andrews, was a bomber man and had long been a proponent of a separate air force. He was ideally suited for a job that called for independent thinking—he'd been so independent in the past that he was one of the few people ever to have been close to court-martial while on active duty and again after retirement. It was typical of Spaatz that he would nominate a man who had been a thorn in Arnold's side to an important position simply because he felt he was the best man for the job.

It turned out to be a happy choice that boded well for the future, for Knerr set up the Air Board so that it would view the USAF as a business, a viewpoint that would put the Board—and consequently Spaatz—exactly on frequency with the first Secretary of the Air Force, Stuart Symington.

Working entirely within the AAF, depending upon a core of personnel that had been involved in aviation almost from the its beginning (Knerr had actually swept out the Wright brothers' bicycle shop in Dayton), Spaatz and his team (which after August 1947 included the formidable talents of Brigadier General William Fulton "Bozo" McKee) were flexible enough to change the organization of their war-winning service to meet the demands they perceived ahead. Forty years later, another team of officers, victors this time in the Cold War, would undertake the same process.

The leaders of the nascent air force had already identified the Soviet Union as the next potential enemy and were planning how to meet the challenge, not realizing that there was a bigger hazard to an independent air force's survival close to home—the United States Navy.

An Organizational Street Fight

At the end of World War II, there were two views on the need for an independent air force: those of the U.S. Navy and those of virtually everyone else. President Truman, General Marshall, General Eisenhower, and most members of the Congress agreed that the results achieved by air power in winning the war justified the existence of a separate, coequal service. In parallel to the obvious requirement for a separate air force was the need for a unity of command, as had been exemplified by the unified commands of World War II. (A unified command is an organization composed of two or more U.S. armed forces; a joint command is an organization that includes an allied force.) Under the pressures of war, the Joint Chiefs of Staff (JCS) had worked together in relative harmony to make unified commands a success. Without that pressure, and subject to the territoriality deriving from competition for the vastly reduced peacetime budgets, few believed that the JCS would still function harmoniously. When the harmony vanished, so would the hope for operating a successful unified command to conduct combat operations.

The Navy was adamantly opposed to both a unified national department of defense and to an independent air force. The Navy believed a unified command, with a single defense chief at the top, would block the access of the Secretary of the Navy to the President. It believed that an independent air force would compete not only for a limited budget, but also for roles and missions, despite AAF disclaimers. They were completely correct in this, for, somewhat euphoric with their success in the war, some AAF leaders, including people as prominent as Lieutenant General Jimmy Doolittle, were openly casting doubt on the future value of aircraft carriers in a world where the only potential enemy had no navy and possessed a huge continental army.

Somewhat hypocritically, given its tendency to deprecate the effectiveness of an air force, the Navy was at the same time lusting for a strategic bombing arm. After the atomic bombs had been dropped on Hiroshima and Nagasaki, this became a naval imperative. It was given tangible expression immediately after the war by the push to create the 65,000-ton super carrier the *United States*, whose size and configuration would have permitted the operation of heavy—i.e., nuclear—bombers.

There was some basic logic in the Navy's argument. The armed forces of the United States had just fought and won the greatest war in history, proving to the Navy's satisfaction that the contemporary organization worked and should not be changed. They relied on what became the popular aphorism "If it ain't broke, don't fix it." The potent counterargument was that air power and nuclear weapons had so altered warfare that it would be foolish to remain tied to World War II concepts.

From 1944 through 1947, there was a veritable storm of special studies, committees, unification bills, and recommendations from senior officers. The conclusions reached were drawn on political lines; those with naval sympathies were against unification, while almost everyone else was for it. An important change in the equation stemmed from the change in administrations. President Roosevelt had been an ardent Navy man, as might be expected from his background, while President Truman came from an Army background. Roosevelt's presidential cousin Theodore had created the Great White Fleet, and he himself had served as Assistant Secretary of the Navy during World War I. Truman had been an artillery captain, and while all of his experience with aviation had not been positive—his committee had been ruthless in exposing waste and inefficiency in the aircraft industry—he still favored an independent air force and unification.

With the style that would bring him tremendous popularity—but only many years after he left office—President Truman intervened decisively for unification. On December 19, 1945, he called for legislation from the Congress that would combine the War and Navy departments into one single department of national defense, with a coequal, independent air force. His words were cogent: "Air power has been developed to a point where its responsibilities are equal to those of land and sea power, and its contribution to our strategic planning is as great. Parity for air power can be achieved in one department or in three, but not in two. As between one department and three, the former is infinitely to be preferred."

Yet the Navy had not lost anything; it was to retain its carrier and water-based aviation and the Marine Corps. The Army, in turn, would retain aviation integral to its operation, i.e., observation, liaison, and intratheater troop transport.

Truman called for a civilian secretary to administer the new Department of National Defense, with a military Chief of Staff of National Defense, the office to be rotated among the three services. All three requirements were anathema to the Navy.

Despite the support of the President, the naval opposition was too strong, and legislation intended to implement his wishes fell short because of the adamant and expert opposition by the Navy. Vulnerable, the Navy felt a terrible sense of ingratitude, for it had done so much to win the war in the Pacific. Now it was fighting for its life, not least because there was no great enemy naval power to offer an obvious rationale for the Navy's existence.

In a series of standoffs not unlike today's budget battles, the Navy fought for its views with its usual political acumen, exploiting its hold on legislators with large shipyards or naval bases in their states or districts and counting on the loyalty of those members of Congress who had actually served in the Navy.

Eventually, it fell to Major General Norstad, Director of the Army General Staff's Operation Division, and Vice Admiral Forrest P. Sherman, Deputy Chief of Naval Operations, to work out an agreement that had eluded everyone else, including Secretary of War Robert P. Patterson and Secretary of the Navy James Forrestal. Like Norstad, Sherman had a brilliant war record. A flier since 1922, he had commanded the USS *Wasp* until her loss in 1942; he then became a leader in planning the naval campaigns that ran from the Carolines to Okinawa. Sherman was considered the best naval strategist in the Pacific, and he became Admiral Chester Nimitz's most trusted adviser. It was Sherman who had convinced Nimitz that by 1944 the Navy's carriers had so grown in strength that they could take on and knock out Japanese land-based aviation, a true revolution in air power philosophy. When things became difficult between the services, Nimitz employed Sherman to negotiate with AAF Lieutenant General George C. Kenney and with the Southwest Pacific Commander in Chief, General Douglas MacArthur. Like Norstad, Sherman was a polished and diplomatic negotiator, and most felt that if these two men could not arrive at a solution, no one could.

Sherman and Norstad put the national interest first, and by compromise provided the basis for a draft of the bill for unification. The House and Senate debated the bill from February through July, with continued Navy opposition, but the National Security Act of 1947 finally became law on July 26. President Truman had signed it under poignant circumstances.

Truman's mother was desperately ill, and he sweltered for almost an hour in the presidential airplane, the Douglas C-54 *Sacred Cow*, at D.C.'s then bucolic National Airport, delaying takeoff until the bill arrived. He signed it, then departed for his home in Independence, Missouri. Sadly, his mother died while he was en route.

The Declaration of Policy of the National Security Act of 1947 provided for the three military services, Army, Navy, and Air Force. All were a part of the National Military Establishment, headed by a Secretary of Defense, who coordinated the departments of the Army, the Navy, and the Air Force, each of which had its own secretary.

The success of the Navy's political infighting was evident in the selection of the former Secretary of the Navy, Forrestal, to become the new Secretary of Defense, and by the careful way that the functions of the Navy were spelled out to include a full range of naval aviation and the Marine Corps. (Forrestal had been Truman's second choice for the position, which he offered first to Secretary of War Robert A. Patterson, who declined.)

The act also created a Joint Chiefs of Staff, a Joint Staff, a Munitions Board, a Research and Development Board, and a War Council.

Within the Department of the Air Force, the United States Air Force was created out of the existing Army Air Forces, which included (still) the Army Air Corps and the Air Force Combat Command.

The act was, on balance, a good effort, for it resolved the major issues of unification and the existence of an independent air force. Yet there were many things to be worked out in relation to the United States Air Force, including its roles and missions, the way it addressed functions that were formerly accomplished by elements of the Army, and, most important, its size and force structure.

President Truman had also signed the short-lived Executive Order 9877, which detailed the function and roles of the three services. There had not been time for his own staff people to discover and correct discrepancies between his Executive Order and the National Security Act of 1947, and these discrepancies were to be the source of controversy for some time. Most of the elements of the Executive Order regarding the Air Force had been provided Truman in draft form by the excellent staff work supervised by Generals Spaatz and Eaker during the prior two years. It hewed closely to the vision of the Air Force that Mitchell, Arnold, and their colleagues had advocated for so long. The specific functions of the Air Force were noted in Section IV of the Executive Order. They are of such fundamental importance to the politics of the time and to the later development of the Air Force that they need to be rendered in full, with their import noted in brackets:

General
The United States Air Force includes all military forces, both combat and service, not otherwise specifically assigned. [This placed a limit on the concessions to the Army and the Navy. The Navy was limited to aircraft for reconnaissance, antisubmarine warfare, the protection of shipping, and air transport essential for naval operations. Included in this were combat, service and training forces, as well as land-based aviation, a crucial loophole.] It is organized, trained, and equipped primarily for prompt and sustained air offensive and defensive operations. [This put an irreducible lower limit on the size of the Air Force and served to blunt the most severe of the postwar budget cuts.] The Air Force is responsible for the preparation of air forces necessary for the effective prosecution of the war except as otherwise assigned, and, in accordance with integrated joint mobilization plans, for the expansion of the peacetime components of the Air Force to meet the needs

of war. [This set the stage for later unified and joint command operations of which the Air Force would often be a principal component, but would rarely command.]

The specific functions of the United States Air Force are:

1. To organize, train and equip air forces for:
   (a) Air operations, including joint operations.
   (b) Gaining and maintaining general air superiority. [This is the key issue about which all Air Force planning revolved: maintaining general air superiority. It automatically conferred precedence in the event of war.]
   (c) Establishing local air superiority where and as required.
   (d) The strategic air force of the United States and strategic air reconnaissance. [This was a vital element, for the strategic component inevitably involved the nuclear, i.e., war-winning, component.]
   (e) Air lift and support for airborne operations. [This would become key in later arguments with the Army over roles and missions.]
   (f) Air support to land forces and naval forces, including support of occupation forces. [Again, critical to later arguments with both the Army and the Navy over the ground support role.]
   (g) Air transport for the armed forces, except as provided by the Navy. [An ambiguity that would not be resolved with the formation of the Military Air Transport Service (MATS) on June 1, 1948, but had to wait until the Military Airlift Command (MAC) was created in 1966.]

2. To develop weapons, tactics, technique, organization, and equipment of Air Force combat and service elements, coordinating with the Army and Navy on all aspects of joint concern, including those which pertain to amphibious operations. [The absence of a specific reference to the development of the nuclear mission as an element of joint concern was important here.]

3. To provide, as directed by proper authority, such missions and detachments for service in foreign countries as may be required to support the national policies and interests of the United States.

4. To provide the means for coordination of air defense among all services. [This would prove to be a continuing problem, one never satisfactorily resolved, as the Army would not give up the antiaircraft role, and the Navy could never subordinate fleet defense to air defense.]

5. To assist the Army and Navy in accomplishment of their missions, including the provision of common services and supplies as determined by proper authority. [The danger implicit here was hidden, and applied to all three branches of the military, for it contained the seed of the current—and, in the opinion of many, dangerous—trend toward single-point Department of Defense research and development and procurement for all services. Carried to its logical extreme, the services would be left only with the people who do the fighting—the "spear-chuckers"—while all the functions of R&D, procurement, supply, etc. would be carried out by civilians in the Defense Department.]

The Army, surprisingly supportive through all of the myriad meetings, worked willingly to arrange a series of implementation agreements that facilitated the separation of the Air Force. These ranged in nature from decisions on tactical and strategic missile responsibility (given to the Army and the Air Force, respectively) to intelligence, housekeeping, service and supply, procurement, and administrative functions.

Even though much was still left unresolved, these two documents were sufficient for the launch of the United States Air Force, which came officially into being on September 18, 1947, with the swearing in of the first Secretary of the Air Force, W. Stuart Symington.

Symington was a charismatic individual who had been a highly successful businessman before serving as Assistant Secretary of War for Air. As Secretary of the Air Force, he sat on the National Security Council, where he had great influence on his fellow Missourian, President Truman. This friendship gave him a leverage over Secretary of Defense Forrestal, who, in any case, saw his role as a coordinator and facilitator rather than as an authoritative leader. Symington would have a tremendously beneficial impact on the newborn USAF, through the introduction of a cost-control consciousness and his firm and unequivocal backing of the integration policies of the President. The degree of Symington's business sense might be found in the small size of his staff: four officers and eleven civilians. (Symington would later take the ultimate step of a great leader who has somewhere else to go, resigning in 1950 when he determined that the then Secretary of Defense, Louis A. Johnson, had reduced the Air Force budget to unconscionably low levels. Fortunately for the USAF, Symington was elected Senator from Missouri, and in that role was able to continue to foster the Air Force's growth and well-being.)

The first Secretary of the Air Force had a long and pleasant relationship with General Spaatz, the first Chief of Staff. Spaatz, regarded by Eisenhower as the best operational airman in the world, was rather stiff in presentations with Congress, so it evolved that Symington became the point man for the Air Force, his information provided by Spaatz. And because of his confidence in Spaatz, Symington was amenable to his suggestions regarding senior Air Force positions, including that of Spaatz's successor. It was here that Arnold's influence continued, for Spaatz followed in his footsteps in selecting for the future rather than rewarding past performance.

The independent air force, conceived in the 1920s, had experienced an abnormally long gestation period, not coming to term until 1943 and struggling in labor until 1947. The organizational street fight that characterized the birthing process would be maintained in the future, often with more intensity and less decorum. The basis for the continued conflicts lay in the ambiguity caused by the general statements in the National Security Act of 1947 and by the comparative brevity of Executive Order 9877. It would have been unreasonable to expect that such a major change in the American military posture, coming as it did at the conclusion of a great world war and in the midst of a cataclysmic demobilization process, could have been brought into being with every element decided.

The ambiguity in the two enabling documents would in the course of the next year call forth further meetings and agreements to refine the decisions. The results of these meetings impelled President Truman to revoke Executive Order 9877 on April 21, 1948, and issue in its place a much more detailed statement of functions of the armed forces and the Joint Chiefs of Staff. This did not resolve all the turf problems, but it reduced them to a few core issues that would continue to flare up over the years. Some of them could never be resolved.

After Independence

When the Air Force Falcon was finally hatched on September 18, 1947, it immediately became obvious that an immense amount of work had to be done to see that the new service had all the necessary elements to be on a completely equal footing with the Army and Navy.

One of the first and most important changes was to secure a separate promotion list for Air Force personnel, taking them out from under the shadow of the long Army list. Spaatz wanted to build a strong officer corps, one that would want to stay in the service regardless of civil enticements. The Air University was to provide an integrated training system, one that would treat flying and nonflying officers evenhandedly, just as the promotion system would. (It would probably be difficult to convince some nonflying officers that this noble sentiment ever became a reality.)

Spaatz placed equal emphasis on creating a stable enlisted force, one that would attract the best possible candidates by providing extensive training, not only in Air Force specialties but in more general courses. Unfortunately, the essential elements of what today is called "quality of life" were largely overlooked; in many respects, the Air Force remained the Army in its lack of consideration for the living standards, pay scales, and respect due enlisted men.

The most difficult task was to persuade the Army that many of the functions that it had performed for the Army Air Forces—such things as medical, commissary, laundry, and salvage and repair—now had to be duplicated in the Air Force. Such duplication obviously flew in the face of economies of scale, but the strong feeling within the Air Force was that it could never be truly independent unless it operated almost all of its services. It couldn't be done all at once, but it had to be done over time.

In the meantime, the Air Force still had to carry out its mission of defending the country amid the confusion of demobilization and the division of ever tighter budgets with the Army and the Navy.

A Critical Time

Even as the emerging Air Force sought to resolve the problems of roles and missions with the Navy and administrative and logistic problems with the Army, it was placed in the unpleasant position of being (although almost disarmed) the principal deterrent to the increasing incidence of Soviet aggression.

With the gigantic Soviet Union now several years gone, is difficult to remember the concern it caused in the early postwar years. Joseph Stalin managed to conceal just how near mortal had been the wounds inflicted by Germany. He did not demobilize on anything like the scale that the Western Allies did; instead of turning the mammoth wartime industry of his country to peaceful use and returning his soldiers to the farms, he maintained a war economy.

Stalin, while not the most doctrinaire of his Communist colleagues, must certainly have felt that the tide of Communism was rising. Just as the effects of World War I had brought Communism to all of Russia, so could the effects of World War II create global Communism. Stalin felt that he could almost immediately parlay his political, military, and territorial advantages into the expansion of Communism. Much of the world seemed ripe for the picking; significant gains were being made in China, and the Communist Party was the largest political party in both France and Italy and could bring down the weak coalition governments that faced them at any time they chose.

The great Soviet leader—for as despotically cruel as he had been and would be, Stalin was as essential to the Soviet Union during World War II as Churchill had been to England—believed in his trusted subordinate Vyacheslav Molotov's observation "What happens to Berlin happens to Germany; what happens to Germany happens to Europe." Like all dictators, he needed a foreign enemy against whom he could rally his people, and he fostered a hatred of the Western Alliance. None of the material evidence of America's friendship, from what Churchill called "the most unsordid act," Lend-Lease, to all of the concessions rendered at Yalta, gave Stalin any doubt that the United States was the next enemy.

And it was against that enemy that he gathered information from German scientists and American and British traitors to rush the technological development of his country. He pushed on every front, including the design of an atomic bomb and a masterfully adapted copy of the Boeing B-29 to carry it. Just as American forces had done, Russian teams ransacked German scientific centers for data on jet engines, swept wings, poison gas, and every other military advance of the Germans.

Stalin also accelerated Communist progress politically, from his domination of the division of power in the United Nations to his overt political takeovers backed by the threat of Soviet power. With Communist regimes in all the governments behind the iron curtain, the Soviet Union engineered a coup in Hungary, taking over on June 21, 1947. Similar techniques brought Czechoslovakia to heel on February 29, 1948. Romania came under complete control in December of that year as King Michael was forced to abdicate.

The momentum of the Soviet expansion was accentuated by the obvious political decline of the Western Allies. During the same interval, the Philippines had been granted independence by the United States, while England was divesting herself of India, Pakistan, Burma, Palestine, and other elements of her once vast empire. Other nations, including the Netherlands and France, also looked to the certain loss of their former possessions.

With the exception of the United States, the Western Allies and the former adversary states of Austria, Germany, and Italy were also in critical economic distress. To offset this, in June 1947 the United States announced the European Recovery Program, which was designed to rehabilitate the economies of Western or Southern European countries, including those occupied by the Soviet Union. This became the keystone of the Truman Doctrine, which stated that the United States would help free people fight Communist aggression. Although Truman was really behind the European Recovery Program, for internal U.S. political reasons it was called the Marshall Plan, in honor of George C. Marshall, now Secretary of State.

The Soviet Union refused assistance, and also forced those countries under its dominance to refuse. The $13 billion in assistance that went to the remaining seventeen countries (Austria, Belgium, Denmark, France, Greece, Ireland, Iceland, Italy, Luxembourg, the Netherlands, Norway, Portugal, Sweden, Switzerland, Turkey, the United Kingdom, and West Germany) proved to be vital. The Marshall Plan was a tremendous success, causing a rise in the gross national product of these countries on the order of 15 to 25 percent. It should be remembered that, to his credit, Truman backed the Marshall plan at a time when he was considered a "lame duck," when his standing was so low that his party asked him *not* to campaign for Democratic candidates, when he lacked a majority in both houses of Congress, and when the drive to achieve a budget surplus coincided with increased spending for welfare programs so that the defense budget was kept at about $13–14 billion.

Recognizing the hazards, the new United States Air Force immediately tried to assemble the forces necessary to protect American interests, but it had an embarrassingly long way to go.

The leaders of the postwar Air Force were faced with some contradictory problems. The Air Force was supposed to be the sword and shield of the United States defense policy, yet it had few aircraft and fewer atomic weapons. At the end of 1948, SAC had 319 Boeing B-29s, 230 North American F-51s, and 120 Lockheed F-80s. (The designation "P" for pursuit had changed to "F" for fighter in 1948.) An air-refueling capability was just in the process of being developed, and there would be no operational tankers until late 1948. There were no very-heavy-bomber bases in Europe. As a result, the presumed enemy, the Soviet Union, lay beyond the USAF's practical striking capability unless one-way missions were used. In effect and by default, the strategy of the United States had sunk to the level of a kamikaze attack. (The one-way attack would not disappear from planning for years to come. In 1953, the author's B-50 crew had the mission to attack targets at Tula, a Soviet city near Moscow, and then to turn southwest in the hope that a successful bailout could be made somewhere in the Ukraine, where, we were told, we might encounter "friendly natives." We were not optimistic about the outcome.)

The specter of American economic intervention combined with the successful coups in Czechoslovakia and Hungary to embolden the Soviet Union.

Its next step was to begin tightening its hold on Berlin, which lay deep within the Soviet occupation zone of Germany.

THE BERLIN BLOCKADE:
FIRST CHALLENGE FOR THE USAF

The presence of French, British, and American troops in West Berlin was a continual irritant to the Soviet Union, from a social rather than a military standpoint. Because it was political dynamite to expose Soviet soldiers and the citizens of East Berlin to the decadent luxuries common to the armed forces of the Western Allies, the Soviets began to squeeze the Allies out of West Berlin by a gradual application of force in the form of restricted road and rail traffic.

On June 22, 1948—not coincidentally the seventh anniversary of the Nazi invasion of the Soviet Union—the gloves came off, and all barge, rail, and road traffic into West Berlin was halted.

The Soviet Union's ground forces vastly outnumbered those of the Allies, having thirty full-strength army divisions—as many as 400,000 troops, well supported by the capable Soviet tactical air force. The United States had about 60,000 troops in Europe, and of these about 10,000 were in the understrength 1st Cavalry Division.

The USAF was in equally poor shape. The 28th Bomb Group flew into Scampton, in Lincolnshire, on July 17, while on the 18th, the 2nd Bomb Group arrived at Lakenheath. On August 8, the 307th Bomb Group, comprising the 270th and 371st Bombardment Squadrons, came to Marham. The B-29s had no atomic bombs at their disposal, but the Soviet Union could not be sure of this. During the third week of July, sixteen F-80s of the 56th Fighter Group at Selfridge AFB, Michigan, set out for Europe. Led by the famous 22.5-victory ace and future in-flight-refueling expert Lieutenant Colonel David Schilling, it flew in stages to Odiham, in Hampshire, and then was transferred to Fürstenfeldbruck, near Munich. It was a token effort, but it was all that the USAF could muster.

The Berlin Blockade proved that committees are not always bad. Lieutenant General Albert Wedemeyer, Deputy Chief of Staff, Operations, and Lieutenant General Henry S. Aurand, Chief of Logistics of the Department of the Army, visited General Lucius D. Clay, the U.S. Military Governor in Occupied Germany, in the spring of 1948. The three men concluded that an airlift was feasible, and so recommended to President Truman. Clay then asked Major General LeMay, now Commander of United States Air Forces in Europe, if the Air Force could supply West Berlin with enough essentials by air to sustain the citizenry until the blockade could be lifted by diplomacy—or other means. LeMay, typically, immediately responded that it could, although preliminary calculations showed that to supply West Berlin at minimum level would require at least 4,500 tons of food, coal, and other supplies a day—far beyond any capability then in Europe.

LeMay immediately brought to bear all his authority and energy to create an initial airlift of 102 Douglas C-47s and two Douglas C-54s, and on June 26, flew in the first 80 tons of supplies, consisting primarily of medicine, flour, and milk. The 60th and 61st Troop Carrier Groups formed the heart of the effort, but C-47s were pulled in from all sides. Two days later, the Royal Air Force began operations with its own fleet of Dakotas (C-47s) and seven squadrons of Avro Yorks, a four-engine sister ship of the Lancaster that had devastated Germany at night during the war.

The initial airlift efforts were not very efficient, being pure improvisation, a tailoring of existing procedures to the mundane task of hauling coal, food, and medical supplies. On July 23, the Military Air Transport System established an Airlift Task Force (ATF) that brought seventy-two C-54s to the task, along with three crews per plane for a twenty-four-hour operation.

The USAF next brought in the world's acknowledged expert in airlift techniques, Major General William H. Tunner, who had made his mark flying the Hump in the massive air supply system of the China-Burma-India theater. After analyzing the situation, Tunner brought about a merger of USAF and RAF operations and organized the airfields, routing, and equipment to provide the maximum efficiency so that the airlift functioned more like a bulk-materials-handling factory than an airline.

Along with a few other miscellaneous types, including (briefly) a Boeing C-97, the giant Douglas C-74, and the workhorse Fairchild C-82, he eventually brought in 319 C-54s, which could carry three times the load of the C-47, and launched them at ninety-second intervals from bases at Celle, Fassberg, Rhein-Main (a base built originally for the giants of another age, the *Graf Zeppelin* and the *Hindenburg*), and Wiesbaden. The aircraft cruised at 170 mph, separated by 500-foot altitude increments and three minutes in time, and landed at either Gatow or Tempelhof, and later at Tegel. All air traffic was strictly controlled, and the ground-controlled approach (GCA) system became increasingly more proficient as time passed. Tunner insisted that all crews land if ceiling and visibility were 400 feet and one mile or better, but threatened court-martial for anyone who went below those minimums.

Efficiency was everything; Tunner reduced ground time to an amazing thirty-minute turnaround, including unloading and refueling. Crews could not leave their planes; refreshments, flight clearances, and weather information were brought directly to them while German civilian crews, many of them women, raced to unload, glad to be helping foil the Reds, happier still to have a job that paid DM 1.20 per hour (about a quarter) and a hot meal.

The results were stunning. Daily airlift totals grew steadily, reaching 2,000 tons per day by July 31 and 5,583 tons per day by September 18, 1948. By October 20, the basic requirement for Berlin was raised from 4,500 tons to 5,620 tons, permitting an increase in rations. By April 15, 1949, 1,398 aircraft brought in 12,940 tons of goods, a figure that convinced the Soviet Union that the blockade was not only broken but had been turned into a smashing political triumph for the Western Allies. When the airlift was finally terminated on September 30, 1949, it had delivered 2,325,000 tons of food, fuel, and supplies to Berlin, a quantity in excess of what would have been routinely brought in by surface traffic during the period. U.S. aircraft had flown 189,963 flights, in 586,827 hours of flying time, for a total distance of 92,061,863 miles—about the distance to the sun.

The $200 million effort had not been without human cost—twelve crashes had claimed thirty-one American lives. The political benefits were incalculable. For the first time since the end of the war, American resolve had halted the Soviet Union in its tracks—with air power and without the use of weapons. Yet had the United States been strong, the blockade would not have been attempted.

THE PRESIDENT'S AIR POLICY COMMISSION

In spite of the cost—and the success—of the Berlin airlift, the Truman administration remained acutely conscious that the American public was focused on reviving the economy. In these times of multitrillion-dollar deficits, it seems strange, but the idea was not only to balance the budget but to create a surplus with which to reduce the deficit caused by the war.

Truman was a believer in air power, and he recognized the hiatus the drastic reductions in spending had caused the U.S. aviation industry. In July 1947, he appointed Thomas K. Finletter to head the President's Air Policy Commission (usually called the Finletter Commission) to provide advice on national aviation policy. (The action was similar to the significant President's Aircraft Board of 1925, which did much to promote the growth of aviation.)

Finletter (later a Secretary of the Air Force) advised that a modern seventy-group air force be developed to counter the foreign (i.e., Soviet) nuclear threat by both defensive means and the threat of offensive action. The commission's report, released in January 1948, said that the military establishment had to be built around air power. Further, it candidly admitted that the Air Force was not capable of fulfilling those duties with its current strength. It recommended a force of more than 12,000 modern aircraft of which at least 700 were to be bombers capable of carrying atomic weapons. The report estimated that it would be 1953 before any aggressor would have nuclear weapons, and also recommended an expansion in naval air strength.

Unfortunately, the Finletter Commission's recommendations ran counter to fiscal realities as perceived by the Truman administration. A limit was put on the fiscal year 1949 budget of $14.4 billion (net)—about one half that projected by the Joint Chiefs of Staff as the absolute minimum necessary. The funds were to be split on an approximately equal basis by the three services.

Secretary Forrestal's health was gradually undermined in part by the disagreements between the Air Force and the Navy over the budget and over roles and missions. After a series of meetings, at Key West, Florida, in March and at Newport, Rhode Island, in August, the strategic issues were parceled out, with strategic air operations awarded to the USAF and control of the sea to the Navy—each service was to assist the other. Unfortunately for the future, no clear mandate was given for missile and space activity. An unexpected but fortunate result of these talks was the creation of the position of Chairman of the Joint Chiefs of Staff.

The strain of the arguments, amplified by the fact that Forrestal himself had insisted that the role of the Secretary of Defense be more that of a coordinator than a decision-maker, took its toll on his mental and physical health. He resigned in March 1949, and was diagnosed as suffering from paranoid delusions, one of them being that he was followed by Israeli agents. On May 22, 1949, he leaped to his death from the nineteenth floor of the Bethesda Naval Hospital. Years later Israel revealed that it had in fact had him followed by its agents.

Forrestal was succeeded by Truman's chief fund-raiser for the 1948 election campaign, Louis A. Johnson. A great table-pounding bear of a man, Johnson was 6 feet 2 inches tall, weighed 250 pounds, and employed intimidation as his principal human relations tool. He alienated every member of Truman's cabinet on many occasions, displaying to them at their meetings the manners of a professional football coach responding to a referee making an adverse call in the Super Bowl.

Yet Johnson harbored his own political ambitions to succeed Truman, and thought these could be furthered by exercising a tight control over military spending. Under his leadership, and against the timid advice of the JCS and his service secretaries, Johnson supported Truman's insistence on keeping proposed defense budgets at a very low level.

The budget dropped to $14.2 billion for fiscal year 1950, reducing the USAF to forty-eight groups rather than seventy. To reach even that modest goal, the USAF had to cut eleven groups from its already too-weak force.

Johnson's economy measures extended to the Navy; in April 1949, he abruptly canceled the USS *United States*, the supercarrier that the Navy wished to use as a step toward a strategic nuclear bombing capability.

He made the decision on the advice of the JCS, but without consulting the Secretary of the Navy, John L. Sullivan, who resigned three days later. The cancellation precipitated what has been called the "Revolt of the Admirals," but what was actually a mutiny. Had some poor enlisted man attempted a similar defiance of order and protocol, he would have been court-martialed without a moment's hesitation. Instead, admirals got away with an unseemly attempt, using crudely bogus anonymous documents, to malign Secretary Symington and the Air Force. Called by Symington "the best hatchet job that I have seen since I've been in town," the documents alleged that the Consolidated-Vultee B-36 did not perform up to specification (future Chief of Naval Operations and then Vice Admiral Arthur W. Radford termed it "the billion-dollar blunder") and that the procurement process was fraudulent, with illegal agreements having been made by Symington and the president of Consolidated-Vultee, Floyd Odlum, often better known as Jacqueline Cochran's husband.

The genesis of the B-36 had been the Air Corps' concern that England might be invaded by Germany, and that the United States would require an intercontinental bomber. Design studies were submitted on April 11, 1941, and Consolidated won the paper competition and was awarded a contract for two prototypes. The war relegated the B-36 to a lower priority, and its first flight did not occur until August 8, 1946.

The huge six-pusher-engine B-36 had problems that might be expected in such a precedent-breaking aircraft, but it also proved to be too slow and unable to attain a sufficiently high bombing altitude. Development continued, and on March 26, 1949, the first B-36D flew. It benefited from a spin-off of jet engine technology; four jet engines, two each in nacelles exactly like those used on the inboard wing stations of the new Boeing B-47, were fitted. The new power imparted a speed and altitude capability that enabled the B-36 to earn its nickname "Peacekeeper." It never dropped a bomb in battle, thus exactly fulfilling its mission of deterrence.

House Armed Services hearings, headed by Congressman Carl Vinson, proved that there was not an iota of truth in the allegations contained in the documents being cited by Congressman Carl Van Zandt (a lieutenant commander in the Navy Reserve). The author of the anonymous documents proved to be Cedric Worth, an aide to the Assistant Secretary of the Navy and a former Hollywood script writer, who, by some imaginative twist, was also the person assigned by the Navy to verify the validity of the documents.

The hearings gave both the Air Force and the Navy a chance to present their cases. In essence, the Air Force contended that the B-36 was not invulnerable, but that enough would get through to achieve its mission. The Navy took the tack that it was essential first to win the tactical air battle and obtain air superiority before committing the bomber forces. It also stated that it would be immoral to carry the war to the enemy homeland's population, as an atomic strike would do, although, somewhat inconsistently, it wanted its own capability to do so. The Navy's view was that "we cannot afford our strategic air component over a long term in peacetime, and we should not plan for intercontinental strategic bombing if any other way is possible." The Navy then found itself in the strange position of extolling the virtues of the Boeing B-47 jet bomber, but only because funding for the B-47 was downstream, when another battle could be fought.

Air Force Chief of Staff General Vandenberg showed the B-36 to tremendous advantage in its defense. He rejected the Navy's argument that the Air Force was seeking a cheap and easy way to wage war, saying, "A prime objective of this country must be to find a counterbalance to the potential enemy's masses of ground troops other than equal masses of American and Allied troops. No such balancing factor exists other than strategic bombing." He also noted that only four of the Air Force's forty-eight groups would have B-36s and that the total cost of the B-36 program was less than $1 billion.

The net result of the committee hearings on the surface was that the Air Force emerged untainted, the B-36 was accepted as a worthwhile weapon system, and the Navy had egg on its face. At these hearings, the Navy's staunchest friend, Representative Vinson himself, moved perceptibly toward the Air Force camp.

Yet the Navy had touched a sensitive nerve. The United States Air Force was not yet capable of carrying out the type of mission called for by the Finletter Commission, or, indeed, by its own charter.

Secretary of the Air Force Stuart Symington was dissatisfied with the defense budgets of 1950 and 1951, and with a becoming courage resigned in protest. He was succeeded by Thomas K. Finletter, who lacked the leadership qualities Symington possessed in abundance. Finletter was not a "people person" as Symington had been, and had difficulty getting along with his Chief of Staff, General Vandenberg, among others. He was not interested in routine details, following larger issues like the newly formed North Atlantic Treaty Organization (NATO), perhaps because they distanced him from interaction with his staff. Yet it was to be Finletter's lot to be Secretary when the pressures of war suddenly opened the money floodgates again. In the meantime, as so often would be the case, the serviceman suffered.

PEOPLE ISSUES

One of the most important considerations of contemporary Air Force leadership is the improvement in the quality of life of service people. While never actually defined, the standard for that quality of life is that of middle America as a whole. To appreciate the change in thinking, we need to understand the baseline of issues regarding the care and welfare of service members in the new United States Air Force. These are reflected in the typical public attitude to the military services in the pre–World War II years, which, sadly enough, ranged from condescension to contempt.

In those days, while an education at West Point or Annapolis was not to be discounted, and while a person of independent means like George Patton or Billy Mitchell might be able to imbue service life with a panache not found at a local country club, service personnel, officer and enlisted alike, were generally looked down upon by their civilian counterparts. In the case of officers, the agonizingly slow promotion system, stultifying social life, and generally hardscrabble nature of living on remote posts was enough to cause civilians to wonder why anyone would put up with it—if one could do something else. "Genteel poverty" was a term used to describe the threadbare existence of military families; the niceties of calling cards, being "at home," and other such relicts of a bygone age served as plaintive substitutes for vacations, good schools, elective orthodontia for children, and other perks enjoyed by civilians in comparable positions.

In the Air Corps, the pilots at least found that flying was an exhilarating compensation for some of the disagreeable aspects of service life. Flying pay was both an inducement and an issue of contention. Fliers loved it, and showed it off with modestly conspicuous consumption, perhaps buying a secondhand Pontiac instead of a secondhand Chevrolet. Nonfliers, naturally enough, hated the very idea of flying pay, and the economy-minded Congress continually sought to eliminate it. The cruel fact of the matter was that the flying pay would, if used solely for the purpose, buy just enough life insurance to offset the loss of pay inherent in the shorter average life span of pilots. It was actuarially, if not emotionally, sound.

But the public on the whole thought that those who stayed in the service, fliers or not, did so because they didn't consider themselves able to compete "on the outside." The term "lifer," although more common now than then, was used contemptuously to characterize servicemen who stayed in for the dubious reward of a minimal pension. Yet when war came, the spectacular performances turned in by a host of the veterans of the deprived years of the Air Corps should have disabused this idea forever.

The civilian view of their life had a reverse resonance within the service, whose members took a grumpy pride in enduring the hardships and regarded those who opted out as "quitters." When the record-setting Jimmy Doolittle, with a growing family, a doctorate in engineering, and a celebrated flying career, found himself locked in place as a first lieutenant for eleven years, with no assurance of promotion in the near future, he decided to leave the service. He found ready employment that included not only higher pay, but promotion and perquisites. Yet when he returned to duty with the advent of World War II, many senior officers, including General Eisenhower, looked at him askance as one who had "jumped ship" when times were tough. It took many months of typically outstanding Doolittle performance to change their minds.

The public opinion of enlisted men (for women were not yet allowed to make their contribution to the ranks) was equally harsh, although somewhat mitigated by the fact that the Depression had caused such widespread unemployment that any job, even a military one, was coveted. In general, however, fathers' faces did not light up when their daughters reported that they were going to date an enlisted man, nor did mothers' hearts leap when sons told them they had been accepted for enlistment. While James Jones's depiction of Army life in his famous *From Here to Eternity* might not be absolutely accurate in the eyes and minds of the servicemen it portrayed, it was on the mark in terms of public perception of those men.

To the extent possible, the cruel public attitude was mitigated within the services. During peacetime, little privileges such as "Wednesday-afternoon athletics" during which officers and men had time off to exercise—or whatever— were common. Officers were careful to maintain a social distance between themselves and enlisted personnel (this gap was narrowest in the Air Corps), but they did help them as much as they could. There were usually baseball fields and other similar low-cost sports facilities, and, of course, the clubs, where beer and liquor were cheap, and drinking was for many a way of life. (Drinking was a costly factor in service life for decades; only in the past thirty years have the services emphasized the virtues of temperance by relating it to performance reports.)

There was not, however, the array of services available to either officers or enlisted men that is common today; it would not have occurred to a private in the 1930s Air Corps that the military would provide him any perquisites such as legal assistance, psychological counseling, job training, or the opportunity for on-base college courses. Even less likely would be the provision of well-equipped fitness centers, bowling alleys, comfortable theaters, or post exchanges with an inventory including household items such as refrigerators, radios, and the like. Most PXs were considered well stocked if they had anything beyond cigarettes, shaving gear, and shoeshine materials.

When war came, all of the stigma of service life fell away, and civilians competed to treat servicemen, officers or enlisted, well. If every able person had to be in service, if there was a war to be won, everything was different. The sounds of young women curtly refusing to date servicemen was suddenly drowned out by the music and laughter of slow-moving dances at the USO, and very often, by wedding bells.

In short, the serviceman in peace and war had everything but adequate pay, housing, and services. Of food there was plenty, in wastefully gargantuan amounts, but often poorly prepared by amateur cooks who ladled their sloppy mixtures into cold and greasy compartmented trays. (Today's mess halls, with salad bars, diet plates, food kiosks, all immaculately clean and staffed by eager professionals, would have seemed half Disney, half Orwell at the time.)

In the vast World War II expansion of the military, bases were thrown up at a staggering rate, all characterized by barren open-bay barracks with group showers and open rows of toilets, complemented by equally austere mess halls, administrative offices, classrooms, and supply huts. It was a temporary world of two-by-four pine and slapdash coats of whitewash, intended to be just good enough to last the war. Larger organizations, like hospitals, were a series of barrackslike buildings strung together with covered walkways. Like most wartime construction, the buildings were rarely insulated. The windows stuck and the heating systems malfunctioned. Elbow grease and hard paste wax kept buildings clean and glistening, but nothing could make them comfortable. They were "temps," temporary buildings, all slated for salvage or replacement in peacetime.

Except, of course, that after the war was over and the services had demobilized, there was no money—nor any apparent reason—to either salvage or modernize the facilities. Many buildings designated "temporary" in 1941 were still doing business thirty years later, improved with some interior walls and some extra coats of paint perhaps, but still just rectangular pine shacks. It was worse overseas. As late as 1948 in Guam, and in many other places, enlisted personnel still had to make do with open-pit latrines, tents, and open communal washbasins.

Families had suffered during the war, for there was no mechanism to care for them, and no incentive to do so in the pell-mell rush of war. A $10,000 life insurance policy was the only thing available to ease the cold distress brought by the standard War Department telegram notifying parents of a lost son.

In the prewar days, base housing had always been limited. It was intended primarily for relatively senior officers. Junior officers and noncommissioned officers were informally counseled to stay single, while at intervals enlisted personnel were officially counseled to do so. When war came, there was little pressure to expand base housing for families, for personnel rarely stayed for more than a few months at any one place. The wives who followed their husbands from base to base, trying to make a home on the economy, found themselves living in everything from modified chicken coops to two-family-per-room boardinghouses.

Because they were young, and because it was wartime, it was somehow not only endurable but—retrospectively at least—romantic. For many years after the war, the situation eased only slightly. Supply and demand took some of the most outrageous accommodations off the market, but in areas where bases remained at full activity after the war, landlords made a killing for as long as they could.

The fledgling Air Force simply had too much on its hands to begin to attend to such fundamental concerns as making life better for the enlisted troops or improving conditions for families. It would be several years—in many cases till long after the Korean War—before commanders would begin to realize how essential these matters were, and to lobby effectively for reform with Congress. As the years passed, however, the Air Force became increasingly aware that such care was not only important, but essential to survival, and as such, one of the most cost-effective ways to use Air Force resources.

Something that should perhaps have been obvious became recognized only in the late 1970s, when it was demonstrated that budget money spent on improving the quality of life was as important as budget money spent on weapon systems, because of its direct and potent impact on job performance.

Even though the Air Force would be in the vanguard of improving the quality of life of its members, compared to the other services, it had to learn the hard way, following old habits and old routines. Some of the lessons would be taught in the three years of bitter fighting in Korea.

3
≋

The First Crucible

*I*n many respects, the United States' bungling, unprepared entrance into the Korean conflict—the wrong war, in the wrong place, at the wrong time, with the wrong enemy, according to the Chairman of the Joint Chiefs of Staff, General Omar Bradley—had many elements of a Greek tragedy. There was hubris on the part of both President Truman and Secretary of State Dean Acheson; there was vision blinded by ambition on the part of Secretary of Defense Louis Johnson; and, for all three, there was an exquisitely painful irony regarding the usually laudable quality of loyalty. To this unhappy mixture must be added the supine acceptance by the Joint Chiefs of Staff and the service secretaries of military budgets that they knew to be inadequate to support the foreign policies of the United States in the face of an increasingly aggressive Soviet Union.

Truman's hubris was certainly human enough, given his triumphant election in which he not only defeated his "man on the wedding cake" opponent, Thomas E. Dewey, but carried the Democrats along to majorities in both houses of Congress. This led to a certain arrogance, unusual for him, expressed by his discounting the counsel of his military experts and preferring instead that of his cronies. His antimilitary bias had deep roots. It began with his inability to qualify physically for a service academy appointment, and was heightened by the traditional disdain a National Guardsman has for regular officers. This attitude was reinforced forever by his experience heading the Special Committee to Investigate the National Defense Program—the Truman Committee—during World War II. The very nature of this task brought Truman into contact with the seamy side of wartime military procurement, coloring his attitude about the service and industry long before Dwight Eisenhower (or, more likely, his speechwriter) coined the term "military-industrial complex" in his farewell address in 1961. As President, Truman undoubtedly saw a parallel in his own military self-education and that of Abraham Lincoln, and became convinced that his judgment was as astute militarily as it was politically. He gave his strongest loyalty downward not to his professional military advisers, but instead to former political cronies like Major General Harry H. Vaughn and to fund-raisers like Johnson. Vaughn, his military aide, had no meaningful military experience, yet was in a position of tremendous influence while serving primarily as a yes-man to corroborate Truman's views.

Nor did upward loyalty work for Truman. Acheson's undoubted loyalty simply reinforced the incompatibility of Truman's foreign policies and his defense budgets, the former expressing a hearty appetite for global interests and support for the United Nations, even as the military was placed on a starvation diet. Misguided loyalty bared its ugly head in other venues as well. It was not in Truman's or the nation's interest when Bradley, the Chairman of the Joint Chiefs of Staff, and the three Chiefs all strongly supported Johnson's defense budgets in public, even though they knew that the low levels of those budgets were destroying the very fabric of the armed services for which they were responsible. It was a time when the heads of the military could better have distinguished themselves by a mass resignation than by a continued service. Truman would undoubtedly have accepted the resignations—if Douglas MacArthur did not intimidate him, Omar Bradley surely would not—but the statement they would have made would have brought them more honor than their past accomplishments in battle.

In Acheson's case, hubris stemmed from his patrician background, the success he had gained as a behind-the-scenes manipulator of White House/ State Department intrigues and the inevitable triumphs he scored when brought to a battle of wits with his political enemies. In his aristocratic view, his devotion to Truman was an affecting idiosyncrasy, an illustration of just how superior he was to the hoi polloi of the Congress and the press, for he and almost no one else could see the true merits of the plebeian President from Missouri. To his credit, it must be noted that late in the game his State Department called for the defense of South Korea and for a rearmament policy that should have emanated from the Secretary of Defense.

The opportunistic Johnson's hubris stemmed from humbler roots; he was simply a loose cannon who blew out the decks of the Defense Department and threatened to demolish the entire cabinet. It took but a few weeks of the Korean War to prove just how wrong his eager enforcement of Truman's budget policies had been, for when the North Korean armed forces launched their surprise attack on South Korea on June 25, 1950, the military forces of the United States were caught dangerously unprepared.

INVITATIONS TO TROUBLE

Although the catalog of Communist aggression was impressive, it still had not been enough to arouse the American public or Congress to face up to the cost of an adequate defense force. The administration's naive belief in the deterrent power of the USAF's "air-atomic" (a catchphrase of the time) monopoly prevented it from working for an adequate military budget.

American diplomatic policy had been as remiss as its defensive efforts. The world in general and the North Koreans in particular had been assured repeatedly that the United States did not regard South Korea as a major interest in a series of self-reinforcing disclaimers, each with a different rationale, beginning at the Potsdam Conference during July 1945. Amid the many towering issues of the day, an offhand decision had been made to allow the Soviet Union to disarm Japanese armed forces in Korea down to the 38th parallel, while the United States disarmed them up to that point.

Although it was not until September 8, 1945, that troops of the U.S. 7th Infantry Division entered Korea, the Soviet Union had moved earlier and more rapidly. Within a few days after entering the war, it had sealed off the border, interrupting rail, river, and road traffic permanently between the primarily industrial North and the primarily agricultural South. The thirty-five years of oppressive Japanese occupation had exploited Korea, but had turned it into a viable economic entity. The Soviet division at the 38th parallel created two half-states, neither fully functional.

The Joint Chiefs of Staff, attempting to deal with the disruptions of complete demobilization, and knowing that the services faced a period of declining budgets, had to make hard choices, apportioning forces on the basis of the nation's greatest interest. On September 25, 1947, the JCS provided a memo to the Secretary of Defense, under the signature of JCS Chairman General Eisenhower, advising that there was little strategic interest in keeping the current level of 45,000 (mostly green) troops in Korea. If military action was required to defend South Korea, it was to be provided by means of air power from Japan.

To offset this lack of support, the United States requested the United Nations to arrange to have elections that would unify Korea under one government and end the artificial division that prevented rational development.

Despite opposition to the idea from the Soviet Union, which never welcomed free elections, arrangements were made for a national election on May 10, 1948. The elections were suppressed above the 38th parallel, but in South Korea, the radical Syngman Rhee, seventy-three years old and difficult to handle, was named president of the Republic of Korea on July 20, 1947.

The Soviet Union contested the validity of the election and in the fall of 1948 proclaimed the existence of the People's Republic of North Korea, with the thirty-two-year-old Kim Il Sung (later hailed as "the Great Leader") as premier. As an apparent sop, the Soviet Union proposed a mutual withdrawal of foreign troops, an arrangement exactly to the taste of the United States.

The American troops were pulled out in stages, dropping first to a 7,500-man regimental combat team and then, after a Korean constabulary had been trained, to a small Korean Military Advisory Group (KMAG) of about 500 men. The Korean constabulary was deliberately equipped with only small arms and given no artillery, tanks, or combat aircraft, for the Americans feared that President Rhee would attack the North if he had the military capability to do so.

The Soviet Union had no similar inhibitions and immediately set about training and equipping a formidable North Korean People's Army (NKPA), fully equipped in Soviet style with artillery, tanks, and planes.

The Central Intelligence Agency, still young, but derived from World War II organizations, had said that it did not believe that South Korea could maintain its independence in the face of the withdrawal of U.S. forces. Yet it did not predict that the North Koreans would attack, despite the imbalance of forces and the gravitational tug ordinary economics exerted to unify North and South. North Korea, with a population of about 9 million, was equipped with a formidable offensive force guided by 3,000 Soviet advisers. Nominally democratic South Korea, with a population of about 21 million, had an ill-trained, ill-equipped constabulary of about 50,000 troops, with the 500-man American advisory group. It should have been obvious that North Korea intended to unite the country by force of arms as soon as possible, before South Korea could build its strength.

By the spring of 1949, Truman agreed to internal military pressure that the Republic of Korea army should be built up with U.S. assistance. This decision—foreshadowing the shabby policy of "Vietnamization" more than twenty years later—was another clear indication that the United States had abdicated a serious military role in South Korea, and it was duly noted by the North Koreans.

Two major events should have precipitated a major American defense buildup, but did not. The Nationalist Chinese forces were at last driven from mainland China and Mao Tse-tung announced the People's Republic of China on October 1, 1949, with Chou En-lai as premier. An event not a week earlier was even more ominous: the detection of a Soviet atomic explosion by a just activated Air Force system. When this discovery was announced on September 23, the administration preferred to believe that it was the first Soviet test and that it would still be some time before the USSR had enough bombs stockpiled for a sneak attack. (Despite the lack of a U.S. stockpile of weapons, Truman did not authorize increased production of standard nuclear bombs until October 17, and waited until the following January 31 to authorize research on the hydrogen bomb.)

Yet the possibility lurked that only the most recent test had been observed, that the Soviets had put the bomb into large-scale production even before its test, and that they might, given the pathetic state of U.S. air defenses, send their Tupelov Tu-4s on one-way missions to a dozen or more major American cities. Once again, the fundamental military rule of preparing for what an enemy could do was put aside for the less expensive strategy of preparing for what it was thought it would do.

The two events did trigger a paper from the National Security Council on December 30, 1949, one that was more an admission of weakness than a statement of policy. The paper, NSC 48/2, called for economic assistance to Asian nations to strengthen their resistance to Communist encroachment and explicitly defined the area that the United States would defend militarily as a line that encompassed Japan, the Ryukus, and the Philippines. South Korea was not mentioned. In a further attempt to limit American responsibilities, on January 5, 1950, Truman stated that the Nationalist Chinese forces on Taiwan would get economic but not military assistance.

If the Soviet Union and the North Koreans were not yet aware of the implications of these policies, Dean Acheson made them crystal clear in a speech at the National Press Club on January 12, in which he once again confirmed that the United States would defend only the perimeter specified in the NSC memo, and that other nations would have to rely upon themselves and upon the United Nations for their defense.

Nonetheless, Acheson belatedly recognized the need for a stronger military that would require more than tripling the 1950 defense budget level of about $13 billion. It was too late. Much blood would be shed, and a war would come within a few days and a few miles of being lost because the United States was so terribly unprepared.

MISNOMERS AND MISJUDGMENTS

Three major powers fought two undeclared wars in Korea, and like passengers nervously watching their respective cab drivers brawl, studiously ignored the fact. The Soviet Union, China, and the United States allowed North Korea's army, with Soviet and Chinese "volunteers," to fight against South Korea's army, allied to a United Nations Command made up largely of U.S. troops.

President Truman called it a "police action" to ease his way around the right of Congress to declare war. The United States was able to get a resolution through the UN Security Council accusing North Korea of unprovoked aggression because of an incredible diplomatic blunder by the Soviet Union. The USSR had boycotted the Security Council in objection to Nationalist China's continued representation, and its representative, Jacob Malik, was not present to veto the UN resolution as he otherwise would have done. The resolution laid the groundwork for the later establishment of the United Nations Command that would prosecute the war.

Each side made catastrophic misjudgments. In the first war, the North Koreans, believing what they had been told, invaded South Korea, confident that the United States would not intervene. They were wrong, and they lost that war.

The United States, under the UN banner, pursued the defeated North Koreans to the Chinese border, in the belief that Red China would not intervene. This time the United States was wrong, and it lost the second, larger, war.

And all the while, the United States, the Soviet Union, and Red China remained nominally at peace.

JUNE 25, 1950

At 0400 hours on the morning of June 25, 1950, six columns of North Korean assault troops crashed through token South Korean defenses. A total of 89,000 men, about a third of whom were veterans of the fighting in China, made up seven divisions, a tank brigade equipped with the formidable Soviet T-34 tanks, and two independent infantry regiments trained for special duties. The North Koreans had a reserve force of some 51,000 men grouped in three divisions and their own Border Constabulary. The North Korean Air Force (NKAF) had 162 combat aircraft, including seventy Yakovlev and Lavochkin fighters (near equivalents to the F-51 Mustang) and sixty-two of the excellent Ilyushin Il-10s, an improved version of the historic Shturmovik attack aircraft.

Opposing them were approximately 100,000 South Korean troops in eight understrength divisions; of these, fully one-third were service and support troops, virtually useless for combat. The Americans had intentionally deprived this constabulary of antitank weapons, for the advisory group had considered Korea—like the Ardennes—not to be "tank country." The ROK Air Force had sixteen aircraft: three North American T-6 trainers and thirteen liaison aircraft—five Piper L-4s and eight Vultee (Stinson) L-5s, none fitted with armament.

The North Korean assault was reminiscent of the later days of the Soviet advance through Germany, using tanks, artillery, and infantry in swift-moving combinations that threatened to achieve its objective, the reunification of Korea by military force, within a few weeks. The Republic of Korea forces were badly beaten, and could not have recovered without outside intervention. On June 26, knowing that Seoul would inevitably fall, U.S. Ambassador John J. Muccio ordered all nonessential U.S. embassy personnel and all U.S. dependents in Korea evacuated to Japan.

STATUS OF THE FAR EAST AIR FORCES

In a war full of many ironies, one of the most important was that the budget difficulties which prevented the modernization of the United States Air Force to the degree required left the Far East Air Forces (FEAF) with exactly the right kind of equipment with which to fight the nonnuclear war in which it found itself. Although more modern aircraft in greater numbers would have been welcome, the 365 Lockheed F-80s, thirty-two North American F-82s, twenty-six Douglas B-26s and twenty-two Boeing B-29s that formed the bulk of the combat forces were well suited to slow down and eventually stop the

North Korean invaders. The F-82s would be withdrawn relatively early from combat because of their small numbers and lack of spares.

This force would be supplemented by North American F-51s called back into service because of their longer range and ability to operate from the primitive Korean airfields. These included ten pulled back from a flight intended for transfer to the ROK Air Force, thirty withdrawn from storage in the theater, and 145 obtained from stateside Air National Guard units. (There were more than 1,500 F-51s still available in the United States, about half in storage and half assigned to Air National Guard units.) In a retrograde step unusual in wartime, six F-80C squadrons were converted to F-51s, a seemingly innocuous measure until you realize that the pilots were going into combat immediately in an unfamiliar aircraft with totally different flying characteristics, and knew that maintenance was performed by mechanics equally unfamiliar with their mounts.

In addition, FEAF had twenty-five RF-80s and six RB-29s for reconnaissance, along with a number of aircraft dedicated to weather reconnaissance and air/sea rescue.

Another irony, less fortunate, would rear its head almost four months after the war's start, when the MiG-15 was introduced. One of the poorest countries in the world, China, had brought in an advanced aircraft clearly superior to all UN aircraft yet in the war. They were purchased or bartered for from the Soviet Union, which had a history of not giving away war material that extended back to the Spanish Civil War, when Loyalist gold paid for the Ratas and Chatos received from Russia. Nonetheless, China had the swept-wing MiG-15 weeks before the United States was able to offset them with the North American F-86. Further, the Chinese would maintain an establishment of MiG-15s that would dwarf the available strength of F-86s throughout the war.

Although FEAF had been starved for aircraft, men and supplies, it had nonetheless maintained a high standard of training for its assigned tasks. FEAF was the air component of the Far East Command (FEC) commanded by General of the Army Douglas MacArthur. Its primary mission was to maintain an active air defense of Japan, the Ryukyus, the Marianas, and American bases in the Philippines.

FEAF was commanded by Lieutenant General George E. Stratemeyer, a personable commander with wide experience in the field, but one whose tactical air experience was insufficient to gain the complete confidence of the Chief of Staff, General Vandenberg, a professional in that discipline.

The principal component of FEAF was the Fifth Air Force, famous for its wartime exploits under General George Kenney, and trained primarily for the air defense role. The F-51s, F-82s, and F-80s of the Fifth would soon be called on to engage in close air support, for which they were not trained, but in which they would do surpassingly well. The Fifth was commanded by Major General Earle "Pat" Partridge, and was equipped as the following table shows:

Unit	Station	Aircraft
8th Fighter Bomber Wing (FBW)	Itazuke	Lockheed F-80C
68th Fighter All-Weather Squadron (F[AW]S)	Itazuke	North American F-82
49th Fighter Bomber Wing (FBW)	Misawa	Lockheed F-80C
35th Fighter Interceptor Wing (FIW)	Yokota	Lockheed F-80C
339th Fighter All-Weather Squadron (F[AW]S)	Yokota	North American F-82
8th Tactical Reconnaissance Squadron (TRS)	Yokota	Lockheed RF-80A
3rd Bombardment Wing (Light) (BWL)	Johnson	Douglas B-26
374th Troop Carrier Wing (TCW)	Tachikawa	Douglas C-54
51st Fighter Interceptor Wing (FIW)	Naha, Okinawa	Lockheed F-80C
4th Fighter All Weather Squadron (F[AW]S)	Naha, Okinawa	North American F-82
31st Photo Recce Squadron° (VLR) (PRS)	Kadena, Okinawa	Boeing RB-29
19th Bombardment Wing (M) (BWM)	Andersen, Guam	Boeing B-29
18th Fighter Bomber Wing (FBW)	Clark, P.I.	Lockheed F-80C
21st Troop Carrier Squadron (TCS)	Clark, P.I.	Douglas C-54
6204th Photo Mapping Flight (PMF)	Clark, P.I.	Boeing RB-17

°Later the 91st SRS

Additional units rounding out FEAF's strength included flights of the 2nd and 3rd Air Rescue Squadrons (ARS), with SB-29 and SB-17 aircraft; the 512th and 514th Weather Reconnaissance Squadrons (WRS), with Boeing WB-29s; and No. 77 Squadron of the Royal Australian Air Force, equipped with North American F-51s and stationed at Iwakuni.

All in all, it was a great World War II air force with which to prevent World War III. Fortunately, although the equipment was obsolete and the mission dictated by the war was not the one they had trained for, the aircrews were both highly skilled and adaptable. More important, they were ready to fight.

THE AIR FORCE INTERVENES

As the North Korean forces swept south, the NKAF strafed Seoul and Kimpo airports, wiping out the light aircraft of the ROK AF and destroying a U.S. Military Air Transport Service C-54—a sufficient *casus belli* in earlier times. FEAF planning had anticipated a requirement to evacuate Korea in an emergency, and General Partridge, who was commanding in General Stratemeyer's absence, designated Colonel John M. "Jack" Price of the 8th Fighter Bomber Wing to be the air task force commander for the evacuation.

Price gathered a force of his own F-80 and F-82 fighters, ten B-26s of the 8th Bombardment Squadron (Light), twelve C-54s, and three C-47s, and prepared to evacuate personnel from Seoul. He was then ordered to provide combat air patrols over the freighters evacuating personnel by sea from Inchon.

The mission distance ruled out the use of the F-80s as combat air patrol, and Price sent out a call for reinforcements of F-82s and F-51s and made the decision to use the B-26s in the convoy protection role as well.

A series of aerial combats followed that would define the Air Force's role in Korea and permit it to act as both sword and shield for the ground forces. General Partridge told the Fifth Air Force, "No interference with your mission will be tolerated." And none was.

On June 27, a mixed bag of five Communist fighters headed for the transports operating out of Kimpo, including a two-seat Yakolev Yak-11, a Yak-9, and a Lavochkin La-7. All of the aircraft were slower than the Twin-Mustangs, but more maneuverable, and each had formidable armament. In the hands of expert pilots, they could have been tough opponents for the five F-82s from the 68th and 339th F(AW)S that ambushed them.

The Yak-11 tried to pick off the number four F-82, but Lieutenant William G. "Skeeter" Hudson and his radar observer, Lieutenant Carl Fraser, slipped behind him and fired the six .50 caliber machine guns; the first burst knocked pieces off the Yak's tail, and the second took the flap and aileron off as it set fire to the wing. It was the first aerial victory of the Korean War; 975 enemy planes were to follow.

A few moments later, Lieutenants Charles "Chalky" Moran, pilot, and Fred Larkins, radar observer, shot down a Yak-9, and an La-7 was destroyed by Major James W. "Poke" Little, Commander of the 339th, and his radar observer, Captain Charles Porter. The eager but inexperienced Yak pilots fled. (Some sources attribute the first victory to Moran, who was lost a few weeks later on a night mission, but official credit has gone to Hudson.)

This first step toward air superiority was reinforced later in the day when eight Il-10s attempted to attack Kimpo; four F-80C jet fighters of the 35th Fighter Bomber Squadron promptly knocked down four of the Ilyushins, and the remainder turned tail. Two victories were credited to Lieutenant Robert E. Wayne, while Captain Raymond E. Schillereff (the 35th's operations officer) and Lieutenant Robert H. Dewald got one each.

The FEAF fighters provided impeccable protection to the transports operating out of Seoul and the freighters operating out of Inchon. C-54 and C-47 transports flew 851 persons safely to Japan, while another 905 were removed, far less comfortably, by water.

It was a good beginning for the Air Force in a bad war. These opening battles also forecast the essential, decisive element of the Korean War—absolute air superiority—that alone sustained the United Nations forces, twice keeping them from ignominious defeat and ultimately convincing the Communist forces that military victory was impossible. Space limitations prevent detailing each of the 720,980 sorties flown by FEAF during the Korean War, but each one, whether flown by a North American F-86 Sabre in a thrilling dogfight over the Yalu or by an aging Curtiss C-46 Commando limping in to deliver vital cargo to trapped troops, represents a level of proficiency and heroism that nurtured the dramatic changes that the war was going to bring about in the Air Force.

The Korean War was not only a conflict against a competent, determined enemy, it was also the birthing process for the professional Air Force, which would grow in size and strength and carry the United States to the rank of sole superpower. The Korean War was not fought the way the Air Force would have chosen to fight it—nor would most later wars be so fought. The Air Force fought the Korean War according to the dictates of Far East Command and the decisions made in Washington by politicians who were dismayed that the magic "air-atomic" shield was not applicable to "local" conflicts and that a World War II–style battle was in process.

These dictates and decisions often contravened sound air power policy, but the Air Force not only carried out its orders, but accomplished its own larger aims of defeating the enemy in the air and interdicting him on the ground. When at last truce talks began, Lieutenant General Nam Il, chief North Korean delegate to negotiations at Kae-song, said, "I would like to tell you frankly that in fact without direct support of your tactical aerial bombing alone your ground forces would have been completely unable to hold their present positions. It is owing to your strategic air effort of indiscriminate bombing of our area, rather than to your tactical air effort of direct support to the front line, that your ground forces are able to maintain barely and temporarily their present positions." (Nam Il considered the bombing of bridges, railroad yards, electrical power stations, and other such targets as "indiscriminate".)

THE PHASES OF THE WAR

The Korean War was divided on the ground and in the air into five phases that were correlated but not identical. The ground phases consisted of the following:

1. *June 25 to September 14, 1950.* UN forces were driven back into the Pusan (Naktong) perimeter and, severely punished, were several times in danger of being driven into the sea. Even when U.S. forces were reinforced and attempted a counterattack, the NKPA was able to handle them roughly.

2. *September 15 to November 24, 1950.* General MacArthur reversed the course of the war with his magnificent (albeit risky and perhaps even unnecessary) counterstroke at Inchon, coupled with the subsequent breakout of the Eighth Army from the Naktong perimeter. The NKPA,

exhausted by battle and depleted by a merciless air interdiction, was driven back across the 38th parallel by UN forces.

On September 27, a decision was made that the UN forces would proceed all the way to the northern borders of North Korea and unify the country politically under the regime of Syngman Rhee. Despite explicit warnings that Communist China regarded Korea as its special sphere of interest, the Joint Chiefs of Staff continued to assume that the North Korean action had not only been sanctioned but masterminded by the Soviet Union. The JCS analysis indicated that the Kremlin would not intervene militarily on behalf of North Korea nor allow Red China to do so. Although General MacArthur normally disagreed with the JCS on practically everything, in this instance he reinforced their stand by stating that China would not intervene with more than a few thousand troops and covert assistance to the North Koreans. He repeated this assurance personally to President Truman at their celebrated (first and last) meeting at Wake Island. The CIA, keeping its Asian forecasting record unblemished, also stated that there would be no intervention to assist the beleaguered North Koreans. The erudite Dean Acheson, who should have known of Korea's two millennia of importance to China as a buffer state, also discounted the risk. Even the first Chinese offensive on October 25 did not register. The wine of the UN victory was too heady for American leaders, who were about to experience a Communist Chinese hangover.

3. *November 25, 1950, to January 24, 1951.* The massive Communist Chinese intervention, begun during the coldest winter in Korea in more than 177 years, caught UN forces completely by surprise and sent them reeling back down the peninsula, suffering heavy losses in the process. The 38th parallel was crossed again on January 1, and Seoul captured for the second time on January 4. Eighth Army Commander, General Walton H. Walker, was killed in a vehicle accident in December, and was replaced by Lieutenant General Matthew Ridgeway, who by sheer force of personality reinvigorated UN troops with the will to resist.

4. *January 25 to November 12, 1951.* Ridgeway adopted a new strategy of attrition that won him acclaim as one of the great combat commanders in history. He no longer had victory and unification as his primary task, but instead negotiation and an armistice. Under his direction, the UN forces fought their way back up the peninsula to a line that ranged beyond the 38th parallel by a few miles. Frequently interrupted armistice talks began at Kae-song on July 10, and were moved to Panmunjom on October 7. The most significant personnel action of the war came on April 11, 1951, when President Truman relieved General MacArthur of all three of his commands—U.S. Far East Command, United Nations Command, and Supreme Commander, Allied Powers. General Ridgeway was named to replace him, and Ridgeway in turn was replaced by Lieutenant General James A. Van Fleet. (Van Fleet would be in command when his own son, Lieutenant James A. Van Fleet, Jr., was lost flying a 13th BS B-26 on a night mission over North Korea.)

5. *November 13, 1951, to July 21, 1953.* As the talks dragged on, there was a long and bitter war consisting of local attacks and counterattacks that caused many casualties but did nothing to assist the peace process. A principal problem was the repatriation of prisoners of war; almost half of the 134,000 Communist prisoners indicated that they did not wish to go back to their homes of origin. The Communists demanded that all POWs be returned, regardless of their personal preferences. A compromise was finally reached that allowed the prisoners who wished to return home to do so within sixty days of signing the armistice; those that did not would be released to do as they wished.

A final large-scale Communist offensive took place in June 1953, but, without air support, suffered a devastating loss of 70,000 troops and ground to a halt. Total casualties in June and July, the last two futile months of a futile war, amounted to more than 160,000 for both sides. The armistice was signed on July 27, 1953, with the battle line becoming the boundary between the two countries.

The UN forces lost 118,515 men killed and 264,591 wounded, with 92,987 captured. U.S. forces lost 33,629 killed and 103,284 wounded. About 3 million South Korean civilians were killed. The Communist armies had about 2 million casualties. A quick calculation shows that there were at least 5.5 million casualties not including civilian wounded, and an incalculable expenditure of treasure, with virtually no change in the boundary or the political status quo.

THE CONDUCT OF THE AIR WAR

All elements of Air Force effort, including air superiority, strategic bombing, reconnaissance, air/sea rescue, cargo (both inter- and intratheater), close air support, and the interdiction of enemy supplies, were constantly adjusted to the exigencies of the ground war, and although frequently praised, were sometimes the subject of criticism, both informed and uninformed.

Praise would have been heightened and criticism diminished if it had been more widely appreciated that the Air Force was fighting its war with aging equipment, inadequate logistics, and a shortage of manpower. And it is rarely acknowledged that the USAF, so strapped for resources in comparison to the demands made on it that General Vandenberg called it "a shoestring Air Force," had a priority higher even than the combat in Korea. This was the creation of a nuclear deterrent force so powerful that it would succeed in preventing a third world war.

Given the limited means available, it is not surprising that many elements of Air Force doctrine, learned at such great expense in World War II, were violated in the conduct of the Korean War, sometimes deliberately, sometimes accidentally, sometimes by force majeure. In the main, however, the Air Force attempted to follow its doctrine wherever possible and revised it only when it was demonstrated that conditions had changed.

Analysis of both the phases of the war and the application of Air Force doctrine can often be illustrated by a discussion of how specific types of aircraft were employed.

AIR SUPERIORITY

The United States Air Force was consistently able to follow only one element of its air doctrine, the maintenance of air superiority. It did this throughout the war, without regard to the numerical superiority of the enemy, the surprising relative qualitative superiority of the MiG-15 fighter, or the pervasive advantage the rules of engagement gave to the Chinese Communist forces. The USAF maintained air superiority through the training and élan of its pilots and the loyal, determined support of its aircraft by members of the ground crews. The latter functioned under conditions of extreme climatic and logistics difficulties, with the young airmen enduring the freezing winters and scorching summers with better grace than they endured the shortages of parts and equipment that made their work so much more difficult.

In brief, air superiority, established with very limited means, saved the United Nations from utter defeat on two occasions, permitted the two "comebacks" by UN forces, and, in the end, provided the climate that drove the Communist forces to agree to armistice terms that were, if not satisfactory, at least acceptable. Oddly enough, the actual battle for air superiority was almost independent of the ebb and flow of the fighting on the ground, except when the F-86s were forced to remove themselves to Japanese bases. For the other elements, including strategic and tactical air, the phases of the ground war had more direct influence on operations.

The greatest threat to American air superiority, the Soviet-built MiG-15s, was first seen on November 1, 1950, when they attacked a flight of Mustangs and a T-6; all managed to escape, but it was immediately obvious that the MiGs were superior to any UN aircraft in the theater.

The MiG-15, like the Zero in World War II, should not have been a surprise. First flown on December 30, 1947 (just ninety days after the first flight of the XP-86A), the MiG was built on the brilliance of the Mikoyan-Gurevich OKB (design bureau) and the utterly stupid—if not covertly treasonous—action of the British government in approving the sale of fifty-five Rolls-Royce Nene engines, in batches of ten, fifteen, and thirty. The first of these were promptly reverse-engineered (as interned Boeing B-29s had been) and put into production by Major General V. A. Klimov as the RD-45. The Nene, with its 5,013 pounds of static thrust, was several years in advance of any contemporary Soviet engine, and the developments it inspired ensured the success of not only the MiG-15 but of a whole series of Soviet aircraft.

The MiG-15 was revealed to the world in a flyover at the 1948 Tushino air show, and, despite the usual Soviet secrecy, it was known to be in quantity production—at least 200 per month—and was supplied to satellite nations. In marked contrast, the first order for F-86s called for a total of 221 planes, only enough to outfit the 1st, 4th, and 81st Fighter Interceptor Wings by June 1950, given the unusually high attrition of the early aircraft. A further 333 F-86As were ordered before the pressure from the Korean War spurred increased production of later models.

Smaller and lighter than the F-86A, the MiG's performance was marginally superior in all performance aspects. The Soviet Union would learn from combat as the USAF did, and the MiG would be improved over time, just as the F-86 would be. The MiG's greatest advantage was in combat endurance, provided not by its design but by geography and the UN rules of engagement, which forbade attacking the MiGs on their home fields. The American fighters had to fly to the vicinity of the Yalu River to engage the MiGs, a distance that left them with perhaps twenty minutes of fuel for combat. The MiGs could take off from airfields protected by the sanctuary given by the Chinese border, climb to altitude, and be ready for combat with almost full loads of fuel.

The first combat with MiGs came on November 7, 1950, when several swept in to attack the F-80Cs of the 51st Fighter Interceptor Wing escorting B-29s in an attack on Sinuiju. The F-80s of the 25th FIS had made three weapons delivery passes on the flak installations; on the third pass, First Lieutenant Wilbur Creech, in "Jungle Jim Blue" flight, was hit; his throttle jammed at 83 percent power—just about enough to keep the F-80 airborne. He jettisoned his tanks and made his way five feet above the Yalu River toward the China Sea.

Creech had just reached the China Sea when his wingman called, "MiGs are coming in at six o'clock" Above them was the "Top Coat Dog" flight of the 16th FIS, with Lieutenant Russell Brown flying in the number two slot. The F-80s peeled off and roared down, with the leader overshooting; Brown slowed up and shot down the first MiG. The MiG's were much faster than the F-80s, but in a pattern that would be maintained throughout the war, their pilots had inferior training to their USAF opponents. Brown hammered the enemy out of control with his one unjammed .50 caliber machine gun, becoming the victor in the first jet-versus-jet air combat in history.

There was later some controversy about the number two man breaking off to shoot instead of covering his leader, but when Brown's picture appeared on the cover of *Time* magazine, all controversy ended.

A cry went up for F-86s to be sent to the theater, and on November 8, General Vandenberg ordered the 4th Fighter Interceptor Wing into action. (He also directed that 27th Fighter Escort Wing of F-84s be sent.) The three squadrons of the 4th (334th, 335th, 336th) with a total of forty-nine F-86s were embarked upon the escort carrier USS *Cape Esperance* on November 29, arriving at Yokusuka, Japan, two weeks later with many of the aircraft damaged from salt-air corrosion. The first orientation flight in Korea was flown on December 13.

FIRST KILLS FOR THE SABRE

On December 17, 1950, Lieutenant Colonel Bruce N. Hinton, Commanding Officer of the 336th, led his Baker Flight of four from Kimpo (K-14), the most modern air field in Korea, on a combat air patrol along the Yalu, 430 miles away. They flew to a triangular sliver of land known as "MiG Alley"— 6,500 square miles of territory bounded on the north by a Manchurian sanctuary teeming with airfields filled with MiGs, and on the south by the long flight home on nearly empty tanks for the Sabres.

Cruising at Mach .62 to conserve fuel (a technique that was soon abandoned), Hinton sighted a flight of four MiG-15s climbing toward them. The Sabres dove toward the enemy, Hinton damaging the MiG leader and then fastening onto the tail of the number two MiG. A series of long bursts from the F-86's six .50 caliber machine guns—1,500 rounds—smashed into the MiG, which went inverted and dove straight in. It was the first F-86 victory over a MiG; Sabre pilots would down another 791 Soviet fighters before the war was ended.

Two tough hombres in air combat scored in the next encounter. After twenty-three victories in World War II, Lieutenant Colonel John C. Meyer was comfortable commanding the 4th Fighter Interceptor Group and led the first of four flights of F-86s to the Yalu on December 22. (In the Vietnam War, Meyer would be Commander in Chief of the Strategic Air Command and oversee the conduct of Linebacker II, the December 1972 attack on Hanoi and Haiphong.) The CO of the 334th FIS, Lieutenant Colonel Glenn T. Eagleston, who had 18.5 victories in World War II, led the second flight. They arrived in the area cruising at Mach .85, a speed permitting only twenty minutes for the patrol or ten minutes of combat, but giving them the best chance of catching any MiGs they flushed. Meyer's sixteen Sabres attacked a flight of fifteen MiGs and shot six down in a swirling dogfight that ranged from 30,000 feet all the way down to the surface. An F-86 flown by Captain Lawrence V. Bach was shot down in flames. Meyers and Eagleston both scored one of the two victories they each would add to their tallies in Korea. (Bach's aircraft was the second F-86 lost. The first had fallen victim on June 17 to a bomb from a wood-and-fabric Polikarpov Po-2 biplane, the famous nocturnal visitor "Bedcheck Charlie.")

By the end of the year, the 4th had shown that the MiG-15 had met its match; in 234 sorties, eight MiGs had been shot down for the loss of one Sabre in combat. The MiG was conceded to be an excellent aircraft, with only a slight performance edge at high altitudes but with a tremendous advantage of almost 8,000 feet in service ceiling. Cruising at 45,000 feet and above, where the Sabres could not reach them, the MiGs controlled where, when, and if fights would take place.

In 1952, the Communist forces also supplied the MiGs with an excellent ground control intercept system that was at once a tremendous advantage and a drawback. It was a plus for the MiG pilots in that it enabled them to know exactly where their F-86 opponents were, but a minus in that the Communists tended to control the activities of the MiGs from the ground, a fatally incorrect tactical doctrine that would persist through the Vietnam War and beyond.

The American pilots would have bet that the MiG was less maintenance-intensive than the more complicated F-86, which demanded many hours of mechanics' time on the ground for every hour of time in the air. Often as many as 50 percent of the Sabres were grounded for maintenance, many of them AOCP (aircraft out of commission, parts), and the 4th was often hard pressed to get sufficient aircraft for missions.

The F-86 was a more stable gun platform, and its armament of six .50 caliber machine guns was superior to the MiG's battery of cannon, which were originally intended for bomber intercept work. If they could have had their wishes, the Sabre pilots would have preferred to have cannons with a fast rate of fire, for it was difficult to keep a MiG in the gunsight for very long—high speed and G forces made deflection shots difficult—and few pilots got more than one chance to shoot.

The nature of the F-86/MiG-15 contest varied over the long months of the Korean War, primarily because of the varying degrees of resolution with which the enemy elected to fight. Often after a series of severe losses the MiGs would be much more difficult to bring to combat, as if they were assessing their defeats and attempting to come up with solutions. By the end of 1951, the enemy reverted to Spanish Civil War practice, using the battleground over the Yalu as a training ground in which successive groups of pilots would be exposed to combat in six-week cycles.

Competition among the American pilots for victories was fierce, and while few made any but joking allusions to the prospect, the desire to become the first American jet ace was intense. A leading contender was Captain James Jabara of the 334th Fighter Interceptor Squadron, who had scored four victories over the MiGs in his first month in combat in April 1951.

The outspoken Jabara was born in Wichita and trained as an aviation cadet. He flew two tours of combat in Mustangs in Europe and was credited with destroying five and a half enemy planes in the air and four on the ground. On May 7, 1951, Jabara's squadron rotated back to Japan, but he stayed on—looking for the next kill.

It came on May 20, late in the afternoon, when thirty-six Sabres and fifty MiGs engaged in a swirling dogfight. One of Jabara's wing tanks would not jettison, but he plunged into combat anyway, shooting down two MiGs to become the first man in history to shoot down five enemy jets. Official policy concerning aces forced his rotation back to the United States, but he returned in 1953.

The overriding importance of air superiority lay least, oddly enough, in the numbers of MiGs shot down, for the Chinese had MiGs aplenty, and the Soviet Union was turning them out in profusion. Air superiority's great gift was that it permitted the F-80s, F-51s, B-26s, and B-29s to operate, if not with impunity, at least without too much interference of mission—the interdiction of enemy logistics. This interdiction was important at all times, as necessary when the lines were stabilized as when the Communists were advancing. The enemy forces recognized this and made major efforts to reverse the situation by introducing new pilots and new tactics and beginning an airfield construction program. Sheer numbers were heavily in their favor, as China marshaled 445 MiG-15s by 1951; this total would steadily increase throughout the war. The United States had managed to get eighty-nine Sabres to the theater, but only forty-four were in combat with the 4th, and of these perhaps only twenty-two would be ready for action. The poorest of the great powers was demonstrating a ten- or twenty-to-one advantage in first-rate equipment over the richest.

THE HONCHOS ARRIVE

The first of the new Communist programs revealed itself in mid-June 1951. On the 17th, twenty-five MiGs, well and aggressively flown, engaged the Sabres, who promptly shot one down and damaged six others. On the next day, forty MiGs tackled thirty-two Sabres; this time five MiGs went down, but the third F-86 of the war was lost. The third day was worse; four MiGs were damaged in a dogfight, but a Sabre was shot down. American pilots recognized the dramatic change in the enemy pilots' skill level by using the term "honcho," or boss, to describe them. The American pilots assumed the newcomers to be veteran aces and instructor pilots from the Soviet Union, this view being validated over time by reports of blond pilots ejecting from damaged MiGs. Less expert pilots were termed "students" and usually provided easy kills.

Apparently gaining confidence despite their losses, the honchos began ranging out from their normal Yalu River fighting area, flying as far south as Pyongyang. The MiGs had a wide variety of markings, from which intelligence officers inferred that the pilots had been seconded from a different unit within the Soviet bloc, and that each pilot had brought his own airplane to war. They also introduced new tactics, including one the Americans named the "Yo-Yo." A formation of MiGs would use their superior altitude capability to orbit over the Sabres, then in pairs make diving attacks that would carry them through the formation and also give them the speed to zoom back up to a height the F-86s could not reach. Later, they would employ what were called "trains," large formations of as many as eighty MiGs in staggered parallel tracks at varying heights, flown down each side of the peninsula. Small units would be dropped off to engage the Sabres patrolling MiG Alley. The two large formations would turn in to join together near Pyongyang, swooping down to attack returning fuel-short Sabres or UN fighter-bombers.

General Stratemeyer had suffered a severe heart attack on May 20, 1951, and Major General Otto P. Weyland was selected to succeed him. Weyland had Vandenberg's complete confidence, for when Weyland commanded the XIX Tactical Air Command in Europe, he protected the Third Army's flank by air, thus earning General George S. Patton's ultimate accolade as "the best damn general in the Air Corps." At the same time, Major General Frank F. Everest was named Commander of the Fifth Air Force.

The new commanders watched the growth in power and aggressiveness of the enemy forces, which they now were forced to credit with the capability of wresting air superiority from the pitifully few F-86s on strength. In these days of sometimes revisionist history, military leaders in the early years of the Cold War are often portrayed as opportunists using a phantom threat from the Soviet Union to back up their wanton budget demands. These views should be analyzed in the light of the F-86 controversy. The leaders of the Air Force were eager for victory in Korea and wanted more than anything to support their ground and air forces fighting there. Yet they considered the threat of an atomic attack by the Soviet Union so genuine that they retained F-86 squadrons for the air defense of the United States instead of sending them to Korea. If the USAF had possessed the 137 to 145 groups that it deemed necessary, instead of the forty the defense budget provided, there would have been adequate forces for both Korean and American air defense.

Fortunately, help was at last on the way, in the form of improved Sabres, creative leadership, and, after anguished deliberation, additional wings of fighters. In the spring of 1953 two more F-86 units would be sent, the 8th and 18th Fighter Bomber Wings. Configured for the ground support role, these Sabres also maintained their interceptor capability.

The first improved Sabre was the F-86E, which featured the all-flying tail pioneered in the Bell X-1 rocket aircraft in which then Captain Charles "Chuck" Yeager broke the sound barrier. (Diehard F-86 fans insist that a Sabre had broken the sound barrier just before Yeager's flight, but that's another story.) The F-86Es were a little slower than the F-86As, but were more maneuverable and easier to maintain. A vastly improved F-86F was put into production at an additional production line in Columbus, Ohio, and was first flown on March 19, 1952. With a more powerful uprated engine, the F model had a new "6-3" wing that was 6 inches wider in chord at the root and 3 inches at the tip, and on which the slats were eliminated and boundary layer fences added. The F-86F reached Korea in the fall of 1952. The Soviet Union was upgrading the MiG as well, and the MiG-15 bis ("bis" indicating a follow-on version) appeared in the Chinese Communist inventory. The F-86F was clearly the superior aircraft, 10 knots faster than the F-86E and able to operate with the MiGs at altitudes above 50,000 feet.

Far more important in both a morale and equipment sense than even new aircraft was the inspiring leadership of Colonel Harrison R. Thyng. A double ace who shot down five airplanes in World War II and five more in Korea, Thyng took over command of the 4th FIW in early November 1951 and soon demonstrated just how vital the qualities of a commander were to his unit. Dismayed by the pilots' frustration with the maintenance on their aircraft and with the even greater anger of the hardworking mechanics, who routinely got the airplanes airborne with patchwork repairs and cannibalized parts, he was outraged at the seemingly insuperable bureaucratic barriers that intervened between the combat area and stateside logistic support. Thyng rebelled and took action on the ground as he had in the air.

From Longstreet at Gettysburg to von Paulus at Stalingrad to Walker in Korea, history is replete with stories of brave military leaders who would risk their lives in combat on a daily basis but would not risk their careers bucking their own superiors. In a stunning gesture defying the established order, Thyng did both, laying his career on the line by going directly to the top, then leaving to lead a patrol to the Yalu. In a message to the Chief of Staff, General Vandenberg, Thyng told him with chilling clarity that he could no longer be responsible for air superiority in the area that had become infamous as MiG Alley. Thyng sent information copies to his direct superiors in the intermediate commands, a gesture that, dependent upon Vandenberg's reaction, ensured either his survival or his removal.

Fortunately for Thyng and the 4th FIW, his message's timing was perfect. On October 22, 1951, Vandenberg had already ordered an additional seventy-five Sabres sent to Korea, which permitted the equipping of an additional wing, the 51st FIW, at Suwon. Only two squadrons of the 51st could be supplied aircraft, but they were all F-86Es and came under the command of Colonel Francis S. Gabreski from the 4th. Gabreski had been the leading American ace in Europe in World War II, with twenty-eight victories; he would score 6.5 more in Korea. The greatest effect of Thyng's message was a massive improvement in the supply of parts and equipment, vastly improving Sabre readiness statistics. This came about in part because of the patriotic enthusiasm of the manufacturers, who supplied the necessary parts without contracts to cover their risk. The joint effort enabled the United States to prevail despite never committing more than six fighter squadrons solely to the single most important factor in the Korean War—the maintenance of air superiority. Despite the odds, those six squadrons of the 4th and 51st FIWs administered an increasingly severe punishment to the Chinese Communist air force over the remaining months of the war, regardless of whether the pilots were honchos or students.

ACE MAKING TIME

At the beginning of 1952, the Communists had lost 339 aircraft; by the armistice, on July 17, 1953, the total had jumped to 954. Of these, 810 were destroyed by F-86s, including no less than 792 MiG-15s.

MiG kills reached their peak when seventy-seven were shot down in June 1953, sixteen being scored on the last day of the month. The eighteen-month-long harvest of MiGs naturally yielded more aces; by the end of the war, thirty-nine F-86 aces had accounted for 305 aircraft. Major William Whisner, who had scored 15.5 victories in World War II in Europe, including six in one day, was the first ace of 1952. Whisner, CO of the 25th FIS of the 51st FIW, raised his score to 5.5 MiGs on February 23, 1952, to become the seventh ace of the Korean War. (Ironically, Whisner, who served his third war in Vietnam, died in an allergic reaction to a bee sting in 1991.)

In the following months, more Sabres with greater capability would allow thirty-two more pilots to achieve five victories. "Gabby" Gabreski became the eighth jet ace on April 1, 1952, and Captain Ivan Kincheloe became the tenth on April 6. Kincheloe went on to become a famous test pilot, only to lose his life flight-testing the Lockheed F-104 on July 28, 1958. The twentieth ace was Major Robinson Risner, who scored eight victories between July 5 and January 21, 1953, those six months representing just 7 percent of the more than eighty-four months he would spend as a prisoner in the Hanoi Hilton during the Vietnam War.

A natural rivalry for the position of top ace developed between Captain Manuel J. Fernandez, Jr., of the 334th FIS and First Lieutenant Joseph Mc-Connell, Jr., of the 16th FIS. Fernandez had started earlier, scoring his 14.5 kills between April 10, 1952, and May 16, 1953. (When the author talked to Fernandez a decade after the war, the ace was convinced that his Hispanic ancestry had worked against him in his fight to become number one.) Scoring sixteen victories between January 14 and May 18, 1953, McConnell survived bailing out over the Yellow Sea and being rescued to emerge as the Rickenbacker of the Korean War. He lost his life in an F-86 accident at Edwards Air Force Base, on August 25, 1954.

James Jabara, now a major, volunteered to fly a second combat tour in Korea, just as he had in Europe; after his return in May 1953, he shot down nine more MiGs, the last on June 15, 1953. His total of fifteen made him the second-ranking ace of the war.

The last ace of the war was Major Stephen J. Bettinger, who scored his fifth victory on July 20, 1953, but was himself shot down and captured, not being released until October 2, 1953. Bettinger had the pleasure of confirming his own fifth victory after he was released. His potential status as an ace had been kept a closely guarded secret while he was a prisoner, in fear of a Communist reprisal.

One Sabre pilot won the Medal of Honor, awarded posthumously. Major George A. Davis was the fifth ace of the war and went on to score fourteen victories, most of them in twos and threes. The leading ace with twelve victories on February 10, 1952, Davis lost his life in a brave attack against a formation of twelve MiGs preparing to attack some F-84 fighter-bombers. Davis destroyed two MiGs and was attacking a third when a fourth enemy fastened on his tail and shot him down.

The war had ended with total UN air superiority, just exactly as its commanders wanted, and just exactly as the ground situation demanded. One reason was the aggressive quality of FEAF's commanders, men like Lieutenant General Glenn O. Barcus, who had assumed command of the Fifth Air Force on June 10, 1952. Unknown to his own superiors, Barcus began flying combat missions, highlighted by a May 1, 1953, attack on Pyongyang. All four F-86 wings participated, the 4th and 51st flying high cover while the 8th and 18th Fighter Bomber Wings worked over the radio station, which had been broadcasting adverse propaganda about the Fifth. Barcus, using a radio frequency known to be monitored by the Communists, announced his presence overhead and promised, "We will be back every time you broadcast filthy lies about the Fifth Air Force."

Despite the MiG-15s' great advantages in number, geography, and rules of engagement, the Sabres had prevailed. While the war was going on, and when it was over, there was a fundamental fact known to the generals, to the aces, to the medal winners, and to the men themselves: none of the successes could have been achieved without the devoted efforts of the noncommissioned officers and men who outdid themselves day after day, month after month, with little reward except the satisfaction of knowing they were appreciated by the men in combat.

AIR TO GROUND: CLOSE SUPPORT AND INTERDICTION

From the earliest days of the war, individual Army commanders and zealous journalists complained that the Air Force did not provide the same degree of close air support to Army units as that provided by the Marine and Navy fliers to Marine units. The messages of gratitude and the praise expressed for Air Force close air support by other Army commanders, including Generals Walker and MacArthur, were often overlooked. More important, the relatively higher effectiveness of interdiction efforts compared to close air support, particularly in the last two years of the war, was never fully understood or acknowledged.

The phases of the air-to-ground war closely followed the changing phases of ground operations, each of which placed different types of demands upon the equipment and the aircrews of FEAF. During the first phase, the rapid advance of the North Koreans down the peninsula resulted in an ad hoc air-to-ground campaign, one that demanded every effort from all FEAF's resources, applied in armed reconnaissance missions against the plentiful targets of opportunity. General MacArthur ordered FEAF to attack and destroy all North Korean military targets south of the 38th parallel; no operations north of that line were to be undertaken except in self-defense. The total forces available were two squadrons of Douglas B-26 Invaders, four squadrons of Lockheed F-80s, and two squadrons of North American F-82s.

Initial results were unsatisfactory because of bad weather and a lack of reconnaissance information. On the morning of June 28, in the first of more than 67,000 reconnaissance sorties conducted during the war, Lieutenant (later General) Bryce Poe II made the first USAF jet combat reconnaissance flight ever in his RF-80A. The results cleared the 3d Bombardment Wing (Light), commanded by Colonel Thomas B. Hall, for an attack by twelve B-26s on railroad yards at Munsan. On the way back, strafing targets of opportunity, one B-26 crashed and two were forced down.

The B-26 attacks were supplemented by a morning and afternoon attack by twenty-four F-80s from the 8th FBW, which took full advantage of roads crowded with North Korean vehicles, troops, and artillery.

Still active—though not for long—the North Korean Air Force sent four aircraft to strafe the airfield at Suwon while General MacArthur was there attending a briefing. As the Supreme Commander watched, a flight of four Mustangs that had been slated for service with the ROK AF attacked and shot down all four enemy planes. Second Lieutenant Orrin R. Fox of the 80th FBS destroyed two Ilyushin Il-10s, while Lieutenant Harry T. Sandlin of the 35th FBS nailed a Lavochkin La-7 and First Lieutenant Richard L. Burns of the 35th FBS got an Il-10. Impressed, MacArthur acquiesced to General Stratemeyer's request to attack North Korean air power at its source.

The next day, eighteen B-26s made the first strike north of the 38th parallel, attacking Heijo airfield at Pyongyang, the North Korean capital, destroying hangars, fuel dumps, and barracks and returning to Japan unscathed. Staff Sergeant Nyle S. Mickley, a gunner on one of the 13th BS (L) B-26s, shot down one of the two Yak fighters that had pressed home their attack. Twenty-five more fighters were claimed destroyed on the ground. The success of the B-26s had depended upon surprise; strong North Korean antiaircraft would soon force them to higher altitudes and then to a night bombing role.

As the North Koreans pressed the UN forces back into the Pusan (Naktong) perimeter, the air-to-ground activity intensified with naval aircraft from Task Force 77, B-29s, and the mixed bag of FEAF's ground attack aircraft participating. Extraordinary measures were taken. The F-82Gs, designed as interceptors and operating out of Yakota and Itazuke, did yeoman work strafing and dropping napalm, the pilots flying mission after mission until they were exhausted.

The F-80Cs also operated out of Itazuke, their weight and footprint ruling out flying from the rough airfields remaining in South Korea. With rockets and tip tanks, the Shooting Stars had an operational radius of about 225 miles with about fifteen minutes time over target (TOT). At a maximum radius of 350 miles, the target had to be found on arrival, as there was zero TOT. An interim solution to gain more range was a field modification generated at Misawa by 49th FBW Lieutenants Edward "Rabbit" Johnson and Robert Eckman. They inserted two center sections of a larger (Fletcher) tank in the center of the standard 165-gallon Lockheed tip tank, creating the "Misawa tank" with a 265-gallon capacity. This extended the F-80C time over target to forty-five minutes, but placed dangerously high loads at the wingtips, resulting in some wing failures. The Misawa tanks were also in short supply for several months.

Perhaps the most unusual improvisation was the reversion to Mustangs. The initial effort became famous through the works of Major Dean E. Hess. Ten Mustangs originally intended to be given to the ROK AF were instead manned by U.S. and Korean personnel and moved to Taegu on June 30, to commence "Bout-One" air-to-ground operations against the enemy. A larger-scale but less-well-known effort was Project Dallas, which pulled thirty F-51s from storage and reassigned F-80C pilots from the 18th Fighter Group (most from the 12th Squadron) in the Philippines to fly them. Organized with the

"Bout-One" planes and pilots into the 51st Provisional Fighter Squadron at Taegu (K-2), the Mustangs flew their first mission on July 15—the same day the United States scraped the bottom of its air power barrel with 145 National Guard F-51s embarked on the USS *Boxer*. The 40th FIS was also converted from F-80s to F-51s, and moved from Japan to Pohang (K-3) on July 16. A few days later, the 40th, their pilots barely familiar with their mounts, flew more than 220 sorties against a North Korean probe toward Pohang, literally stopping it in its armored tracks.

The National Guard F-51s from the *Boxer* were used to equip the 18th FBG's 67th FBS, the latter commanded by Major Louis J. Sebille, a veteran of sixty-eight combat missions in Martin B-26s in Europe, where he won two Distinguished Flying Crosses and twelve Air Medals. Sebille led his unit out of Ashiya, Japan, to attack an enemy troop concentration on a river near Hamchang. One of his two 500-pound bombs failed to release on his first attack, but he returned for another strafing attack and his aircraft was severely damaged by heavy antiaircraft fire. Sebille knew how vulnerable the Mustang was, but he also knew what a threat the 1,500 well-armed troops were to the Pusan perimeter. Ignoring advice to make an emergency landing at Taegu, he dove to his death on the target. Major Sebille was awarded the Medal of Honor, the first of four posthumously awarded to American airmen in the Korean War.

(Major Charles A. Loring would win his Medal of Honor under strikingly similar circumstances. On November 22, 1952, flying an F-80 from the 80th FBS, Loring was hit after leading a four-plane element against an enemy artillery emplacement. After he was hit Loring turned and deliberately dived into the target, destroying it but killing himself.)

Despite the improvised nature of FEAF's initial response to the invasion, it was effective against the ever-lengthening North Korean supply lines, and the Pusan perimeter held. Standard USAF practices were put into effect as soon as possible, especially in regard to targeting. By mid-July, a joint operations center (JOC) and tactical air control center (TACC) were in place at Taegu, and no less than eighteen tactical air command posts (TACPs) were in the field, controlling strikes. These would be enhanced immeasurably after July 10, when Lieutenant Harold E. Morris demonstrated the utility of the North American T-6 Texan trainer as an airborne controller. The airborne controllers first call signs were Mosquito Able, Mosquito Baker, and Mosquito How, and Mosquitoes became the generic term for the Texans. The Mosquitoes were especially valuable to the F-80Cs, making the most of their short loiter time.

A complex but swift and flexible methodology of calling for and executing a close air support strike was quickly established. A strike request from a TACP would be passed through divisional and corps headquarters to the TACC/JOC, which would clear the fighter-bombers to scramble. They would report in to the airborne controller, who would point out the target. After the attack was made, the fighter-bombers would give a strike report and return to land. Total time from request to landing was often as short as forty minutes.

This grunt-level cooperation was reflected in command operations; General Partridge established his own command post in Taegu, where General Walker's headquarters had been established. One of the first tasks agreed upon was the establishment of PSP (pierced-steel-plank) runways in Korea to permit quick turnaround of F-51 operations. As the F-51 units came on line, they were given the close support work, while the F-80s were interdicting the supply lines.

Long debates raged about the wisdom of using the F-51s in the ground support role, given that their liquid-cooled engines were so vulnerable to ground fire as well as to the dust-laden atmosphere of the airfields. The sole reason that they were used was their availability as "war surplus," a costly economy dictated by the improvidently low peacetime defense budget. The increasingly tired Mustangs were used until 1953, carrying heavy loads of ordnance over a great range, and even operating from Japan when the Communist advances made it necessary.

Some pundits argued against the F-80s, saying that the jet's high speed made it less useful as a ground attack airplane. They were wrong. The high speed was useful in getting to the target with a greater degree of surprise, and the jet was less vulnerable to ground fire than the Mustang. The lethal firepower of the F-80 was demonstrated on July 10, 1950, when yellow-nosed Shooting Stars of the 8th FBS used machine guns and rockets to attack a target of opportunity, destroying 117 trucks and thirty-eight tanks. Tank-busting became a science for the F-80 pilots, who would dive at the T-34s at an angle of 45 to 60 degrees at an airspeed of 400 to 500 mph, aiming where possible at the more vulnerable rear of the tank. The delay fuse on the 5-inch HVAR (high-velocity aircraft rocket) was set so that the rocket would shatter against the armor before exploding, providing a shrapnel effect to kill the tank crew. The F-80 had to break off at least 500 feet above the ground to avoid debris from the target.

During the retreat to the Pusan perimeter, F-80s were given credit for 75 percent of the destruction inflicted by air on the enemy. One important reason was ground crews who worked night and day to keep the airplanes flying, achieving an 84 percent in-commission rate, excellent for a peacetime operation, but brilliant under the conditions at hand.

The long series of extemporaneous responses to North Korean aggression had bled the invaders dry, permitting the defenders at Pusan to build up their strength for a counterattack that coincided with the invasion at Inchon. The North Korean forces could move only at night, and their front lines were wracked as the tempo of close-support sorties built up, almost 7,500 being flown in August. As the North Koreans made a last effort to drive UN forces into the sea in September, Fifth Air Force fighter-bombers and light bombers mounted as many as 683 sorties a day. When the Eighth Army counterattack began driving the North Koreans back, General Walker stated that the Eighth had received excellent air support, and that had it not been for the Fifth Air Force, it would have been forced to evacuate Korea.

THE FIRST MOVE UP THE PENINSULA

The breakout from Pusan coincided with an increase in FEAF strength in the Korean theater obtained by drawing in units then tasked with the air defense of Japan, Okinawa, and the Philippines. The tempo of Fifth Air Force activity picked up in September, with as many as 361 sorties a day flown. Mustangs with napalm proved especially effective against entrenched enemy positions; when the North Koreans broke and ran, they were pounced on by F-80s and B-26s. By September 22, most North Korean tanks had been destroyed and the Communist forces were beginning to reel under the air attack. Enemy morale dropped to the point that on September 23, 200 NKPA troops surrendered to a T-6 Mosquito flown by Lieutenant George A. Nelson, who had dropped a note signed "MacArthur," ordering them to lay down their arms.

Where before the interdiction effort had been to prevent troops and supplies reaching the front, it now was directed at keeping them from getting to the rear. The North Koreans moved primarily at night in their retreat, and B-29s and B-26s worked together, the Superfortresses dropping huge M-26 parachute flares while the Invaders worked over the troops illuminated below.

FEAF's air power, improvised as it had been, had imposed a stunning defeat on enemy forces. Analysis of prisoner interrogations revealed that air power had inflicted losses on the NKPA of 47 percent of personnel, 75 percent of tanks, 81 percent of trucks, and 72 percent of artillery. The inference was drawn by the USAF—if not by MacArthur—that the unending airground action had defeated the NKPA—and not the landing at Inchon.

The Fifth Air Force swarmed to South Korea as soon as airfields could be built or improved to the level necessary for operations. The pursuit of the North Koreans turned into a rout, and the fighter-bombers now sometimes found themselves returning with ordnance unexpended for a lack of targets, a condition that changed dramatically with the intervention of the Chinese Communist forces.

FIGHTING BACK DOWN THE PENINSULA

In response to the Chinese intervention, General Vandenberg had ordered the Strategic Air Command's 27th Fighter Escort Fighter Wing (FEW) from Bergstrom AFB, Texas, to action. The 27th flew its Republic F-84D and E model Thunderjet fighters to San Diego, where they were too hurriedly embarked on the escort carrier USS *Bataan*. Once again inadequate preventive measures resulted in severe corrosion problems that required a week of work to overcome.

The 27th was sent to Taegu (K-2), where its first mission would be led by Colonel Don Blakeslee. During World War II, he had been a member of the famous RAF No. 121 "Eagle" Squadron, and later he had commanded the famous 4th Fighter Group. In his 1,000 hours of combat and 400 sorties in Europe, Blakeslee had scored fifteen victories. On December 7, 1950, the veteran Blakeslee led a four-aircraft flight on an armed reconnaissance south-west of Pyonyang. Using HVAR rockets and machine guns, the F-84s knocked out locomotives and started raging fires in a marshaling yard—a good effort for a unit trained for escort fighter duties and with no F-84 experience in ground attack work. It was a textbook start for an intensive first month in which the 27th flew 972 combat sorties, of which 275 were ground support. It was also symbolic of the F-84 during the Korean War, during which it would participate in almost every major campaign, complete more than 86,000 sorties, drop more than 50,000 tons of bombs and 5,500 tons of napalm, fire 22,154 rockets, and shoot off literally miles of belted .50 caliber ammunition. A total of 153 Thunderjets were lost, eighteen of them to MiGs. The F-84s managed to shoot nine of the faster MiGs down. Despite this monumental effort, the F-84 garnered far fewer headlines than the more glamorous F-86. Did its pilots and ground crewmen mind? Absolutely!

Although the straight-wing F-84s were faster than the P-80s, with a top speed of 613 mph, they were no match for the MiG-15s in air combat at high altitude. If they could lure the MiGs into fighting at 20,000 feet and below, they stood a better chance. In a low-level battle on January 21, 1951, Lieutenant Colonel William Bertram of the 523 FES demonstrated what a good pilot in an F-84 could do, shooting down a MiG-15 in flames. But everyone recognized that the Thunderjets were better suited to the ground support role, where they excelled, like their ancestor the fabled Republic P-47 Thunderbolt. Capable of carrying up to 4,000 pounds of bombs, the heavy (18,000-pound) F-84s sometimes needed a JATO boost to get off the short Korean airfields.

The F-84s had many moments of glory. Early on the morning of January 23, 1951, thirty-three Thunderjets of Colonel Ashley B. Packard's 27th FEW attacked Sinuiju, the North Korean airfield always protected by the flights of MiGs stationed at Antung, a few miles away across the Yalu.

Before the MiGs could be scrambled, the first eight F-84s made a firing pass across Sinuiju, climbing quickly up to 20,000 feet to provide a cover for the successive flights of attacking Thunderjets. Thirty MiGs dove to the defense of Sinuiju, and though faster, could not turn with the F-84s at that altitude. Lieutenant Jacob Kratt quickly shot down two MiGs, and Captains Allen McGuire and William W. Slaughter each claimed another. (McGuire's claim was not confirmed; Kratt would get his third and final victory three days later, destroying a Yak-3.)

Other UN fighter-bombers distinguished themselves in supporting the 3d Marine Division and the 7th Infantry Division as they retreated from the Chosin Reservoir area, saving them from annihilation. For the next month, the disheartened UN forces gave up ground almost as quickly as they had gained it, moving from one designated "stand-fast" line to the next with demoralizing rapidity. Pyongyang fell to the enemy on December 5, Seoul was lost on January 4, and Inchon on the 5th. The pell-mell retreat had forced President Truman to declare a national emergency on December 15.

The Chinese offensive quickly overran many of the UN's airfields, forcing American units to retire to Japan. By February, the Sabres' limited range kept them from reaching farther north than Pyongyang, effectively giving the enemy complete control of MiG Alley. Fortunately, the Communists were not proficient in the use of air power, and despite an immense airfield rehabilitation campaign, were never able to give UN forces the kind of brutal air-to-ground punishment they continued to receive. That punishment would bleed the Chinese offensive dry, as troops, trucks, trains, and supplies were destroyed by the combined efforts of UN fighter-bomber and bomber units.

The constant deprivation of supplies, exacerbated by the requirement to move surviving equipment and troops to secondary roads and land trails to avoid the bombing, vitiated the Chinese offensive. Then the resurgence of the UN forces under General Ridgeway's leadership slowly reversed the course of the war. The continuous retreat was turned into a slow advance by January 6; by March 14, Seoul had been recaptured. In recognition of the vulnerability of the lengthy Communist supply lines, FEAF fighter-bombers were assigned armed reconnaissance areas. Pilots became thoroughly familiar with their own area, so that the slightest change would indicate where the targets were.

Ridgeway's emphasis on inflicting the maximum number of casualties on the Chinese, whether in defensive or offensive operations, slowly began to pay off. The Chinese clung to their doctrine, launching no less than six major offensives, before they finally realized that they could no longer hope to overwhelm the UN forces and drive them into the sea. With the realization came a change of objectives on both sides; neither sought victory any longer, and instead each looked for the best deal possible in a face-saving way out. In July, the Great Leader, Marshal Kim Il Sung, responded favorably to Ridgeway's broadcast announcement that the UN was willing to discuss armistice terms. The long series of talks at Panmunjom would ultimately result in an armistice. For the remaining two years of the war, the fighting on the ground was stabilized at a fierce, bloody, but low-order stalemate. The air war would rage on with ever greater intensity, as it was the only means the United Nations had to ensure that the Communists would continue to negotiate.

In the course of the UN forces' struggle to just beyond the 38th parallel in the summer of 1951, the 27th FEW was replaced by the 49th and 136th Fighter Bomber Wings. The experienced 49th, with the 7th, 8th, and 9th FBS, happily gave up their war-weary F-80Cs in exchange for F-84s. The 136th, an Air National Guard unit that had just transitioned from Mustangs into F-84Es, was composed of the 111th and 182nd Squadrons from Texas and the 154th from Arkansas. Another Air National Guard F-84 unit, the 116th, arrived in late July with squadrons from Georgia (158th), Florida (159th), and California (196th) and took over the air defense of Japan.

This combined force of Mustangs, Shooting Stars, and Thunderjets would (with the B-26 and B-29 bombers) effectively throttle the enemy—but not the critics of air power. Each battalion commander wanted to see "his" aircraft on "his" front, every hour of the day, especially since (at least anecdotally) Marine air seemed to service the Marine ground troops on that basis. When the front was fluid, on either the advances or the retreats, the very nature of the battle provided a rationale for any absence of local air power, for in a rapidly changing situation it was plausible that aircraft were occupied elsewhere. But when the front settled down to the static fighting of the last two years of the war, when both enemy and UN probes were confined to discrete sections of the line, it seemed irrational not to be able to call in all the air power required, all the time.

The problem lay in the very nature of static warfare. The Communist forces were superb diggers, with deep entrenchments, heavy bunkers with tunnels to connect them to underground storage dumps, and heavily protected, well-concealed artillery and mortar positions. These were so well done that anything less than a direct hit from close air support was not only nonproductive, it was a far more expensive process than conventional artillery fire. Statistical analysis of the first years of fighting had revealed that interdiction far behind the lines yielded greater results, even though lacking the morale-building character of close air support.

Some rough rules of thumb evolved. It was demonstrated that if air power was concentrated on the front line during an offensive, reinforcements would add to the momentum of the attack and allow it to continue. If, however, sufficient forces were dedicated to long-range interdiction, the lack of replacement troops and equipment would gut the attack and it would grind to a halt. The longer the supply line, the more this was true.

The effect of long-range interdiction upon the war became even more evident as the truce talks continued, and the United Nations elected to keep the pressure on the Communist negotiators by the application of air power. Daytime attacks by fighter-bombers forced the enemy to take to the road at night, providing targets for the bombers; working together, they kept pressure on the enemy.

THE LIGHT BOMBERS

The Douglas B-26 Invader—"the little racer" to some of its pilots—has the distinction of having flown the first and the last bombing missions of the Korean War. As noted above, the first was flown on June 28, 1950, led by Captain Harrison Lobdell, Jr., of the 13th BS. The last was flown on the evening of July 27, 1953, when a B-26C of the 34th BS of 17th BW, piloted by First Lieutenant Herbert L. Atkins, used Shoran (short-range navigation) to drop its bombs on Wonson at 8:59 P.M. It landed at 11:55—five minutes before the cease-fire was to take effect.

In the long, hard thirty-seven months intervening between those two missions, a mere handful of B-26s made life hell for the Communist forces, carrying the war to the North on long solo missions that added miserable weather and rock-filled clouds to the danger of low-level attacks on enemy convoys. The 3rd Bombardment Wing and its attached 731st (later 90th) Bombardment

Squadron (Light-Night Attack) were based at Iwakuni Air Base on southern Honshu and never in the course of the war had the full complement of seventy-two aircraft on hand. The "rival" B-26 units—the 452nd BG (L) and its successor, the 17th BG (L)—were equally short-handed. (The 452nd was an Air Force Reserve unit that flew its B-26s overseas from George AFB, California.) By the end of the war, B-26s were in such short supply that the Air Force was desperate enough to send older models with flat canopies that could not be flown with standard winter flying equipment and survival equipment. The two units were also short on replacement crews for most of the war.

The effectiveness, inventiveness, and heroism of the B-26 crews can be typified by the actions of Captain John S. Walmsley of the 8th Bombardment Squadron. On the night of September 12, 1951, Walmsley had tested new tactics employing a huge 80-million-candlepower searchlight, illuminating a convoy and destroying sixteen trucks in ten passes. The searchlight gave the B-26s the awesome appearance of an oncoming train dropping out of the sky at 300 mph, terrifying some truck drivers into careening off the road. Two days later, Walmsley was at it again, this time attacking and locating a train. He had run out of ammunition and summoned another B-26 in to finish the train off. Walmsley continued to illuminate the train with his searchlight, and, an ideal target, was shot down and killed. He was posthumously awarded the Medal of Honor. Use of the searchlights was subsequently dropped.

With only a handful of aircraft, the B-26 outfits were routinely assigned to fly thirty-eight missions each night, against all sorts of targets. When the situation demanded, this was raised to forty-eight missions, which could be achieved only by having some of the aircraft fly two missions a night, one as an intruder on armed reconnaissance, and the second in the ground support role. It made for an exhausting evening, for, pleasant as the B-26 was to fly ordinarily, it was a handful to muscle around when pulling out of an attack with the airspeed indicator red-lined at 420 mph.

A single mission was grueling. Flying out of Iwakuni, the B-26s would stage through Taegu airfield, where both fuel and ordnance were often in short supply and the pierced-steel-plank runways played havoc with the tires. At various stages of the war, B-26s also flew out of K-1 (Pusan West), K-8 (Kunsan), and K-16 (Seoul). Takeoffs were made at one-minute intervals with the crews flying alone to assigned areas over the main North Korean supply routes. Targets were difficult to find, and the Communists used guards along the road to warn road traffic of approaching aircraft. Usually trucks would switch off their lights within fifteen to twenty seconds of learning of an intruder. Trains ran without lights and were adept at speeding rapidly from tunnel to tunnel.

The B-26s would try to surprise the enemy, searching for targets while flying at about 2,000 feet above the ground. If flares were carried, they were dropped upwind, about 3,500 feet above the terrain. If there were no flares, the B-26s often dropped their napalm first to light up the area, then tried to block off the convoy's advance and retreat. If the bombs stopped a train or a convoy, strafing attacks would be made until ammunition was exhausted, the pilots sometimes descending to as low as 200 feet to get results.

Mountains were often cloaked in clouds, and the aircrews had to have nerves of steel to descend through the weather into—they hoped—a valley where they could make an attack. It was physically demanding as well, with the noisy pounding of the big R-2600 engines, the constant changes in night adaptation going from full blackness into the blinding light of flares dropped by "Firefly" C-47s, and the ordinary psychic demand of combat, where flak traps set in the valley walls could bring an aircraft down in an instant.

As many as six aircraft might be scheduled to patrol the same area during the night, and timing was crucial. In one incident, a B-26 pilot felt a faint jar as he was breaking away from a run on the target and assumed it was either flak or a bird strike. The next morning, his crew chief found a bomb shackle smeared with the same color paint found on the wingtips of another squadron. A quick check of wingtips in that squadron confirmed just how near a miss it had been.

Despite their age and limited numbers, the B-26s soldiered on, destined to fight yet another war in Vietnam. By the summer of 1952, doubts arose as to the economic return of night intruder missions, and General Barcus redirected their effort in daylight formation raids against hostile communication centers in areas not ordinarily defended by MiGs. Only the most experienced crews remained on night-intruder duty, grouped into the 13th Squadron of the 3rd BW and the 37th Squadron of the 17th Wing. Some aircraft were also equipped to do Shoran bombing.

The B-26s flew more than 55,000 sorties in Korea, 44,000 of these at night. One of the great dissatisfactions of the B-26 crews was that they could not get verification of their work, and in the heat of battle could only make rough estimates of the destruction they caused. A compilation of their claims reveals that they destroyed almost 40,000 vehicles, 35,000 railway cars, 406 locomotives (their destruction was more dramatic and thus easier to verify), 168 bridges, and seven enemy aircraft.

HEAVY BOMBERS: THE EARLY DAYS

Although North Korea was industrialized compared to South Korea, the bulk of its war materials originated in the Soviet Union and were supplied through Communist China. Therefore, a strategic campaign against the North had finite limits both as to range of targets and effective payoff. Further, during the two periods of Communist advances, FEAF's B-29s were often placed into the close support role, doing carpet bombing. Nonetheless, the B-29s— and their crews—did a superb job throughout the war, even though many of the aircraft were pulled from mothballed status and many of the crewmen were reservists, their lives disrupted by another tour of combat duty.

FEAF's initial strategic force was the 19th Bomb Group (Medium), based at Andersen Air Base, Guam, and the only B-29 unit outside the control of the Strategic Air Command. The 19th was given orders to move immediately to Kadena, and four aircraft went into action on the afternoon of June 28, primarily as a show of force, hitting rail lines and roads leading to Seoul.

On July 8, 1950, General Stratemeyer established the Far East Air Forces Bomber Command (Provisional), and Major General Emmett "Rosie" O'Donnell, Jr., was dispatched to command it. O'Donnell, who commanded the 14th Bombardment Squadron in the dark days of 1942 in the Philippines and led the first B-29 attack on Tokyo, had gone on to become commander of SAC's Fifteenth Air Force after a distinguished combat career.

O'Donnell was thus uniquely qualified to lead FEAF's Bomber Command, as it was supplemented by units from SAC. The training and discipline of SAC showed to good advantage when the 92nd Bombardment Group (M) at Spokane (later Fairchild) AFB, Washington, and the 22nd Bombardment Group (M) at March AFB, California, were alerted for combat duty in Korea. Nine days and 8,000 miles later, they flew their first combat mission against Wonsan, a tribute to the flyaway kits, the resilience of the aircrews, and SAC theories on mobility. The experience of the move enabled later groups to move even faster. The 98th and 307th Bombardment Groups (M), from Spokane AFB, Washington, and MacDill AFB, Florida, respectively, were summoned to FEAF on August 1. The 98th (stationed at Yokota) flew its first combat mission only six days later, while the 307th (operating from Kadena) took seven.

During the early days of the war, as the initial B-29 effort was directed against delaying the enemy by destroying bridges and railways, a jurisdictional dispute over the selection of targets broke out between MacArthur's GHQ (General Headquarters) Target Group and the professional airmen of the FEAF Target Section. Grudging agreement was given to permit FEAF to select targets, although MacArthur would insist on his organization's designating interdiction targets for the B-29s under certain "special conditions" that he would define.

FEAF conducted a hurried analysis that identified five primary strategic targets in North Korea, including the capital, Pyongyang, plus Wonsan, Hungnam, Chongjin, and Rashin.

The first major effort, three strikes over a five-day period, was against a huge chemical industry site at Hungnam specializing in explosives and would set the precedent for subsequent strategic operations. Forty-seven B-29s used APQ-13 radar sets to bomb the Chosen nitrogen explosives factory complex on the morning of July 30, 1950, destroying 30 percent of it and heavily damaging another 40 percent. On August 1, forty-six B-29s used the Norden bombsight to destroy the Chosen nitrogen fertilizer factory and on August 3, thirty-nine B-29s struck the Bogun chemical plant, using radar bombing. The Hungnam complex was destroyed for all practical purposes for the remainder of the war.

FEAF Bomber Command worked around the clock to sustain the bombing offensive, which soon exceeded World War II records by delivering 30,136 tons of bombs between July 13 and October 31. The tried-but-true B-29s averaged 8.9 sorties each per month, a tribute to the maintenance personnel, who were operating 8,000 miles from their normal equipment and supplies.

The efficiency of the FEAF missions was enhanced by the use of an airborne commander who preceded the bomber force on weather reconnaissance and then, based on conditions at the target, directed whether the attack would be by radar or visual means and decided whether to use formation or individual bombing techniques. The airborne commander also assessed the effects of the bombing and directed any required changes in technique on the spot.

The pattern of success was repeated against a variety of targets through September, including a controversial attack against the Fusen hydroelectric plant and the naval oil storage areas at Rashin. The latter case was a harbinger of future troubles in Vietnam, for the State Department demanded that Rashin not be a target because of its proximity to the border of the Soviet Union.

During this early period of the war, the enemy was unable to employ fighters against the B-29s, and his antiaircraft defenses were weak, but the B-29 crews faced extreme hazards from the weather, often having to return to Japan for instrument approaches with minimum ceilings and visibility.

The decision to allow UN forces to cross the 38th parallel caused the JCS to halt attacks against strategic targets in North Korea on September 26. During October, Bomber Command facilitated the North Korean rout by hitting troop concentrations, bridges, and other tactical targets, but the advance of UN troops rapidly removed the requirement even for these operations. As a result, a decision was made to return the 92nd and 22nd Bombardment Groups to the United States.

A NEW BALL GAME

The movement was premature, for when the Chinese Communist forces intervened, FEAF's B-29s had much more fighting to do under conditions far less benign than during the first five months of combat. Necessity again forced FEAF to go back to its first targets, bridges, but this time in a very particular way, and against a new and formidable opponent.

The Communist ground strategy was well conceived. The preliminary offensive, which began on October 25, was limited in strength and raised questions as to the intent of the Chinese, whom some optimists saw as perhaps seeking only to stake out a buffer zone south of the Yalu. The response of the Eighth Army, buoyed by its successful run up from Pusan, was to prepare for a "final" offensive itself.

In the face of the flood of Chinese Communist volunteers he had insisted would not be forthcoming, a chagrined and chastened General MacArthur demanded increased bombing, including incendiary attacks against North Korean arsenals and communications centers, but especially against the "Korean end" of all the bridges across the Yalu. MacArthur was still unaware that the bulk of Chinese forces were already across the Yalu, having filtered in over the previous month. In an un-Solomon-like decision, the JCS authorized the use of bombers against the bridges—but with no violation of Manchurian territory. Bridges from Siberia to Korea were off-limits.

The incendiary attacks were relatively easy and quite effective. The attacks against the bridges were something else, for a number of reasons. First was the strength of the bridges themselves. Built by the earthquake-wary Japanese to withstand natural disasters, they proved able to shake off damage from 500-pound bombs. Then, because they could not violate Chinese airspace, the FEAF bombers had to attack in an east-west direction on an axis parallel to the river and perpendicular to the bridge. Forced by antiaircraft fire to fly at altitudes above 18,000 feet while on the bomb run, they were subject to jet-stream crosswinds in excess of 100 miles per hour. Hitting a pinpoint target—and a 50-foot-wide bridge is less than a pinpoint from 18,000 feet— was stretching the B-29's capability. In addition, they had to endure attacks from Communist aircraft, slow Yaks and swift MiGs, which made use of their sanctuary by climbing to altitude and then making diving passes through the bombers and their escort.

The first major attack came against Sinuiju, on the south shore of the Yalu, directly opposite the Manchurian city of Antung, home base of the MiG-15s. On November 7, 1950, the FEAF attack on Sinuiju was preceded by Fifth Air Force F-80s and F-51s in a flak-suppression attack, during which Lieutenant Brown got the first MiG. Seventy B-29s dropped nearly 600 tons of incendiary bombs on the city, while nine other B-29s dropped their 1,000 pounders on the approaches to the two international bridges.

When the strike photos were interpreted, the bridges were still standing, revealing just how daunting the task of bombing one end of a bridge was. The Chinese were the victors in the battle of the bridges against both B-29s and Navy fighter-bombers. By the end of November, only about half had been cut, and the Chinese proved to be adept at supplementing them with easily repaired pontoon structures. Nature then helped them by freezing the river, reducing the need for bridges.

A dramatic, if imperfect, answer to the destruction of bridges came in the form of updated World War II technology, with the introduction of 12,000-pound Tarzon radio-guided bombs in January 1951. The Tarzons were guided in both range and azimuth to the target by the bombardier and were extremely powerful, able to take out whole sections of any bridge they hit. Unfortunately, they required a high degree of proficiency on the part of the crew and possessed an inherently dangerous feature in that they could not be salvoed safely. Their use was discontinued after eight months; thirty had been dropped, six bridges were destroyed, and one was damaged. True "smart bombs" were still a few years in the future.

REVERSAL OF FORTUNE

When the Chinese Communist forces, 180,000 men in eighteen divisions led by General Lin Piao, intervened, they had played their hand cleverly, allowing the UN forces to move almost to the limits of the North Korean border before springing their trap. The Eighth Army, preparations for its final "Home for Christmas" offensive complete, attacked on November 24, 1950. The Chinese responded on the night of November 25 with a tremendous attack on both the ROK forces and Eighth Army, and followed this with an assault on November 27 against U.S. Marine units.

A week after the Chinese offensive started, the Eighth Army had retreated more than 50 miles; by December, it was dug in below the 38th parallel, 120 miles from its starting point. B-29s were employed in incendiary attacks, including two raids on Pyongyang on January 3 and 5 which destroyed 35 percent of the city.

Brigadier General James E. Briggs took over FEAF Bomber Command on January 10, 1951, when General O'Donnell returned to the United States. Briggs was faced with a tremendous challenge, not only because the MiG-15s were becoming both more numerous and more aggressive, but also because the B-29s' 1944 vintage fire control system, designed to combat 400 mph fighters, made them increasingly dependent upon their escort fighters for protection. Unfortunately, most of the escort fighters available were straight-wing Republic F-84s and F-80s, the latter almost 200 miles slower than the MiGs. Adequate defense budgets would have provided the missing 300 F-86s that would have evened the odds.

With the spring thaw of the Yalu River, destruction of the bridges again became a priority for the B-29s, which in turn became a priority for the MiGs. Employing greater numbers and using new tactics, some MiGs would engage the F-86s in dogfights, while others brushed past the F-84s and F-80s to attack the bombers. The MiGs' armament of one 37mm and two 20mm cannons was lethal, a few hits being enough to bring down a bomber. Sitting on the "perch" high above the bomber formation, they would dive through the escorting fighters to strike the bombers, then pull out before they could be attacked. On April 12, in yet another attack on the railroad bridge at Sinuiju, two B-29s were shot down and six others were badly damaged.

The Red Chinese signaled their further aggressive intentions by building airfields in the recaptured territory. This was a crucial threat, for if they were able to operate their MiGs from the new airfields, the hard-won UN air superiority would be lost, and with it the ground war. Bombing the airfields became an essential—and costly—task for the B-29s.

The week of October 21, 1951, signaled a revolution in the aerial war. In a series of attacks on airfields and bridges, five B-29s were lost and eight suffered major damage. There were sixty-seven B-29 personnel casualties, including fifty-five dead or missing and twelve wounded. General Vandenberg summed up the situation sadly: "Almost overnight, Communist China has become one of the major air powers of the world." One of the fiercest air battles of the war occurred on October 23, when B-29s attacked a newly constructed airfield at Namsi. Eight B-29s of the 307th BW rendezvoused with fifty-five F-84s of the 49th and 136th FBW acting as escorts. As the B-29s began their run-in, fifty MiGs attacked in classic swooping pursuit, the F-84s performing close escort being unable to intervene. The lead B-29, commanded by Captain Thomas L. Shields, pressed on despite damage and dropped on the target. He bailed his crew out at the coast, but lost his life when his aircraft crashed. Two other B-29s were shot down, and the remainder were damaged. One F-84 was lost, and four MiGs were claimed destroyed, three by B-29 gunners.

The dreary scenario was repeated the next day when eight B-29s of the 98th BW attacked a railway bridge at Sunchon. As many as seventy MiGs attacked the Superfortresses, brushing aside the escort of ten F-84s and sixteen Gloster Meteor F.8s of the Royal Australian Air Force. One B-29 was shot down, but its crewmen were rescued. (The twin-engine F.8 Meteor was an improved model of the Royal Air Force's first jet fighter, comparable to the F-80C in performance. The RAAF used the Meteor primarily for close air support in Korea; it lost forty-eight Meteors while shooting down three MiGs.)

The MiGs reacted strongly again on the 27th, almost a hundred attacking eight 19th BW B-29s. This time, there were sixteen Meteors and thirty-two Thunderjets flying escort, and the MiGs were not as aggressive; only one B-29 was heavily damaged.

Brigadier General Joseph W. Kelly had taken command of FEAF Bomber Command on September 30, 1951, and soon recognized the inevitable: the B-29s were obsolete for daylight operations in the jet age. The Superfortresses flew one more daylight mission in Korea without incident, and then were assigned to night operations, using Shoran bombing techniques, which gave them a temporary respite until the Chinese Communist forces could build up their radar, searchlight, and antiaircraft defenses in critical areas.

For the remainder of the war, the B-29s, their undersurfaces painted black, would continue to hammer the enemy, striking at the few remaining strategic targets when warranted, but concentrating on supply centers and railway junctions.

On March 15, 1952, Brigadier General Wiley D. Ganey assumed command of Bomber Command. He had the benefit of some administrative changes which, for once, resulted in a nominal increase of strength to an average 106 airplanes instead of the authorized ninety-nine. The difference was small but critical, as it permitted an increase in training in the theater.

Ganey also contributed FEAF Bomber Command's share to the "air pressure" tactic designed to ensure that the Communists remained in good faith at the negotiating table. Strongly advocated by FEAF Commander General Otto P. Weyland, it was an air campaign to obtain an armistice. The enemy had come to the conference table knowing the only weapon in his arsenal was a bloody war of attrition in which he was able to inflict an unacceptable level of casualties on UN forces even though he had to sacrifice his own troops in far greater numbers.

The air pressure campaign denied the Communist forces this capability. Called at various times Operation Strangle and Operation Saturate (terms Weyland hated), the combination of bombers, light bombers, and fighter-bombers inflicted enough damage to prevent the enemy from launching an offensive that would bleed the UN forces.

One element of the air pressure campaign began in the summer of 1952 and ran through March 1953, as FEAF and NavFE (Navy Forces, Far East) joined forces in a massive effort to attack the electrical generating facilities at Suiho, Fusen, Chosin, and Kyosen, which serviced Communist industries on both sides of the Yalu. Thirty-one B-29s participated in follow-up attacks on September 12, 1952, on Suiho, placing 2,000-pound armor-piercing bombs on the target for more than two hours and forty minutes. Damage assessment photos indicated that it would take the Communists years to place the systems back in operation.

The final tasks for the B-29s came in the late spring of 1953. Anticipating an armistice that would limit the number of arms allowed to be brought into North Korea, Bomber Command turned to destroying bridges and airfields once again. FEAF was determined not to allow the Communists to bring an air force into North Korea before the armistice was signed. Bomber Command mounted a systematic campaign, neutralizing one airfield after another, returning night after night to reverse the remarkable repair efforts of the enemy. Just seven hours before the cease-fire deadline on armaments movements, a 91st Strategic Reconnaissance Squadron RB-29 toured North Korea and brought back photos to prove that all the airfields they had been concentrating on were unserviceable.

The B-29's performance was remarkable for an aircraft designed in 1940 and built during War II, particularly considering that many of them were mothballed for years in the Arizona desert. During that period, when it seemed the B-29s would never be used again, the men and equipment required to maintain them disappeared. When war came, it was undertaken by a force never greater than an average of 106 aircraft. Headquarters USAF, concerned about the paucity of spares and replacements, sometimes placed arbitrary sortie limits on the B-29s, often as low as twelve per day. Nonetheless, they flew all but twenty-one days out of the three years and one month of the war. In their 21,000 sorties, they dropped 167,100 tons of bombs and claimed sixteen MiGs and seventeen other fighters shot down. Sixteen of the B-29s were shot down over North Korea, but perhaps three times that many were lost or written off in crash landings when they returned to base.

The airplane that had been called "the billion-dollar gamble" proved itself once again.

RECONNAISSANCE

Of all the false economies of the military downsizing that occurred after World War II, none was more costly than the virtual elimination of USAF reconnaissance capability. The art of photographic reconnaissance had been raised to new heights by the USAAF in World War II, then promptly thrown away, the baby with the bathwater, with the cutback of the American military.

When the United States government declared that it would fight in Korea, FEAF did not have a reconnaissance system; it did have two squadrons of aircraft, the RB-29s of the 31st Strategic Reconnaissance Squadron of Kadena, two flights of RB-17s for photo-mapping purposes, and the RF-80As of the 548th Reconnaissance Technical Squadron at Yokota. However, it lacked the personnel, the equipment, and the methods to convert any photos that this ill-assorted collection of aircraft might take into usable intelligence.

And this is the great double-edged sword of budget cuts, which often retain some military hardware while gutting the system that was designed to use it meaningfully. The men who had done such a sterling job of aerial reconnaissance in World War II were discharged or reassigned; the photo labs they had worked with had long since been sold for surplus. Everything had to be reinvented in Korea. In the beginning, the reinvention was done incorrectly, with an irrational administrative setup, inadequate equipment, and ad hoc procedures.

To provide some semblance of a reconnaissance capability, all available resources were pulled in, and by November 1950, the 45th Tactical Reconnaissance Squadron began to receive RF-51s for visual reconnaissance, flying out of Taegu (K2). General Stratemeyer knew that leadership was the key, and he asked that Colonel Karl L. "Pop" Polifka be assigned to the task. Polifka had flown 130 missions in Europe and then commanded the 8th Reconnaissance Squadron, flying Lockheed F-4s in the Pacific. Brigadier General George Goddard, the father of USAF aerial photography, called Polifka the most outstanding reconnaissance pilot of World War II, one who quite literally wrote the book on technique.

In Korea, "Pop" Polifka was almost forty years old, heavyset and absolutely without nerves, still flying the tough missions he did not want to assign to younger men in the 67th Tactical Reconnaissance Wing he commanded. His knowledge and leadership quickly whipped the wing into shape, normalizing the procedures for getting the photo coverage that Fifth Air Force intelligence demanded. Under his guidance, the 67th maintained a periodic surveillance of enemy airfields, did bomb damage assessment, and, just as had been done in World War I, took photos of the front.

On July 1, 1951, he scheduled himself for a dangerous long-range mission into North Korea. On his run-in over the target, his RF-51 was hit by ground fire and he bailed out, only to have his parachute snag on his tail assembly. His loss was a severe setback to the already troubled reconnaissance program.

Polifka's loss in a World War II Mustang symbolized the USAF reconnaissance effort in Korea, which was never given the equipment and the personnel commensurate with the demands placed upon it by the Eighth Army and by the Fifth Air Force. The primary units involved included the 67th Tactical Reconnaissance Wing and the 31st (later the 91st) Strategic Reconnaissance Squadrons. They used a wide variety of aircraft, including North American RF-51s, Lockheed RF-80As and RF-80Cs, RF-86s, RF-26s, a few four-jet North American B-45s, and RB-29s. Each of the aircraft types had limitations which inhibited the performance of the mission.

The RF-80s were red-lined at Mach .8 and thus vulnerable to the MiG-15 and unable to operate in MiG Alley without a heavy F-86 escort. In addition, their cameras had been designed for piston-engine aircraft speeds, so that they had to slow down when taking the required photos. The RF-51s were hopelessly obsolete, and, as Polifka's mission had shown, very vulnerable to ground fire. The RB-26s operated at night, and the artificial illuminating systems they used were not only often defective but had to be operated at a low 3,000-foot altitude where the antiaircraft fire was deadly. The RB-26s eventually reverted to using standard photoflash bombs, which gave good results from higher altitudes. The RB-29s also needed heavy F-86 escort to enter MiG Alley, as did the faster RB-45s. The latter were not useful at night, for opening their bomb bays to drop the photoflash bombs caused an unacceptable buffet. The RF-86s, although fast enough to operate in MiG Alley, had a totally inadequate camera system.

Despite the sorry state of its equipment, USAF reconnaissance in Korea outpaced World War II efforts by a large margin. In April 1945, a record number of 1,300 sorties had been flown in Europe; in May 1952, the 67th Reconnaissance Group, then commanded by Colonel Edwin S. Chickering, flew 2,400 sorties, and it averaged close to 1,800 sorties per month for the year. Even this was not enough, however, for the Eighth Army complained bitterly that only 75 percent of its reconnaissance requirements were met. The F-86s had been able to prevail over superior numbers of MiGs because of the superior training of their pilots. The reconnaissance pilots were faced with a different situation; there was no individual enemy to intimidate, defeat, and thus overcome. Instead, there were simply too many targets to hit and too many square miles of ground to cover. The tight prewar budgets had denied them the capability to do their job, and there was no way that innovation or courage could completely compensate for the equipment they lacked.

KOREAN AIRLIFTERS

The USAF fought the cargo war in Korea with new ideas and, for the most part, old aircraft. Originally called the FEAF Combat Cargo Command (Provisional) and later the 315th Air Division (Combat Cargo), the basic organization for hauling troops and supplies in the Korean theater originated with the master of the Berlin Airlift, Major General William H. Tunner. Tunner simply wanted to bring to a logical conclusion the concepts he had developed in the China-Burma-India theater, flying the Hump, and later amplified in the Berlin relief operation. He believed firmly that the theater air commander should have continuous centralized control over subordinate transport units. With this control established, a single airlift command could meet all commitments, from carrying cargo and dropping parachutists to supplying cutoff units by air or bringing home tired GIs for R&R.

Tunner did not attempt to allocate airlift priorities—he left this to the Joint Airlift Control Organization. Instead, he concentrated on how to best execute the airlift duties he was assigned. By centralizing scheduling—which naturally involved maintenance, crew assignments, and all other aspects of an effective airlift—Tunner was able to accomplish miracles with a small number of aircraft. His fleet averaged about 210 airplanes, and it was made up of a ragtag collection of Douglas C-47s, Curtiss C-46s (many of which had flown the Hump under Tunner), Douglas C-54s, Fairchild C-119s, and small numbers of the huge-for-the-time Douglas C-124s.

The variety of aircraft posed problems for maintenance, parts, supplies, crew coordination, and crew replenishment; it also gave the 315th the flexibility it needed to fight a war under the most primitive conditions. The C-47s—considered "ancient" at the time, but destined to soldier on for another twenty years—were useful in landing supplies at small airfields with limited runway capacity. After ten years of use, the C-46s, great hogs on the ground and not much better in the air, were still reliable, and the 315th pleaded to keep them in service long after the time when they should have been salvaged. The C-54s served as they had in Berlin and everywhere else—completely reliable, lovely to fly, and limited only by their numbers. The C-119s turned out to be a maintenance nightmare, with fragile landing gear and fracture-prone propellers. They were themselves a case study in management—how do you get a decent utilization from aircraft that are regarded as so dangerous to fly that they were forbidden to carry passengers? How do you get aircrews to fly planes that were allowed to drop parachute troops only because it was considered that a paratrooper would know how to bail out quickly in an emergency? The grave limitations of the C-119s precipitated at least two crises in which the 315th Air Division's airlift capability almost collapsed, and with it the Fifth Air Force's logistic support. With all its limitations, the ground-loving C-119 eventually became a useful aircraft. The Douglas C-124, in later years the beloved "Old Shaky," was new on the scene, and went through the usual growing pains of a new aircraft, including massive fuel leaks, lengthy groundings, and reduced operating limitations.

In spite of the limited capabilities of the aircraft on hand, the 315th managed to do a remarkable job, flying more than 200,000 sorties and lifting more than 300,000 medical air-evacuation patients, carrying 2.6 million passengers, and lifting almost 400,000 tons of freight.

Besides these logistic duties, the 315th also participated in combat, dropping both troops and supplies and participating in two of the biggest airborne assault operations of the Korean War. The first took place on October 20, 1950, when the war was scarcely underway. Both General MacArthur and General Tunner were airborne, watching seventy C-119s and forty C-47s pump out 2,860 paratroopers and 300 tons of equipment to drop zones at Sukchon. It was a near-perfect operation and validated Tunner's contention that airlift and paratroopers did not have to train together continuously to be effective. The second operation was at Munsan-ni on March 23, 1951, when seventy-two C-119s and forty-eight C-46s dropped 3,447 paratroopers and 220 tons of equipment, enabling the 187th Regimental Combat Team to assist in closing up the U.S. I Corps drive to the Imjin River.

Despite the fact that the 315th functioned well with its ill-assorted equipment, there were continuous demands by the Army, Navy, Marines, and Air Force for their own dedicated transport capability. These demands might have been blunted if the 315th had been equipped with an all-purpose aircraft that could have handled any of the tasks, from C-47s dropping into tiny fields to C-54s carrying wounded soldiers to Japan to C-124s carrying outsize equipment. The right aircraft for the job, the Lockheed C-130, would fly for the first time a year after the war ended.

AIR/SEA RESCUE

If any one organization typifies the American spirit, it is the Air Rescue Service, which introduced the concept of making rescues even far behind enemy lines. Building upon tactics first used by the Royal Air Force in World War II, the air rescue squadrons quickly established procedures that enabled them ultimately to save 170 U.S. airmen and eighty-four airmen from other UN air forces who had gone down in enemy territory. Operating in flights at various bases and following the front lines as they moved forward or back, the 3rd Air Rescue Squadron (later 3rd Air Rescue Group) of Detachment F had the usual grab bag of World War II equipment like the converted RB-17s and the Vultee L-5 Sentinel liaison plane, plus a pathetically small number of modern aircraft like the Grumman SA-16 Albatross and the Sikorsky H-5 helicopter.

The rugged Albatross made an immediate impression, for it could operate in waves up to 5 feet high, and soon after the war started, it began making saves off the coast of North Korea. Gutsy pilots began landing them close to the enemy coast, ignoring enemy fire to pick up downed aviators. The helicopters were given multiple roles, one of which was immortalized by the M*A*S*H* television series, carrying critically wounded soldiers from the front to the mobile Army surgical hospitals. By the war's end, they had carried more than 8,500 wounded to the MASH units. Their primary task was the penetration of enemy territory to pick up downed airmen, but they were also used to pick up agents from behind enemy lines.

September 4, 1950, set the pattern for the Air Rescue Service's future. Captain Robert E. Wayne, who in his F-80 had shot down two Il-10s at Kimpo on July 27, had just made his twelfth strafing pass (not a good idea, he says today) when his F-51 burst into flame from antiaircraft hits. (His 35th FBS had converted to Mustangs in January.) Wayne bailed out, landing in a rice paddy, where he buried himself facing south, hoping to see a helicopter on its way from Pusan while fourteen of his buddies formed a ResCAP (rescue combat air patrol) overhead.

Wayne waited for an hour and fifty minutes before the welcome clopping sound of H-5 helicopter blades came from behind—the chopper had flown wide of his position on the first pass. He jumped up, tore off his flying suit to get at his clean T-shirt, and began waving it. First Lieutenant Paul W. Van Boven had flown to Hanggan-dong, penetrating about 8 miles inland from the coast, to drop his line to the grateful Wayne. This, the first behind-enemy-lines rescue by the Air Rescue Service, was a harbinger of the magnificent efforts that would be undertaken in the Vietnam War.

The immediate success of the rescue efforts brought about an expansion of the organization to new airfields and a moderate amount of additional equipment. The larger Sikorsky H-19 proved immediately useful, for its range of 120 miles gave it a 35-mile advantage over the little H-5. To maximize their utility, the aircraft were spotted at various fields close to hospitals and command posts. When it became evident that the Chinese Communist air force had an aversion to over-water operations, detachments were moved to Chodo, a small island off the coast of North Korea. A communications center there housed the tactical air controller operating under the familiar call sign DENTIST, and the H-5s stationed there were able to work effectively to rescue damaged aircraft returning from missions over North Korea. Speed was of the essence, especially in the winter, when a downed airman could expect to live for only a few minutes in the gelid waters of the Yellow Sea.

The effectiveness of the helicopter operation resulted in many demands being placed on it. Flash floods in July 1952 isolated hundreds of UN troops, who were saved by the helicopters in an intense series of missions that strained the air rescue resources. The helicopter crews usually made the headlines only with the rescue of a famous ace. In September 1952, Major Frederick C. "Boots" Blesse, the leading Sabre ace at the time and a fighter tactician, was saved by an SA-16 after he had run out of fuel after combat in MiG Alley. In April 1953, Captain Joseph C. McConnell, Jr., was saved by an H-19 so that he could go on to become the leading ace of the Korean War.

OTHER FACETS OF THE WAR

Despite the fact that FEAF had begun the war with inadequate resources and had to compete with the nuclear deterrent mission of the stateside Air Force, it grew in strength and proficiency, and in July 1953 it was a much more effective fighting force than when the war began. By the time of the armistice, FEAF had nineteen groups with a total of 1,536 aircraft. It could sustain a daily sortie rate of eighty-five B-26 sorties, 181 F-84 sorties, 171 F-86 fighter-bomber sorties, and 143 F-86 fighter interceptor sorties.

Underlying this offensive strength was a remarkably complex and diverse defensive organization that had grown up in Korea and Japan. The Lockheed F-94Bs night fighters of the 68th and 319th Fighter Interceptor Squadrons operated out of Japan and Korea. The F-94B scored its first victory on January 30, 1953, when Captain Ben L. Fithian and Lieutenant Sam R. Lyons shot down a piston-engine Lavochkin La-9 fighter. The F-94Bs went on to shoot down several jet and "Bedcheck Charlie" aircraft, the latter actually proving more hazardous. The F-94 had to slow down to the point of stall to attack the slow-flying Po-2s and ran the risk of either stalling out or overrunning the target and colliding with it.

A huge apparatus grew up to support the combat units, to include an extensive radar network, sophisticated radio communications, airfield construction, and weather forecasting, along with myriad other support units. General Barcus's directive to keep 75 percent of all aircraft ready for combat forced changes in maintenance procedures and resulted in the development of REMCO (rear echelon maintenance combined operations) maintenance methods. All maintenance personnel above those necessary for ordinary pre-and postflight services were concentrated in REMCOs. Spare parts were also centralized with the REMCO units. There was some initial resistance to the concept—unit commanders liked to keep tabs on their own maintenance and complained about the flying time involved in getting aircraft to and from the REMCO organizations in Japan—but the higher in-commission rates spoke for themselves.

Although FEAF stood shoulder to shoulder with the U.S. Navy and with UN forces in the battle against the Communists, aerial warfare in the Korean War was *not* conducted as a unified command. The primary reason was the decision by the United Nations Command/Far East Command not to have a joint headquarters for most of the war. The activities of the naval and marine units of NavFE were coordinated with FEAF, but not controlled by it. The situation was analogous to that which would prevail in Vietnam, where it was more convenient to designate broad areas of territory to the Navy for general operations than to attempt to have an actual unified command. Coordination was undertaken with good results on specific missions, but the degree of cooperation that had been achieved in joint operations with the USAAF and the RAF in World War II was never matched between FEAF and NavFE.

The Royal Australian Air Force, first flying F-51s and then Gloster Meteors; the South African Air Force, with its F-51s; the Royal Navy, with Supermarine Seafires, Fairey Fireflies, and Hawker Sea Furies; and the Royal Air Force, with its Short Sunderland flying boats, all made important contributions, but they were also coordinated with, rather than controlled by, FEAF. The Royal Thai Air Force and the Royal Hellenic Air Force contributed C-47 units that became a part of the 374th Troop Carrier Wing. Like them, the Republic of Korea Air Force units were controlled by FEAF.

Despite the lack of true unified control, the monumental air effort in Korea resulted in some staggering statistics. Following is a brief tabular distillation of the many months of hard work, danger, injury and death that were the currency of the air effort:

Item	Total UN Effort	Total FEAF Effort	%FEAF
Effort			
Total sorties	1,040,078	720,980	69%
Tons ordnance delivered	698,000	476,000	68%
Aircraft lost	1,986	1,466	74%

Although air power did not win the war for the United Nations, it certainly prevented defeat. In essence, the United Nations was not willing to commit the millions of troops that would have been necessary to win a ground war and relied upon air power to contain the Communist forces. For their part, the enemy realized that without air superiority, it could never win the war through its massive advantage in numbers on the ground. Air power had brought about the tactical stalemate that permitted the armistice to be negotiated.

The effects of the Korean War would be felt throughout the world for many years, not least in the very organizational makeup of the United States Air Force.

General Henry H. "Hap" Arnold steps out of his staff car to be greeted by the famous designer Jack Northrop.

It was Hap Arnold's vision that led to the establishment of the vitally important Scientific Advisory Board. Here, assembled for its first full meeting in June, 1946, are some of the nation's most distinguished scientists.

The end of one era and the beginning of another. President Harry S. Truman presents the Distinguished Service Medal to General Carl Spaatz, Chief of Staff, USAF.

The exact moment that the USAF officially became an independent service as the first Secretary of the Air Force, Stuart Symington, is sworn in by Chief Justice Fred Vinson on September 18, 1947.

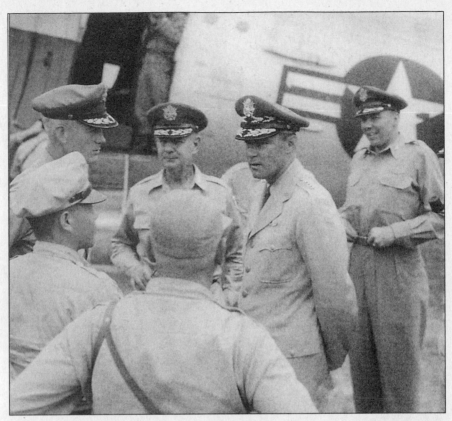

Within days after the North Korean invasion of South Korea, General Hoyt Vandenberg, the second USAF Chief of Staff, flew to the scene of the action.

The North American F-86 Sabre of the 4th Figure Interceptor Wing taxi out to do battle with the MiG-15s to be found in abundance in "MiG Alley."

A pair of aces. Colonel (and later General and CINCSAC) John C. Meyer, CO of the 4th FIG, congratulates James Jabara on his third MiG kill. Jabara would later go on to be the second-ranking ace of the Korean War.

First Lieutenant Joseph McConnell, Jr., is shown after shooting down his fifth MiG. Portrayed by Alan Ladd in the Hollywood film, *The McConnell Story*, Captain McConnell would lose his life in the crash of a North American F-86D at Edwards Air Force Base, California, in 1954.

Eugene M. Zuckert was a powerfully influential Assistant Secretary of the Air Force to the first Secretary, Stuart Symington, and led the fight for racial integration of the USAF.

The first jets that operated in Korea were the Lockheed P-80 Shooting Stars.

Heavily laden Republic F-84Es of the 49th Fighter Bomber Wing head for targets in North Korea.

Major William T. Whisner, Jr., of the 51st Fighter Interceptor Wing, scored 15.5 victories in World War II and added 5.5 more in Korea. Ironically, Whisner would die in 1989 in a reaction to the sting of an insect.

Air Force Chief of Staff John P. McConnell pins Major General Benjamin O. Davis, Jr.'s third Legion of Merit award. Davis, who led the Tuskegee Airmen in World War II, was the Air Force's first African-American general officer.

Two of the most important men in the United States' ballistic missile program. On the left, Trevor Gardner, Assistant Secretary of the Air Force (Research and Development); on the right is Major General Bernard A. Schriever, who masterminded the development of three ballistic missile systems and one intermediate range system in less than eight years.

4

≡

POWER AND PEACE:
1953–1961

FRIENDLY LEADERS

*F*ortunately for the Air Force, President Eisenhower had a benign view of air power. Despite his Army background, Eisenhower emphasized strategic forces that would respond to enemy challenges by massive retaliation, which led to budget decisions favoring the Air Force and the Nation.

This naturally suited the Air Force Chief of Staff, General Nathan F. Twining, who grasped both the possibilities and the limitations of nuclear weapons, bombers, and ballistic missiles. Twining needed all the help he could get, for it happened that the next two men who occupied the position of Secretary of the Air Force, Harold E. Talbott (February 4, 1953, to August 13, 1955) and Donald A. Quarles (August 15, 1955, to April 30, 1957) were less than powerful leaders. Their performance, however well intentioned, was hampered by external events and by the general shift in power from the service secretaries to the Secretary of Defense.

Distinguished in name and appearance, General Nathan Farragut Twining became the third Chief of Staff on June 30, 1953. Twining was blessed with the kind of quiet, pleasant personality and commonsense style that made people like him without their realizing just how bright he was, qualities that served him well as Chief of Staff and later as the first Chairman of the Joint Chiefs of Staff from the Air Force. He had a remarkable combat career, having commanded the Thirteenth, Fifteenth, Allied Mediterranean, and Twentieth Air Forces. He spent six harrowing days on a raft in the Pacific after his plane ditched. In true Rickenbacker survivor style, Twining, a marksman, defied superstition and provided some of his crew's scant food by shooting an albatross with his pistol.

As Commander of the Twentieth Air Force, he, like some others, opposed dropping the atomic bomb, and he insisted on a written order for its use. Later, he changed his mind, realizing that the deterrent value of nuclear power depended in great measure on the enemy's conviction that the United States would use it if required. Twining was an advocate of the use of nuclear weapons to save the French at Dien Bien Phu and saw to it that tactical nuclear weapons became a part of the Air Force arsenal. Yet his views on the use of atomic power were more sophisticated than most; he believed that it could be used to disarm the enemy completely, rather than destroying him.

He was well prepared to understand the thunderbolt announcement of the explosion of a Soviet thermonuclear device on August 20, 1953, less than a month after the Korean armistice had been signed. In response, the National Security Council issued NSC-162, a paper giving guidance to the Joint Chiefs of Staff on the nature of any war that it now might be expected to fight. The NSC directed that the nation's first line of defense should be an air atomic strike force, one that would deter the Soviet Union from attacking. President Eisenhower announced the NSC policy in his State of the Union message to Congress on January 4, 1954, stating that there would a permanent professional corps of trained officers and men. Air power would be backed up by a mobile strategic reserve that could be deployed to meet local emergencies. Continental air defense was to be upgraded, and industry was to be ready to mobilize to a full wartime status. Funds were to be supplied to meet the additional requirements of the Air Force, while both the Army and the Navy would make reductions in their forces to provide offsetting economies. Eisenhower's policy was called the "New Look," and the Congress provided a defense budget of about $30 billion, divided to give roughly $12 billion to the Air Force, $10 billion to the Navy, and $8 billion to the Army. (The term "New Look" is sometimes attributed to an incongruous derivation from a contemporary phrase for the latest in women's fashions, but actually went back in military history to 1948.)

Five days after Eisenhower's State of the Union address, Secretary of State John Foster Dulles defined the concept of deterrence as "instant, massive retaliation." The capability to inflict such retaliation by a bomber force had long been the goal of General Curtis E. LeMay. An extension of that capability, the creation of the intercontinental ballistic missile (ICBM), was being made simultaneously under the leadership of one of the most famous heroes in Air Force history, General Bernard A. Schriever. Schriever pulled off a managerial coup by fielding no less than three generations of intercontinental ballistic missiles almost simultaneously, a feat in many ways more daunting than the Manhattan Project. In addition, he masterminded the Thor intermediate-range ballistic missile (IRBM) system, instigated the Lockheed U-2 aircraft, and essentially created the managerial, engineering, and administrative basis for the U.S. space program, for which he is rarely given credit.

The growth of the Strategic Air Command and the successful creation of a retaliatory ICBM fleet were so important in creating the deterrent sword and shield that still defends the free world that they justify going back a moment in time to tell the story. That story is also a demonstration of just how prescient Hap Arnold's emphasis on research and development proved to be.

THE PATH TO TRUE AIR POWER

Even while the Korean War raged, and long before the "New Look" called for it, the U.S. Air Force was firmly focused on the task of making the Strategic Air Command's bomber fleet so powerful that it would deter the Soviet Union from aggression. Hard internal decisions were made relative to the scope of effort to be applied in Korea, the allocation of resources to be spent on air defense, and, perhaps the most contentious of all the problems, the amount to be spent on the Tactical Air Command (TAC). Matters not only of judgment but of personality came into play, for General Vandenberg and Lieutenant General Elwood R. Quesada, TAC Commander, had a thinly veiled hostility of long standing.

Amid all this, USAF leaders were looking to the future, aware that the Soviet Union not only was building a sophisticated bombing force, but also was pushing the development of the intercontinental ballistic missile, against which there was no technically feasible defense. The bomber threat could be met with a sophisticated warning system and an adequate interceptor force, both expensive but achievable. Antiballistic missile weapons were very far in the future. The only way the ICBM threat could be countered was with our own system of ICBMs in such strength that the leaders of the Soviet Union would be deterred from attempting a first strike, knowing that our surviving retaliatory power would still be sufficient to destroy their homeland.

The decisions that put these policies into effect were costly in both economic and psychic terms. For the first time in its history, the United States was going to maintain professional armed forces in peacetime, rather than relying, as it had done in the past, on cadres of professionals upon which to build essentially civilian armed services. The oceans were no longer relevant protection for our shores, and the warning time for crises had dropped from the years provided in World Wars I and II to a matter of minutes. Further, the creation of a system of ICBMs meant the virtual establishment of a second kind of air force, one totally different from anything known in the past. The differences went far beyond the obvious ones of hardware; the missile force structure was made up of personality types new to the Air Force. They were non-aircrew, and even seemingly trivial questions like the creation of a missile badge to rank with the coveted pilot's wings insignia were filled with controversy. The typical "missilier" was very technically oriented and implicitly posed a threat, albeit long-range, to the existence of the flying Air Force.

The ICBM also required a new industry. Although the standards of the American aviation industry were very high, the new requirements for accuracy, reliability, and precision for missile propulsion, navigation, guidance, and weapon systems were an order of magnitude greater. As these requirements were just beginning to be perceived, a basic battle raged as to how the missile force should be divided among the Army, the Navy, and the Air Force, and had to be fought out in Congress and the White House.

The introduction of new missiles was only a part of the upsurge in the demand for funds to maintain professional armed services. Personnel costs loomed large in the equation, as did the purchase of new weaponry. Those responsible for the defense budget were thus confronted with huge increases in the projected costs over an indefinite, indeed, unlimited, number of years. As usual, the first approach to dividing the defense budget was met at every level, from the Pentagon to the Congress to the White House, by the time-honored (if inefficient and inequitable) proportionate distribution of funds, raising each service by about the same amount. Over time, however, the Air Force received an increasingly larger share of the pie.

Fortunately, two factors favored the establishment and growth of professional armed forces, including the expensive new missile arms. The first was the general sense of patriotism the populace felt after the triumph of World War II, which was not significantly diluted by the stalemate in Korea. More important, the buildup of the military budget was coincident with an unprecedented period of economic growth in the United States. (The coincidence was so marked that an argument can be made that the growth in the military spending was the engine that drove the economy during this halcyon period of expansion, rather than being a motor driven by economic growth.)

In retrospect, one thing is quite clear. The economy could not have grown, nor the nation have survived, had it not been for the absolute military dominance so quickly established by the Strategic Air Command, a military superiority that was used wisely and justly by the United States. It should be remembered that in all previous history, any nation with a demonstrable military dominance used its strength to increase its territory, whether it was Spain conquering the New World, England extending its empire to every corner of the globe, or the abortive smash-and-grab tactics of Germany and Japan. The nuclear striking power of the Strategic Air Command alone dwarfed the power of every other nation alone or in combination and was vastly greater in both absolute and relative terms than that of any of the great nations of the past. It reflects well upon the United States that this massive superiority in armed might was used only to deter war.

CURTIS LEMAY AND THE STRATEGIC AIR COMMAND

The year 1948 had seen two significant changes of command. The first came on April 30, 1948, when Hoyt Vandenberg succeeded the grand old man, Carl Spaatz, to become the second Air Force Chief of Staff. Vandenberg's tenure would be characterized by crisis, and then by emergency-forced growth. His personal reputation as a leader varied; some thought him the right man at the right time, carefully orchestrating the birth of the independent Air

Force, while others were less charitable, believing that he had used politics and his connection with his uncle, the powerful Senator Arthur Vandenberg, to offset certain limitations in his intellect and industry. Given Senator Vandenberg's political views, which were certainly not rabidly promilitary, this seems unlikely.

The second came on October 19, 1948, when Lieutenant General Curtis E. LeMay became Commander in Chief of the Strategic Air Command. LeMay would hold this position until June 30, 1957—the longest tenure for any military force commander since General Winfield Scott. The USAF and the country were extraordinarily fortunate that, over time, LeMay would be in a position to execute the nuclear strategy conceived of by the brilliant leader and future Chief of Staff General Thomas D. White.

No other U.S. military force commander so imprinted his personality and ideals upon his organization as did LeMay. SAC became LeMay personified— but only after tremendous effort on his part. There were no criticisms of his intellect or industry, nor any suggestion of patronage, but the hard, and often seemingly cold, manner in which he drove SAC gave rise to many stories about him, most of them apocryphal. In 1951, at the age of forty-six, he was confirmed as a full four-star general, the youngest since Ulysses S. Grant. LeMay was "the Iron Eagle" to his admirers, and simply "Iron Ass" to detractors who feared him. Some of his seemingly tough demeanor probably stemmed from a deadened nerve that left his face immobile and unsmiling. In practice, LeMay took better care of his troops than anyone else in the Air Force, and his tenure at SAC was filled with achievements such as improved housing, pay, recreation, promotion, medical care, and other vital personnel requirements. The most important assessment of LeMay was defined by the loyalty and the high morale of the people he commanded.

After his retirement in 1965, LeMay ran as a Vice Presidential candidate in George Wallace's 1968 third-party bid, a move that tarnished his reputation in the eyes of many. One time later in his life, he was in the company of several other retired four-star generals, including his former aide David C. Jones, himself a former Chief of Staff of the Air Force and Chairman of the Joint Chiefs of Staff. The evening had been mellowed with some drinks, and the conversation took a daring turn—for retired or not, LeMay was still LeMay— to the question of why the general had supported Wallace.

Jones recalls LeMay saying that he had not run because of political ambition—he had none, and knew that Wallace could only lose—but because he feared the direction the country would take if the Democratic candidate won. LeMay told the little group of intimates, friends for many long years, "Don't tell me about George Wallace. I know all about George Wallace. I knew he had no chance of winning. But I ran with him anyway because I thought he could take enough votes away from Humphrey. Humphrey would have been a disaster for this country as President." Always the strategist, LeMay wanted to add enough strength to Wallace's ticket to split the Democratic vote and thus defeat Humphrey. In essence, LeMay was making a last great sacrifice, his political reputation, to serve his country's cause as he saw it.

If his politics offended some, there could be no censure of his military record. No one, friend or foe, doubted for a moment that he made SAC into an elite force, capable of strategic operations on a scale never before conceived and conducted at a level of proficiency that became the standard for the USAF. Inevitably the USAF became a benchmark to which the Army and the Navy, not to mention many foreign armed forces around the world, would aspire. It marks the second phase of USAF professionalism, both in terms of proficiency and in terms of its view of the quality of life within the service. LeMay was a genius at organization, and the Management Control System (MCS) he installed at SAC Headquarters (and which was replicated at lower levels of command) is but one example of his style. The MCS gave LeMay the capability to spot every breakdown or potential breakdown within the SAC system, and because lower-echelon commanders were aware of his system and used it themselves, potential breakdowns were usually detected and corrected before they occurred.

LeMay also had the capacity for choosing good subordinates, delegating authority to them and then letting them do their job. Not all of his choices were popular. His deputy and later successor at SAC, General Thomas S. Power, had a reputation for cold-hearted efficiency that many considered bordering on sadism.

LeMay knew that Power was tough—but he also knew that he got his job done, and that was what counted.

Not a West Pointer, LeMay graduated from Ohio State University in 1927 as a civil engineer before entering Air Corps aviation cadet training. Commissioned a second lieutenant as a pilot in 1929, he labored quietly in the background for years, learning a variety of skills that would benefit him in the future. One of these was navigation, and he later became the proponent of multirated officers, qualified as pilots, radar observers, and bombardiers/navigators.

During World War II, LeMay became one of the all-time great combat commanders, leading from the front to raise standards of professionalism and proficiency in the Eighth Air Force in Europe, and with the XXI Bomber Command in the Pacific. He drove his airmen hard, making them practice formation flying, gunnery, and bombing to a degree many considered fanatic. But his classic "sweat instead of blood" methods saved lives even as they molded forces that would fight hard.

At SAC, he was at war again, this time with the Soviet Union, an enemy he never underestimated. Since the fall of the Soviet Union, it has been shown that LeMay's assessment of their intentions was exactly on the mark, even if intelligence reporting led him occasionally to overestimate their capabilities. LeMay followed the sound military dictum that you should always prepare for what the enemy can do, not what you think he will do. The maxim sounds almost trite, but failure to observe it is costly, as was demonstrated many times during World War II—in Norway and the Low Countries in 1940, by Rommel in North Africa, by the Japanese at Pearl Harbor, the Philippines, and Savo Bay, and by the Germans in the Ardennes in 1944, to name just a few. From the early 1950s on, the Soviet Union *could* have inflicted grievous harm upon the United States and the world. That it did not may be attributed to the fact that it had realistic rather than rogue leaders, hard-bitten survivors of the Soviet competition who were aware of SAC's capability and knew that it would be used to its fullest extent. Neither LeMay nor anyone else could have counted on this, and he flogged SAC continually so that there would never be what script writers term a worst-case scenario.

When LeMay arrived to take over command, he was disappointed but not surprised at what he found—senior Air Force officers were aware that the Strategic Air Command in 1948 was woefully lacking in proficiency, discipline, and professionalism. He went to work immediately to correct things, using on-the-spot leadership to do so.

David Jones, LeMay's aide and pilot at the time, recalled an early incident in which LeMay called him and told him to prepare their command Boeing C-97 for takeoff on a Sunday. When LeMay flew, he flew in the left seat and made all the takeoffs and landings. They landed, without warning, at McDill AFB, Florida. When the startled wing commander, dressed in golf clothes, met him, LeMay said merely, "Execute your war plan." (Their war plan was to load their aircraft up with the specified flyaway kits, fly to England, and fly strike missions from there.)

A Keystone Kops pandemonium followed: squadron commanders could not be found, keys to equipment storerooms were lost, aircraft were not ready. It was, in short, a shambles. LeMay canceled the exercise and flew back to Omaha.

He took no disciplinary action, painfully aware that a similar landing at any other base would produce the same results. Instead he called his staff together and told them what had happened. It was all that was necessary. Six months later he repeated the exercise at Hunter AFB, Georgia, and everything went like clockwork.

Quiet acceptance was not his usual style; he relieved commanders immediately if he found they were not meeting his standards. But he rewarded competence with equal swiftness, once getting into enormous bureaucratic difficulty because he promoted an able young lieutenant to the rank of captain on the spot, illegally ignoring the customary—and cumbersome—Air Force promotion procedure. A mad scramble of communications ensued between SAC and Air Force Headquarters to legalize the promotion retroactively. This was an antecedent if not the inspiration of the controversial, much envied, but highly effective "spot promotion" system that he inaugurated in SAC. Crews were ceaselessly evaluated, and had to meet exacting standards to become "lead" or "select" crews. If all the members of a select crew performed to the highest standards, meeting all their obligations under the intensified SAC training program, having excellent bombing, refueling, and navigation results, they were placed in competition for spot promotions to the next rank. Some exceptional individuals—a relative few—actually had the highest accolade, a "spot on a spot," in other words, a promotion two ranks beyond their normal levels. Needless to say, it was a tremendous incentive program within SAC and the subject of bitter envy outside it.

LeMay's style was to have his best crews set the highest standards, then provide more than adequate training and flying time for other crews to reach those standards of proficiency. He also insisted on scrupulously accurate records and very demanding evaluation procedures, knowing that he had inherited an air force that had reflexively gone from the rigors of war to the pleasures of a really well equipped flying club, one that paid you for belonging. It was a long process, for SAC was expanding rapidly. When the author joined the Strategic Air Command in January 1953, as a green second lieutenant freshly graduated from flying school, he was puzzled by the flying club atmosphere. Flying the big Boeing B-50s was done as a sport, radar bombing, navigation, and gunnery scores were fudged, and the principal occupation seemed to be playing hearts in the briefing room. Then one bright day LeMay's inspection team came in. Heads rolled, rigorous standards were introduced and enforced, and reporting became squeaky clean. Oddly enough, everyone still retaining his head was happier with the new system.

LeMay arrived at SAC Headquarters in Omaha, Nebraska, with a certain knowledge of what he wished to create: an enormous fleet of long-range bombers with sufficient atomic weapons at their disposal to devastate the Soviet Union within the first few days of World War III. LeMay did not wish to judge the margin closely; he wanted overkill on a scale that would terrify his opponents. He knew that he would have to make do with the B-29s on hand and the Boeing B-50s (a B-29 upgraded with larger engines and better systems) just coming into service, in addition to the few squadrons of the controversial B-36s. But he was planning for the future—the Boeing B-47 was showing much promise (as well as many development problems), and the even larger, more capable Boeing B-52 was in the pipeline.

LeMay's management of the jet bomber demonstrates his independent judgment and vision of the future as much as any other facet of his career. After the B-47 had entered widespread service there was a strong call from both within and outside the Air Force to economize by canceling the B-52 and instead modernizing and improving the B-47. LeMay fought this concept tooth and nail, believing that the B-47 was not sufficiently advanced technologically, and more important, that it was not big enough. He instead insisted on the purchase of the B-52, whose introduction would be accompanied by a phased retirement of the B-47.

His judgment was right on two counts. The B-47's structure proved to have a limited fatigue life because of the metallurgy and metal construction techniques current when it was built. It was also far too small to adapt to the many additions of equipment that have given the B-52 an unprecedented longevity performing an incredible variety of missions. It was first flown in 1952, and current estimates have it in the force structure until the year 2040—almost ninety years. If it does stay in service until then, it will certainly be the longest-lived aerial weapon system in history, and second in seniority overall only to the USS *Constitution*.

LeMay in Action

In both the European and Pacific theaters, LeMay had learned that getting bombs on targets involved far more than a crew, an airplane, and the bombs. Equally important were training, maintenance, morale, health, logistics, pre- and postflight briefings, and a host of other factors. He had also found that training was rarely rigorous enough, and so it was at SAC. When he arrived, practice missions were being flown at "comfortable" altitudes where engines were not taxed and the pressurized cabins worked well. Radar operators helped out in establishing visual bomb runs for bombardiers, while the bombardiers often monitored—and corrected—the radar runs. There was little concern about crew integrity—people flew with those they liked, time hogs (pilots who wanted to build up their flying time to get the coveted Senior Pilot and Command Pilot wings) flew all the time, and shirkers got by putting in their four hours per month for pay purposes.

The changes required in crew training were but the tip of LeMay's problem iceberg. To represent the scope of his predicament, it might be well first to show something of the growth of SAC during his tenure, in terms of size and equipment, and then explore the complications of managing and directing that growth.

The Growth of the Strategic Air Command

LeMay came to SAC, on October 19, 1948, and remained until mid-1957, when he left to become Vice Chief of Staff. On his arrival, he found the following force of 837 combat aircraft:

35	Consolidated B-36 Peacemakers
35	Boeing B-50 Superfortresses
486	Boeing B-29 Superfortresses
131	North American F-51 Mustangs
81	North American F-82 Twin Mustangs
24	Boeing RB-17 Flying Fortresses
30	Boeing RB-29 Superfortress reconnaissance planes
11	Douglas C-54 Skymasters
4	Beech RC-45 Expeditors

In addition, there was a ragtag collection of support aircraft, including Douglas C-47s and North American B-25s. (For a pilot, one of the great advantages of the period was the unlimited flight time available; pilots were literally begged to take the support aircraft on personal weekend cross-countries, to build proficiency and to "burn up the gasoline," an administrative holdover from World War II when future gas allocations were based on past use.)

LeMay commanded a force of 51,965, including 5,562 officers, 40,038 airmen, and 6,365 civilians. At the time of his departure, SAC personnel had grown fivefold, to 258,703 people, including 34,112 officers, 199,562 airmen, and 25,029 civilians. And he had at his disposal the following fleet of 3,040 aircraft:

22	Consolidated B-36 Peacemakers
380	Boeing B-52 Stratofortresses
1,367	Boeing B-47 Stratojets
176	Boeing RB-47 Stratojets
182	Boeing KC-135 Stratotankers
789	Boeing KC-97 Stratofreighters
51	Douglas C-124 Globemaster IIs
54	North American F-86 Sabres
19	Martin RB-57 Canberras

The growth in numbers from 1948 to 1957 was remarkable, but did not compare with the growth in technology. The B-36 had earned its name, Peacemaker, never having to drop a bomb in anger, but was a 1940 design. A large, complicated aircraft, it was difficult to maintain and had its performance raised from disappointing to marginal for Cold War standards by the addition of four jet engines. It was phased out in 1957.

In gleaming contrast, the Boeing B-47 Stratojet, which was built in greater numbers (2,032) than any other postwar bomber, was the most important multiengine aircraft of the jet age, one that launched not only the B-52 and KC-135 sister ships but the entire fleet of Boeing civilian aircraft from the 707 to the 777. With its swept wings, six jet engines, and almost perfect streamlining, the B-47 had a sensational performance in all respects but range, and this was totally compensated for by in-flight refueling. Compared to the piston-engine aircraft that it replaced, the B-47 was easy to maintain, but it was demanding to fly and unforgiving of mistakes. It was difficult to land, compared to its predecessors, for it was so streamlined that it was slow to decelerate, requiring both an antiskid brake system (known in today's automobiles as ABS) and a brake chute for landing. Go-arounds were difficult because its jet engines were slow to accelerate. A drag chute was fitted for use during the approach, enabling the pilot to maintain a higher power setting to permit faster acceleration. The most dramatic of its challenges could be found on takeoff, when an incorrect response to its roll-due-to-yaw characteristics could turn a simple outboard engine failure into a catastrophe.

The B-47 fleet, in combination with a steady growth in the nuclear weapon stockpile, established SAC as the most powerful military force in the history of the world. The B-47 had a protracted development, entering operations in

1951 on a very small scale. It became the mainstay of SAC, and then was re-tired in 1967 as an economy measure. Its successor, the B-52, combined the revolutionary advances of the B-47 with the lessons learned in its operation. The most important difference besides its larger size was the use of a thick wing instead of the B-47's very thin wing. The thick wing was easier to con-struct, stronger, and less fatigue-prone, and it provided room for enormous fuel tanks. As a result of the larger wing and the knowledge gained in operating the Stratojets (as they were almost never called), the B-52 was generally less de-manding to fly than a B-47. In a rough automotive comparison, the B-47 was a sports car while the B-52 was a pickup truck.

The improvement in aircraft performance had to be matched by an im-provement of onboard equipment. The great fleets that LeMay had com-manded in Europe had flown in vast formations to the target and on the formation leader's signal dropped their conventional 500- and 1,000-pound bombs en masse to provide the concentration necessary to damage targets. The B-47s and B-52s flew their missions as single aircraft or in small "cells" of three aircraft. Each one had to be of "lead aircraft" quality. Consequently, hundreds of millions of dollars were devoted to the development of electronic equipment to improve navigation, bombing, and electronic countermeasures. Both aircraft demonstrated their proficiency by winning SAC's annual bomb-ing competition with better results than their piston-engine predecessors. However, the size of the B-52 permitted it to continue development for a much longer period, incorporating equipment that allowed it to completely change its mission's flight regime from dropping nuclear weapons from high altitude to being a low-level penetrator capable of carrying cruise missiles of uncanny precision armed with conventional warheads. The result has been a reduction in fleet size from almost 2,000 bombers in 1961 to about 200 today.

THE TANKERS

All the improvements in bombing capability would have been almost meaningless without the very risky and mostly unsung development of aerial refueling. General Spaatz had seen the need for a refueling capability as far back as 1926, for the need to stretch aircraft range was always present, and the only practical solutions were forward bases and aerial refueling. (In the post–World War II period, research into nuclear-powered aircraft determined that they posed too many problems in terms of weight, radiation hazard, and disaster in the event of an accident.) Forward bases were the only alternative during World War II, despite early Air Corps experiments with aerial refueling. It was left to the British to make the first practical advances with Flight Refuel-ing Ltd.'s trailing-hose system, first patented in 1935, which used a hauling line that trailed behind the tanker and a contact line that trailed behind the receiver. The two aircraft made a cross-over maneuver, enabling grapnels on the end of the contact line to engage the hauling line. The receiver then winched both lines in, the contact line was detached, and the hauling line pulled the hose from the tanker. The nozzle of the hose was manually seated in the receptacle, and refueling was begun.

This obviously cumbersome and time-consuming system was followed by a probe-and-drogue system. The tanker reeled out a hose equipped with a funnellike female connector into which the receiver inserted a probe. In SAC, this system was adapted to use by the KB-29, the KB-50, and even two B-47s modified for test.

Both the hose and the probe-and-drogue systems required good weather conditions and took so long to transfer fuel that they were essentially impractical for use with jet aircraft. Boeing proposed a new method called the flying boom, which revolutionized in-flight refueling. (The probe-and-drogue system would prove to be more suitable for refueling fighters and was adapted as standard by TAC on its KB-29 and KB-50 tankers.)

The flying boom consisted of a long telescoping transfer tube with two V-shaped control surfaces, called "ruddervators." The refueling plane flew above and in front of the receiver, which flew in close formation, permitting the boom operator to fly the boom into the receiver's receptacle. With the boom system, fuel could be transferred under high pressure so that refueling times were reduced.

Despite the advantages of the flying-boom system, the disparity in performance between the new jet bombers and the piston-engine tankers (KB-29Ms and KC-97s) greatly compromised the technique. The jet aircraft had to descend to the refueling aircraft's altitude, slow down to almost stalling speed, refuel, and then climb back to its own mission altitude, greatly reducing the net yield of the fuel transfer. The introduction of the Boeing KC-135 Stratotanker, which made its first flight on July 15, 1954, obviated the performance differences and raised in-flight refueling from an arcane art to an industry.

Aerial refueling almost immediately became routine, being adopted first by bombers, then fighters, and later even by cargo planes and helicopters. Well publicized nonstop round-the-world flights informed the public—and the Soviet Union—that aerial refueling conferred almost unlimited range on SAC bombers. Later, aerial refueling permitted the concept of the continuous airborne alert of a portion of the SAC fleet. The full measure of its technically and personally demanding nature is still not generally appreciated. Aerial refueling requires a high level of airmanship, conducted under conditions of imminent danger. Consider the case of the KC-135 and the B-52. The KC-135, cruising at 250 knots indicated and weighing some 300,000 pounds (of which perhaps 150,000 was highly flammable jet fuel), was physically connected, by a tube that averaged about 40 feet long, to the B-52, itself weighing more than 450,000 pounds, flying at the same speed, and perhaps carrying a number of nuclear weapons. Add to this picture of mass and momentum such factors as the dark of night, turbulent weather, and the prospect of combat, and you have all the elements of a genuine white-knuckle drama.

OTHER ELEMENTS OF SAC AVIATION

The requirement for instant mobility had led to the creation of strategic support squadrons, equipped with Douglas C-124s, as an integral element of SAC. These aircraft facilitated the deployment of wings to forward bases. The growth of aerial refueling, combined with budget pressures, resulted in the transfer of the function to the Military Air Transport Service (MATS) in 1961.

The lessons of World War II had also predisposed SAC to require fighter escort units, reaching a peak of 411 Republic RF/F 84 Thunderjets in 1955. Some attempts were made to develop parasite fighters like the unfortunate McDonnell XF-85 Goblin, or the only slightly more practical Republic RF-84K FICON (fighter conveyor), as a means of escorting the B-36s. The in-flight-refueled jet bomber changed the conditions of warfare; B-47s or B-52s penetrating deep within enemy territory individually or in small cells of three aircraft were almost impossible to escort. The fighter escort squadrons were transferred to TAC in 1956.

As some of the support units became obsolete, others were introduced, including North American F-86 fighters to defend SAC bases in Spain, and such exotic aircraft as the Martin RB-57 and Lockheed U-2 reconnaissance planes.

Whatever their function or size of the support units, and regardless of whether they were waxing or waning, they received the full LeMay scrutiny and had to meet LeMay's standards. They were, like the bomber units, most immediately characterized by their equipment. Yet, as LeMay was quick to point out, procuring equipment was but a part of the equation in building a fully capable Air Force.

LESS OBVIOUS ASPECTS

LeMay always insisted that people were the most important factor in building an effective Air Force. His views were echoed by virtually every other Air Force leader, from the first Chief of Staff to the present. In spite of this, for much of the Air Force's existence, circumstances have placed a higher priority on funding weapons with their attendant fuel, spares, and maintenance than on personnel. In effect, the people of the United States Air Force, particularly its enlisted personnel, have subsidized the service by accepting lower pay, inferior living conditions, and inadequate equipment, even as they compensated for other shortages by their own extra effort. Yet this extra measure of service seemed to inoculate the noncommissioned officers and enlisted men of the Air Force with the desire and the ability to take on more responsibility and discharge it with a highly professional competence.

As important as people were to LeMay, was grappling with even greater challenges caused by the rapid progress of technology. For most of the history of military aviation, aircraft procurement had been a relatively simple matter involving the purchase of an airframe from one manufacturer and engines from another. Systems were simple and common to most types. It was almost a rule to create a new aircraft type around a proven engine, and to test new engines in a proven type.

The jet age greatly complicated this process. New aircraft and new engines were often required simultaneously to meet new specifications, and there were many other factors to integrate, including electronic suites, new materials, new methods of flight control, vastly more sophisticated weapons, and the like. And where in the past maintenance had been performed by mechanics largely divided into airframe and engine specialties, a host of new maintenance types emerged to cater to the new equipment.

Bringing an aircraft (missiles were even more demanding) into operational service thus meant the simultaneous management of the procurement, training, maintenance, and logistic processes, among a great many other requirements. For an aircraft as large as the B-52, there were additional considerations, including the building of runways adequate to take the landing weights, hangars large enough to house them (at Castle Air Force Base, California, the first operational B-52 base, the hangars were so large that miniature weather systems, including rain clouds, formed *within* the hangars), new fuel farms for swift refueling, larger munitions storage areas, enormous simulator buildings, and very specialized dollies, tools, and maintenance stands, to name but a few of the unique requirements.

There were massive changes not only in procedures but in processes. Where maintenance was once the fiefdom of the hardworking crew chiefs who "owned" an airplane and were both authorized and competent to do anything from a preflight run-up to an engine change, maintenance systems became industrialized, with successive levels of work being performed on the line, in base shops, and (over time) in various levels of depot maintenance. Some elements of maintenance became so esoteric that building-block units were exchanged rather than repaired when they malfunctioned—a "black box out, black box in" approach. Training naturally became equally rigorous and ever more demanding. SAC flight crews received training at special centers, the 3520th Flying Training Wing (later the 4347th Combat Crew Training Wing [CCTW]) at McConnell AFB, Kansas, for B-47s, the 4017th CCTW at Castle AFB for B-52s, and the 4397th Air Refueling Wing at Randolph AFB.

Huge training centers for each of the specialized tasks were required, and everyone—flight crews, mechanics, specialists—needed almost continuous updating. Where aircraft like the F-80 had flight manuals of less than an inch thickness, with a few maintenance manuals as backup, there now came entire libraries of manuals, the updating of which became a demanding science—configuration control reached out of the factories down to the manuals.

Flight time became so expensive—and the possible hazards so great—that remarkably realistic simulators were provided for additional procedural training. Crews could practice approaches to strange airports, navigation, bombing, and every conceivable emergency situation. Qualified crew members acting as simulator instructors were almost sadistic in the way they would torque up the pressure, adding one malfunction after another to create emergency situations that would have been impossible to attempt in the air. The simulators themselves required dedicated technicians doing specialized maintenance in a climate-controlled environment. Other specialized gear such as test equipment, the calibration equipment for the test equipment, high-altitude physiological training chambers, ejection seats, and personnel equipment had similar needs for their own personnel, who also had to be continuously trained and evaluated.

Underlying all of the requirements was the need for standardization, so that people and equipment could be freely interchanged and anomalies quickly identified.

This brief description merely touches on the scope of the task facing everyone from commanders to mechanics throughout the Air Force. All of these needs and many more had to be met within very short time periods with limited budgets, and with an overwhelming need for secrecy, not only in SAC but in TAC, the Air Defense Command, MATS, and elsewhere. That these needs were met was in large part due to LeMay's SAC management attitude and techniques spilling over to the rest of the USAF even before he became Vice Chief of Staff in 1957 and then Chief of Staff in 1961.

This spread of LeMay's influence was right for the time; his centralist form of management brought the necessary discipline and training that the Air Force required. However, his Air Force–wide influence would be transcended when he became Chief of Staff by the Department of Defense–wide influence of Secretary of Defense Robert Strange McNamara. McNamara would push the pendulum of authoritarian, centrally controlled management to its limits, going far beyond anything LeMay had done and virtually paralyzing the services with paperwork. This phase would, however, set the stage for the third, more liberal and people-oriented management phase of the Air Force that began in the mid-1970s and continues to this day.

All that LeMay had done to field SAC's incredibly powerful bomber force was possible only because of the research and development encouraged by Hap Arnold, which had become part of the fabric of Air Force thinking, and which was made manifest in the establishment of the Air Research and Development Command in 1950. This R&D effort had also laid the foundations for the nation's intercontinental ballistic missile systems, which, as we will see, were under tentative development for an extended period. However, the true incentive for the United States' need for space dominance would come as a total surprise, in the innocuous-sounding yet utterly ominous beeping of the world's first artificial satellite, the Soviet Union's Sputnik I, on October 4, 1957.

LEADERSHIP, R&D, AND MISSILES

The interservice rivalry over the control of air power was but a prelude to the long fight to control the development of missiles. The Army insisted that missiles were a form of artillery and hence within its province; the Air Force disingenuously protested that missiles were nothing more than unpiloted bombers and therefore within its province; the Navy, at odds as usual with both the other services, insisted that naval requirements made it necessary to control all forms of seaborne missiles, whether from ships, submarines, or aircraft. This quarrel differed somewhat from previous situations. In the old days of the "Revolt of the Admirals," the Army and the Air Force were usually allied. In the battle for missile roles and missions, the Army and the Navy most often aligned against the Air Force.

All the normal bureaucratic means to resolve the dispute were tried—existing boards were used; a National Guided Missiles Program was created; committees were formed (one with the classic initials GMIORG for Guided Missiles Interdepartmental Operational Requirements Group); Chrysler's K. T. Keller (whose automotive engineering expertise was exceeded only by his abysmal taste in styling) was appointed as Director of Guided Missiles, i.e., a "Missiles Czar"—but each service stubbornly continued on its own path, refusing more than token coordination or cooperation.

President Eisenhower chose a former president of General Motors, Charles E. "Engine Charlie" Wilson, to be his Secretary of Defense, the first to come from the automotive world. (Wilson, whose domineering personality and budget-cutting tactics had an adverse effect upon R&D, has been called a Saint John the Baptist presaging the arrival of the Messiah, Robert S. McNamara, in the Department of Defense.) As it had fallen to Louis Johnson to implement President Truman's budget cuts, so did it fall to Wilson to implement President Eisenhower's cuts, which reflected his view that "a sufficiency" of weapons was now the desired standard because Soviet advances had made achieving an overwhelming superiority in the number and size of weapons impossible. To sell the President's ideas, Wilson stated that the military expenditures of the Soviet Union were primarily for defensive purposes—an argument the Communists had refuted at the 1955 Tushino air show, where they had proudly demonstrated three potent bombers, the twin-jet Tupelov Tu-16 Badger, the four-jet Myasishchyev M-4 Bison, and the four-turboprop Tupelov Tu-20 Bear.

Aware of the indecision over the question of roles and missions for missiles, Wilson issued a memorandum on November 26, 1956, that addressed a number of issues, including the size of the combat zone in time of war (100 miles behind the line of contact), the limits on Army aircraft (5,000 pounds for fixed-wing and 20,000 pounds for rotary-wing), but most important, assigned departmental responsibility for missile systems among the three services. Wilson established that the Army was given point defense surface-to-air missiles; the Navy and Marine Corps were assigned responsibility for weapon systems to carry out their functions; and the Air Force was to have area defense missiles.

In addition, Wilson allowed the Army to continue development of surface-to-surface missiles with a range of about 200 miles; the Navy was assigned responsibility to employ ship-borne intermediate-range missiles (IRBMs), while the Air Force was confirmed in its operational employment of intercontinental ballistic missiles (ICBMs) and given the joint assignment with the Army of land-based IRBMS. The apportionment was not clean-cut, but it sanctioned the extensive work already undertaken by the Air Force.

Wilson continued the trend of concentrating power in the office of the Secretary of Defense, automatically reducing the influence of the service secretaries. Unfortunately for the Air Force, the two men who occupied the position of Secretary during the critical 1953–1957 era, Harold Talbott and Donald Quarles, while very much committed to the development of the intercontinental ballistic missile, could not stand up to Wilson. Talbott had done an excellent job overseeing the post-Korea expansion of the Air Force to a 137-wing force and worked diligently to improve the lot of the enlisted man, improving pay, housing, and educational opportunities. Unfortunately, his tenure was cut short by a conflict-of-interest controversy. A scientist, Quarles found himself at odds with the Air Force military leadership as well as with Wilson. Both Talbott and Quarles backed the development of the ICBM and of space satellites; sadly, both suffered deep disappointments while serving as Secretary, and both died within two years of leaving office.

It was well that the civilian leadership muddle was compensated for, at least in part, because the United States Army Air Forces had as early as 1946 begun an orderly and systematic missile development program. The most important effort was a study contract for $1.9 million to the Consolidated Vultee Corporation (later unofficially called Convair until the name was made official in 1954) for Project MX-744. A very refined development of the German V-2, Project MX-744 was placed under the leadership of a Belgian émigré, Karel J. Bossert. His rocket featured a thin, single-wall, pressurized monocoque structure that had to be kept rigid with nitrogen under pressure until fuel was introduced. Unpressurized, it would collapse under its own weight. Bossart also proposed a nose cone that would detach (with its warhead) from the main rocket structure and swiveling rocket engines. Project MX-744 was one of the first victims of the postwar budget cuts and was canceled in 1947, but Convair continued to spend its own money on the project.

Within the USAF, there was an inherent organizational bias against missiles, one freely acknowledged by both Twining and LeMay, and this, coupled with the tendency to hedge development bets, led to a preoccupation with surface-to-surface winged missile types that themselves were never brought to fruition. The aviation industry, which had soared to new heights in World War II, seemed strangely unable to cope with unmanned missiles, most of which fell years behind in their schedules. Among them were the winged, subsonic turbojet missile, the Martin Matador, the rocket-boosted, winged, supersonic ramjet-powered North American Navajo, and a long-range supersonic turbojet, the Northrop "Boojum." An even greater variety of air-to-surface, surface-to-air, and air-to-air missiles were also under development.

Budget cuts and the Korean War caused a decline in funding for missile research, which was probably not disadvantageous in the long run, for it tended to quickly weed out missiles whose performance seemed doubtful, and those whose development cycle was taking too long. Convair was rewarded for both its engineering skill and its patience on January 23, 1951, with a contract for Project MX-1593. The $500,000 contract called for Convair to outline a vehicle with a maximum over-the-target speed of Mach 6, a range of 5,000 nautical miles carrying an 8,000-pound warhead, and the ability to strike a target with a CEP of 1,500 feet. (CEP stands for circular error probable, a bombing-range term that defines the radius of a circle within which half the weapons targeted for the center of the circle can be expected to land.)

Convair reported that the desired missile was feasible and could either be a ballistic or a glide-rocket type. The ballistic missile would be twice as large as the glide rocket and would require a cluster of five to seven of North American Aircraft's 120,000-pound-thrust alcohol–liquid-oxygen engines that were decendants of the booster engines used on the Navajo Project. The MX-1593 project was also given the name Atlas, although the Air Force, concerned about roles and missions, continued to call it the XB-65 pilotless bomber for a considerable time.

Many within the Air Force were properly concerned about spending scarce resources on long-range missiles that might not be technically feasible. However, these concerns were soon overshadowed. Reports from the Soviet Union in late 1951 of huge rockets of more than 250,000 pounds of thrust lent urgency to the development of the Atlas.

The principal impediment to proceeding with the Atlas was the size and weight of the warhead, which, on an almost exactly proportionate basis, dictated the size and weight of the rocket required to launch it to the target. (There were many other difficult problems to be solved, including the high heat associated with atmospheric reentry, guidance and control, engine staging, and so on, but these seemed to be more amenable to resolution than the fundamental question of the warhead weight.)

The first experimental hydrogen fusion device was detonated at Eniwetok on November 1, 1952, as a part of "Operation Mike." This device used cryogenic D (deuterium) and T (tritiom), and weighed about 50,000 pounds. It clearly seemed impractical for delivery by bombers, even though SAC immediately began framing requirements for an aircraft that could carry a 25-ton bomb. At a 1953 Scientific Advisory Board meeting in Florida at Patrick Air Force Base, two of the most brilliant scientists of the day, Dr. Edward Teller and Dr. John von Neumann, gave independent presentations which indicated that a "dry" thermonuclear bomb weighing about 1,500 pounds could be built. General Schriever in a recent interview recalled, "I almost came out of my seat in excitement, realizing what this meant for the ICBM."

Then a colonel, Schriever sought confirmation and credibility for this new development. Upon the recommendation of retired Lieutenant General James H. Doolittle, a nuclear weapons panel headed by Dr. von Neumann was established by the Air Force Scientific Advisory Board. The results of this panel's deliberations were more conservative than the original presentations, indicating that a thermonuclear warhead weighing only 3,000 pounds and yielding .5 megaton was possible.

This effectively solved the warhead weight problem, simultaneously reducing the projected weight of the Atlas rocket from 440,000 pounds to 220,000 pounds. Spurred on by this finding, one of the most influential men in the missile program, Trevor Gardner, the Special Assistant for Research and Development to the Secretary of the Air Force, established another committee, called the Strategic Missiles Evaluation Group. The committee followed the spirit of Hap Arnold's direction, with von Neumann as chairman, and with stellar members Simon Ramo, Dean Wooldridge, Clark B. Millikan, Hendrik W. Bode, Louis G. Dunn, Lawrence A. Hyland, George B. Kistiakowski, Charles C. Lauritsen, Allen E. Puckett, and Jerome B. Weisner. These men, quite literally the cream of crop of industrial and academic scientists, placed themselves at the service of their country. (It is sad to reflect that in a similar emergency, an equivalent national need, we might not in these liberal days be able to count on an equivalent representation from academia.) Schriever served as the committee's military representative.

After a series of meetings, the "Teapot Committee" (no one seems to be able to recall why it was so called) recommended that the Atlas program be revised to take double advantage of the new lightweight, high-yield warheads by reducing the stringent CEP requirements from 1,500 feet to between 2 and 3 nautical miles.

Gardner took the recommendation and ran with it. An impulsive, sharp-tempered, sometimes acerbic man, Gardner was convinced of the imminent danger of the Soviet ballistic missile program and was determined to ramrod a counter to it. Gardner was fortunate in having the backing of Secretary of the Air Force Talbot and Generals Twining and White in a plan that called for $1.5 billion expenditure over the next year to achieve, by July 1958, two launch sites and four operational missiles, and by 1960, twenty launch sites and a hundred missiles.

Designated as Talbott's direct representative in all aspects of the Atlas program, Gardner pressed ahead with a no-holds-barred attitude toward spending. By July 1, 1954, an ARDC field office, called the Western Development Division, was established in Inglewood, California, with now Brigadier General Schriever as commander. It was the best possible appointment.

THE RIGHT MAN IN THE RIGHT PLACE AT THE RIGHT TIME

Bernard Adolph Schriever was born in Bremen, Germany, in 1910, and was perhaps destined for aviation and space when as a boy he saw the Kaiser's Zeppelins flying over his home, returning from their raids on England. By an unusually provident act of war, the ship upon which his father served was seized in New York harbor in 1916 and its crew interned. Bernard, his mother, and his brother Gerhard came to America in the following year.

Young Bernard's life was difficult, but molded his character and his interests. His father died in an industrial accident in 1918, and his mother supported the family. He worked hard at school and at play, graduating from San Antonio High School with honors. A caddy early in life, Schriever became an expert golfer. He graduated from Texas A&M in 1931 with a degree in architectural engineering; his success in ROTC gave him a chance to become a flier, and he won his wings at Kelly Field, Texas, in June 1933.

The next six years were varied, including flying the airmail for the Army, commanding a Civilian Conservation Corps camp, flying for Northwest Airlines, playing a lot of golf at the professional level, and finally returning for a regular commission when the Air Corps at last began to expand in 1939. During World War II, he flew sixty-three combat missions with the veteran 19th Bombardment Group.

After the war he worked with von Karman, who recognized his talent and made certain that Schriever would carry on the liaison Arnold had established with the scientific community. Schriever's technical expertise led him to push technology to obtain advances in weaponry; his methods were not traditional, and he sometimes experienced opposition within the Air Staff, often with the preeminent pragmatist Curtis LeMay. (It was Schriever who had pushed for the B-47 instead of the B-52.) However, he had the patronage of both Generals Twining and White—necessary to deal with LeMay on an equal basis—and was obviously destined for a leadership position in a technologically advanced Air Force.

In his new assignment as Commander of the Western Development Division, Schriever had been able to handpick his staff. Further, against much opposition, he was permitted a radical new organizational scheme, in which his WDD would retain system responsibility for the Atlas program, employing the Ramo-Wooldridge (later TRW) firm to do the systems engineering and technical direction, with Convair being awarded an airframe and assembly contract. Later, Schriever was made assistant to the Commander of the Air Research and Development Command for Ballistic Missiles, a second hat that enabled him to direct ARDC to assist WDD's efforts, rather than having to solicit help from the ARDC staff. A dotted line is a subtle distinction on the organization chart, but it was a time-saver in developing the ICBM, and Schriever recalls that many seemingly impossible bureaucratic problems were solved by simple telephone calls.

Trevor Gardner had received the backing he wanted for the ICBM with the advent of a report from the President's Technological Advisory Committee in February 1955. Called the Killian Report, after its chairman, James R. Killian, Jr., the president of the Massachusetts Institute of Technology, the report noted the progress in Soviet rocket technology and stressed the vulnerability of the United States to attack. Schriever, Gardner, and von Neumann briefed President Eisenhower and virtually his entire cabinet on the potential of the ICBM. Eisenhower directed that the ICBM program become a nationally supported effort of the highest priority.

With Gardner securing top priority for the ICBM, with separate funding for the Atlas, and under Schriever's leadership, the Western Development Division was well launched. Its furious pace of work began the development not only of the Atlas missile but of what would become the aerospace industry. The new specifications were challenging, and some companies had to be induced to compete for systems unlike any they had ever created. Schriever fostered a competitive climate by subcontracting out every major subsystem to separate contractors and using the WDD as the integrating facility. Schriever never doubted that the Atlas would be a success, but was concerned that the requirement that the missile be fueled and launched in twelve minutes might be difficult to achieve.

The Soviet threat and the momentum of the Atlas program spawned other missiles. A decision was made in May 1955 to develop an intermediate-range ballistic missile (IRBM), the Thor, from the Atlas, and an award was made to the Martin Company in the fall of 1955 to develop the Titan missile as an alternative to the Atlas.

In brief, Schriever found himself at the head of an organization that dwarfed the Manhattan Project in terms of scientific difficulty, spending, number of employees, and, most significant, urgency. For the last year of the Manhattan Project, the United States was confident that neither Germany nor Japan had the capacity to build or deliver an atomic bomb. In striking contrast, there was every reason to believe that the Soviet Union was far more advanced than the United States in rocketry, and its successes with both atomic and hydrogen bombs clearly created an imminent threat. Gardner directed that listening posts to be established in Turkey to monitor Soviet rocket experiments; it would not be long before the hard evidence of their success would be trumpeted to the world by Sputnik.

Schriever's system of concurrent development, production, and operations was risky, but paid off. Within a short time, he was calling on the talents of 18,000 scientists, seventeen prime contractors, 200 subcontractors, and 3,500 suppliers, all employing an estimated 70,000 people, to build the more than 100,000 individual components of the Atlas. Five hundred million dollars was spent on new test facilities. The WDD grew correspondingly, and became the Ballistic Missile Division (BMD) on June 1, 1957. (One spin-off of Schriever's management methods was the growth of what came to be called systems management, an approach to business that was widely adopted in industry.)

Building the Atlas was only part of the task. General Thomas D. White, Vice Chief of Staff, called on the resources of SAC, the Air Training Command (ATC), the Air Materiel Command (AMC), and the Air Research and Development Command to prepare for the advent of the missiles.

The Atlas program was fraught with difficulties and, even after the Killian Report, subject to budget cuts that caused delays in the schedule. There were also the usual development problems. The first launch of an Atlas took place at Cape Canaveral on June 11, 1957; it suffered the classic American first-launch syndrome when an engine failed, sending it into wild maneuvers that called for detonation by the range officer. A second launch suffered the same fate, but the third, on December 17, 1957, was very successful. By then the one-two punch of Sputnik I and Sputnik II in the fall of 1957 had spurred Congressional budget largess again, for it was obvious to everyone that the Russian rocket that orbited a satellite could also propel an ICBM warhead to the United States.

Schriever did his work well. The Strategic Air Command activated its 704th Strategic Missile Wing on January 1,1958. On September 9, 1959, the first Atlas, a D model, was launched from Vandenberg Air Force Base by a SAC crew. Deployment of the missile was complicated in 1959 when advances in Soviet ICBM capability forced a crash program to place a large part of the Atlas force in huge underground silos, 174 feet deep and 52 feet wide.

By 1963, SAC had thirteen Atlas missile squadrons, with 127 missiles deployed. All of the money, effort, and management expertise had paid off—the Soviet threat had been met. By then the Atlas missile was obsolescent, already overtaken by Titan and Minuteman follow-on missiles. The Atlas force was deactivated during 1965, with many of the missiles going on to serve splendidly as launch vehicles for satellites and other space programs.

Missiles are more difficult to describe than aircraft. The Atlas was produced in a number of models, of which the D is typical. It was 82 feet 6 inches long and 10 feet in diameter and had a launch weight of 260,000 pounds. The D's 10,360-mile range was made possible by its five engines, a LR 89 sustainer rated at 57,000 pounds of thrust, two LR 105 engines each rated at 150,000 pounds of thrust, and two small vernier rockets to provide final "trim" velocity. All five engines were fed the combination of kerosene-like RP-1 fuel and LOX (liquid oxygen) from the two stainless-steel, thin-wall pressurized balloon tanks.

The success of the Atlas program alone was an astounding managerial achievement, raised an order of magnitude by the fact that the Thor IRBM and the Titan and Minuteman ICBMs were all being successfully developed in parallel with it.

The Thor was a direct spin-off of the Atlas, with a liberal helping of technology from the rival Army IRBM, the Jupiter. As a result, it had a very successful test program and the shortest interval between contract signing and the IOC (initial operational capability) of only 3.9 years—less time than General Motors was taking to develop and introduce a new-model Chevrolet. With a length of 65 feet, a diameter of 8 feet, and a launch weight of 105,000 pounds, the Thor had a range of 1,976 miles, which made Great Britain the perfect basing point.

The Titan was a far more sophisticated missile than the Atlas, and it was quickly improved with the Titan II. With a rigid rather than a pressurized structure, the Titan used two stages of liquid-fueled rockets to achieve its mission goals. The first stage was an Aerojet LR87-1 engine with two gimbaled chambers rated at 150,000 pounds of thrust each; the second stage was an Aerojet L91-1, rated at 80,000 pounds of thrust, which was ignited in the vacuum of space. A huge vehicle, the Titan was 98 feet long with a 10-foot-diameter first stage and a launch weight of 220,000 pounds. It had a range of 8,000 miles with a huge 4-megaton-yield warhead. The Titan was stored upright in a silo, but had to be lifted to the surface and fueled before firing.

The Titan II was even larger, 103 feet long and 10 feet in diameter and with a launch weight of 330,000 pounds. Its range was over 9,000 miles. It had the significant advantage of being designed for storable fuels and could be launched in position from the bottom of the silo. The LR87-5 engine provided 216,000 pounds of thrust from each of its two chambers, and the second-stage LR91-5 engine put out 100,000 pounds of thrust. The missile could be fired on two minutes' notice.

As advanced as the Titan II was, the Minuteman series was an even more radical and much more satisfactory approach to the problem. Boeing won a hotly fought contest with fourteen competitors to secure the contract, and delivered a three-stage solid-propellant rocket fired from its underground silo. The less complex nature of the Minuteman greatly reduced crew requirements, only two persons being necessary to control a flight of ten silos. Like its predecessors, the Minuteman was continuously refined, resulting in the Minuteman II and III series. Firing was virtually instantaneous.

Each stage of the Minuteman was assigned to a different company, to reduce risk. Thiokol's first stage had 200,000 pounds of thrust, Aerojet's second stage had 60,000 pounds of thrust, and Hercules' third stage had 35,000 pounds of thrust. Launch weight was 64,815 pounds and range was over 6,000 miles.

Every one of the follow-on programs, with their many changes and improvements, confronted Schriever and his rapidly expanding organization with a degree of challenge the same as or greater than the original Atlas had. All four programs were compressed into an amazingly tight schedule, as the following table indicates:

Event	Missile			
	Atlas	Thor	Titan	Minuteman
Contract award	1/55	12/55	12/55	10/58
First launch	6/57	9/57	2/59	2/61
Initial operating capability	9/59	6/58	4/62	11/62

Thus from January 1955 to November 1962, just short of eight years, Schriever's organization guided the United States Air Force and the missile industry to four complete missile systems. Titan and Minuteman had a remarkable longevity, and the first three systems—Atlas, Thor, and Titan—went on to more than pay for their investment through their contribution to the space age as launch vehicles. As early as February 1957, General Schriever announced that about 90 percent of the developments in the ballistic missile program could be applied to advancing in space and to satellites and other vehicles. He viewed ballistic missiles as but a step in a transitional process leading to flights to the moon and beyond, and now sees his contributions to the space program as being even more satisfying than the creation of the ICBM fleet.

There is no governmental or industrial counterpart to Schriever's management feat. In government, the Manhattan Project is most nearly comparable. In industry, one can look to the widespread chaos of Howard Hughes's empires, or to the prairie-fire-swift expansion of Ross Perot's interests, or to the seemingly limitless prospects of Bill Gates software world, and in none of these is there anything near the magnitude of the challenge, the scope of the work, or the rapidity of achievement that can be found in the USAF intermediate and ballistic missile programs. Yet General Schriever, while fully acknowledged within the service and industry for his talents, was never accorded his full measure of fame by the civilian community. This was due in part to his informal manner and modest nature, in part to the security considerations of the time, and in largest part to the fact that the sheer magnitude of the challenges he faced and overcame were beyond most people's ready comprehension. General Schriever died on June 20, 2004, at the age of ninety-two. In 1998, Schriever Air force Base at Colorado Springs was named for him—a signal honor generally reserved for deceased heroes.

An aircraft comparison is revealing. During the same time period, the Convair F-102 fighter, admittedly an advanced design, took ten years to go from the establishment of the requirement to the completion of the program, and it cost $2.3 billion. By any measure—scale, scope, importance, technological advance—the ballistic missile and space programs initiated and carried to a successful conclusion by Schriever were far more complex, yet were executed in a shorter period.

One of Schriever's strong backers, General Thomas D. White, a brilliant intellect, had succeeded General Twining as Chief of Staff on July 1, 1957. In November he made a prescient speech at the National Press Club in which he stated, "Whoever has the capability to control the air is in a position to exert control over the land and seas beneath. . . . I feel that in the future whoever has the capability to control space will likewise possess the capability to control the surface of the earth. . . . We airmen who have fought to assure that the United States has the capability to control the air are determined that the United States must win the capability to control space."

White's words were a ringing challenge in the post-Sputnik atmosphere, when the embarrassment of the United States led to spin-control language in the legislation creating the National Aeronautics and Space Act, which President Eisenhower signed in July, 1958. The act provided that activities of the United States in space would be "devoted to peaceful purposes for the benefit of all mankind" in contrast to the ominous, absolutely secret activities of the Soviet Union, which clearly had a military purpose. The act did not abolish military space research and development, and it assigned "those [research and development] activities peculiar to or primarily associated with the development of weapon systems, military operations or the defense of the United States to the Department of Defense." Neil H. McElroy, Secretary of Defense from October 1957 to December 1959, and not a dynamic leader, did have the foresight to assign to the Air Force the responsibility for the development, production, and launching of space boosters. It was a small start to what would become a supremely powerful space presence.

In the Shadow of SAC Between Korea and Vietnam

The emphasis given here to the Strategic Air Command is in part because of its importance and in part because the SAC story is also the story of much of the rest of the Air Force. The growth and change in SAC was reflected sooner or later in similar growth and similar change in the major and minor commands.

The physical creation of the bomber and missile fleets was not done in a vacuum; indeed, the Air Force, SAC included, was in a period of continuous turmoil in the years between its involvement in Korea and Vietnam. Fortunately, the rising budget tide lifted all boats. Air Force R&D budgets grew from $62 million in 1950 to $814.8 million in 1959. During the same period, expenditures for aircraft and related equipment grew from $1.2 billion to $7.1 billion, permitting TAC, ADC, and other commands to expand their operations, receive new equipment, and grow—but always in the shadow of SAC. The great support commands without which nothing could be done, Air Training Command, Air Force Logistic Command, and Air Force Systems Command, in all their varying names over time, made it possible for SAC, TAC, and ADC to operate by providing the planning, the training, the equipment, and the maintenance facilities.

SAC's size and prestige received much attention, as did its innovative means of meeting a mounting threat. The war-making qualities that U.S. airmen had demonstrated in World War II and Korea (and would demonstrate again in Vietnam and the Gulf War) were evident in the aggressive, assertive manner in which SAC forces amplified their strength by means of tactics. One common characteristic of these tactics, however, was that they usually came at the expense of the crew members in terms of extra hours at work and extra months away from home.

In 1955, SAC began rotating entire combat wings and their air refueling support to overseas bases in North Africa, England, and later Spain. It was known that the Soviet missile threat was growing and that SAC air bases were a primary target. To offset this, SAC began an alert system, in which one-third of its aircraft were "cocked," i.e, fully armed, fueled, preflighted, and ready to take off on fifteen minutes' notice. The system was extremely costly to operate and played havoc with the family life of crews, who had to spend days in alert facilities, away from their families, and then many more days flying to meet their training requirements.

There was an added psychic embellishment, one that spoke directly to the insanity of modern war. In times past, warriors would go away to war to chance death while their families waited in relative safety at home. Under the new threat of a missile exchange, the warriors would go away to battle with at least a chance of surviving, very much aware that the families they left behind would be killed by the thermonuclear attack on their base. The instructional films of the period, now dated and so often the butt of jokes, showed the measures to be taken in the event of a nuclear attack—crawling under desks, closing eyes, etc.—and were no comfort to crews who knew too well what the full measure of the carnage would be.

Yet the decisive importance of the alert force was demonstrated during the 1958 Lebanon crisis when President Eisenhower sent 5,000 Marines to Beirut to help quash an internal rebellion. The Soviet Union rattled its sword, attacking the move as "open aggression." In response, Eisenhower put SAC on alert, and within a few hours, 1,100 aircraft were poised ready for takeoff. Soviet rhetoric cooled rapidly, and after a few days the alert was called off.

As the gravity of the Soviet missile threat increased, the alert concept was continuously refined. As noted, SAC had achieved an alert status that required one-third of its crews to be ready to launch within fifteen minutes. In 1961, President John F. Kennedy called for a 50 percent alert, and this was later supplemented by an airborne alert, in which nuclear-armed bombers were kept airborne on station on a continuous basis. The stultifying hours spent in the alert shelters (as they would later be spent in underground missile control rooms) were used by crew members to take correspondence courses for classes at Squadron Officers School, Air Command and Staff College, and the War College, along with civilian courses. The crews studied hard, even though they recognized that while the correspondence courses filled a square in their career checklist and demonstrated their seriousness of purpose, in no way did they confer the prestige or, more important, provide the networking to be gained by actual attendance at the schools in question. Thus SAC crew duty actually became a career penalty for many officers.

The monotony was somewhat broken by a continuous flow of new equipment and new concepts. To offset the incredible buildup of the Soviet radar, antiaircraft, and missile defenses into the most sophisticated defense system in history, the B-52 units adopted new low-level tactics. The importance of electronic countermeasures was stressed, and the newest G and H models were designed to carry the AGM-28A Hound Dog cruise missile. This exceedingly sophisticated Mach 2 cruise missile, which had the capacity to go in at high or low altitudes and later in its life even had a terrain-following feature, remained in service until 1975. The crews loved the idea of suppressing enemy air defenses with the Hound Dog's atomic capability.

The general progress in technology meant that new aircraft would come along at greater intervals and smaller numbers. The gorgeous Convair B-58 Hustler, the world's first supersonic bomber, became operational on August 1, 1960. The delta-wing Hustler could fly at twice the speed of sound and featured an unusual jettisonable pod to carry weapons and fuel. Expensive, difficult to maintain, and cursed by a high accident rate (twenty-six of the 116 built were destroyed in accidents), the Hustler was withdrawn from service in 1970.

No one understood better than the leaders at SAC the value of dramatic flights in improving public relations. From January 16 to 18, 1957, in Operation Power Flite, three SAC B-52Bs flew around the world nonstop, completing the 24,325-mile trip in forty-five hours and nineteen minutes. In June 1958, in Operation Top Sail, two KC-135s broke the speed record for New York to London and return, making the round trip in just under twelve hours. In 1971, a B-58 Hustler flew from New York to Paris in three hours, nineteen minutes, and forty-one seconds, setting a transatlantic record.

Such risky flights were not without their cost. A KC-135 crashed on take off during Operation Top Sail, killing its crew. Like many aircraft accidents, the crash was the result of a seemingly innocuous problem, the precession of the aircraft's attitude indicator as a result of acceleration on takeoff. The instrument showed a higher than-actual climb attitude; unnecessary corrections by the pilot flew the plane into the ground. The B-58 that set the transatlantic record later crashed at the Paris air show, scene of many crashes. Yet there was no doubt that record-setting flights sent messages to the Soviet Union with carbon copies to the Congress and the public.

R&D + POLITICAL FACTORS = CONVERGING CAPABILITIES

The flowering of research-and-development efforts in the 1950s combined with the changing tide of political strength (fostered by the Defense Reorganization Act of 1958) to alter significantly the nature of the United States Air Force and, for the first time, set the capabilities and missions of the Strategic Air Command and the Tactical Air Command on converging paths. The functions of the two commands would become thoroughly homogenized in the next decade, during the Vietnam War, when SAC bombers conducted close support operations and TAC fighters conducted long-range strategic bombing missions. Yet strategic, budget, and turf considerations would make it necessary for another twenty years to elapse before the logical conclusions of this development were drawn. Only after the dramatic highlights of winning both the Persian Gulf War and the Cold War would TAC and SAC be disestablished, and two new commands—Air Combat Command and Air Mobility Command—be formed.

TECHNOLOGICAL CHANGE IN THE 1950S

The foundation for technological change came with four important breakthroughs: the thermonuclear weapon (especially its smaller versions); improvements in electronics and communications, which raised command and control to a new level; the successful development of the long-range intercontinental ballistic missile; and the concept of space systems, which called for extreme improvements in electronics, especially miniaturization.

Of these the development of the Atlas caused a change in both scientific and budgetary accounting attitudes in weapon development. In the past, research and development had been considered an area in which many failures were routine in the process of achieving periodic successes. The urgency of the Atlas program had required that a series of unknown scientific discoveries and technological breakthroughs occur on a rigid schedule. By monumental effort of the best minds in the country, these did occur, with the inadvertent result that on later systems, the "bean-counters" in both the Department of Defense and Congress began to regard such achievements as routine, like obtaining a new muffler for an automobile. The long-term effect of this raised level of expectation would prove to be stultifying to research-and-development programs, for failure was no longer considered an option. Congressional and DOD scrutiny virtually demands certainty of success, cost, and schedule. It takes an exceptionally able and brave advocate to guarantee an R&D project's success in all three areas, so that it will survive the review process.

This sense of "anything is possible" was also fostered by the seemingly endless series of ever more capable aircraft under test at Edwards Air Force Base. The confluence of powerful new jet and rocket engines, advanced aerodynamics, including swept wings and area-ruled (Coke-bottle-shaped) fuselages, and new electronics brought forth a series of "X-planes" to lead the world in scientific advances as well as set records. Along with the X-planes came a proliferation of fighter and bomber prototypes to be tested by the most competent professional test pilots in history.

The first, and perhaps most memorable, X-plane achievement had occurred earlier, on October 14, 1947, when then Captain Charles "Chuck" Yeager exceeded the speed of sound in the Bell XS-1. The swept-wing Bell X-2, which followed it, was designed to reach speeds at which the problems of aerodynamic heating would come into play. The X-2 had a troublesome early development period, including an explosion while being carried by its Boeing EB-50A mother ship that cost the life of Bell test pilot Jean Ziegler and an observer on board the EB-50A, Frank Wolko. After three years of intensive tests, on July 23, 1956, Captain Frank Everest flew the X-2 to a speed of 1,900.34 mph (Mach 2.87), earning Everest instant identification as "the fastest man alive."

Korean ace Ivan Kincheloe took over X-2 test duties from Everest, setting an altitude record of 126,200 feet on September 7, 1956. Twenty days later, Captain Milburn Apt would fly the Bell X-2 on its thirteenth powered flight, pushing it to and beyond its maximum capabilities. Apt reached a speed of 2,094 mph (Mach 3.196) before the aircraft plunged out of control, carrying him to his death.

The most advanced and perhaps most successful of the X-planes, the North American X-15, made its first glide flight on March 10, 1959, with the company test pilot, the irrepressible Scott Crossfield, at the controls. Three X-15s were built, and in 199 flights, they set many records, including a peak altitude of 354,200 feet and a speed of 4,534 mph (Mach 6.72). The X-15 also served as a test bed for dozens of experiments to explore the hazards of high-altitude hypersonic flight.

While these and other X-aircraft were tested, the test pilots of Edwards had a field day with new aircraft (year of first flight in parentheses), including the Boeing B-52 (1952), Convair YF-102 (1953), Convair F-106A (1956), Convair B-58 (1956), Douglas B-66 (1954), Lockheed F-104 (1954), Lockheed U-2 (1954), Lockheed C-130 (1954), McDonnell F-101 (1954), North American YF-100 (1953), North American YF-107A (1956), Northrop T-38 (1959), Republic YF-105 (1955) and a host of less well known planes.

These new weapon systems would provide SAC, TAC, and ADC tremendous new capabilities, the latter two commands especially benefiting as they expanded their capabilities despite roller-coaster budgetary changes.

The Defense Reorganization Act of 1958

President Dwight Eisenhower had spent thirty-nine years in the military, rising from West Point cadet to Chief of Staff, and then after retirement being recalled to serve as the first Supreme Allied Commander, Europe, from 1950 to 1952. As a two-term President, he also enjoyed eight years as Commander in Chief, for what might be said to be a total of forty-nine years of combined military service. There is some irony in that it was this quintessential military commander who contributed two vital elements to the basic idea of civilian control over the military. The first of these was his often misconstrued remarks about the military-industrial complex, which have been a rallying cry for the antimilitary over the intervening years.

The second, and far more significant, was his insistence upon the Department of Defense Reorganization Act of 1958. Eisenhower was influenced by a number of factors, the most important of which was his belief that joint and unified commands had worked so well during World War II that some means had to be created to make sure they would be used by the U.S. military in a new emergency. Unremitting interservice rivalry had convinced him that the contemporary DOD arrangement would not ensure this. Disturbed by a series of reports that indicated that the growth of technology had obscured military service roles and made them more competitive than complementary, disgusted by the turf wars indulged in by the members of the Joint Chiefs of Staff, and alarmed by the fact that the Secretary of Defense had been more preoccupied with resolving these turf wars than with initiating military policy, Eisenhower tasked his new Secretary of Defense, Neil H. McElroy, to reform the DOD.

McElroy employed a committee that included the incumbent JCS chairman General Twining and drafted legislation that Congress quickly passed and Eisenhower signed into law on August 6, 1958. In essence, the reorganization placed few limits on the powers of the Secretary of Defense, removing the service secretaries, the Army and Air Force Chiefs of Staff, and the Navy Chief of Naval Operations from the line of operational command. This was an unprecedented departure from American military tradition. The operational chain of command now ran from the President to the Secretary of Defense and through the JCS to the unified and specified commanders. In effect this meant that the Chief of Staff of the Air Force no longer commanded its combat forces, but instead was responsible for their creation, training, and support. When combat forces were to be used in a specific military mission, they were to come under the command of the unified or specified commander, who had full operational control, and who might or might not be an Air Force officer. Reasonably enough, this element of the reorganization has been difficult for many of the succeeding Chiefs and CNOs to accept; for some their lack of acceptance has bordered on denial. The nonoperational line of command flowed from the President to the Secretary of Defense and then to the service secretaries.

Some of the impact of the 1958 act was obscured in its language. Although the original unification act of 1947 had called for three military departments to be separately administered, the 1958 act specified a Department of Defense with three military departments—a subtle but significant difference that handed the keys of the castle to the SecDef. The Secretary of Defense was authorized to establish single agencies to conduct any service or supply activity common to two or more of the services.

This seemingly logical organizational construct paved the way for a loss of control by the military services over several areas of critical *war-making* importance. Only four examples will be given here. The first was the establishment of the control and direction of military research and development in the Director of Defense Research and Engineering in the DOD. This action is now considered by many to be a potentially fatal flaw in the defense establishment, for it removes the military services from the essential decision-making process and places it in the hands of civilian appointees who may or may not have an adequate understanding of future military requirements. The general effect is to circumvent all R&D efforts whose results cannot be guaranteed as to effectiveness, schedule, and cost. Its direct effect is to inhibit managers from attempting projects, no matter what their ultimate value, if they cannot guarantee success—the antithesis of normal R&D philosophy.

A less obvious deficiency came to light during the Vietnam War. In the process of consolidating authority in the DOD, the mission of targeting was removed from the Department of the Air Force and passed to the Defense Intelligence Agency, where it promptly (and naturally) withered and died from a lack of interest—it was no one's rice bowl in that organization. The analysis of enemy logistic capability somehow defaulted to the CIA, so that when the Air Force entered the war in Vietnam, it had very limited tools to assess the enemy's logistic flow for determining interdiction missions.

As another example of good intentions gone awry, the next chapter will show how Eisenhower's desire for a unified command was, under the direction of the Secretary of Defense, carried to such extremes in Vietnam that it forced the abandonment of the air doctrine learned at such great expense in World War II and Korea.

The fourth and most all-embracing deficiency in the Defense Reorganization Act of 1958 was that, in the words of Colonel Harry C. Summers, Jr., USA (Retired), the military had become "merely a logistics and management system to organize, train and equip active duty and reserve forces." The concept of the senior military professional, schooled in warfare, dedicated to a life of service to his country, and using his wisdom to create policy, was abandoned to the hands of competent, well-meaning civilians who are appointed to the role of assistant secretary for this or that as a part of their civil career progress, but who usually do not claim military expertise and cannot be expected to have the ability to craft the long-term goals of the services in planning or to direct them in execution. This policy of political displacement is not followed in industry. Willing and able novices do not substitute for experienced chief surgeons, for large fund managers, for automobile company presidents, or for any position in any progression where experience, specialized education, and long-term devotion are required.

Neither McElroy, who served from October 9, 1957, to, December 1, 1959, nor his successor as Secretary of Defense, Thomas S. Gates, Jr., who served from December 2, 1959, to January 20, 1961, had the personality or the tenure in office to make full use of their newly conferred powers, although McElroy did a great service in establishing the Advanced Research Projects Agency (ARPA) within the Department of Defense. ARPA (later DARPA) was chartered to provide unified direction of antimissile programs and outer-space projects. The newly elected President, John F. Kennedy, had originally asked the veteran and admired Robert A. Lovett to be his Secretary of Defense. Lovett, who had done brilliantly in all his work at the Pentagon, including his 1951–1953 stint as Secretary of Defense, demurred because of his age. And while Lovett did not know him personally, a vigorous young man named Robert McNamara had been recommended to him. Lovett passed on the recommendation to Kennedy, saying that what the Pentagon needed was a good administrator rather than a man on horseback who would be intent on making all the decisions. In a career characterized by good calls, Lovett had finally made a bad one. On January 21, 1961, McNamara became Secretary of Defense, and he would use the Defense Reorganization Act of 1958 like a master puppeteer, gathering the strings of control to his office on an ever-increasing basis. He would also be, in the opinion of a great many senior officers of all the services, the absolute worst Secretary of Defense in the nation's history. Extremely bright and able to spot instantaneously the most innocuous error on a briefing chart, he hammered subordinates and assistants alike with his amazing grasp of detail and memory so that he could win arguments even when he was wrong. His insistence on having the last word would lead to major debacles like the attempted procurement of the TFX fighter to serve both the U.S. Navy and U.S. Air Force, and major tragedies like Vietnam, where his philosophy of graduated response would simultaneously nullify air power even as it immured ground power.

Throughout the 1950s, the decline in the power of the office of the Secretary of the Air Force continued undiminished. It was mitigated in part by the conciliatory efforts of James H. Douglas, Jr., who had been Under Secretary for four years before taking office on May 1, 1957. Douglas had managed to soothe relations between his office and the Air Staff and was able to fight the budget cuts demanded by Secretary of Defense Wilson. Douglas's successor, Dudley S. Sharp, served just over a year, from December 11, 1959, to January 20,1961, and, as a lame duck, was unable to provide vigorous leadership.

Both Douglas and Sharp benefited from their association with the Air Force Chief of Staff, General White. White, an intellectual, was fluent in Chinese, Greek, Italian, Portuguese, Russian, and Spanish. Understated and refined, he understood the implications of nuclear strategy and the impending space age as well as he knew the grass-roots politics of the Pentagon. His successor, General LeMay, would work shoulder to shoulder with the next Secretary of the Air Force, Eugene B. Zuckert, in their joint—and losing—battle with Secretary McNamara.

THE UNDERDOGS: TAC AND ADC

As organizations, the Tactical Air Command and the Air Defense Command (later Aerospace Defense Command) suffered from the roles in which fate had cast them. Although they executed their respective missions with panache, they were, for quite different reasons, always the bridesmaid but never the bride when it came to budget allocation and to their basic wartime roles. The role of SAC, as a specified command, was always well defined, and even when in the Korean and Vietnam wars SAC sent units into combat, the basic integrity of its deterrent mission was preserved. TAC's mission was preparing units for war; when war came those units were assigned to a unified command, such as the Pacific Air Force. ADC units were also seconded to PACAF during the Vietnam War, and when the fighting was over these individual ADC and TAC units came back covered with a glory that did not directly transfer to their parent commands. ADC's principal difficulty, however, was that the threat for which it was created diminished just as ADC came into full flower.

TAC'S·MISSION AND OPERATIONS

From the start, the mission of the Tactical Air Command was to command, organize, equip, train, and administer the forces assigned and attached to participate in tactical air operations. Like SAC, it was established on March 21, 1946. Unlike SAC, it was soon reduced in status, when in January 1949 it was absorbed into the Continental Air Command. The advent of the Korean War caused it to be reestablished as an active command in August 1950. It built up rapidly during the war, as a host of new responsibilities were added to its mission, including electronic countermeasures, increased airlift capability, and air-ground operations.

After the Korean War, the defense budget had been increased and the technological advances of the previous ten years were being translated into hardware. In an incredibly short period of time, TAC went from its essentially World War II origins to a force equipped with jet fighters and bombers. Tactical nuclear weapons became available, and at about the same time, supersonic fighters. Aerial refueling turned TAC into an ocean-hopping force, and new transports, including the versatile and long-lived Lockheed C-130, came into the inventory. Among the advanced aircraft and missiles were the century-series fighters, including the Lockheed F-104, North American F-100, Mc-Donnell F-101 and RF-101, and Republic F-105. New bombers included the Douglas B-66 and RB-66 and the Martin B-57. The Matador surface-to-surface missile came into use with the 1st and 69th Pilotless Bomber Squadrons, the designation another artless attempt to make the roles and missions more palatable to the Army. About 1,000 Matadors were built; they were succeeded by the Mace, which was retired in 1966.

General Otto P. "Opie" Weyland, fresh from his duties as Commander, FEAF, and just after reorganizing Japan's air defense and aircraft industry, assumed command of TAC on May 1, 1954. Weyland appreciated the new weaponry, but his long experience in limited warfare put him in advance of his time as he advocated the development of the "Composite Air Strike Force" (CASF), an organization that foreshadowed the development of composite units in today's Air Force. The CASF was very mobile and self-contained, with its own fighters, bombers, and transport aircraft. Weyland tested the concept in 1956, during Operation Mobile Baker, which sent fighters, reconnaissance planes, bombers, and tankers from a variety of U.S. bases to Europe.

The first real test of the CASF came on July 15, 1958, with the Lebanon crisis. President Eisenhower ordered the CASF into action, and within twelve hours, it was arriving in at Adana, Turkey. Just over a month later, during the week of August 23, another CASF strike force was sent to reinforce U.S. forces during the Quemoy-Matsu crisis.

TAC used its mobility and its cargo capacity in a succession of humanitarian missions, including delivering medical equipment to Argentina and iron lungs to Japan. Such missions of compassion would become increasingly important in the years to come.

Weyland and his immediate successor, General Frank F. Everest, had maintained TAC's capability and its personality. Under their leadership, TAC remained a freewheeling outfit, mission-oriented but not as concerned with conformity and centralized control as SAC. Yet while TAC's mission had remained much the same, the world of air combat was maturing. The new equipment, and especially the tactical nuclear weapons, demanded a more rigorous environment. It came, in what many veterans of the period describe as the best thing ever to happen to TAC, in the person of General Walter Campbell Sweeney, Jr., who took command on August 1, 1961.

In the phrase of the time, TAC was "SACimcised," for Sweeney was a tough but fair commander who had fought the Pacific war from the battle of Midway on and led the first low-level attack on Tokyo. He became the Commander of SAC's Fifteenth Air Force; in 1954 he led a flight of three B-47s in history's first nonstop jet bomber flight across the Pacific, an event depicted in the Jimmy Stewart film *Strategic Air Command*. In the next few years, TAC's élan and skill would be enhanced by Sweeney's discipline and rigor, enabling its units to operate well under the Pacific Air Command in the tough years to come in Vietnam. Inevitably, however, Sweeney's centralized and authoritarian methods, supercharged by the methods emanating from McNamara's DOD, would be carried to an extreme. Change would come in the 1970s.

THE AIR DEFENSE COMMAND: OPERATIONS AND MISSION

Like TAC, the Air Defense Command did not have the same visibility as SAC, in part because of external events. When the Cold War began, the Soviet threat was limited, but so were ADC's capabilities in terms of aircraft, radar defenses, and weapons. Over the decade of the 1950s, ADC grew strong in all these areas, with new supersonic aircraft being supported by an increasingly sophisticated radar network and equipped with armament of ever-greater lethality.

Even as ADC flourished, the Soviet threat began to shift from bombers to ICBMs. There always remained the possibility that the enemy would launch a bomber attack, in conjunction with, or as a follow-up to, an ICBM attack, but the bomber threat was relatively less terrifying. In January 1958, ADC's name was changed to Aerospace Defense Command, to reflect the dual threat. Unfortunately, any budget conflict within the Air Force began to be resolved with cuts for ADC, and it would diminish in size and importance over time until in 1980 it was inactivated and its assests absorbed by TAC and SAC.

Duty for ADC personnel, officers and enlisted, aircrew and ground crew, was very difficult. Most of the ADC bases were located in the north, and weather conditions were often miserable. The bases had few amenities, and the alert crews had to stay next to their airplanes in drafty hangars and be ready to launch no matter how bad the weather was outside. It was not uncommon in far-northern bases for the snow to be so deep that an aircraft taking off could not be seen until it lifted above snow walls lining the runways. After an intercept—usually of some errant passenger liner—the approach to a landing was often made with the low-fuel warning lights winking red, so that the pilot knew he had only one shot at the snowbound runway. With a minimum ceiling and visibility and frequently gale-force crosswinds, it was enough to make even an expert's palms sweat. It was also enough to make pilots who survived experts.

Other factors impacted adversely on morale. The original ADC equipment—North American F-51s, North American F-82s, and Lockheed F-80Cs—barely had the performance to intercept Tu-4-type aircraft under visual conditions. The first generation of all-weather fighters—the Lockheed F-94 and the Northrop F-89—had a series of problems that took years to resolve. The early airborne radars were difficult to keep in operation, and as the F-94 had no deicing equipment, it was unsuitable for many weather conditions. The F-89 had a very long development process, with several crashes because of structural and engine failures. While both of these aircraft were adequate to meet the Soviet piston-engine threat, their capability against jet bombers was marginal. It was extremely difficult for them to intercept B-47s, especially when the six-jet bomber would turn into them, and even the older B-36 flew at altitudes beyond their reach.

The Air Defense Command recognized the problem and sought to field an array of advanced aircraft and missile weapon systems. The aircraft requirement was to be met by the emerging century series of fighters, which offered promise of extraordinary performance but, as might be expected, suffered teething problems that delayed their entry into service. The medium-range Convair F-102 Delta Dagger—always called the Deuce—entered service first, in mid-1956, and was the first USAF fighter to be designed as a missile carrier. The Deuce, which saw service in thirty-two squadrons, was succeeded by the Convair F-106 Delta Dart—always called the Six—which equipped fourteen squadrons beginning in 1959. The F-106 was fitted with the Hughes MA-1 electronic guidance and fire control system and was designed to operate with the SAGE (Semiautomatic Ground Environment) defense system. The McDonnell F-101B long-range interceptor also entered service in 1959, and eventually equipped seventeen squadrons. The Lockheed F-104 was pressed into small-scale service on two occasions by ADC, but it was not an all-weather fighter by any standards and was clearly an interim measure.

All of the century-series aircraft had the equipment and the performance to intercept incoming bombers, but all were beset by maintenance difficulties that kept too large a percentage of aircraft out of commission. The art of shooting down a bomber had changed considerably, calling for equally new weapons that themselves were beset by development problems. ADC pilots regarded the 2.75-inch-diameter unguided rockets as virtually worthless and derisively called the early Hughes GAR-1 (later AIM-4) Falcon the "sandseeker" (as opposed to heat-seeker) because it was so unreliable. The most impressive of the missiles was the MB-1 (later AIR-2) Genie, an unguided rocket with a 1.5-kiloton nuclear warhead that had a lethal radius of over 1,000 yards. The 822-pound Genie had a range of about 10 miles and could be fired from a pull-up maneuver that placed even high-flying aircraft in range. In operation, the fire control system (which varied with the aircraft) tracked the target, designated the missiles to be used, instructed the pilot to arm the warhead, fired the missile, then pulled the aircraft into a tight turn to avoid the resulting nuclear blast. The first live Genie was fired by a Northrop F-89J over Yucca Flat,

Nevada, on July 19, 1957. Despite the demonstration, aircrews felt a natural apprehension about using a nuclear weapon to explode a bomber carrying nuclear weapons—they could not be sure that they would elude the resulting blast. (It is not generally known that ADC fighters of the period were often launched to unfamiliar civilian airfield dispersal sites with nuclear Genie missiles fully armed and ready to fire. TAC fighters of the period on full alert presented the same hazard. A similar undertaking today would be met with the full fury of environmentalists and homeowners' associations along with full-page headlines in the press.)

While development of a short-range missile system for base defense was left to the Army because of the basic issue of roles and missions, ADC's interceptors were complemented by the Bomarc medium-range interceptor missile. The Bomarc was intended for fixed-base defense of a huge area and was created through the cooperation of Boeing and the Michigan Aeronautical Research Center at the University of Michigan. Fifty-three Bomarc squadrons were initially specified, and thirty-six were ultimately acquired. The Bomarc was the world's first surface-to-air missile with an active homing system. Originally designated the XF-99 (once again maintaining the fiction of "it's just another airplane"), it was subsequently called the IM-99A and later the CIM-10A. Looking like a very sleek airplane, the Bomarc was vertically launched (on thirty seconds' notice from SAGE) by a 50,000-pound-thrust solid-fuel Thiokol rocket motor. In flight, two 12,000-pound-thrust Marquardt ramjets took over to boost it to its cruising height of 65,000 feet and cruise speed of Mach 3. At about 10 miles distance, the Westinghouse DPN-34 radar locked on the target. Conventional or nuclear warheads could be carried. The Bomarc was highly successful against many high-speed drone targets, and 570 were built. Its first flight occurred on September 10, 1952, and it served for twenty years.

Both missiles and aircraft depended upon ground radars to detect incoming bombers, and mammoth amounts of energy and money were spent by the United States and Canada during the decade to create an enormous early warning system. When the Soviet bomber threat was first recognized in 1950, there were no radar warning lines, and World War II radars were pulled out of storage and installed in makeshift buildings in an operation aptly named Lashup. The Ground Observer Corps, which could trace its origins to Great Britain in World War I, was revived and thousands of volunteers vigilantly watched the northern skies to supplement the fragile radar network of forty Lashup stations.

The threat was so great that a decision was made to build the Distant Early Warning (DEW) line stretching more than 3,000 miles across the 69th parallel, spanning the frozen north from Barter Island, Alaska, to Thule, Greenland. These DEW Line stations called for highly skilled, dedicated technicians who could man them without regard to weather, absence from family, and the almost overpowering boredom. These, like many ADC and related assignments, were truly hardship postings that often created long-term problems, for alcoholic drinks were very inexpensive and often the only apparent anodyne. The DEW Line began test operation in 1953 and was completed by 1955.

Canada was a full partner in the defense effort, and besides supplying nine squadrons of the excellent Avro CF 100 fighter, it cooperated in building the Pinetree Line of aircraft control and warning centers about 1,000 miles farther south than the DEW line. A third line—the Mid-Canada Line—was built across the 55th parallel from Labrador to Hudson Bay. The Mid-Canada Line was to confirm the attack and scramble the interceptors, while the Pinetree Line and permanent radar stations in the United States would guide the interceptors to the bombers. The radar network was further supplemented by the construction of Texas tower sites in the North Atlantic and by the operation of airborne early warning aircraft.

All of this monumental effort—century-series aircraft, supersonic missiles, and extensive radar warning nets—was designed to culminate in the Semiautomatic Ground Environment (SAGE) system. Designed by the Lincoln Laboratory, an organization founded under an agreement by the Air Force with the Massachusetts Institute of Technology, SAGE harnessed giant computers to process the masses of incoming data from the radar sights, Ground Observer Corps, aircraft, and elsewhere. Threats were identified and fighters were designated to intercept the bombers.

ADC had divided the United States into eight air defense regions, each one with a SAGE combat operation center, and thirty-two air defense sectors, each with a SAGE direction center. SAGE installations were elaborate structures of concrete, some designed to withstand nuclear blasts and keep functioning. The first SAGE direction center became operational at McGuire Air Force Base, New Jersey, in 1957, and by 1963 the system, inevitably reduced by budget cuts to twenty-two air defense sectors in the United States and one in Canada, was complete. SAGE was supplemented by the Back-Up Interceptor Control (BUIC) system and would serve until it was replaced by the Joint Surveillance System (JSS) in 1983.

Initially, SAGE's reach for automation exceeded the contemporary level of computer grasp, and controllers were often forced to resort to the manual direction of interceptors, just as controllers had done during the Battle of Britain. The general consensus in the early days of operation was that in the event of an actual attack, the automatic system would be bypassed and directions given manually. Like all such sophisticated systems, it was improved over time and came to serve very efficiently.

Ironically, all of the great ADC effort—fighters, missiles, and SAGE—culminated in an increasingly efficient fighting machine just shortly after the time when the perceived threat of bombers began to decline. Premier Khrushchev had stated that the Soviet Union was emphasizing ICBMs rather than missiles, and it was eventually determined that the USSR indeed had not built as many Bison or Bear aircraft as it could have. General LeMay himself recognized that ICBMs were the major threat, noting that the principal task of the early warning radars was to give SAC sufficient time to scramble its bombers. To ensure that this time—now down to a virtually irreducible fifteen minutes—would be available, two Ballistic Missile Early Warning Sites were begun in 1959, one in Point Clear, Alaska, and the other in Thule, Greenland.

ADC's organizational road was as rocky as its equipment development had been. It became a part of the North American Aerospace Defense Command (NORAD) in 1957, and over time its strength was reduced as its tasks were assigned to Air National Guard and Air Force Reserve units. As previously noted, it was ultimately deactivated in 1980, with its remaining assets divided between TAC and SAC.

Because of the nature of its duties and the placement of its far-flung bases, ADC developed a culture of its own. The isolation and the sense of frustration at flying a mission in which the best possible outcome was that nothing would happen led to discipline and morale problems. The offset, as far as pilots were concerned, was the demanding nature of the flying, for when the word came to scramble, they had to be prepared to go no matter how bad the weather was. This rigor weeded out inferior pilots, and when ADC pilots were assigned to other duties such as close air support, they did exceptionally well, despite a lack of training in the specialty.

CONCLUSION

The United States Air Force had mastered a number of challenges in the eleven years from 1950 to 1961. It became a professional military force in being amidst a technological revolution of unprecedented scope and complexity. Even as its forces were massively expanded, it achieved ever higher levels of proficiency. The great expense and greater lethality of the new weapon systems were recognized, and met with corresponding improvements in procedures, doctrine, and discipline. The combat capability of SAC, TAC, and ADC units was made possible only by a corresponding growth in size and efficiency of supporting commands that will be covered in the following chapters.

Commanding General of the U.S. Far East Air Forces, Lt. General Otto P. "Opie" Weyland, meets with Lt. General Laurence S. Kuter in Japan.

Colonel Harrison B. Thyng is shown receiving five awards—a cluster to his Silver Star, another cluster to his Distinguished Flying Cross, the Legion of Merit, and two more clusters to his Air Medal.

1957 brought a series of promotions to these four gentlemen. From the left, Under Secretary James N. Douglas would become Secretary of the Air Force; Secretary of the Air Force Donald A. Quarles would become Under Secretary of Defense; Chief of Staff General Nathan Twining would become Chairman of the Joint Chiefs of Staff; and Vice Chief of Staff General Thomas D. White would become chief of staff.

November 13, 1957: The legendary commander of the Strategic Air Command, General Curtis E. LeMay, emerges from the Boeing KC-135 tanker, which he has just flown from Buenos Aires to Washington, D.C., in the record-setting time of eleven hours and five minutes. Secretary of the Air Force James Douglas shakes his hand while Air Force Chief of Staff Thomas D. White looks on.

From left: Lieutenant General Archie J. Old, Commander, 15th Air Force; Lieutenant General John P. McConnell, Commander, 2nd Air Force; General Thomas E. Power, Commander, Strategic Air Command; and Lieutenant General Walter C. Sweeney, Jr., Commander, 8th Air Force.

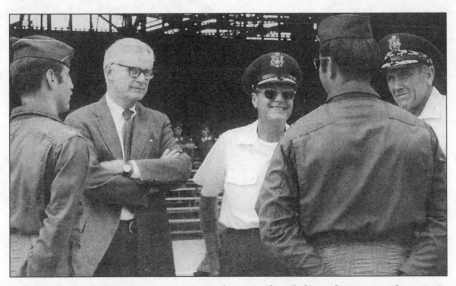

Telling the brass how it is. Captain Joel P. Gordes (left) and Captain Thomas G. Dorsett of the 363rd Tactical Reconnaissance Wing at Shaw Air Force Base, S.C., explain a McDonnell RF-4C mission to (from left) Secretary of the Air Force Robert C. Seamans, General William H. Momyer, Commander, Tactical Air Command, and General John D. Ryan, Chief of Staff, USAF.

The most advanced airplane in the world at the time of its debut in 1956, the supersonic Consolidated B-58 Hustler bomber provided SAC with a dramatic supplement to the B-52. A high accident rate and soaring maintenance costs forced its withdrawal from service in 1970.

Necessity forced the Air Force to develop some unusual weapon systems. This is a Republic F-84F hooking up as a FICON aircraft to a Consolidated B-36.

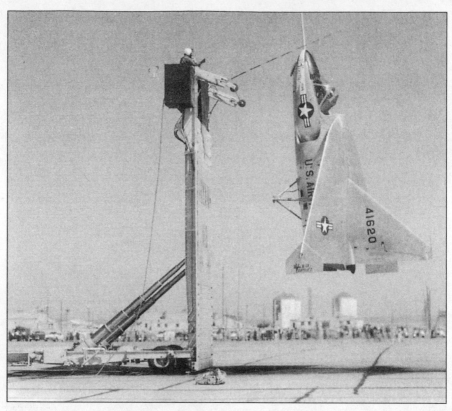

The desire to have vertical takeoff aircraft raged for a number of years; among the more successful was the Ryan X-13 Vertijet of 1956. The problem of all vertical take-off jets until the advent of the McDonnell Douglas AV-8B Harrier was first the land-ing, and second the small payload.

Jack Northrop's jet-flying wing looked like the wave of the future, but an aging design and stability problems kept it from becoming operational.

The author's squadron and their aircraft at Kirtland Air Force Base, circa 1959.

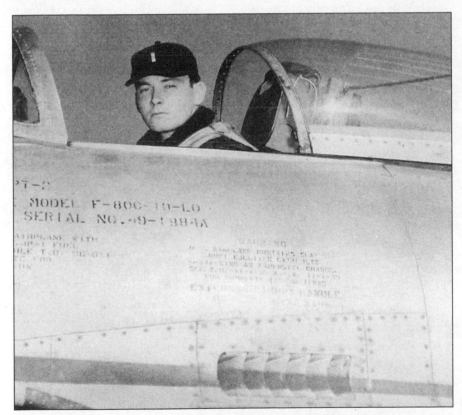

First Lieutenant Bill Creech in the cockpit of his Lockheed F-80C in Korea. Creech would ultimately fly a total of 280 combat missions in Korea and Southeast Asia. *(Courtesy of General W. L. Creech)*

Twenty years ago, no one would have thought that a four-engine transport and helicopters would be crucial to dominating the battlefield. Yet the Lockheed Martin MC-130P Combat Shadow/Tankers were used with spectacular effect by the 1st Special Operations Wing, refueling Sikorsky MH-53 Pave Low helicopters. *(Courtesy of Robert F. Dorr)*

History is made when Jeanne M. Holm, Director of Women in the Air Force, became the first woman general in Air Force history.

5

THE MANY FACETS OF WAR

In his 1960 Presidential campaign, Senator John F. Kennedy railed against the Eisenhower administration for being so dependent upon the concept of massive retaliation that it was incapable of fighting limited wars. Kennedy also charged that an inept Department of Defense policy of "sufficiency" in nuclear arms had created a "missile gap" such that the Soviet Union's ICBM capability now exceeded that of the United States. He made a number of strong recommendations for a new defense posture, including increasing defense appropriations, improving conventional forces as well as their airlift capability, and greatly increasing civil defense measures. After his election, Kennedy tasked the members of his new administration to rectify the problems he had pointed out, work that the new Secretary of Defense, Robert S. McNamara, eagerly embraced.

During the campaign, Kennedy had endorsed the views of retired Army General Maxwell D. Taylor on the strategy of flexible response. The concept had been first articulated in 1954 in *The Realities of American Foreign Policy* by the respected George F. Kennan. As Army Chief of Staff, Taylor had expanded on the idea, and some of his concepts were expressed in Army Field Manual 100-5, even though they were not wholeheartedly accepted by the Department of Defense. A glamorous wartime figure who led the 101st Airborne Division into France on D-day, he remained prominent after his retirement in July 1959, putting forth his ideas in his well-received book *Uncertain Trumpet*. In essence, Taylor stated that the bomber and missile fleet had become too large and possessed an "overkill" capability. It was now time for the Army and the Navy to build up their forces so that they could react decisively to limited wars. He called for a process, later adopted in modified form by Secretary

McNamara, to revise the military budget so that it would show in terms of operational forces how much defense money bought.

In Taylor's view, the policy of massive retaliation had been appropriate only when the Soviet Union did not possess a comparable nuclear power. When the Soviet Union attained a relative parity in nuclear strength, the strategy of massive retaliation became a handicap, reducing conventional forces, like NATO units in Europe, to mere trip wires for setting off a nuclear exchange. Taylor held that unless the United States was willing to unleash a massive attack on the enemy—and suffer the consequent counterattack—a massive retaliation strategy meant that we could not intervene in any limited war the USSR might instigate. In effect, the enemy could nibble away at the rest of the free world, taking one country at a time through intimidation, Sputnik diplomacy, limited brushfire wars, and fomenting of internal revolutions.

His belief that a tenfold "overkill" of nuclear power currently existed was attractive to the Army and the Navy, and they supported his call for finite— and reduced—bomber and missile forces. These would still be sufficient to provide a nuclear deterrent, the all important "air-atomic" shield under which the revitalized conventional Army and Navy forces could handle limited war. Admiral Arleigh "30-Knot" Burke, who had been the Navy's point man in the infamous "Revolt of the Admirals," embraced the argument, for the concept of a finite deterrent capability with its corollary of "controlled retaliation" helped make the case for the Polaris submarine fleet. The banner of the new strategy thus bore a haunting resemblance to the old turf wars.

As might be expected, General White expressed the USAF's disagreement with Taylor's ideas rather eloquently, while General LeMay was more direct. White's superior intellect gave Taylor pause and might in fact have intimidated him. White argued, "Our strategic objective in the event of global war is to eliminate an enemy's war-making capacity in the minimum period of time. In determining the force requirements to do this, we must take into account not only the number, location and vulnerability of the targets, but the reliability, accuracy and warhead yield of our weapons as well as countless operational variables, and the evaluation of the expected enemy defense."

LeMay's view was: "A deterrent force is one that is large enough and efficient enough that no matter what the enemy does, either offensively or defensively, he will still receive a quantity of bombs or explosive force that is more than he is willing to accept." White and LeMay did not agree that the strategic air and missile forces should be reduced to an as yet undefined "finite" limit.

Despite the demonstrated success of the USAF strategy through 1960, Taylor's views would ultimately prevail, especially after Kennedy called him out of retirement in October 1962 to become the Chairman of the Joint Chiefs of Staff. Even before he assumed his new office, Taylor's doctrines provided the most important impetus to the long series of events that propelled the United States into a full-scale military involvement in Vietnam. There "flexible response" became a conveyor belt that ultimately fed 536,100 troops to fight what every thinking American military man had always abhorred: a land war in Asia.

Five American Presidents, from Harry Truman to Richard Nixon, would be involved in the problems of Vietnam. The Kennedy administration's entanglement began on a low-key basis in 1961, a year that was studded with a series of crises that validated Taylor's views that limited warfare would dominate the future. The first erupted in Laos in the spring of 1961, when the new President supported a coalition led by Prince Souvanna Phouma that was resisting a Communist takeover. Following his stated policy of coordinating military and diplomatic action, Kennedy showed the flag with Marine and naval air forces while working with members of the Southeast Asia Treaty Organization (SEATO) to reach an agreement in negotiations at Geneva. The crisis subsided, even if Communist subversion did not; it would continue throughout the Vietnam War.

The second emergency was in the beleaguered nation of Congo, where an emergency airlift, Operation New Tape, had been underway since July 1960. (It would last until January 1964 and be a true testimony to the skill and courage of the men flying the Military Air Transport Service Douglas C-124s, Douglas C-133s, Lockheed C-130s, and Boeing C-135s. It was a long and difficult trip from England or Germany to Elizabethville, with primitive landing fields, few en route navigational facilities, and no weather reporting.) Kennedy made a speech at the United Nations on September 25, 1961, taking note of Communist support of disturbances—called "wars of liberation" by Premier Nikita Khrushchev—in the Congo, Laos, South Vietnam, India, Berlin, and elsewhere, and challenged the Soviet Union to a "peace race" instead of an arms race.

Khrushchev was euphoric because of the repeated Soviet space achievements, stating with his customary diplomatic finesse, "We have bombs stronger than 100 megatons. We placed Gagarin and Titov in orbit, and we can replace them with other loads that can be directed to any place on earth." The Soviet premier, brought up in the hard knocks of the Communist Party and World War II, discounted Kennedy as young, inexperienced, and vulnerable to pressure. Intending to torque up the tension, he announced that he intended to sign a separate peace treaty with the German Democratic Republic (East Germany), believing that this action would force the three Allied powers out of Berlin, as the Berlin Blockade had been intended to do.

Kennedy—a clutch player despite Khrushchev's assessment—was determined not to yield, even though he was aware that NATO forces were inadequate for a conventional war, and that if conflict came he might have to toss Taylor's ideas out the window and resort to nuclear war to defend Europe. The Strategic Air Command accelerated its ground alert, and six wings of B-47s scheduled for inactivation were retained on duty. Thirty-six Air National Guard and Air Force Reserve airlift squadrons were activated and the inactivation scheduled for a number of older tactical fighter and transport units was canceled.

The East German civil populace, particularly its youth, voted with its feet in a 4,000-person-per-week-exodus through Berlin. This drain of the best and the brightest was an intolerable commentary on life in a Soviet satellite and was met with a Soviet solution: the erection of the Berlin Wall.

The tension over Berlin had just begun to subside when President Kennedy was confronted by a new and potentially even more explosive crisis. His administration, somewhat to its embarrassment, had become aware soon after the election that the issue of a missile gap had been a fabrication, and that the United States was qualitatively and quantitatively superior in terms of ICBMs. (The Eisenhower administration had been unable to refute Kennedy's charges during the election without revealing its intelligence sources.) This very superiority, however, had untoward effects in 1962, even as Khrushchev smarted from his rebuff over Berlin.

The so-called missile gap had failed to materialize because the Soviet Union had rightly been dissatisfied with its cumbersome and vulnerable first generation of liquid-fueled ICBMs and had virtually ceased their production, turning instead to medium-range and intermediate-range ballistic missiles (MRBMs and IRBMs). Adequate for deployment in Europe, they offered no threat to the continental United States, which had deployed similar-range Thor missiles in England, Italy, and Turkey. To close its own missile gap and offset its ICBM deficiency, the Soviet Union courted the newly established regime of Fidel Castro in Cuba, with the aim of placing IRBMs there. Thus, in the way of politics, the spurious Kennedy "missile gap" inverted itself to become a genuine crisis in Cuba.

THE CUBAN MISSILE CRISIS

Cuba was extraordinarily attractive to the Soviet Union, for it offered a base within 90 miles of the United States and a conduit for intervention in Central and South America. The United States had been maladroit in its treatment of Fidel Castro's new regime, established in January 1959, and looked on helplessly as the bearded leader turned more and more to Moscow for support. The situation was exacerbated in April 1961 when a newly inaugurated President Kennedy sanctioned the abortive Bay of Pigs invasion by Cuban counterrevolutionaries.

In keeping with its policy of applying pressure at peripheral points all over the globe, the Soviet Union began supplying Cuba with arms at an ever-increasing rate, sweetening the deal with subsidized sugar purchases to maintain the island's economy. When the United States protested the heavy influx of tanks, artillery, antiaircraft guns, SA-2 surface-to-air missiles, and more than fifty MiG-15, -17, and -19 fighters, Khrushchev replied that these were purely defensive weapons. His reply was inadequate, for the Soviet presence represented an incursion on the long-standing Monroe Doctrine, which excluded foreign bases from the Western Hemisphere. The threats to the U.S. naval facility at Guantánamo Bay and to the Panama Canal were intolerable—as was the loss of American prestige.

On October 14, two expert U-2 pilots of the 4080th Strategic Reconnaissance Wing, Majors Richard S. Heyser and Rudolf Anderson, Jr., brought back photographs that proved conclusively that the Soviet Union was building offensive missile sites in the San Cristóbal area. Follow-up flights revealed Ilyushin Il-28 Beagle medium bombers being assembled, and evidence that there were both 1,000-mile-range MRBM and 2,000-mile-range IRBM sites being prepared. Succeeding flights showed that work was proceeding at a feverish pace, and that within two weeks, the Soviet Union would have in place as many as two dozen missiles capable of reaching almost every heavily populated area in the Western Hemisphere.

Aware of the utter gravity of the situation and determined not to have a repeat of the Bay of Pigs fiasco, the President established an Executive Committee of the National Security Council and made his brother Robert, the Attorney General, chairman. Other members included Secretary of Defense McNamara; Secretary of State Dean Rusk and his Under Secretary, George Ball; Director of Central Intelligence John McCone; National Security Adviser McGeorge Bundy; and Chairman of the JCS General Taylor.

Normal diplomatic relations with the Soviet Union proceeded as the Ex-Comm, as it was called, met in secret, reviewing all the intelligence data and seeing what it could select to solve the basic problem of getting the Soviet offensive capability out of Cuba without starting World War III.

The full array of flexible response options was considered. The primary proposal by the military was to take out the missiles, bombers, and air defenses in a large-scale attack. Other options included a "surgical" air strike, an invasion by conventional forces, or a blockade. (Historically, "surgical" air strikes have always been a favorite choice of nonfliers. Fliers have been less optimistic, aware that such surgery was difficult to carry out effectively and without collateral damage. Today, with precision-guided munitions, it is a better option.) The blockade—termed a "naval quarantine" because no state of war existed—was finally accepted as a means of applying a carefully calculated degree of power under the shield of America's nuclear capability.

The Strategic Air Command was placed on full alert, with almost seventy B-52s airborne, fully armed, and prepared to fulfill their war mission on an instant's notice. In addition, almost every ounce of existing U.S. strength was mobilized and sent south, to Florida if there was room, or to other Southern states if there was not.

In perhaps the most stirring address of his career, President Kennedy informed the American public of the existence of the Soviet missile threat in a television broadcast on the evening of October 22. He made it clear that the United States was not dealing with Cuba, but with the Soviet Union, and that it was prepared to use the entire weight of its war-making capability to ensure that the missiles were removed. Kennedy announced that the naval quarantine would be lifted only when the Soviet offensive weapons were removed and the missile sites destroyed. The full measure of his determination was revealed when he said, "It shall be the policy of this nation to regard any nuclear missile launched from Cuba against any nation in the Western Hemisphere as an attack by the Soviet Union on the United States, requiring a full retaliatory response upon the Soviet Union."

The first sign of the possibility of a resolution came when Soviet ships bound for Cuba reached the 500-mile boundary line of the quarantine, stopped, then turned around. On October 26, Premier Khrushchev sent a long, lugubrious message to President Kennedy, followed the next day by a shorter, much tougher one that demanded the removal of U.S. missiles from Turkey. Both letters, however, indicated that the Soviet Union recognized the danger of a miscalculation leading to a nuclear exchange, and that it sought a way out. The slight relief in tension was clouded on the 27th, however, when Major Anderson, one of the heroes of the October 14 discovery, was shot down and killed in his Lockheed U-2 by an SA-2 missile.

The possibility of disastrous miscalculation was uppermost in the minds of both American and Soviet leaders as tension built. A considerable degree of private diplomatic activity went on, with the Soviet Union being assured first that the United States did not intend to invade, and second that the missiles in Turkey were going to be removed routinely later in the year. In a formal response to Khrushchev's first letter, Kennedy stated that the United States would guarantee not to invade Cuba if all work was stopped on offensive missile sites immediately, all offensive weapons were removed, and the site destruction was monitored by United Nations observers.

Khrushchev accepted the terms, and the crisis was over, although additional pressure had to be applied to secure the removal of the Il-28s. Secretary of State Rusk made a public statement to the effect that the two nations had stood eyeball to eyeball, and that the Soviet Union had blinked. Red China also took note of the Soviet backdown, which it regarded as cowardly, and relations between the two countries took a turn for the worse in succeeding years.

In later analyses of the crisis, two very unlikely characters found themselves in agreement on the reason for the American success, General LeMay and Premier Khrushchev. Both emphasized the importance of the American nuclear shield, with Khrushchev pointing out, "About 20 percent of all Strategic Air Command planes, carrying atomic and hydrogen bombs, were kept aloft around the clock." LeMay stated that he believed the success of the American efforts was due to SAC's superior strategic nuclear power and President Kennedy's obvious willingness to use it.

Although it defused the situation and avoided a nuclear exchange, the crisis revealed some deficiencies in the American military machine. As the Tactical Air Command had insisted for years, it had an insufficient number of fighter and cargo aircraft. The Army's general-purpose forces were not adequate for the occasion, and the Navy was desperately short of ships to enforce the quarantine. All of these would be addressed in the months to come, but the success in Cuba had another and more serious result. The members of the

ExComm, most particularly Defense Secretary McNamara, National Security Adviser Bundy, and JCS Chairman Taylor, all came away with an inflated sense of their value in the management of the situation, and, more important, of the techniques used to respond to the behavior of the enemy. Their conclusions would be applied in far greater measure in Vietnam, only to result in a long, agonizing, divisive catastrophe.

THE TRAGEDY OF SOUTHEAST ASIA

Many mistakes were made during our involvement in Southeast Asia, but the most egregious mistake of all was Secretary McNamara's basing U.S. strategy for the conflict on the obvious fiction that the war was merely a civil insurrection within South Vietnam. McNamara knew full well that North Vietnam was behind the elaborate problem facing South Vietnam, yet he insisted on this fabrication as a basis for his military policy, for it enabled him to apply his concept of graduated response. In effect, the SecDef planned to educate the North Vietnamese, giving them a logical reason to accede to American policy. The education was simple carrot and stick: concessions for good behavior and varying degrees of punishment for varying degrees of bad behavior. This arrogant philosophy, a by-product of the administration's complete ignorance of the psychology of the North Vietnamese, derived more from Dr. Spock than Clausewitz. McNamara's powers of persuasion were so great that both Presidents Kennedy and Johnson acquiesced in the basic deceit of fighting a war as an insurrection, and thus abandoned traditional military strategy.

Their acceptance is difficult to understand, given that North Vietnam was totally involved in South Vietnam, from deploying regular army troops there to its maintenance of multiple logistic routes through Laos and Cambodia. Everyone was aware of the open, committed support the Soviet Union and China gave North Vietnam. The two giant powers overcame their contemporary mistrust of each other to unite in furnishing equipment, supplies, and training to North Vietnam. It was an excellent investment, for they traded relatively inexpensive military material for the total absorption of the United States into a war disapproved of by its allies, its media, and ultimately a significant and vocal proportion of its population. Nonetheless, McNamara continued to insist to the very end of his tenure that the war was a simple rebellion best handled within the confines of South Vietnam. McNamara was convinced that the North Vietnamese were a mere pawn for Red China, with which he (wisely) wanted to avoid war at all costs. (All through the war, McNamara's interpretation of intelligence data on China was overly pessimistic and did not realistically assess China's means or intentions.) A cursory study of the history of the area would have made it clear that North Vietnam had its own agenda, one that went back to its earliest roots, and that Ho Chi Minh was determined to unite Vietnam for his own regime and not for any other.

As a consequence of this flawed assumption and his acceptance of General Maxwell Taylor's advice, McNamara let the conflict develop into a ground war, with an eventual commitment of more than 500,000 troops—and 56,000 fatalities. Worse, from the point of view of the leaders of the USAF, he forfeited an opportunity to end the war in 1965—and many times thereafter—by the full application of air power. Instead he created a muddled framework of command and control that jettisoned all of the experience gained in the employment of air power in World War II and Korea. (The contention that he forfeited an opportunity to end the war in 1965 might be a matter of debate, were it not for Linebacker II in late December 1972, which proved that it could have been done.)

With the fussy precision of a dedicated executive accountant, McNamara declaimed that the war in Vietnam was to be a laboratory for handling counterinsurgency. McNamara's all-knowing attitude permeated his staff, who came to regard advice from senior military officers as a mere irritant. A rather typical incident captures this attitude of omniscience and contempt for military planners. In early 1965, the North Vietnamese began building surface-to-air missile sites around Hanoi and Haiphong. General William C. Westmoreland, Commander of U.S. Military Assistance Command, Vietnam (COMUSMACV), and Major General Joseph H. Moore, Commander of the 2nd Air Division, asked permission to knock them out. Their request was refused, and Assistant Secretary of Defense John T. McNaughton admonished them, "You don't think the North Vietnamese are going to use them! Putting them in is just a political ploy by the Russians to appease Hanoi." McNaughton had a tremendous influence on McNamara's thinking throughout the war. Early on he was an enthusiastic advocate of ground-troop increases and even of bombing the north; by 1966, he had revised his position and become an advocate of abandoning the Vietnamese.

With occasional exceptions, each succeeding year of the war in Southeast Asia transformed the concept of "flexible response" into a rigid belief that warfare was better executed in terms of clever signals, subtle increases of pressure, and "punishment" applied in delicate gradations, rather than in classic military terms of mass, surprise, and selection of the most appropriate objective. Many thousands of miles away from the scene of combat, the leaders in the Department of Defense and in the administration persisted in regarding the enemy as errant schoolboys who could be reformed by a combination of inducements and punishments. Yet it should have been obvious to McNamara and Johnson, of all people, with their massive resources for research and intelligence, that the North Vietnamese and their National Liberation Front auxiliaries were formidable opponents who had been fighting for twenty-five years, battling the tough Japanese and defeating the French.

It should have been seen that the North Vietnamese national psychology did not recognize the size, technological advantage, or national interests of the United States; it recognized only the Vietnamese ability to fight without respite under the most difficult of circumstances. To assume that the North Vietnamese leaders would be sensitive to subtle signals and give up their lifelong tradition of fighting to the threat of "escalation" betrayed an unforgivable intellectual arrogance on the part of the American political leaders. What was presented to the American public as a cool and cautious approach to fighting turned out to be merely a means of training an already capable enemy. The restrictions on bombing and the bombing halts were not perceived by the North Vietnamese as invitations to negotiate, but as weaknesses to be exploited. Unfortunately, the rarefied academic decisions on policy in Vietnam were different from similar decisions in a corporation or a classroom, for they carried with them the weight of daily death in battle of young Americans, as well as the prolongation of suffering and the ultimate sellout of the South Vietnamese government and people.

BACKGROUND TO INVOLVEMENT

The United States had been involved in Indochina at some level since 1945, usually in support of an eventual loser. During the Eisenhower administration, the United States had almost acceded to French requests for direct military support at Dien Bien Phu, where American C-119 Flying Boxcars of the 50th Troop Carrier Squadron were hastily painted in French markings and used to airdrop supplies. The use of a nuclear weapon had been discussed, and preliminary arrangements had been made for conventional bombing by a fleet of ninety-eight Boeing B-29s. Unable to obtain approval from Great Britain and some of Vietnam's neighboring states, the United States declined. Dien Bien Phu fell, and France was ejected from the colony it had exploited for almost a century.

After the fall of Dien Bien Phu, the interim fate of Vietnam was determined at a conference in Geneva, Switzerland, in the summer of 1954, where representatives of France, North and South Vietnam, Cambodia, Laos, Red China, and the Soviet Union hammered out a Korea-like agreement. Secretary of State John Foster Dulles had attended the conference, but left because he felt that the outcome would not be favorable to the United States. A decision was reached to divide the country roughly in half along the 17th parallel, with the northern portion going to the Communist-sponsored Vietminh, led by Ho Chi Minh, and the southern to the Emperor Bao Dai, who was sponsored by the French. As a by-product of the conference, the final dissolution of the French empire in Indochina was ratified by the recognition of Laos and Cambodia as independent sovereign states.

The agreement at Geneva was followed by a mass movement of more than a million refugees from North Vietnam to South Vietnam. The Vietminh soldiers in the South moved to the North, but many thousands were ordered to stay behind to form cadres for local guerrillas who would become known as the Viet Cong. The name comes from "Viet Nam Cong San"—a phrase reportedly coined by South Vietnamese President Ngo Dinh Diem in 1959. Diem did not wish to dignify the Communists with their party title National Front for the Liberation of South Vietnam or National Liberation Front (NLF). The NLF was controlled by the People's Revolutionary Party (PRP) and directed by the Central Committee of the North Vietnamese Communist Party.

Although the United States agreed to observe the Geneva agreement, it was not optimistic, and in September 1954, in a meeting in Manila, it fostered the creation of an Asian counterpart to NATO. The Southeast Asia Treaty Organization (SEATO) included the United States, Great Britain, France, Australia, New Zealand, the Philippines, Pakistan, and Thailand.

U.S. Military Involvement

As the French withdrew from Vietnam, the South Vietnamese turned to the United States. A 325-man Military Assistance Advisory Group was in place by 1955, to train the Army of the Republic of Vietnam (ARVN). They found the situation in Vietnam in 1955 had an amazing similarity to the situation in Korea in 1950. Both Korea and Vietnam were peninsulas, bordered on the north by Communist China. Both had been divided artificially in the middle. In both countries, the northern half was Communist and more heavily industrialized, while the southern half was primarily agricultural and governed by only nominally democratic regimes that did not have the full backing of their people. In both instances, the northern army was well trained and well equipped by the Soviet Union, and openly intended to conquer the southern half by force of arms. In both instances the southern army was trained by U.S. forces and badly equipped.

There were many significant differences in the two conflicts besides the temperature contrasts, of which several stand out. In Korea, while the front was sometimes fluid, there was always a front, while the war in South Vietnam was the "war without a front." In Korea, the peninsula was bounded on the west by the Yellow Sea and on the east by the Sea of Japan, both of which were controlled by UN naval forces. Vietnam shared its peninsula with Laos, Cambodia, and Thailand. Although the U.S. Navy could operate off the eastern coast, the North Vietnamese would exploit the unsettled political conditions in the western portion to establish supply routes—the Ho Chi Minh Trail—through Laos and Cambodia, along with sanctuaries to support an insurgent movement in South Vietnam. U.S. activity began in Laos in 1960, in the form of reconnaissance and covert military action; it would turn into an eight-year interdiction campaign.

Another significant difference in the two situations was the adroit way that the North Vietnamese leaders, particularly Ho Chi Minh and his military leader, General Vo Nguyen Giap, took care to build up a guerrilla movement in the South. The soon-to-be-infamous black-clad Viet Cong, built up from cadres left behind by the Vietminh, were indistinguishable to foreigners from the indigenous population, and over the years they built up a fluid organization that could come together to fight a battle in the city or in the countryside and then quickly disappear. Its techniques of terror were more effective in producing recruits, taxes, foodstuffs, and supplies than were the more conventional techniques of the South Vietnamese. The Viet Cong and the population loyal to the South Vietnamese government were not merely two opposing political parties, Tories and Loyalists, with equal claims on the affection of the people.

The Viet Cong would take over a village and terrorize its residents, forcibly recruiting them and threatening death if they did not support the North Vietnamese cause. If they were forced to leave by the arrival of ARVN or U.S. troops, the Viet Cong would do so with the warning that they would return to punish any defectors. The American media never recognized the fundamental terrorist function of the Viet Cong.

FIRST USAF INVOLVEMENT

The initial entrance into Vietnam was utterly innocuous. President Kennedy authorized sending a mobile control and reporting post to Tan Son Nhut Air Base, near Saigon. By October 5, 1961, a detachment of the 507th Tactical Control Group from Shaw AFB, South Carolina, was in operation. On October 11, the President raised the ante, transforming Air Force operations in Vietnam from a purely advisory capacity to a limited combat role by authorizing deployment of a detachment that became known as Farm Gate. Formed by a portion of the 4400th Combat Crew Training Squadron, which had been activated at Eglin AFB on April 14, 1961, the original unit of 155 officers and airmen had four Douglas RB-26s, four Douglas SC-47s, and eight North American T-28s. The B-26s and C-47s had already proved their worth in World War II and Korea, while the T-28 was a trainer masquerading as a fighter-bomber. The troops and the T-28s and SC-47s began arriving at Ben Hoa Air Base in November 1961, with Colonel Benjamin H. King, CO of the 4400th, bringing the first SC-47 in from Hurlburt Field in just over seventy-five hours of flying time. The RB-26s came the next month. All aircraft were in the yellow-and-red South Vietnamese Air Force markings. Their nominal charter was training the South Vietnamese Air Force and "working out tactics and techniques." Operational flights were authorized provided a Vietnamese was on board for purposes of receiving combat or combat support training. Circumstances and natural disposition soon saw the Farm Gate crews flying combat missions, including reconnaissance, close air support, and surveillance. The C-47 crews provided no direct training to the Vietnamese and instead developed an increasingly higher level of proficiency in bringing supplies to short, rough fields and in air drops.

The 2nd ADVON (Advanced Echelon) was established by the Thirteenth Air Force on November 15, 1961, with four numbered detachments, three located in South Vietnam and one in Thailand. It was deactivated on October 8, 1962, and replaced by the 2nd Air Division. Brigadier General Rollen H. Anthis was the initial commander for both units, in addition to operating as Chief, Air Force Section, Military Assistance Advisory Group (MAAG).

Farm Gate's main training effort was to teach twenty-five Vietnamese Air Force (VNAF) pilots to fly the T-28s in the newly organized 2nd Fighter Squadron, with instruction in bombing, rocket firing, and gunnery. Although somewhat beefed up for their fighter-bomber role, the T-28s would, like the B-26s, prove to be structurally inadequate for the task. Both aircraft were also very vulnerable to the increasingly heavy ground fire they encountered.

The SC-47s were used for psychological warfare and for dropping flares to permit night strikes by the VNAF. These were highly successful, and a directive by McNamara permitted equipping 520 villages with radios to call for help if the Viet Cong attacked. One of the psy-war SC-47s was the second aircraft to be lost in Vietnam, crashing on February 11, 1962, with six Air Force and two Vietnamese crew members killed.

The Farm Gate unit—officially designated Detachment 2, 4400th Combat Training Squadron—was well received by its Vietnamese colleagues, who proved to be apt students. Under the command of Lieutenant Colonel Miles M. Doyle, Farm Gate operations had built up rapidly to 275 officers and men and more than forty aircraft by early 1963 and had been expanded to other South Vietnamese bases. Then attrition and increased enemy defenses took their toll of the aging equipment. Two B-26s were shot down in February 1963, and a wing failed on a third on August 16. All three crews were killed. Unable to obtain reinforcements, by October the unit was down to nine serviceable T-28s and twelve B-26s, the latter operating under flight restrictions intended to avoid wing stress—hardly reassuring when plunging into battle. Once again, the richest country in the world was letting its fliers fight an Asian war with old, inadequate, and dangerous equipment.

The unit was renamed the 1st Air Commando Squadron in mid-1964, and reequipped with the far sturdier if still Korean War–era Douglas A-1E Skyraider, the formidable "Spad." A second Farm Gate squadron, the 602nd Air Commando Squadron, was organized in October 1964, an indication of the success of the down-and-dirty operations of the unit, despite the miserly way it had been provided resources.

OPERATION MULE TRAIN

As soon as the intervention ice was broken, additional material began to flow at an increasing rate. On January 2, 1962, Lieutenant Colonel Floyd K. Shofner, CO of the 346th Troop Carrier Squadron, brought in the first of sixteen Fairchild C-123 Providers as a part of Operation Mule Train. The first mission was flown the following day. The C-123 had started life as the Chase XC-20G combat glider, designed by Michael Stroukoff, and was transformed into the XC-123 Avitruc by the addition of two Wright R-2800 engines. Fairchild took over development and built 302 for the USAF and twenty-four for Allied countries. In 1966, the C-123K was fitted with a 2,850-pound-static-thrust General Electric J85 turbojet under each wing. The combination of improved performance and increasing pilot proficiency turned the Provider into an excellent short-field and para-drop aircraft, with a capacity of as many as sixty troops or 8,000 pounds of cargo.

C-123 operations were distinguished by their daring, ingenuity, and high morale. Despite abysmal working conditions—changing engines by the light of jeep headlamps was not uncommon—the relationship between flight crews and maintenance personnel was excellent.

Contact with the enemy's guerrilla forces was difficult to maintain, calling for an increased reconnaissance capability. Aircraft and training were provided to create the VNAF 716th Reconnaissance Squadron. The request for a jet reconnaissance plane, the Lockheed RT-33, had been denied, and McNamara instead authorized a handful of Beech RC-45s, Douglas RC-47s, and North American RT-28s—not exactly fast movers. These were supplemented by a flight of USAF McDonnell RF-101Cs operating out of Don Muang in Bangkok, Thailand.

OPERATION RANCH HAND

The C-123 reached its peak of fame—or notoriety—early in its Vietnamese career by its work as part of Operation Ranch Hand, the painful operation that began as a "limited, three-phased defoliation plan" authorized by Secretary McNamara on November 3, 1961, and ran for nine years. Operation Ranch Hand is a heart-rending example of how good airmen can be forced to do unpleasant work when it is determined that the war effort demands it.

The purpose of the three phases was to defoliate a portion of the Mekong Delta area where the Viet Cong were known to have bases, to destroy manioc groves that the Viet Cong used for food, and to destroy mangrove swamps to which the Viet Cong fled to take refuge.

The British had used defoliants in Malaysia, and this was cited as a precedent in the arguments prepared for President Kennedy's consideration. The full approval of President Ngo Dinh Diem and his government had been received. The risk of charges of biological warfare was recognized and accepted.

The USAF had developed a limited capacity for the work in the course of public service activities. After World War II, a Special Aerial Spray Flight (SASF) had been established at Langley Air Force Base to undertake mosquito-control tasks with aerial spraying. The unit was alerted for duty in Southeast Asia, and six C-123s from TAC were sent to the enormous depot at Olmsted AFB, Pennsylvania, to be equipped with the 1,000-gallon MC-1 spray tank and additional necessary equipment. Somewhat to his surprise, Captain Carl W. Marshall, SASF commander, had no trouble getting volunteers who were willing to fly aircraft without U.S. markings, wear civilian clothes, and be on temporary-duty status (TDU) for extended periods of time. They understood that they would not be acknowledged as USAF members if captured. That there was no shortage of volunteers is a comment on their patriotism—and the powerful appeal of flying a dangerous mission.

After a grueling flight from Pope AFB, the six C-123 spray aircraft arrived in South Vietnam in January 1962 to begin Operation Ranch Hand. Over the next nine years, the unit, expanded over time, would spray more than 20 million gallons of herbicide over almost 6 million acres in South Vietnam. The decision to fight a ground war within South Vietnam changed the use of herbicides from a trial run to a military necessity, for, unlike many experiments in Vietnam, defoliation worked.

In the first year of its use, the number of ambushes dropped dramatically. It was extremely effective in exposing hidden enemy camps and supply dumps, and it was the only means to peer beneath the jungle canopy at the traffic on the Ho Chi Minh trail. In late 1962, the Ranch Hand charter was extended to the destruction of crops that the Viet Cong depended upon, so as to increase their supply difficulties.

Operation Ranch Hand grew in size until it had nineteen UC-123Ks to support the expanded mission. With the strategic hamlet concept in full flower, Army bases ranging in size from large semipermanent fortified establishments to tiny outposts now dotted South Vietnam. The bases were also quite successful, and by 1970 had established control over large areas and prevented the Viet Cong from returning. But almost all the bases were surrounded by the ever-encroaching jungle, and their perimeter fences and minefields had to be kept clear of growth to function. Herbicides kept them clear.

The immediate protests by the North Vietnamese on the use of defoliants were echoed by the Soviet Union and China. By 1965 many civilians in the United States were objecting to their use on humanitarian and environmental grounds. Three years later, President Nguyen Van Thieu of South Vietnam believed that herbicide use had become counterproductive. A 1968 report prepared by a committee appointed by the American ambassador to Vietnam, Ellsworth D. Bunker, indicating that the three principal herbicides, Agents Blue, Orange, and White, were *not* harmful had an inflammatory reverse effect upon the media, and pressure mounted to stop the practice of defoliation. (Later studies repudiated the report. Eventually, suits against the manufacturers of the herbicides were settled for $180 million, the funds going to some 250,000 claimants. The Veterans Administration also settled claims for 1,800 U.S. veterans. Studies still go on, a recent one showing a link between children born with spinal bifida and their fathers' exposure to Agent Orange. Oddly enough, the people with most exposure to the defoliants, the aircrews, have displayed few symptoms characteristic of Agent Orange poisoning.)

Despite a wistful lingering belief that the herbicides being used in South Vietnam posed no permanent hazard, Secretary of Defense Melvin Laird informed President Nixon on December 22, 1970, that future herbicide use in Vietnam would conform to the standards for herbicide use in the United States. On January 7, 1971, the last Ranch Hand mission was flown.

Once launched, the large-scale ground war had required the use of herbicide to save the lives of troops in the field. The Air Force Ranch Hand units flew their mission with honor, skill, and dedication, often under heavy antiaircraft and small arms fire, and with more intimate exposure to the herbicides than anyone else. The first USAF aircraft lost in the Vietnam War was a Ranch Hand C-123, which crashed on February 2, 1962, killing all three crew members, Captain Fergus C. Groves, Captain Robert D. Larson, and Staff Sergeant Milo B. Coghill.

Time has not dimmed the controversy. A Fairchild UC-123K that had received more battle damage than any of its kind and still flew was brought to the magnificent USAF Museum at Wright Patterson Air Force Base in Dayton, Ohio. There it became the center of a dispute with the local environmentalists, who demanded its removal, decontamination, and destruction, despite the fact that it offered no hazard to anyone. The "compromise" was to have the aircraft cleaned once again, and then totally sealed. It now sits virtually mummified, an eloquent testimony to the basic mistakes that brought about its use in the first place.

AN UNPRODUCTIVE STRUGGLE

From 1962 to 1965, the struggle McNamara chose to wage in Vietnam closely emulated the fate of the Edsel automobile. Both the war and the car had been introduced with a publicity gasconade; both proved to be resounding failures. The crucial difference was that the Edsel was canceled.

McNamara's management style, however difficult, might have been salvageable if he had been willing to use the talents and advice of the men who had fought wars in the past or were engaged in fighting the war in Vietnam. Despite the fact that any operation in Southeast Asia would be undertaken beneath the umbrella of SAC's air-atomic shield, McNamara refused to take the advice of the man who had created that shield: Curtis LeMay. As Air Force Chief of Staff, LeMay had been unable to convince the DOD of the efficacy of an all-out air attack on North Vietnam.

Parenthetically, it is interesting—and sad—to note that as Chief of Staff, LeMay could not influence the Air Force as he had been able to influence the Strategic Air Command when he was its commander. There were a number of reasons, of which the Defense Reorganization Act of 1958 was one: LeMay no longer commanded a combat force. The Air Force was also large, and its bureaucracy in the Pentagon was less cooperative than his staff had been at SAC. The Secretary of the Air Force, the energetic and capable Eugene M. Zuckert, was similarly hamstrung, for despite a good working relationship with LeMay, who kept him fully informed on JCS activities, he was out of the DOD decision loop. In the course of his four years of service, he saw his influence on policy drop to zero. It was an especially bitter pill for Zuckert to swallow, for he had served as an assistant secretary under Stuart Symington, who had enjoyed almost equal influence with the Secretary of Defense.

Despite the best efforts of these two strong men, Zuckert and LeMay together could not stop the flow of power into the hands of McNamara, who used his authority over the DOD budget to withhold funds for projects of which he did not approve, sometimes thwarting the wishes of both the military services and the Congress. In McNamara's view, the day of the heavy bomber had passed; he canceled the B-70 and refused to allow additional B-52s to be built. LeMay was outspoken about McNamara, in one instance comparing him to a hospital administrator who practiced brain surgery on the side.

There is an interesting side note. The centralist, authoritarian management style that LeMay had used so well to bring SAC to its peak of power and that now permeated the rest of the Air Force was amplified by a power of ten by McNamara. The Secretary of Defense carried the style to an extreme, creating a climate of fear and placing a value on reporting rather than on doing. Numbers—whether it be of bodies counted, sorties flown, bombs dropped, or Congressional delegation visits—became the be-all and end-all of the military system McNamara shaped. Report numbers were transformed from black dots on a page into truths that were quantified, defended, and extrapolated from. Whole bureaucracies sprang up to create the numbers, challenge them, defend them, and mold them into new requirements for more numbers. In short, McNamara carried LeMay's methods to an extreme from which a rebound was inevitable. When the rebound came, it proved to be immensely beneficial.

McNamara, Zuckert, and LeMay were still at loggerheads by the fall of 1963, a time of momentous political changes. The most important of these without doubt was the assassination of President Kennedy on November 22. His death had been preceded by that of President Ngo Dinh Diem, who was killed by his own generals on November 2. The junta was led by General Duong Van Minh, who took over as leader. Minh was himself replaced in January by another coup, this one led by Major General Nguyen Khanh. One result of the November coup was the appointment of then Colonel Nguyen Cao Ky to become head of the Vietnamese Air Force. Two years later, Ky, with Major General Nguyen Van Thieu and General Ming, would form a triumvirate to run the country. Ky, a dashing, courageous pilot, was vehemently anti-Communist and much liked by U.S. politicians and airmen alike. Under his leadership, the VNAF grew proficient even as it expanded, a situation that conformed to the official U.S. policy of withdrawing American forces as South Vietnamese forces became capable.

But despite that proficiency, and the increased demands made upon the small U.S. military contingent, the Viet Cong and North Vietnamese were steadily increasing their influence. USAF forces were still suffering from a lack of support and the implicit disadvantage of war-surplus equipment. By February 1964, all Farm Gate B-26s were grounded, two T-28s had been shot down, and two other T-28s had suffered catastrophic wing failures. The 1st Commando Squadron was ultimately forced to borrow nine T-28s from the VNAF to stay operational.

Secretary McNamara's reaction to the growing disaster was to authorize a second Air Commando squadron equipped with Douglas A-1E Skyraiders. However, in May 1964 he ordered that Air Force pilots could no longer fly combat missions, even with Vietnamese observers on board. As an alternative, he authorized expansion of the VNAF by two squadrons and made Douglas A-1H aircraft available for a total of six squadrons.

If at this time the McNamara team had reexamined their basic assumptions and compared them with the known events in South Vietnam, they would have found that as the United States escalated its level of effort, so did the North Vietnamese. Instead of being intimidated and admitting that a small country like

North Vietnam could not defeat a large country like the United States, they simply fought harder. In essence, the North Vietnamese had the measure of the American psychology and knew that the public would not support a long war. The American leaders, McNamara chief among them, had no grasp of North Vietnamese psychology and perhaps never attempted to attain one. They believed, almost to the end of their time in office, that their superior gamesmanship would automatically force the North Vietnamese to conform to U.S. wishes. When they finally realized that this was not the case, they changed their position, advocating leaving South Vietnam in the lurch and departing the country.

The North Vietnamese and the Viet Cong responded to increased U.S. pressure as if it were a tonic. The Viet Cong increased their efforts and the North Vietnamese dispatched regular army troops over the Demilitarized Zone (DMZ) to the south. Even as late as 1964, much might have been saved if a reinforced USAF contingent had been directed against the North Vietnamese while the VNAF was left to deal with the Viet Cong. Instead, the USAF was dedicated to counterinsurgency.

The North Vietnamese successes led to the hotly disputed incident in the Gulf of Tonkin on August 2, 1964, when enemy torpedo boats were reported to have attacked the destroyer USS *Maddox*. On August 4, there were two additional attacks reported, against the *Maddox* and the USS *C. Turner Joy,* another destroyer. (Much doubt has since been cast on the truth of these attacks, particularly the second one.) On President Johnson's orders, a retaliatory raid was made on North Vietnamese bases, and on August 7, 1964, the Congress passed the Gulf of Tonkin Resolution, authorizing the President to commit the armed forces to defend South Vietnam's independence and territory. The die was cast.

TIT FOR TAT TAG LINES

The year 1965 saw an abrupt transition for the United States forces in Vietnam. The advisory role was retained and training went on, but the American forces were now there to fight. The number of U.S. military personnel in Southeast Asia in 1964 totaled 23,310. It jumped to 184,314 in 1965 and continued to rise steadily, reaching a peak of 536,134 in 1968. Simultaneously, forces in Thailand had grown from 6,505 to 47,631. During the same time period, the number of USAF aircraft in South Vietnam grew from 84 to 1,085, with an additional 523 planes in Thailand, a full 28 percent of the approximately 6,000 planes in the USAF inventory.

The sugar that made this medicine of massive growth in troop strength go down was an original intent to use U.S. and other Allied forces from Australia, New Zealand, the Philippines, and South Korea to guard enclaves and thus release South Vietnamese troops for combat duty. The good intentions soon gave way to reality as the Allied forces were progressively engaged in combat.

With all pretense of adherence to the Geneva agreement gone, jet aircraft began to arrive—a few Martin RB-57s, thirty-six B-57s, McDonnell RF-101s, small numbers of Convair F-102s, several squadrons of North American F-100s, and a lesser number of Republic F-105s. The numbers were relatively small and the crews were rotated every ninety days, but they clearly presaged the future, as did the first use of SAC tankers to support operations. On June 9, 1964, four KC-135s operating out of Clark Air Force Base (and stylishly nicknamed Yankee Team Tanker Task Force) refueled eight F-100 fighters on their way to strike Pathet Lao antiaircraft defenses on the Plain of Jars in Laos. Yankee Team Tanker Task Force (subsequently renamed Foreign Legion) grew into the "Young Tigers" who kept the air war running with their incredible refueling capabilities.

The Viet Cong reacted predictably to the reinforcements; in a November 1, 1964, mortar attack on Bien Hoa Air Base, they destroyed five B-57s and damaged fifteen others, along with four VNAF Douglas A-1s. Four Americans were killed and seventy-two wounded. This was followed by an attack on an American barracks, the Brink Hotel in the heart of Saigon, this time killing two and wounding seventy-one Americans. These actions served to unite the JCS to recommend U.S. bombing of the infiltration trails in Laos and North Vietnam. However, it was election month, and President Johnson did not wish to jeopardize what proved to be his landslide victory over that redoubtable veteran and friend of the USAF Senator Barry Goldwater of Arizona. (The Senator later always jokingly referred to his 486-to-52 defeat in the electoral college as a "real cliff-hanger.")

The U.S. leadership, if unable to come up with a winning strategy, excelled at dramatic names for operations, which began with the Operation Flaming Dart series of attacks. These were "tit for tat"—a phrase used in the official DOD communications—for the twin February 7, 1965, raids on Pleiku and Tuy Hoa, which destroyed five helicopters and killed eight Americans and wounded more than a hundred others.

Secure in his position, President Johnson ordered the limited Flaming Dart retaliatory strikes against troop concentrations and staging areas in the southern portion of North Vietnam. The Navy attacked first, sending forty-nine aircraft against Dong Hoi; the USAF attacked the following day with F-100s and A-1s striking NVN barracks at Chap Le.

Another portent of the future occurred in February 1965 when B-52s of the 7th and 320th Bomb Wings were deployed to Andersen Air Force Base, Guam. General John P. McConnell, who became Chief of Staff on General LeMay's retirement on February 1, 1965, put forward once again the list of ninety-four targets in North Vietnam against which he wished to conduct a massive air offensive, with the aim of destroying North Vietnam's ability to wage war. This list was substantially the same as the list previously nominated by the JCS and Admiral Ulysses S. Grant Sharp, Commander in Chief, Pacific (CINCPAC), and constituted the heart of North Vietnamese industrial and logistical capability. The targets could have been taken out quickly, with a minimum loss of life, but the suggestion was always overruled.

In its place would come applications of limited force in nondecisive areas with lilting names like Barrel Roll, Steel Tiger, and Tiger Hound in northern,

central, and southern Laos, respectively, and Rolling Thunder, which super-
seded Flaming Dart in Vietnam. Each of these missions put a premium on the
skill and daring of the aircrews, whose equipment was often ill-suited to the
task and who had to fly under rules of engagement as adverse as the abominable weather conditions. The basic flaw in the design of the missions was
that they were striking the wrong end of the North Vietnamese/Viet Cong
supply lines, sometimes losing $3 million aircraft in attacks against trucks
worth 6,000 rubles and carrying bags of rice. The frustration of Air Force
leaders and the men flying the missions was extreme; they understood perfectly well that it was far less risky and far more efficient to sink a ship in
Haiphong harbor carrying 300 trucks with one mission than to have to spend
1,000 missions to try to destroy those same trucks on the Ho Chi Minh Trail.
This was hard intelligence, obtained by the pilots who flew north, were shot at,
and returned with an acute awareness of the risk versus return. Unfortunately,
they were never able to communicate this intelligence to the highest levels in
DOD, because the communication lines ran in one direction only: down.

ROLLING THUNDER

Among the multitude of ironies in the Vietnam War, not least is the fact that
tactical aviation, overshadowed for twenty years by the prodigious demands for
resources by the Strategic Air Command, now had to bear the brunt of fighting
the war, with 50 percent of the fighter units trained and equipped by the Tactical Air Command being permanently assigned to Southeast Asia in 1965. The
irony was heightened when SAC bombers, whose leaders were champing at the
bit to strike North Vietnam a decisive blow, were instead relegated to bombing
Viet Cong base camps and conducting close air support missions in South Vietnam, even as the tactical force went north. The attack against North Vietnam
was a tribute to the courage and determination of both the USAF and U.S. naval
air; the air attack was conducted relentlessly, regardless of weather or the ever-
increasing enemy opposition. Day after day, the attacks went on, never deterred
by losses nor by the sickening realization that their effort was not appreciated by
the American public, and often not by their own leaders.

Rolling Thunder was a campaign intended to complement the interdiction
work in Laos and in South Vietnam. And, as with McNamara's entire Southeast Asia strategy, it was to avoid provoking either China or the USSR, despite
the daily evidence that those two countries furnished North Vietnam with almost all its matériel. To ensure that the war remained sanitized, target selection was controlled directly by the White House in a request/approval network
that still defies belief. In theory, target requests were to be assessed by the
combat wings and forwarded to Seventh Air Force. In practice, the combat
wings had no say in the targets. On those few occasions when they were able to
nominate targets, the action next went to the Pacific Air Force (PACAF); they
then went to CINCPAC, to the Joint Chiefs of Staff, to the Secretary of Defense, and then to the President, who could, if he chose, get advice from the

Department of State, the Secretary of Defense, and the National Security Council. The approved target then went back down the same list, five headquarters, before it reached the wings. Even in an age of swift electronic communication, the method took far too long for effective use of tactical air strikes. Instead of getting target lists from the people doing the bombing, most targets were generated at Tuesday luncheons at the White House (from which military men were excluded until 1967), where the President and the Secretary of Defense dictated the choice of targets, tactics, timing, number of aircraft, and ordnance to be used. To say it was ludicrous would be kind; it was an insane way to fight a war, a cold exercise of power for the sake of the exercise, and without regard to the expenditure of life or materiel involved.

The question of unified action by USAF and naval air assets, which President Eisenhower had sought to resolve with the 1958 Defense Reorganization Act, reverted to the same uncooperative setup that had occurred in the Korean War, and for which a similar Band-Aid solution was devised. The Navy insisted that carrier operations, with their inherent rigid requirements for aircraft carriers to position themselves into the wind for both launch and recovery operations, were unable to respond to a centralized control system. Naval air was not the only problem. The command-and-control confusion was exacerbated because the B-52 bomber and KC-135 tanker force remained under SAC's control, where during the Korean War, the B-29 force had been controlled by FEAF. Targeting request/approval for the bombers was equally complex, and even more vulnerable to the time elapsed in getting a decision, for the Viet Cong were extremely mobile, able to remove themselves from a target area within minutes of notice of an impending attack—which they often had.

Rolling Thunder began on March 2, 1965, when twenty B-57s and twenty-five F-105s struck just north of the DMZ at Xam Bong, blowing up an ammunition dump. By September, almost 4,000 strikes had been made by USAF and Navy aircraft against targets mostly in the south of North Vietnam. These included radar sites, rail lines, marshaling yards, highways and trails, and bridges—the last proving to be a difficult target.

The momentum Rolling Thunder had gained in 1965 was brought to a shuddering stop when President Johnson ordered a bombing halt from December 24, 1965, to the end of January 1966, during which both Christmas and the Buddhist festival of Tet were celebrated. Johnson's intent was to have his opposite number, Ho Chi Minh, interpret the gesture as a signal to negotiate. The Vietnamese used it instead as a time to build up their defenses.

The Rolling Thunder attacks seemed to be successful in terms of destroyed vehicles, burned-out tank farms, and other material losses. A report by Secretary McNamara estimated these to be $150 million for the period February 1965 through October 1966. There were two problems, however. One was that the campaign seemed to encourage rather than deter the North Vietnamese, who simply asked their giant benefactors, China and the USSR, for more oil, more SAMs, more everything. The other was that it was costing the United States $250 *million per month* to sustain the entire interdiction effort by the Air Force and Navy in North and South Vietnam and Laos. Despite this expenditure, Secretary McNamara perceived "no significant impact on the war in South Vietnam."

To McNamara, the answer was obvious: spend more money on increasingly less significant targets. Rolling Thunder sorties rose from 55,000 in 1965 to 110,000 in 1966, while costs rose from $460 million to $1.2 billion as aircraft and munitions became more sophisticated and expensive. The sorties went on until November 1, 1968, when President Johnson halted the bombing of North Vietnam north of the 19th parallel. He also called a halt to his own administration, announcing that he would not run for reelection. This time the bombing halt lasted for *four years*—longer than the entire time the United States was engaged in World War II. As long as this sounds in comparative terms, it was nothing compared to how long it seemed in actual days, weeks, months, and years to the American prisoners of war who, during those four years, continued to be tortured, starved, and killed. During the same long period of the bombing halt, American soldiers in the field continued to be wounded and killed at a rate that reached hundreds per week. Bombing in the North was not resumed until North Vietnamese intransigence at the negotiating table in Paris made it necessary.

Secretary McNamara left office on February 29, 1968, to be succeed by Clark Clifford. A report by one of McNamara's top men, Dr. Alain C. Enthoven, who headed the Office of Systems Analysis in the Pentagon, must have given the new Secretary something to reflect on. Enthoven reported that the gradual escalation of the war to inflict an unbearable attrition on North Vietnam had failed. The war was costing $10 billion per year, and despite more than 500,000 troops, millions of tons of bombs dropped, and an estimated 200,000 enemy troops killed in action, the control of the South Vietnam countryside was about the same as in July 1965. Ironically, it was during this same period that Maxwell Taylor, now in Saigon as the American ambassador, reversed his previous position on fighting a ground war with U.S. troops and instead began to advocate the bombing of the North. When Taylor was forced out of the country by then President Nguyen Cao Ky in 1967, he returned to Washington as a special adviser to President Johnson. He advised sending more troops and bombing the north, thus neatly covering his past positions.

Unfortunately, the situation had changed drastically in North Vietnam. In July 1965, the North Vietnamese air defenses were relatively unsophisticated. By 1968, they had grown into a highly effective network of radar systems, missiles, antiaircraft guns, and interceptors under ground control. The SAMs—surface-to-air missiles—were a new threat.

DEFEATING THE SAM

The North Vietnamese learned to use the new tools of war quickly, increasing their antiaircraft and surface-to-air missile capability and scoring the first SAM kill on July 24, 1965, when a MIGCAP (MiG combat air patrol) McDonnell F-4C was shot down 55 miles northwest of Hanoi. The SA-2 Guideline missile, coupled with the radar system designated "Fansong" in NATO terminology, seemed like an implacable foe at first, as several aircraft were lost. Its effectiveness was enhanced by the DOD-imposed rules of engagement that, in retrospect, defy belief. SAM sites within 10 miles of Hanoi and Haiphong were off-limits; within 30 miles of Hanoi, SAM sites could be hit only if they were preparing to fire and were not located in a populated area or on a dike of the widespread irrigation system. These became the very places that the North Vietnamese chose to install their weapons. Perhaps the most ludicrous limitation was the prohibition on attacking a site while it was being built, or even of attacking when the operational site was in the radar-surveillance mode. They could be attacked only when they switched to track, i.e., seconds before launching. In the American West of a century ago, a similar rule of engagement would have forbidden the sheriff to draw his weapon until the hammer had dropped on the outlaw's pistol.

But with the courage and determination that Air Force crews would repeatedly demonstrate in Southeast Asia, ways were found to counter the SA-2. Going in low would have been the best solution, but fuel economy, the increasing strength of antiaircraft artillery, and the massive small-arms fire ruled this out. Fighter strikes eventually entered the combat area at medium altitudes calculated to best avoid the particular combination of MiGs, SAMs, and antiaircraft defenses around a specific target. If an aircrew were fortunate enough to see a SAM launched or in flight toward them, the best defense was to attack it as if it were an enemy aircraft. Breaking toward the missile and executing a high-G maneuver would cause the stubby-winged SAM to attempt to follow it; the SAM could enter a high-speed stall or break up from the forces involved. With the best of luck it crashed back on those who had fired it.

The most dramatic antidote, however, was the introduction of the Wild Weasel system, created in an emergency seminar chaired by Brigadier General K. C. Dempster. Like most weapons introduced to meet emergencies, the first Wild Weasels were a lash-up. What became the APR-25 RHAW (Radar Homing And Warning) system was installed, along with an IR-133 panoramic receiver to analyze the radar signals and indicate whether they originated from an antiaircraft battery, a ground control intercept radar, or a SAM site. The third element was the APR-26 Launch Warning Receiver, which picked up the increased power of the SA-2 guidance system upon launch. All of these (in their experimental versions) were packed into four aging two-seat North American F-100F fighters. Armament initially varied, but always included the 20mm cannon and 2.75-inch rockets. Later, the AGM-45 Shrike missile, tuned to home in on radar frequencies used by the Fansongs, was introduced. A quick reaction development effort led by Major Garry Willard and a dedicated team of air and ground personnel brought the initial Wild Weasel unit into combat.

By November 21, 1965, the first of seven F-100F Wild Weasel aircraft arrived at Korat Royal Thai Air Force Base, Thailand. The unit became the 6234th TFW (Wild Weasel Detachment), operating as part of the 388th TFW, and its mission became known as Iron Hand strikes. Not surprisingly, given the experimental nature of the work, a Wild Weasel crew was lost before a success was scored. On December 20, Captains John Pitchford and Robert Trier were shot down about 30 miles northeast of Hanoi. Both men ejected. Pitchford was shot while being captured after he landed, and spent more than seven years as a prisoner of war. Trier was killed by his captors, who alleged that he had resisted.

The first Wild Weasel success came soon after, when Captains Al Lamb and Jack Donovan took out a site during a Rolling Thunder strike on the railyard at Yen Bai, some 75 miles northwest of Hanoi.

The men flying the F-100s were the pioneers in the effort; like them, their successors had to be good, for only the very best pilots could handle the Wild Weasel mission. They were complemented by equally skilled electronic warfare officers (EWOs), who had to withstand the stress of high-G maneuvers in a potentially fatal environment while operating the new radar warning systems. (It was the custom of the service at the time to treat nonpilots with hearty denigrating humor; thus the EWOs became GIBs, for guy in back, or Bears, for trained bears. It was terribly unfair, for while a pilot flying the aircraft was used to high-G maneuvers, the EWO was often a navigator or electronic countermeasures officer accustomed to SAC's straight-and-level flying. Yet the success of the mission depended upon the EWOs managing the systems, spotting the SAM site or the SAMs, and ultimately allowing the pilot to position the aircraft to launch his ordnance.)

The F-100s were flown in four-ship flights, often with F-105 escort. The two aircraft were not compatible in terms of their speed capabilities, and the F-100 Weasels, whose motto was "First In—Last Out," were vulnerable to the high density of antiaircraft fire in the area of SAM sites. The effect of the first F-100 Wild Weasel attacks was deadly—but at great cost. Seven attacks took out seven SAM sites, but two aircraft were shot down and five were damaged beyond repair.

Nonetheless it was the proper antidote, and the crews flying the F-100s built up a repertoire of techniques that would benefit the F-105s and F-4s that replaced them. New aircraft were no panacea, however. For example, a group of seven F-105 Wild Weasel aircraft arrived at Takhli RTAFB on Independence Day 1966; thirty-eight days later, all seven aircraft had been shot down. Yet the real reason for Wild Weasel's success was the proficiency built up by the crews in the crucible of a very uneven battle. They established a tradition which would endure through the Persian Gulf War and beyond, the EWOs finding the targets (whose defense techniques became increasingly sophisticated) and the pilots managing to maneuver the aircraft well enough to survive attacks and kill the SAMs.

F-105 Iron Hand flights usually consisted of two F-105G two-place Weasel aircraft, each with a conventional single-seat F-105D on its wing. The Wild Weasels routinely preceded the attack force into the target area by about five minutes, just far enough in advance that the SAMs would have to be turned on and warming up. When the instruments showed that a SAM site had activated, the Wild Weasel would attack, early in the war with a Shrike, later with a Standard Arm missile. This would be followed by a general attack using conventional weapons from the remainder of the Iron Hand flight.

The intent was as much to keep the SAMs off the air as to destroy them. The North Vietnamese adapted their tactics. They would sometimes alternate having the radar on between supporting SAM sites, to keep track of the F-105s; sometimes they would leave their radar off so as not to present a target, then salvo the SAMs in barrage fire at the strike force. Close cooperation was maintained with antiaircraft artillery and any MiGs that were available.

And just as a shut-down site measured the effectiveness of the Wild Weasel operation, so did the forced alteration in bombing tactics measure the effectiveness of the SAMs. Loss rates per missiles fired were relatively low. In 1965, eleven U.S. aircraft were shot down for 194 SAMs fired—a 5.7 percent rate. It took 3,202 SAMs to shoot down fifty-six aircraft in 1967, a 1.75 percent rate. In 1972, at the relative height of effectiveness of both U.S. countermeasures and SAM proficiency, 4,244 missiles destroyed forty-nine aircraft, for a 1.15 percent rate. But no matter how low the loss rate, the SAMs could not be ignored—they had to be suppressed.

COUNTERING THE MiG THREAT

With memories of the great aces of World War II and Korea firmly in their mind, American fighter pilots, USAF or naval air, wanted most of all to come to grips with enemy MiGs to establish air superiority—and perhaps, just incidentally, to become aces.

The situation proved to be totally different from Korea, however, particularly in end results. The North Vietnamese Air Force was not interested in mixing it up in dogfights except on rare occasions, and the impediments in equipment and rules of engagement kept the USAF, in particular, from running up the scores the pilots wanted.

The most significant lack of equipment was an airborne radar warning and control system (AWACS) type of vehicle to provide coverage in eastern areas of North Vietnam. The Disco Lockheed EC-121Ts (the familiar Super Constellation with a big radar dome and lots of tracking equipment) operating over Laos and the Gulf of Tonkin were invaluable, as was Red Crown, the radar warning and control vessel in the Gulf of Tonkin.

The first American and only all-MiG-21 ace, then Captain and now Brigadier General Steve Ritchie, stated emphatically that every time he was successful he had received good information from either Red Crown or Disco. Navy fighters almost always operated under Red Crown's beneficent eye, and as a result were able to gain the crucial seconds necessary to create and win an encounter with the MiGs.

North Vietnam's greatest advantage lay in its integrated defense system, which provided its MiG fighters with comprehensive information on the air battle. They had the advantage of knowing the customary inbound tracks of the USAF fighters (the radar was assisted by visual guidance made easy by the twin trails of smoke emanating from the F-4s) and could launch their attacks with every advantage.

North Vietnamese defense was aided by American jurisdictional problems. As it had done in Korea, the Navy stoutly refused to participate in a central control system. The Navy's official position did not seem entirely arbitrary. The Commander in Chief, Pacific (CINCPAC), Admiral Ulysses S. Grant Sharp, Jr., was a dedicated supporter of the war who wanted to keep pressure on North Vietnam. However, the official line was that aircraft carriers were less suitable for centralized control because they had the burden of fleet defense. In practical truth, the Navy wished to control its own war.

As a result, just as in Korea, a decision was made to carve out a geographical area for naval air operations to ensure minimum interference. Vietnam was divided into seven segments known as route packages, starting in the south along the 18th parallel, with what the pilots called Pack One. Packs Two, Three, and Four were horizontal east-west slices stacked farther north. At the top of the stack, stretching to the Chinese border, were Pack Five in the west and Pack Six A and Pack Six B, which included Hanoi and Haiphong.

The intensity of enemy opposition increased as the Pack numbers rose, with Packs One through Five known as "the Easy Packs" and Packs Six A and B known as "Downtown"—where the action was.

Little attention was paid to who owned what rights within the Easy Packs, which, prior to President Johnson's termination of Rolling Thunder, were primarily either routine targets of less importance or alternate targets for the strike forces primed to go Downtown.

Pack Six A, including Hanoi, was designated USAF turf, while Six B, including Haiphong, was assigned for Navy targeting. The imaginary line between the two sections of Pack Six were important at the staff and planning level but disappeared for the airborne fighter pilots whenever they needed to cross, or when they could assist their friends on the other side.

The route package system was a clumsy expedient that avoided the issue of centralized control of air assets that presumably had been established for all time during the North African campaign of World War II. The cost was an implicit limitation on the optimum use of the aircraft available.

THE ENEMY DEFENSES

The North Vietnamese Air Force was small. Its pilots had received basic training in the Soviet Union and advanced training in China. In general, they did not compare in skill, aggressiveness, or training with their American counterparts. This was both offset and exacerbated by the integrated air defense system that exercised rigid control over North Vietnamese pilots at every stage of the flight; they were told when to take off, when to engage, when to disengage, and when to return to base. Their excellent Atoll missiles were fired on command from the ground. The downside of this overcentralized control for the North Vietnamese was the consequent lack of initiative on the part of pilots accustomed to be told what to do and when to do it.

As with the SAM sites, the skewed rules of engagement put Communist airfields off limits for most of the war. Nothing pained USAF pilots so much as seeing juicy targets sitting on the ground, some with engines running, ready to attack moments later, and being unable to take them out in the classic tactic of aerial warfare, a strafing attack. MiG strength rose from about fifty in 1965 to more than 200 by 1972. Losses were quickly replenished from both Chinese and Soviet stores.

MiG Data

The MiG-17 Fresco was an evolutionary development of the MiG-15 that had been such a dramatic surprise in Korea. USAF and Navy pilots both were shocked to find the relatively unsophisticated MiG-17 to be a formidable adversary at low altitudes, where it often loitered to engage U.S. aircraft returning from a mission low on fuel. Called a "vicious, vicious little beast" by the famed Colonel Robin Olds, it had gun armament similar to the MiG-15's, with two 23mm cannons and one 37mm; later in the war, the AA-2 Atoll missile was carried on outer pylons. The Atoll was a virtual copy of the U.S. AIM-9B Sidewinder heat-seeking missile.

The MiG-19 Farmer (the Chinese designated their version the Shenyang J-6) was reportedly the world's first supersonic production fighter, first exceeding the speed of sound in early 1953. (Fans of the North American Super Sabre dispute this claim on the basis that the YF-100A set an official speed record of 755.149 mph on October 29, 1953.) Heavily armed with three 30mm cannon and an array of missiles, the MiG-19 was a tough customer.

The pride of the North Vietnamese fleet was the delta-wing MiG-21 Fishbed, which played to the McDonnell F-4 as the MiG-15 had to the F-86. The tiny MiG-21 had an empty weight of only 13,500 pounds (the F-4 weighed in empty at about 30,000 pounds) and had a speed in excess of Mach 2. The aircraft's rate of climb and acceleration were excellent, although it did not have as high a sustained turn rate as the Soviets desired.

North Vietnamese ground control tactics used the MiG-21's capabilities perfectly, vectoring them in pairs behind an incoming strike force. The Fishbeds would be positioned behind the formation, accelerate to supersonic speed, fire their heat-seeking Atoll missile, and zoom up and away from the strike force. Their mission was accomplished if they got the F-4s and F-105s to jettison their bomb loads, but on too many occasions, the Atoll scored a victory.

Tactics like these, combined with the general reluctance of the NVAF to participate in dogfights, kept the kill ratio of the USAF at a troublingly low level—between 3.5 and 4 to 1 for the period 1965–1967. During the next year, the ratio fell to 2 to 1. The low kill ratio has been attributed to a number of factors, most important the lack of an AWACS-type aircraft, as previously noted.

However, the rules of engagement also required U.S. pilots to have a positive visual identification before firing, and the time required for that put American planes at a severe disadvantage. Until the appearance of the McDonnell F-4E in 1968, the escorting Phantoms had only Sidewinder and Sparrow missiles with which to fight, sometimes supplemented after 1967 by gun pods. The missiles were originally designed for use against bombers. Using them in a dogfight was agonizing, for in a battle where combined head-on speeds could exceed 1,500 mph and where a flick of the stick could remove a fighter from the sights, the setup, arming, and firing of the missiles took hours-long seconds to accomplish. The tracking and turn capabilities of the missiles were also geared to bombers rather than to nimble fighters, and many a MiG got away that should have fallen to cannon or a specialized-for-fighters missile.

There was another even more important factor that still has repercussions today. The experience of combat made it evident that prewar USAF training was not rigorous enough. Safety considerations and the tendency to train against similar types of aircraft had vitiated the training programs. The solution was training programs like the Air Force's famous Red Flag and the Navy's Top Gun that were far more realistic, even at the price of some flying safety considerations.

OPERATION BOLO

The combination of U.S. rules of engagement and the avoidance tactics employed by the North Vietnamese was frustrating to USAF fighter pilots, who wanted to come to grips with the enemy in a classic dogfight. In the relatively few combats that occurred during 1966, nineteen MiGs had been shot down, with the loss of five USAF crews. (The Navy had shot down four and lost four.) The MiG-21s' rear attacks with Atoll missiles were achieving North Vietnamese goals by causing the F-105 formations to jettison their bombs before reaching their targets.

Because the MiG airfields were off-limits, a decision was made by Seventh Air Force Commander General William W. "Spike" Momyer to plan a fighter sweep for just after the year-end stand-down. The North Vietnamese Air Force always reacted more strongly after a quiet period, having more aircraft in commission and their tactics refined by training sorties.

The fighter sweep was deliberately designed to appear exactly like a normal F-105 mission, with the standard call signs, designations, and other indicators to show that the mission was just like the many that had preceded it. The difference was that F-4s armed for air-to-air combat would be flying instead of F-105s. The most important deception was the use by the F-4s of ECM pods previously used only by F-105s. Altitudes and airspeeds were flown exactly as if it were the daily strike force.

Operating out of Udorn Royal Thai Air Force Base, on the morning of January 2, the 8th Tactical Fighter Wing, the "Wolfpack," launched Operation Bolo. Led by veteran ace Colonel Robin Olds, who had thirteen aerial and eleven ground victories in Europe during World War II, twenty flights of F-4s and F-105 Wild Weasel aircraft were used to set the trap. Despite some adverse weather, the MiGs reacted exactly as had been hoped, coming up from behind to attack what they presumed to be bomb-laden F-105s. To their surprise, they encountered F-4s, with their tanks already jettisoned and ready to fight. In the space of a few minutes, seven MiGs were shot down, Olds himself getting two of his four victories in Southeast Asia. When the North Vietnamese lost two more MiGs on January 6, and began an immediate stand-down to analyze the situation, the American fighter pilots, if not content, were a little happier.

Operation Bolo illustrates the indisputable fact that the USAF dominated the airspace, north and south, albeit at great cost. The enemy was unable to attack our airfields, our carriers, our front lines, or our most vulnerable asset, the KC-135 tankers that always plied their trade next to—and sometimes in—the combat zone.

The USAF had air superiority, even if not in the overwhelming degree that it might have had if its leaders had not been hobbled by DOD constraints. As will be detailed later, the basic strength of the USAF was demonstrated in a number of ways besides basic combat. One of the most compassionate was the magnificent level of effort of its search and rescue (SAR) service. A less obvious, yet equally important manifestation of inherent organizational strength was the remarkable manner in which support commands like the Air Force Logistic Command (APLC) and the Air Force Systems Command (AFSC) proved their worth with rapid fixes to seemingly insoluble problems, including the vexing requirement for visual identification of enemy fighters.

THE PRINCIPAL FIGHTERS:
THUDS AND PHANTOMS

THUDS

The workhorse of the war, the Republic F-105 Thunderjet (affectionately, if sardonically, nicknamed the Thud for the sound it reputedly made when crashing) had originally been designed as a supersonic long-range nuclear strike fighter, capable of carrying bombs of 1 to 20 megatons yield. The name "Thud" was an unfair call, for the F-105 had a sculptural beauty, even after years of modifications had disfigured it with various bumps to hold equipment. Republic, noted for building heavy "ground-loving" fighters, experienced trouble during the development of the aircraft, which flew first on October 22, 1955 (exceeding Mach 1 on its first flight), but did not reach operational status until May 1958. The F-105 was a lovely aircraft to fly, and despite having far too few access panels for ease of maintenance, it had a remarkably high in-commission rate. The first F-105 squadron to deploy to Thailand stayed for five months and racked up 2,231 sorties with an 85 percent availability rate.

Eventually, seven squadrons of F-105s belonging to the 355th (based at Tahkli RTAFB) and the 388th Tactical Fighter Wing (based at Korat RTAFB) bore the brunt of the Rolling Thunder campaign, taking heavy losses in the process. Out of the original 833 production aircraft, some 350 F-105s were lost to combat or other operational causes. Despite their relative lack of maneuverability and the manner in which they were employed, the F-105s managed to shoot down 27.5 MiGs, often off the tail of another F-105 being attacked. (A .5 score means a victory shared with another type of aircraft, probably in this case an F-4.) The attrition rate, combined with a refusal by DOD to reopen the production line, ultimately forced replacement of the Thud by F-4s.

PHANTOMS

Perhaps the greatest tribute to the USAF aircrews and ground crews is the manner in which they took aircraft designed for one mission and flew and maintained them so artfully that they excelled in many others.

The McDonnell F-4 Phantom II was originally conceived as a naval fleet defense fighter, armed solely with missiles and using a Weapons System Officer to manage its complex fire control system. A huge aircraft, grossing 61,650 pounds on takeoff, the Phantom was propelled at Mach 2.27 speeds by the brute force of its twin afterburning General Electric J79 turbojets, the incredible contribution of Gerhard "Herman the German" Neuman to engine technology.

In a decision almost without precedent, the USAF decided in March 1962 to adopt the Phantom as its standard fighter with absolutely minimum changes from the Navy production version. The original F-110 designation was changed to F-4 when the DOD instituted a tri-service designation system on October 1, 1962.

The big Phantom's in-flight refueling capability conferred a number of advantages on it—increased range and the ability to take off with less fuel and more bombs, and then refuel after takeoff and, thanks to the bravery of the tanker crews, use up most of its fuel in combat, knowing that KC-135s were on station to refuel them for the trip home.

Yet the Phantom's basic USAF requirement was to fly long distances, heavily laden with ordnance, sometimes to engage in dogfights, sometimes to bomb. When the Phantoms made their long journey to the battle arena in North Vietnam, they naturally found the smaller MiG-21s and their older point-interceptor brethren far more maneuverable, with quicker acceleration and a faster turn rate. The F-4's size and weight put it at a disadvantage in the conventional dogfight, as did its lack of an internal gun, for its missiles were originally designed for bomber interception. (SUU-16 gun pods were supplied as an interim answer, but were not entirely satisfactory.) The answer was found in energy maneuverability tactics, which forced a change in dogfighting from an essentially horizontal plane into a vertical plane. The sheer speed and the power of the Phantom could put it rolling up and down in maneuvers that combated the MiG's greater turning capability. (Energy maneuverability is a complex concept; reduced to its essence, it is the proper management of the combined positional and kinetic energy of an aircraft to extract the maximum maneuverability from it.)

The Phantom's greater speed was offset by the requirement for their protective flights, often loaded with ordnance, to cruise at the same speed as the bomb-laden Republic F-105s they were escorting. To cope with an attacking MiG flight required the Phantoms to be alerted, jettison their bombs, and accelerate to build an energy margin that would take them above the potential attacker to a point from which a diving attack could be made. All this depended upon early warning—and early warning depended upon radar surveillance, not always available. The lack of coverage forced the rapid development of the airborne warning and control aircraft, the giant, rotating-dome Boeing E-3Bs that would be so effective in the Gulf War and elsewhere.

But given the slightest warning, the superior ability of American aircrews allowed them to transform their Phantoms into dogfighters, a transformation later enhanced when the F-4E was fitted with an internally mounted 20mm M-61A1 Vulcan multibarrel Gatling gun.

THE INVERSION OF TACTICS

The fundamental fallacy of DOD strategy was never more evident than in the comparative use of F-105s and B-52s. The F-105s daily flew north with bomb loads of 4 or 5 tons, greater than World War II B-17s could carry, but small relative to the need. A "basic package" of F-105s would be sixteen aircraft in four flights of four, with two flights of F-4s positioned fore and aft for fighter cover. The Wild Weasel force would precede the task force on entry and follow it out after the strike. Radar suppression was supplied by Douglas EB-66s orbiting on each side of the target area. The Thuds would drop a tight pattern of perhaps a hundred Mark 117 bombs (750 pounds each) and break off to return.

As will be shown below, while the Thuds were hitting precisely defined and often carefully limited targets in the north with their 75,000 pounds of bombs, three cells of three B-52s might be operating in the south, dropping as much as 560,000 pounds on *suspected* Viet Cong locations. Thus did DOD directives turn basic commonsense tactics upside down.

The most frustrating mission for the F-105s proved to be North Vietnamese bridges, especially the Thanh Hoa (Dragon's Jaw) bridge spanning the Son Ma and the eighteen-span Paul Doumer bridge across the Red River north of Hanoi. The Thanh Hoa bridge seemed indestructible, shrugging off 871 USAF and U.S. Navy air attacks that cost eleven aircraft. The periodic bombing pauses were used to repair it, and in the four years after the November 1, 1968, pause, the bridge was substantially improved as an artery south.

The Paul Doumer, built by the French, was a massive structure 8,467 feet long that rested on eighteen huge concrete piers. The narrow bridge supported a railroad down the center, with roadways on each side. Each day it carried twenty-six trains and hundreds of trucks, funneling an average 6,000 tons of supplies south. A vital artery connecting Hanoi to the rest of the country, the bridge was heavily defended by 300 antiaircraft guns, including radar-directed 85mm flak batteries, and eighty-one SAM sites. All of the MiG interceptors could be quickly scrambled in the event of an attack. The F-105s flew 113 missions against the bridge, losing two aircraft in the process.

A preliminary success came on August 11, 1967, when the bridge was attacked by a flight of twenty F-105s, each fighter carrying two 3,000-pound Mk 118 bombs. The Deputy Commander of the 355th, Colonel Robert H. White, who had won both the Harmon and the Collier trophies along with his astronaut's wings flying the X-15 research airplane, led the flight. The attack was preceded by the usual Wild Weasel and flak suppression unit and brushed through a flight of MiGs that made an unusual Luftwaffe-style head-on attack.

White rolled in for the attack from 13,000 feet, diving through a curtain of flak to release his bombs at 8,000 feet. The rash of explosions obscured the bridge momentarily; when it cleared, three spans were knocked down into the river. Two 105s were damaged, but all returned to base. The resilient North Vietnamese had repaired the bridge by October 3, even as they delivered material by other means, including a ferry. Succeeding bombing attacks also caused damage that was quickly repaired.

It was not until Linebacker I in 1972 that the issue was resolved by F-4s carrying Paveway I laser-guided bombs. Both the Thanh Hoa and Paul Doumer bridges were at last dumped into their respective riverbeds, with no U.S. aircraft destroyed.

The Paveway series of precision-guided munitions were standard bombs combined with guidance and control kits. In an attack, the target was illuminated by a laser. When the bomb was dropped, a microprocessor fed signals to the bomb's guidance fins to place it on target. These precision-guided munitions were precursors of weapons to come and a tribute to the AFSC/AFLC teams that developed them.

BUFFS IN ACTION

The Boeing B-52, SAC's dependable long rifle, was designed to carry thermonuclear bombs to the heart of the Soviet Union. First flown in 1952, the Buff (an acronym standing for, in its bowdlerized version, Big Ugly Fat Fellow) is probably going to be in service well into the twenty-first century still the most flexible aircraft in military history. Despite its versatility, it will probably still be engaged in the impromptu role into which it was cast in Vietnam, dropping massive quantities of conventional ordnance.

B-52 operations began on June 18, 1965, in the first of what were called the Arc Light strikes, when twenty-seven aircraft from the 7th and 320th Bombardment Wing flew from Andersen AFB, Guam, to a suspected Viet Cong base north of Saigon. The enemy base camp was spread out, and previous air attacks had not been effective.

Neither the Air Force nor the Army was optimistic about the effectiveness of the B-52s performing the same role that artillery had in World War I, grinding up miles of terrain in the hopes of hitting something. One Air Force leader termed it "swatting flies with a sledgehammer." The first mission was further tainted when two B-52s collided in midair because of insufficient distance between their planned refueling tracks. Nonetheless, General William Westmoreland, USMACV Commander, soon became convinced that the B-52 was an indispensable weapon, and Arc Light operations escalated.

The utility of B-52s was expanded when they began actual close support operations in November 1965, driving off Viet Cong attackers at Plei Mei in the Central Highlands. In December, support was provided to the III Marine Amphibious Force, under the command of Lieutenant General Lewis W. Walt III. The Marines were and are notoriously demanding about close support efforts, and Walt was delighted by the scope, scale, and accuracy of the B-52s' effort.

The B-52Fs then being used could carry fifty-one 750-pound bombs, twenty-seven internally and twenty-four externally. In 1965, a massive program called Big Belly was established to increase the internal capacity of B-52Ds for 500-pound bombs from twenty-seven to eighty-four, and for 750-pound bombs from twenty-seven to forty-two. External pylon racks could carry twenty-four bombs, either 500 or 750 pounds. Maximum bomb load was thus increased to 54,000 pounds when carrying all 500-pounders and 49,500 pounds when carrying 750-pounders. Fittings for mines and glide bombs were also installed, and just as a precaution, up to four nuclear gravity bombs could be carried as well.

The B-52s were not bombing typical easily defined targets such as cities or military bases, but rather acres of heavily canopied jungle, so new delivery techniques were needed. Mobile ground radar guidance units were installed to provide Combat Skyspot, a ground-directed bombing system. Radar controllers would direct the B-52s along a course and signal when to drop the bombs. Combat Skyspot was flexible, permitting quick switches in targeting, and was also extremely accurate. The automaton-like nature of the raids on the vast stretches of impenetrable jungle was frustrating, and the overworked crews sardonically referred to themselves as "coconut knockers."

Yet the Viet Cong came to hate the B-52s more than any other weapon, for the first warning of the B-52s presence was the explosions of bombs in a huge corridor. One year after commencing operations, the B-52s were dropping 8,000 tons of bombs per month on Viet Cong targets. Sortie rates rose to 1,800 per month, with aircraft flying out of Andersen and the newly established Royal Thai Naval Air Base at U-Tapao, Thailand. U-Tapao was only two to five hours away from its targets, while aircraft from Andersen faced twelve-hour missions, with at least one refueling in route.

The scale of the logistic effort was massive, as millions of pounds of supplies ranging from food for the guard dogs to bombs for the Viet Cong were trundled across a 12,000-mile pipeline, all scheduled on various milestone charts around the world to keep inventories low—in many ways the system anticipated the modern "just in time" inventory method. The work was demanding for air and ground crews, who worked seventy-two hours a week routinely and eighty-four hours during surge periods. The troops doing the dangerous munitions work were under constant pressure to keep the Big Bellies filled with bombs. The constant heat, the loneliness, and the lack of amenities did nothing to improve their lot. Aircrews were assigned for 179 days of TDY, because a 180-day tour would have had to have been considered a permanent change of station, which was expensive and administratively disruptive. Their life degenerated into an unending cycle of brief, fly, sleep, brief, fly, sleep.

While the B-52 made invaluable contributions throughout during the war, there were two defining missions that stand out, Operation Niagara and Operation Linebacker II. The first of these was early in 1968 at Khe Sanh, where 6,000 Marines and their Vietnamese allies were surrounded by 30,000 North Vietnamese regular soldiers, who hoped to make Khe Sanh for the Americans what Dien Bien Phu had been for the French.

General Westmoreland approved Operation Niagara, which began on January 22, 1968. Even here command and control was difficult, for the Navy and would not agree to subordinate itself to Seventh Air Force for the operation, and the area around the camp had to be divided into small zones, miniature route packages, to avoid interference.

Using Combat Skyspot techniques, six aircraft struck the target every three hours. The B-52s pummeled the North Vietnamese for days, flying 461 missions and 2,701 sorties and dropping 75,631 tons of munitions before the operation ended on March 31. Initially, the B-52 was limited to dropping bombs no closer than 3,300 yards to friendly positions; as the North Vietnamese sappers pushed their trenches forward, the safety zone was dropped to less than 300 yards. This distance may seem like a lot to someone running a race, but to a GI looking out at bombs bursting across a field of barbed wire, it is minuscule. Considering it as the end result of bombs dropped from a bomber flying a course line at 500 mph at 30,000 feet with crosswinds varying as much as 180 degrees in heading and 100 mph in speed throughout the length of the drop, it becomes a remarkable testament to the skill of the bomber and Skyspot crews and the trust of the troops on the ground. In a post-battle assessment, General Westmoreland noted, "The thing that broke their back basically was the fire of the B-52s."

LAOS AND IGLOO WHITE

The sideshow war in Laos was as much a secret war as the war in South Vietnam was a civil insurrection. The DOD tried to pretend that the multibillion-dollar effort to prop up Souvanna Phouma by means of aid, troop training, covert CIA intervention (including running its own airline and using the services of 20,000 Thai troops) and the virtual dedication of Seventh Air Force air assets to the U.S. ambassador in Vientiane simply was not happening.

Yet during the period from 1960 to 1968, American airmen engaged in a life-or-death struggle trying to stop the infiltration of men and material to South Vietnam and support the troops loyal to the central Laotian government. In Laos, as in South Vietnam, the equipment allocated to the task was at first World War II–vintage RB-26s and B-26s and North American T-28 trainers.

And as in Vietnam, as the U.S. effort increased, so did the North Vietnamese response. Over time, the equipment dedicated to warfare in Laos was improved, and by 1968, all of the modern aircraft operating in North and South Vietnam were also at work in Laos, including the deadly Lockheed AC-130 Spectre gunship. The AC-130 had low-light-level television, infrared, radar, and 20mm side-firing Gatling guns. Later versions had 40mm guns and even laser designators. Phantoms were used to provide flak suppression and attack work—extremely nerve-wracking dives into the pitch-black night to deliver bombs that cost just about as much as the trucks they were aimed at. B-52s were brought in to work under Combat Skyspot control. They were once again a success, and a call went up for an expansion of their work, for the quantity of explosive power they could deliver dwarfed that of all the artillery available in the area.

The desperate effort to use the highest possible technology to sustain the fiction that the war in Southeast Asia was a mere insurrection reached its peak of fantasy in what was first known as the Barrier Concept and was code-named Igloo White.

An original proposal by Professor Roger Fisher of the Harvard Law School to build a 60-mile anti-infiltration barrier across the DMZ and into Laos caught the attention of Assistant Secretary McNaughton. The idea was expanded to become an on-the-ground barrier reaching 180 miles across South Vietnam and Laos—a miniature Great Wall of China—approximately along Route 9 to the Mekong. The concept had been passed to the Jason Committee of the Institute of Defense Analysis, which was serving as McNamara's private Strategic Bombing Survey group. The Jason committee viewed bombing the North as futile and endorsed the concept of an air-supported barrier system that would use electronic sensors to detect enemy activity, which would be countered with air strikes.

The JCS argued against the concept on the basis of its costs and the probable ease with which the enemy might counter it. McNamara ignored the JCS comments and appointed General Alfred Starbird to head Joint Task Force 728 (later given the cover name Defense Communications Planning Group [DCPG]) within the OSD Directorate of Defense Research and Engineering. Thus in a remote-control war can an idea be transmuted by a small committee into a concept for a large committee whose work makes it possible for an even larger committee to turn it into a travesty. The famous aphorism "A camel is a horse designed by a committee" was transcended; what became known as Igloo White was a hydra-headed monster that devoured close to $3 billion, cost many lives, and only marginally disrupted the flow of traffic. The JCS repeatedly attempted to fight the program on the grounds of cost, efficiency, and other more pressing needs, but McNamara had seized the concept as his own and was determined to ramrod it through without regard to advice opposing it. It truly became the McNamara Line.

The DCPG had decided upon a highly advanced air-supported anti-infiltration system consisting of sophisticated air-dropped acoustic and seismic sensors (along with sensors attuned to pick up the ordinary odors of hard-working humans). Read-out aircraft were posted to receive the transmitted signals, which were in turn sent to an enormous central processing facility at Nakhon Phanom (NKP) Royal Thai Air Force Base near the Laotian border. There a huge computerized ground assessment center received, analyzed, and reported enemy logistic flow activity (the grinding of a truck, the pounding feet of rice-bag-laden porters, the acrid scent of urine). This information was relayed to an airborne battlefield command and control center (ABCCC) over Laos, which would then direct attack aircraft to the site. Unfortunately, the North Vietnamese reacted to a limited increase in pressure in Laos exactly as they had to limited increases elsewhere: they fought harder. They seized upon President Johnson's bombing halt to extend their radar control into Laos, moving in SAMs and AAA. For the first time, MiG fighters began making incursions in Laotian airspace; these were quickly checked by Phantoms working with Disco or Red Crown.

By the end of 1968, despite improvements that had been made in slowing the rate of infiltration, there was still a sense of general dissatisfaction, for too many trucks were still getting through. General Momyer pointed out that all previous successful air interdiction campaigns had three components. The first was against the production sources of enemy supplies; the second was the interdiction of movement between the enemy's heartland and his forces in the field; the third was forcing the enemy into active ground warfare so that he would consume his supplies faster than he could replace them. In Laos and in Vietnam, only the second element was being executed.

The JCS once again went on record as advising the closing of North Vietnam's major ports and against continuing Rolling Thunder, since it was not very effective because it was limited to the area south of the 20th parallel.

All of the factors coincided with the initial operational capability of the Igloo White system—which, if it worked as hoped, would be the answer for stopping infiltration. Igloo White was tested in a series of interdiction programs during the dry seasons.

New tactics were used in conjunction with assessments made of intelligence provided by the sensors. Roads would be cut by laser-guided bombs, then land mines powerful enough to destroy heavy vehicular equipment would be emplaced. To inhibit mine clearing, antipersonnel land mines were sown. (McNamara had approved a production schedule of 3.5 million Dragontooth antivehicular mines and 10 million antipersonnel land mines per month.) The results of this effort were then to be monitored by sensors on each side of the mined area, which would determine how many trucks were getting through. A small air force supported Igloo White, including twenty-one Lockheed EC-121 ABCCC planes, eighteen McDonnell F-4Ds, eighteen Douglas A-1Es, twelve Sikorsky CH-3 helicopters, and thirty-four Cessna O-2 Skymasters. Additional aircraft could be called in.

Despite the expenditure of enormous effort and billions of dollars, Igloo

White proved to be a spectacular failure for a variety of reasons, the most important being that it was totally the wrong system for existing conditions. North Vietnamese infiltrators streamed down the multiple trails like rain running down a window pane; Igloo White corresponded to a finger placed on the window to halt the runoff. Almost as important was the resilience of the North Vietnamese, who could repair road damage about as fast as bombs could cause it. The North Vietnamese understood immediately what the sensors were for and quickly devised active and passive ways to nullify them or spoof them. They would cover some of the devices with the ubiquitous wicker baskets to mask their sensors; others they would trick by sending a truck back and forth to simulate a convoy. Anything from croaking frogs to thunder could trigger the sensors used to detect vehicles, resulting in traffic reports that made the Ho Chi Minh Trail seem like rush hour on a Los Angeles freeway. The final obstacle was the geography, climate, and vegetation of the area targeted for Igloo White use. The heavy jungle canopy made it impossible for a FAC (forward air controller) to get visual identification of trucks that the sensors might have picked up, and bad weather often prevented strikes.

Despite brilliant efforts by the almost 5,000 people who tried to make it work, Igloo White was a world-class failure. And in the way of Washington politics, its failure went almost unnoticed because the program was so highly classified. McNamara never had to acknowledge the utter bankruptcy of the scheme he had bullied through over all objections, for in November 1967, he had decamped to become the head of the World Bank.

CAMBODIA

Cambodia had long been a haven for North Vietnamese and Viet Cong, and President Nixon authorized air strikes in early 1969 to destroy the sanctuaries. Under the code name Operation Menu, B-52s made almost 4,000 sorties and dropped almost 110,000 tons of bombs in Cambodia.

On April 30, 1970, the combined factors of the American drawdown and the growth in Communist supply storage areas in the Cambodian sanctuaries caused Nixon to authorize an invasion of Cambodia by both U.S. and ARVN troops. The area of the invasion was limited, and the time of occupation was set at eight weeks. B-52s were again used to bomb.

The military effort was largely successful and reduced the threat to Saigon, but domestic reaction was violent, especially on college campuses, and led to Congressional legislation that forbade future introduction of U.S. ground forces into Thailand, Cambodia, or Laos.

The bombing went on at a reduced rate until August 1973. By 1975, the Communist Khmer Rouge forces of Pol Pot had seized control and begun the genocidal slaughter of 1.5 million of their countrymen.

LINEBACKER II

The abominable term "political air superiority" characterized the gift made by Secretary McNamara to the armed forces of North Vietnam by placing inane restrictions on U.S. air activities in the form of unreasonable rules of engagement. The North Vietnamese retained political air superiority until the enunciation of the Nixon Doctrine by the newly elected President Nixon. Nixon was faced with a dilemma. He had to recognize that after seven years of McNamara's leadership, the war was lost politically no matter what happened in the field. The most vivid example of this anomaly was the Tet offensive, a crushing disaster for the North Vietnamese and the Viet Cong, who had lost 45,000 killed of the 84,000 troops participating. Yet this victory in the field for American and ARVN forces was turned into an agonizing defeat by the U.S. media.

Nixon had to find a way out, and he found it in the Nixon Doctrine, which promised that the United States would honor its treaty commitments and would provide its nuclear shield to any nation vital to U.S. interests. It would furnish arms and economic assistance, but it expected any threatened nation to provide the manpower for its own defense. Nixon's Secretary of Defense, Melvin Laird, coined the term "Vietnamization," meaning training and arming of South Vietnamese forces to defend themselves, while U.S. forces were systematically and rapidly pulled out. The policy was announced in a speech on Guam on July 25, 1969, without consulting the South Vietnamese government of then President Nguyen Van Thieu. The troops had another phrase for Vietnamization, calling it the "bug-out." Political air superiority now became intolerable, for the North Vietnamese reaction of increased military pressure indicated that they wished to secure a military victory rather than permit the United States to save whatever face remained in a negotiated withdrawal. Much would happen between 1969 and 1972, and eventually, air power would be called upon not to achieve the victory that it might have produced, but to persuade the North Vietnamese to accept a negotiated peace. On March 29, 1972, powerful North Vietnamese mechanized units drove across the DMZ into South Vietnam, in a flagrant violation of the negotiation talks then being conducted. The routed South Vietnamese fell back in disarray, and the task of stopping the advance fell to the South Vietnamese Air Force and to the remaining units of the depleted USAF forces. On May 8, President Nixon recognized the inevitable, called a halt to the peace talks, and authorized Operation Linebacker, which extended the strikes that had been going on since April above the 20th parallel and included the mining of Haiphong and other ports.

The North Vietnamese had immeasurably strengthened their air defenses during the four-year bombing layoff, but the efforts of AFSC and AFLC had provided the USAF with some new weapons too. In November 1968, the McDonnell F-4E had begun to arrive in Southeast Asia. Equipped with more advanced engines, an internal 20mm M-61A1 Vulcan multibarrel cannon, leading-edge maneuvering flaps to improve its dogfighting ability, and (by 1972) TISEO (Target Identification System Electro Optical), a telescope in the leading edge of the left wing, the F-4E was a formidable MiG killer. Electronic warfare capability had been upgraded, and both EB-66s and EC-135s had new jammers to quell enemy electronics. New ways had been found to dispense chaff effectively from F-4s. (Chaff was perhaps the simplest of the jamming devices, being simply sheets of what resembled tinfoil cut into lengths. It jammed certain of the enemy radar frequencies. It had been first used in 1943 in the calamitous Royal Air Force raid on Hamburg.) The electro-optical and laser-guided munitions were available in increased supply, and aircrews had gained familiarity in their use.

Linebacker was an apparent success, for the North Vietnamese returned to the negotiating table. President Nixon ordered a halt to the bombing above the 20th parallel in response, and the enemy as usual interpreted this as a sign of weakness and increased its military pressure. By December 18, the United States was in a desperate position; the North Vietnamese negotiators now obviously believed they would achieve the military victory for which they had been fighting since 1945.

President Nixon then chose to use air power as had been advocated by General LeMay in 1963 and subsequently by Generals McConnell, Earle Wheeler, and John Ryan along with Admiral Sharp. That use was the full application of air power at its highest intensity in all its forms directly against the key political and military targets of the enemy to force a settlement of the war.

Linebacker II had to be conducted during the height of the monsoon season, so it was built around an all-weather force of B-52s, General Dynamic F-111s (which, after a disappointing debut, were proving to be deadly efficient), F-4s (which had replaced most of the F-105s), Vought A-7s, EB-66s, KC-135s, and naval aircraft. The B-52s were designated to hit airfields, supply depots, and railroad marshaling yards; F-4s, using precision guided munitions, were to hit the Hanoi electrical power plant, Radio Hanoi, and specialized sections of the railyards. F-105 Wild Weasels had their usual dangerous task of SAM suppression. The F-111s were to attack SAM sites, airfields, and marshaling yards, while the A-7s were to attack Yen Bai airfield. These were backed up by KC-135 refuelers, Search and Rescue C-130s and Sikorsky HH-53 Jolly Green Giant helicopters, EC-121s, and a host of others.

The initial attack called for 129 B-52 sorties: forty-two B-52Ds from U-Tapao and fifty-four B-52Gs and thirty-three B-52Ds from Andersen. The second day would have ninety-three sorties and the third ninety-nine; after that, pressure would be applied continuously at the highest level possible. The Andersen Buffs would form the first and third waves over Hanoi, the U-Tapao aircraft the second.

The first takeoff took place at 2:51 P.M. local time at Andersen, on December 18, 1972. Instructions from SAC Headquarters were that the aircraft were *not* to take evasive action from either SAMs or MiGs during the long run in from the initial point (IP) to bombs-away. This rigid—and ultimately costly—requirement was because SAC wanted to make sure that only military targets were hit and to preserve the electronic countermeasure integrity of the three-ship formations.

As soon as the bombers penetrated the target area, SA-2 Guideline missiles filled the sky, their firing signaled by a city-block-square flash on the ground, followed by streaks through the sky that exploded into a huge mushroom-shaped halos of expanding light. As many as forty SAMs were in the air at the same time, often tracking a single cell of three B-52s. The Guidelines were the size of telephone poles, and many passed close enough to B-52 cockpits that a newspaper might have been read in their rockets' glare. Others hit aircraft, knocking them from the sky. More than 200 SAMs were fired, and three B-52s were lost, two from Andersen and one from U-Tapao.

On the second night, six B-52s were shot down out of the ninety-three that were launched, even though permission had been given the second and third waves to take evasive action on their run in from the IP. The heroism was incredible; one pilot on one crew calmly announced that they were going to be hit by a SAM just as they released their bombs. They were, and they were shot down.

The 7 percent loss rate was unacceptable, yet General John J. Meyer, a twenty-six-victory ace in World War II and now Commander in Chief of SAC (CINCSAC), made the tough decision to press on, calling for SAM sites and storage sites to be heavily hit.

His decision was correct, for the enemy had been hurt too, and was running out of SAMs. For three successive days, no B-52s were lost. The MiGs proved to be no threat at all, with two being shot down by B-52 gunners.

Christmas Day saw a stand-down, a pause to get a signal from Hanoi, and to give the crews, particularly the ground crews, much needed rest. The stand-down was a mistake, because it gave the North Vietnamese a respite in which to recover and restock their SAM sites with missiles.

On December 26, entirely new tactics were tried. All Andersen aircraft, seventy-eight B-52s, were to arrive simultaneously over Hanoi, arriving in four waves from four different directions. Three other waves went on to strike Haiphong at the exact same time.

With 110 support aircraft in the air, the attack went off with precision, with only two B-52s being lost despite the heavy increase in SAM firings. In three attacks, the B-52s had established an ascendancy over Hanoi, which at that moment in 1972 had the most heavily defended airspace in the world. All USAF commanders—and all aircrew members—ground their teeth in fury as they recognized what could have been done in 1965 with such ease and with no losses. The argument that had been made time and again—and ignored every time but once—was proved to be correct: air power could dominate North Vietnam and make it comply with U.S. wishes.

The remainder of the eleven raids went off flawlessly, except for the sad, unavoidable loss of two more B-52s. By the eleventh day, the North Vietnamese were out of SAMs, their MiGs were shut down, their radar and communication links disrupted: they were at the mercy of the United States. In the miserable prisons in which they were held, American prisoners of war experienced an unimaginable elation at seeing their brutal captors frightened and suddenly polite.

Pragmatists, the North Vietnamese signaled that they wished to return to the negotiating table in Paris, and the raids were immediately halted. Many observers within and without the U.S. military knew that this was a mistake; had the raids been continued, the North Vietnamese would have had to accept total military defeat. Instead they secured a victory at the peace tables, one they translated in three years to the full conquest of South Vietnam.

In Linebacker II, SAC had flown 729 sorties, dropped 15,000 tons of bombs, and sustained fifteen losses—less than a 2 percent loss rate. About 1,240 SAMs had been fired. The result of Linebacker II was exactly what had been predicted for the total application of air power in North Vietnam: quick military victory. Had it been applied in the first years of the war, the lives of millions of people would have been spared, hundreds of billions of dollars would have been saved, South Vietnam would not have been ravaged, Cambodia would have not had to endure the Khmer Rouge, and the United States would have not had the ugly experience of the disaffected 1960s and 1970s with the continuing cynicism of the media and the general disaffection of the populace with its government.

Secretary of Defense Robert S. McNamara (left) confers with General Earle G. Wheeler (right), USA, Chairman, Joint Chiefs of Staff, and John T. McNaughton, Assistant Secretary of Defense (International Security Affairs).

In an April, 1965, briefing, Secretary of Defense Robert S. McNamara points to the notorious Thanh Hoa Bridge, known as "the Dragon's Jaw."

Colonel David "Chappie" James, Jr., after he assumed command of the 7272nd Flying Training Wing at Wheelus, AB, Libya. James went on to become the first African-American four-star general.

One of the great gentlemen of the Air Force, General Russell E. Dougherty, Commander in Chief of the Strategic Air Command, chats with (from left) General George S. Brown, chairman of the Joint Chiefs of Staff, Secretary of Defense James R. Schlesinger, and General David C. Jones, Air Force Chief of Staff.

A veteran of three wars and an eight-victory ace in Korea, Lt. Colonel Robinson Risner was commander of the 67th TFS, flying Republic F-105s, when he was shot down in 1965. Rescued, he was shot down again, this time to remain seven long years in the infamous "Hanoi Hilton." Here, he stands defiant before the enemy camera.

Robinson endured and prevailed, and is here commended by PACAF CINC General Lucius D. Clay, Jr. The joyous welcome homecoming prisoners of war received was perhaps the high point of an otherwise frustrating war.

Despite the DoD-imposed disadvantages of the rules of engagement, the USAF fought well. Here are two of the three USAF aces of that conflict, Captain Charles Debellevue, a Weapons Systems officer, and Captain Steve Ritchie, pilot.

Older aircraft fought well in South Vietnam against the Viet Cong. Here, a pair of Douglas A-1E Skyraiders attack a Viet Cong barracks near Thanh Minh.

Colonel Robin Olds marks up three victories over MiG-17 aircraft on May 20, 1967. From left: Major Philip P. Combies, First Lieutenant Daniel L. Lafferty, Major John R. Pardo, First Lieutenant Stempen B. Croker, and First Lieutenant Stephen A. Wayne.

Necessity drove technology to reach unbelievable heights. A Lockheed HC-130P tanker refuels a Sikorsky HH-3B helicopter over the Gulf of Tonkin in June, 1970.

Major (later Lieutenant Colonel) Leo K. Thorsness was awarded the Medal of Honor for his courageous actions on April 19, 1967. After silencing a surface-to-air missile site, Thorsness's wingman was shot down. Thorsness circled the crew members to protect them, and in the process shot down one MiG-17 and damaged another. He was later shot down and captured, spending almost six years in Vietnamese prisons.

There was no sweeter sight to a downed aircrew man than a Jolly Green Giant helicopter hovering overhead, lowering its jungle-canopy penetrator for the quick ride home. The penetrator, whose bullet shape was transformed into three petal-like supports, could lift three men at the end of its 240-foot hoist cable.

When the battle situation did not permit a landing, the Lockheed C-130s used a Low Altitude Parachute Extraction System (LAPES) to get their cargo where it was needed. Here a Hercules crew delivers a load of supplies to the U.S. Army's 1st Cavalry Division (Airmobile) at An Khe, Vietnam.

General Lew Allen, Jr., served from 1978 to 1982 as Chief of Staff, United States Air Force. A command pilot with four thousand flying hours, General Allen was also highly respected as a scientist in both the military and civil communities.

6

≡

THE LONGTIME PILLARS
OF THE AIR FORCE: FOUR
MAJOR COMMANDS

*D*uring the Vietnam War, especially the eleven days of Linebacker II, the ferocious capability of fighter, bomber, electronic warfare, tanker, and other combat aircraft was the sharp cutting edge of the United States Air Force's sword.

But the cutting edge of any sword is made lethal only by the reach, heft, balance, and control of the entire weapon and the skill of the person wielding it. During the Vietnam War, the cutting edge of tactical and strategic units' combat capability in the field existed only because of the reach provided by the airlift capability of the Military Airlift Command. The heft of the USAF combat sword was provided by the towering supply support of the Air Force Logistic Command. The balance and control were the result of years of work by the Air Force Systems Command, and the skills of the units wielding the Air Force sword were the product of the Air Training Command.

Over the course of time, the names and sometimes the missions of these organizations have changed.

THE AIRLIFTERS: ANYTHING, ANYWHERE, ANYTIME

The history of the United States Air Force military airlift units has been one of years of absolute drudgery punctuated by periods of shining glory. Despite being understrength and often using inadequate equipment, military airlift units traditionally have been on call at any time, to fly anywhere in the world in every sort of weather. The "trash haulers," as they ruefully came to call themselves, have been decisive in war and, in a much larger way than is generally recognized, in the preservation of peace. Unfortunately, the ability of the United States Air Force military air transport units to solve problems

ranging from the Berlin Airlift and the relief of Israel in the Yom Kippur War
to the thousands of compassionate missions flown to people in need is often
forgotten—especially during the appropriations process.

The very first demonstration of American military airlift took place at Fort
Myer, Virginia, on September 9, 1908, when Lieutenant Frank P. Lahm briefly
rode with Orville Wright on the Wright *Military Flyer*. Since then, the story
has always been the same, from officers bumming rides from Orville Wright to
frantic calls for McDonnell Douglas C-17 support: military airlift is so attractive compared to other means of transport that demand always exceeds supply.

Certain other factors have been consistent throughout the history of military airlift. The first of these has been the desire of expert airlift specialists to
keep all airlift assets in a single organization. The first call for organizational
unity was made by Brigadier General Augustine W. Robins, Chief of the Air
Corps Materiel Division from 1935 to 1939. Opposition to the idea has been
just as pervasive, for the desire of most organizations, from the U.S. Army and
U.S. Navy down to individual squadrons, has consistently been to maintain
their own dedicated airlift capability. In the matter of control, the principal argument over time has been the management of intratheater airlift, including
the combat airlift mission, where the transports go in under hostile fire to land
or to drop troops and supplies.

Another uniform element in airlift history was the inevitable reduction in
airlift resources at the end of every emergency. This has always resulted in a
frantic scramble to manage the next crisis on an ad hoc basis, with the concomitant crash program to rebuild airlift capability. An expensive way to do
business, it is not confined solely to airlift units. Yet, as will be shown, a contradictory combination of advances in technology and reduced military budgets
has served to bring most of these arguments into congruence, if not harmony.

FROM FERRYING AIRCRAFT TO WINNING THE WAR

The first great impetus to military airlift was the requirement to ferry aircraft all over the United States and to the United Kingdom and the Soviet
Union. The Air Corps Ferrying Command served for fifty-five weeks and fer-
ried more than 14,000 aircraft before being succeeded by the Air Transport
Command (ATC). Created on June 20, 1942, following the recommendations
of L. Welch Pogue, then Chairman of the Civil Aeronautics Board, the ATC
assumed responsibility for air transportation of personnel, material, and mail
for all War Department units, except for those served by Troop Carrier Command units. The most vital function was assigned almost as an afterthought:
ATC was responsible for the control, operation, and maintenance of facilities
on Army Air Force air routes outside the United States. This was to be the beginning of a vast network of airfields, runways, meteorological stations, radar
stations, and other facilities that would serve as the basis for postwar civil and
military international air routes.

In four years of war, the ATC delivered 282,437 aircraft and flew 8.1 billion

passenger miles along with 2.7 billion ton-miles of freight. By war's end, an
ATC aircraft was crossing the Atlantic every thirteen minutes and the Pacific
every thirty-seven minutes, twenty-four hours a day. The ATC was started
with 11,000 personnel; by August 1945, it employed 41,705 officers, 167,596
airmen, and 104,677 civilians. Among the most important of these—in societal terms—were the pioneering WASPs. Of the 1,830 women who gained admission to the Women's Air Service Pilots, 1,074 graduated as pilots. They
delivered 12,652 aircraft in the course of the war. It took until 1979 for the
country to come to its senses and grant these brave women the military status
they had earned and deserved.

Many lessons were learned in the ATC, including the need to have larger,
more reliable aircraft. At the peak of its operations, the ATC had 3,090 aircraft, of which 1,341 were DC-3s. The rest were a mixture of Curtiss C-46s,
Consolidated C-87s, and Douglas C-54s. The C-46s were disappointing because
of mechanical problems, and the C-87s had the natural disadvantage in cubic
capacity of a bomber converted to a transport. The C-54s were a blessing, a remarkable aircraft loved by their air and ground crews and capable of flying
great distances with large loads.

Some of the flying performed by the ATC was as dangerous as combat
over Europe—a trail of aluminum-spread crash sites punctuated the perilous
route over the Himalayan Hump, from China to India. During the war, the
command suffered 1,229 accidents, or 5.57 per 100,000 flying hours.

Troop Carrier Command had an equally distinguished record and grew in
strength as the war progressed. On D-day some 900 aircraft were used; by the
time of Operation Market Garden, the abortive September 17, 1944, airborne
attack on Arnhem, more than 2,000 C-47s alone were used, along with 600
gliders. Airlift had come along way from Fort Myer.

After the war, the ATC suffered from the same drastic demobilization as the
rest of the services, despite a "Dear Tooey" letter from retiring General of the
Army Hap Arnold, who advised his successor, General Spaatz, "During times of
peace we are apt to retain our combat units and sacrifice the essentials to their
successful deployment and immediate operation. We must retain our bases and
our means of deployment." His words are as valid today as they were in 1946.

Disturbed by claims of wasteful duplication of Air Force and Navy transport units and spurred by the recommendations of the Finletter Commission,
the new Secretary of Defense, James V. Forrestal, literally forced the merger of
ATC and the Naval Air Transport Service in June 1948, to create the Military
Air Transport Service. Grumbling, the Navy "loaned" MATS three squadrons
and kept most of its transport capability intact for "fleet support."

As related previously, MATS was given an immediate challenge by the
1948 Berlin Airlift, which it met with enormous success, acquitting itself with
distinction. It was confirmed there for all time that fewer large transports were
more effective than more numerous small ones, and the first great step in that
direction was the Douglas C-124 Globemaster II. The C-124 was rarely called

the Globemaster—to most it was "Old Shaky" for the orchestrated vibrations that strummed through it from the moment the engines were started until the flight was completed.

The C-124 was huge for the time, with a 174-foot wing that looked thin and quite inadequate to support the cavernous fuselage. Unlike its gallant C-54 predecessor, the C-124 was designed with hauling freight in mind, with clamshell nose-loading doors and a ramp up which tanks, trucks, and other vehicles could be driven. First flown on November 27, 1949, the C-124 would be used by several commands but would find its niche with the MATS.

When the Korean War came, the hero of the Berlin Airlift, Major General William H. Tunner, was called upon to organize a Combat Cargo Command (Provisional) for the Far East Air Force. There once again the missions of strategic and tactical airlift converged in a combat role, as indicated in a previous chapter. The Combat Cargo Command made history with its air drops to Marines defending their positions at the Chosin Reservoir, dropping 140 tons of ammunition in two days in late November 1950. When the Marines hacked an airstrip out of the frozen soil, Combat Cargo Command aircraft landed 221 times, bringing in 273 tons of supplies and airlifting out 4,600 wounded.

MATS support grew steadily during the Korean War; early on, it averaged 2.5 tons per day, and by the end of the war the average rose to 106 tons. The combined total of 212,034 MATS sorties, tactical and strategic, delivered more than 391,763 tons of cargo, 2.6 million passengers, and 310,000 medevac patients.

It was an incredible effort, rewarded immediately after the war by another catastrophic drawdown and a continued forced reliance on piston-engine transports. In peacetime, MATS was placed under attack by civil airlines that wanted to carry passengers and freight for the services. Congressman Daniel Flood, a flamboyantly mustachioed Democrat from Pennsylvania, attacked MATS as a "billion-dollar boondoggle" and offered legislation thoughtfully prepared by the Air Transport Association—a lobby for the airlines—to remedy the situation. Flood called for 40 percent of all military passenger traffic and 20 percent of military cargo to be carried by civil airlines. The self-interest of the legislation was exposed almost immediately, for when the Taiwan crisis occurred in 1960, the airlines refused to bid on augmenting MATS service because the emergency had occurred during the profitable tourist season. The lesson was obvious: don't have emergencies during the tourist season.

Fortunately, farsighted Congressmen such as Representative L. Mendel Rivers, a Democrat from South Carolina, became advocates for MATS, particularly its requirement for modern aircraft. The combination of this congressional backing and the demonstrated ability of MATS to respond to crises in the Middle East, the Far East, and Africa resulted in the first articulation of a national military airlift policy in February 1960. A report entitled "The Role of Military Air Transport in Peace and War" by Secretary of Defense Neil McElroy resulted in nine "Presidentially Approved Courses of Action." The most important of these provisions directed that the Civil Reserve Air Fleet (CRAF) program would augment the military's airlift capability. MATS was required to provide the essential "hard-core" element of air transport. To accede to the demands of the civil carriers, MATS agreed to reduce its regularly scheduled fixed routes (much to the dismay of "space-available" travelers!) and to procure additional airlift through negotiated contracts. Another provision, the closest to the hearts of Air Force leaders, required that MATS begin to modernize its fleet of aircraft. An effective compromise, McElroy's report defined civil and military airlift responsibilities for the next twenty-seven years, until President Ronald Reagan's National Airlift Policy statement of June 1987.

A congressional subcommittee headed by Representative Rivers directed the Air Force to modernize its airlift. In the 1959 appropriations bill, Public Law 86-601, of July 1, 1960, required that the USAF spend $310.7 million for airlift, including $140 million for Lockheed C-130Es. The C-130 Hercules would set records for production longevity, utility, versatility, and dependability.

The expansion of MATS and the acquisition of modern aircraft coincided with the Kennedy administration's adoption of the concept of "flexible response." Clearly, if highly mobile armed forces were to be used on a moment's notice in any part of the world to stem Communist aggression, a strong MATS was required. Fortunately, the modernization came at a time when airlift technology, profiting by the power of the jet engine, the aerodynamics of jet aircraft, and the relatively new discipline of materials handling, had made it possible to create superlative aircraft with matching ground facilities.

The first foray by MATS into turboprop aircraft had not been happy. The Douglas C-133A Cargomaster, the first USAF turboprop-powered transport, made its initial flight on April 23, 1956. A handsome shoulder-wing aircraft with four Pratt & Whitney T34-P-3 turboprop engines, it could carry 52,000 pounds of cargo over a 4,000-mile route at a 323-mph cruising speed. Fifty aircraft were procured, but they were troubled by a series of enigmatic accidents and, after only ten years of service, by chronic fatigue problems.

Fortunately for MATS, when funds for expansion were made available, a trio of outstanding aircraft were in the wings, all strongly supported by Secretary McNamara. He advocated obtaining ninety-nine Lockheed C-130Es and thirty Boeing C-135s and initiating development of what would become the Lockheed C-141, another stellar performer.

These clearly superior aircraft raised MATS capability to a new level. When the full complement of 284 C-141 StarLifters had entered the fleet by 1968, MATS airlift capability was tripled, literally revolutionizing intertheater airlift. As a simple comparison of the inherent advantage of a swept-wing jet over a conventional piston-powered aircraft, the venerable Douglas C-124 required ninety-five vibrating flying hours to go from Travis Air Force Base, California, to Saigon and return, carrying a 20,000-pound cargo. The C-124 had a limited daily utilization rate, and thus required thirteen days to make the trip. In contrast, the comfortable C-141, flying above the weather, could carry the same amount of cargo the same distance in thirty-four flying hours. In addition, the C-141 used the innovative 463L materials handling system, which permitted the aircraft to off-load 68,500 pounds of cargo, refuel, and reload in less than one hour. Old Shaky drivers could only shake their heads in amazement and hope that their squadron would soon reequip with StarLifters.

AIRLIFT IN VIETNAM

MATS entered the Vietnam war with an obsolete fleet equipped as follows:

Equipment	Number of Squadrons
Boeing C-135 Stratofreighter	3
Lockheed C-130 Hercules	7
Douglas C-133A Cargomaster	3
Douglas C-124 GlobemasterII	21

There were two keys to the strategic and tactical modernization programs. The strategic key, the Lockheed C-141, became operational in April 1965. In terms of sheer performance, the Boeing 707 was 50 mph faster, carried 20,000 pounds more cargo, and flew 500 miles farther, but the C-141 was designed to carry outsized cargo and to work with the 463L system. The result was a superb transport that arrived just in time to support the endless logistic requirements of the Vietnam War.

The tactical key was the Lockheed C-130, which went from strength to strength as it was employed in greater numbers for a wider variety of tasks. Other aircraft were also useful. The C-123, with its short-field capability, was a real workhorse. The de Havilland C-7 Caribou, which had been accepted from the Army only because it was part of the package of assuming the theater airlift responsibility, was also well liked by its crews because it was rugged and simple to maintain. Even the venerable Douglas C-47 did yeoman work, particularly in the dreary task of resettling Vietnamese civilians into protected enclaves. But in the exacting matter of moving tons of equipment over short and medium ranges, too often right into the teeth of enemy flak, there was nothing to compare with the C-130. In terms of efficiency, 100 C-130s could do the work of 1,500 C-47s.

The prototype YC-130 made its first flight at Burbank, California, on August 23, 1954, and the first operational C-130A was delivered to the Air Force on December 8, 1956. The basic design was excellent, and it has been improved over the years as newer models came into service. It is still in production and more than 2,100 have been sold to operators in sixty-five countries. Versatility is the hallmark of the C-130, and its roles include those of personnel and cargo transport, gunship, remote piloted vehicle and drone launcher, search and rescue, tanker, satellite recovery, weather reconnaissance personnel recovery electronic countermeasures, Antarctic delivery, airborne communicator, relay, medical evacuation, maritime patrol, special operations, hurricane hunting, and many more.

The performance of the Hercules was remarkable at the time of its debut, and is still so competitive today that the replacement aircraft for the C-130 seems to be destined to be another, the C-130J. A typical C-130E has a top speed of 384 mph, a service ceiling of 23,000 feet, and a range of 2,420 miles.

And as the C-130 was the right plane for the task, so was General Howell M. Estes, Jr., MATS/MAC Commander from 1964 to 1969, well qualified to meet Vietnam's airlift challenge. With a breadth of experience that ranged from the cavalry to the atomic tests at Eniwetok, Estes drew on the lessons learned in World War II and Korea to establish quick turnaround procedures (called "Quick Stop" and "Quick Change") along with specialized maintenance teams to get aircraft airborne using waivers, one-time flight permits, and even the bête noire of maintenance officers, the cannibalization of parts from spare aircraft. "Red Ball Express" methods used by General Tunner on the Hump airlift were revived; MATS guaranteed that important parts and equipment would move to their destination within twenty-four hours of reaching their aerial port of entry (APOE). Estes also established a command post system that formed an integral chain from MATS Headquarters to the lowest field echelons to monitor airlift operations.

MATS was redesignated Military Airlift Command (MAC) in January 1966, an important step that would lead in eight years to establishing a genuine unified control over airlift assets, strategic and tactical. In October 1966, the 834th Air Division was formed under Seventh Air Force at Tan Son Nhut Air Base, Saigon. The 834th's Airlift Control Center (ALCC) used fourteen airlift control elements (ALCE) at fourteen different aerial port locations. In addition, there were forty smaller detachments located throughout the country. The result was a far firmer control of airlift assets and the more rapid movement of troops and freight. The timing was perfect, because in Vietnam the strategic and tactical missions again converged.

STRATEGIC AIRLIFT IN VIETNAM

The war in Vietnam was odd in many ways, none more so than the standard mode of travel of soldiers to the war. In World Wars I and II, soldiers went to war jammed in the holds of freighters for endless days; the average soldier going to Vietnam went in style. No less than 91 percent of the personnel arrived in Southeast Asia aboard commercial jet transports, complete with reclining seats, helpful flight attendants, cocktail service, and, most of all, airconditioning. All who stepped off the aircraft gasped in disbelief at the humid air smoking from Tan Son Nhut's blistering tarmac. The airliner represented home and comfort; the stinging heat of Vietnam symbolized war with all its loneliness and discomfort.

Yet, such was American air superiority that the civil airliners could operate in Saigon with virtual impunity. The Tan Son Nhut flight line was crowded with the airlines of many nations; sometimes an airliner from Honolulu would be parked next to another whose flight had originated in Hanoi.

Commercial airlift also carried about 24 percent of the cargo. The rest was brought in by MAC aircraft, supplemented by Air National Guard and Air Force Reserve units flying venerable Boeing C-97 Stratofreighters, Fairchild C-119 Flying Boxcars, Lockheed C-121 Super Constellations, and, of course, the ubiquitous C-124s. The excellence of their effort influenced in no small way the thinking of President Nixon's Secretary of Defense, Melvin Laird, the proponent for the "Total Force" concept of bringing Guard and Reserve forces to the same level of skill, responsibility, and utilization as the regular Air Force.

Timing is everything, and the C-141 was introduced just in time to provide the capacity to execute President Johnson's decision to begin a massive buildup of American ground forces. Traffic to Southeast Asia grew from a monthly average of 33,779 passengers and 9,123 tons of cargo in 1965 to 65,350 passengers and 42,296 tons of cargo monthly at the height of the buildup in 1967.

At the same time, strategic airlift proved itself capable of combat operations, beginning in 1965 with Operation Blue Light, when 2,952 troops and 4,749 tons of equipment of the 3rd Infantry Brigade, 25th Infantry Division, was airlifted from Hickam AFB, Hawaii, directly to Pleiku. The movement, conducted by a mixture of C-141s, C-133s, and C-124s, cut days off the planned schedule and was called by General William Westmoreland "the most professional airlift I've seen in all my airborne experience." It was the precursor of even more ambitious efforts that would continue even after the war was long lost.

The next major intervention of strategic airlift into tactical warfare occurred with Operation Eagle Thrust in November 1967. This time MAC carried 10,365 troops of the 101st Airborne Division, along with 5,118 tons of equipment, directly from Fort Campbell, Kentucky, to Bien Hoa Air Base, South Vietnam. The steady stream of C-141s used all of MAC's new tactics, including expedited off-loading that reduced average C-141 ground time at Bien Hoa to an incredible 7.4 minutes. All told, in 1967 MAC strategic airlift carried out tactical missions to transport 141,113 tons of cargo and 345,027 passengers within the Southeast Asian theater.

During the January 1968 Tet offensive, MAC handled a double emergency with Operation Combat Fox. Troops and supplies were airlifted to South Korea from the United States, Japan, and Southeast Asia in response to the USS *Pueblo* crisis and to South Vietnam because of the Tet offensive. The combined effort was more than twice the size of Eagle Thrust.

There were four additional major requirements for strategic airlift in Southeast Asia, three terribly sad and one poignantly heartwarming. The first was during the North Vietnamese Easter offensive in 1972, when TAC made the largest single move in its history to provide air power to stiffen the faltering

South Vietnamese forces. The second was rapid withdrawal of American and Allied personnel and equipment after the Paris cease-fire had been signed.

The third was the memorable Operation Homecoming, the repatriation and rehabilitation of the prisoners of war. No one who was there, or who watched on television, will ever forget the expressions on the faces of the men returning to their homeland and to their families after years of torture and neglect. If MAC had existed for no other reason but this one mission, it would have been sufficient. The Military Airlift Command functioned so well in Homecoming that it received the McKay Trophy from the National Aeronautic Association.

The fourth major airlift event was Operation Frequent Wind, the evacuation of American and foreign nationals (as well as loyal South Vietnamese citizens) from Saigon in late April 1975 as the North Vietnamese approached. Operation Babylift, the evacuation of South Vietnamese orphans, was a subset of this mass turmoil, and was marred by tragedy on the very first mission. The new Lockheed C-5A Galaxy had been introduced into the Southeast Asian conflict in May 1972. A C-5A was designated to transport 314 orphans from Tan Son Nhut to Clark Air Base in the Philippines on April 4, 1975. A massive decompression critically damaged the aircraft, causing it to make an emergency forced landing in which 138 adults, children, and babies perished. It was without doubt the nadir of MAC effort in Southeast Asia. The remaining Babylift sorties went off without a hitch, and by May 1975, 1,794 Southeast Asian orphans had been transported to their new American families in the United States.

The versatility, speed, and capacity of the combination of three Lockheed transports—C-130, C-141, and C-5A—drove home the point again that there was no real difference in the missions of tactical and strategic airlift, but it still would take much agonizing to finally cut the cord and combine the two missions.

If final evidence of the capability of strategic airlift to participate in tactical operations were necessary, it came in the fall of 1973. The wind-down in Southeast Asia had been proceeding more or less on schedule when the Yom Kippur War erupted in the Middle East on October 6. A simultaneous attack on Israel by Syria and Egypt caused a massive loss of Israeli materiel. President Nixon directed that an aerial resupply operation, Operation Nickel Grass, begin on October 13. For thirty-two days, MAC C-141s and the new C-5As carried everything from tanks to missiles directly to Lod International Airport at Tel Aviv, the veritable heart of the battleground. Just a few hundred miles away, a similar stream of Soviet Antonov An-12 and An-22 cargo planes was bringing in similar material to Egypt and Syria. It was an airlift air race, with MAC planes faced with a 6,500-nautical-mile one-way distance, compared to only 1,700 miles for the Soviet aircraft.

MAC won the race handily, the C-5s turning the tide of war by bringing in quantities of vitally needed M-60 and M-48 tanks, 175mm cannons, 155mm howitzers, CH-53 helicopters, and even McDonnell Douglas A-4 fuselages. Israel's prime minister, Golda Meir, described the grateful feelings of her people when she said, "For generations to come, all will be told of the miracle of the immense planes from the United States."

In its first major exercise, the huge Galaxy had carried 50 percent of the cargo while flying only 25 percent of the missions. It was an aircraft whose time had come, despite the controversy that had swamped it almost since the December 22, 1964, announcement by Secretary McNamara that it would be built.

The competition for the contract had been held under the McNamara-inspired Total Package Procurement (TPP) concept. Manufacturers were required to compete for an entire program, including R&D, testing, evaluation, and production, with clearly established price, delivery schedule, and performance commitments. It seemed a marvelous way to "hold the contractor's feet to the fire," but it was impossible to manage in practice. Air Force–dictated changes in requirements, equipment, performance, and, most especially, in the congressionally driven numbers procured played havoc with the program. There were other typical problems as well—overoptimistic cost estimates on the part of the contractor, a rise in the consumer price index, and an inflexible approach to the TPP contract resulted in a huge cost overrun and an aircraft whose maximum life was only about 25 percent of that forecast. When Lockheed saw that the weight of the aircraft was going to exceed specifications, it offered design changes that would not lower the weight, but would increase performance to meet the specifications. The Air Force, perhaps intimidated by Secretary McNamara's insistence on the matter, was unwilling to buck the DOD interpretation of Total Package Procurement policies and refused. The result was a massive engineering effort to reduce weight by means as primitive as drilling "lightening" holes in wing members. As a result, the C-5A emerged with an estimated 8,000-hour life, rather than the desired 30,000-hour life. The changes to the wing structure, which included a new method of riveting, resulted in fatigue cracks early on. The C-5A became the whipping boy of critics of defense procurement, who used it to symbolize government procurement ineptitude and waste, much as the Boeing 747 in its early days was criticized by airline commentators for being too big, too costly, and designed for a market that did not exist.

Nonetheless, from December 17, 1969, when MAC received its first C-5A, through the many modifications that have led to the present fleet of C-5Bs, the aptly named Galaxy has more than paid its way. With its 222.7-foot wingspan and basic mission weight of 712,000 pounds, the C-5 could cruise at 440 knots for 5,500 miles, and an in-flight-refueling capability gave it unlimited range. Despite the vicissitudes of its development, the C-5 was the right aircraft for military airlift, just as the 747 was the right aircraft for the airlines.

TACTICAL AIRLIFT IN VIETNAM

The efforts begun with Operations Farm Gate and Muletrain were soon extended to a massive tactical airlift system that covered Southeast Asia. Quantitatively, the tonnage handled by the aerial port system grew from 30,000 tons per month in early 1965 to a peak of 209,000 tons per month in March 1968. In 1967, the assistance given by strategic airlift diverted to tactical missions freed up a number of C-130s to be employed as virtual assault transports. The C-130s proved to be so effective in short-field work that they became the basic fixed-wing foundation for General Westmoreland's airmobile tactics against the Viet Cong.

The versatile Hercules could haul entire units, with their complete equipment and thousands of tons of supplies, into forward airstrips for large, long-term search-and-destroy missions. The major offensives conducted in Cambodia in 1970 and Laos in 1971 were possible only because the C-130s could deliver supplies—and especially fuel for the helicopters—to forward sites. The requirement for parachute assault, with all its attendant problems of wind and communications, gave way to fixed-wing and rotary-wing airlift operations.

The tactical airlift capacity served as a mobile fire brigade and permitted a far better utilization of troops. Because the transports could provide quick reinforcement, the defensive garrisons in many areas of this "war without fronts" could be minimized. The C-130s resupply capabilities were tested to the maximum during the December 1967 Khe Sanh airlift. Employed to land amid heavy Communist fire, the C-130s would wend their way down through the clouds into the canyon where Khe Sahn was situated, using a steep approach to minimize the time the North Vietnamese had to fire on them. Later, when the North Vietnamese had approached so close to the Khe Sanh perimeter defenses that it was impossible to land, the C-130s used airdrops and the low-altitude parachute extraction system to deliver critical supplies.

The tactical airlifter's work was hot, hard, and heavy. It was also extremely dangerous. In the eleven years from 1962 to 1973, fifty-three C-130s, fifty-three C-123s, and twenty C-7s were lost in action, with 269 crew members killed or missing in action.

MAC aircraft provided a thousand other services, from ferrying troops to embarkation ports at the end of their tours and bringing in supplies to Montagnard villages to carrying Bob Hope's annual shows. In a way, the variety of duties, from death-defying to pleasure-bringing, was a symbol of MAC's resourcefulness. Blessed with new aircraft, it promptly overworked and undermaintained them to get the job done; blessed with good people, it drove them till they dropped from fatigue, for there were never enough air or ground crews to meet the demands of the war. Yet the airlift job in Southeast Asia was done superbly, and the people who did the work were justly proud.

MAC SEARCH AND RESCUE IN VIETNAM

Despite excellent experience gained in Korea in search and rescue techniques, the many years of low budgets had reduced USAF search and rescue capability to the minimum level needed for local-base rescue efforts. The war in Southeast Asia put a strain on resources, so during the first two years of operations in Laos, there was no integral USAF capability and reliance was placed on the men and planes of the CIA airline, Air America. They did excellent work, even though search and rescue for the Air Force was not in their job description.

In South Vietnam, the first USAF air rescue team arrived at Tan Son Nhut on January 10, 1962; it consisted of three officers, three airmen, and no aircraft. Part of the Military Air Transport Command, it received equipment and men slowly. By January 1965 there were helicopter detachments at Ben Hoa and Da Nang in South Vietnam and Udorn, Nakhon Phanom, Takhli, and Korat in Thailand. The fixed-wing aircraft allocated to the task were old Douglas C-54s as airborne command posts and Grumman HU-16 Albatross amphibians. The duty helicopters were the tiny Kaman HH-43B, which had a very limited range.

The requirements for rescue accelerated with the tempo of the war, and Rolling Thunder greatly increased the need. In January 1966, the 3rd Aerospace Rescue and Recovery Group at Tan Son Nhut became the primary rescue agency in Southeast Asia, with responsibility for planning, organizing, coordinating, and controlling rescue operations from the Joint Search and Rescue Center at Seventh Air Force.

A basic technique was developed. On receipt of word that an airman was down, a Douglas A-1E Sandy was dispatched to the area to search for the downed pilot. Meanwhile, another Sandy was sent to escort two helicopters to the scene. One helicopter would go in to make the recovery while the other stood by to lend aid if required. Fighters accompanied the formation to act as ResCAP (rescue combat air patrol). When the first A-1E located the survivor, it determined whether he was in a hostile area. If so, the A-1Es and the fighters worked the area over with cannons, rockets, and other ordnance to neutralize it. When it was safe, the helicopter went in to rescue the downed airman.

Over time, both the techniques and the equipment were radically improved. The Sikorsky HH-3E Jolly Green Giants and their successors, the HH-53 Super Jolly Green Giants, were specially tailored for the role, having the range, armor, firepower, and rescue equipment to do an outstanding job. The faithful Lockheed Hercules was adapted as the HC-130, able to fly swiftly over long distances carrying the electronic gear with which to find the downed pilots. To almost everyone's amazement, the HC-130s and HH-3Es were adapted to permit in-flight refueling. Night-rescue equipment was developed, including low-light-level television, night-vision goggles, automatic Doppler navigation systems, terrain avoidance radar, and electronic location finders.

Individual crew members were given instruction on rescue during their survival school training, learning how to use special flares to penetrate the jungle canopy, setting up the special personal codes based on information only the crew member would know (e.g., "What was the name of your father's favorite dog?") that would be used to verify their identity when they were located, and training on escaping on the jungle penetrators let down by cable from the helicopters.

The annals of the search and rescue efforts are studded with achievements, ranging from routine extractions near ground bases in South Vietnam to the sensational recovery of Captain Roger C. Locher after he had spent twenty-three days on the ground evading enemy capture, and the gallant, if fruitless, rescue attempt at Son Tay.

Operation Kingpin was planned by the JCS in the summer of 1970, with the aim of rescuing a hundred or more U.S. prisoners of war held at the Son Tay prison 23 miles west of Hanoi. Fifty-six volunteers were selected from the Special Forces and the Rangers; their task was to overwhelm the prison force in a swift assault. Helicopters would be standing by to pick up the rescued prisoners while USAF and Navy aircraft attacked Hanoi to divert attention.

The raid was carried off on the night of November 20–21, with precision and élan; the North Vietnamese forces, including Soviet or Chinese troops there for training, were quickly overcome. Unfortunately, reconnaissance photos had not revealed that massive flooding had caused the POWs to be moved from the prison the previous July 14.

Only one man was injured in the raiding party, and although no prisoners were rescued, the raid had a positive effect on prison life. The morale of the prisoners of war was improved, and the North Vietnamese, suddenly aware of their vulnerability, permitted prison conditions to improve somewhat. The biggest dividend was that many prisoners who had been confined alone or in small groups throughout the country were concentrated together in the prison called the Hanoi Hilton. As atrocious as the conditions were there, the prisoners now had others to encourage them throughout their confinement.

During the years from 1964 to 1973, the search-and-rescue team functioned superbly, as it saved 3,883 lives, including 2,807 from the U.S. military, 555 allied airmen, and 521 civilians. In the process, forty-five rescue aircraft were destroyed and seventy-one crew members killed.

The very concept of search and rescue is noble, but even this nobility was transcended by the way in which the entire USAF gave a search and rescue effort first priority, breaking off wartime missions to assist, if necessary. The knowledge that no effort or expense would be spared to rescue them gave downed crewmen a sense of hope that enabled them to survive in otherwise desperate situations.

UNITY AT LAST

All the empirical evidence on airlift in the war in Southeast Asia pointed to the inescapable conclusion that centralized control of all airlift assets was essential. On July 29, 1974, Secretary of Defense James R. Schlesinger directed that the Air Force consolidate all military airlift forces under a single manager by the end of the fiscal year 1977, and that this single manager provide airlift for all the armed services, including the Navy and the Marine Corps.

One month later, Air Force Chief of Staff General David C. Jones announced that the Military Airlift Command would be the designated single manager. The Tactical Air Command transferred all of its airlift units, including Air National Guard and Air Force Reserve units, immediately making MAC into the world's largest single airlift organization. The expanded size and the demonstrated importance of the airlift operations in Vietnam and Israel resulted in the Joint Chiefs of Staff authorizing MAC to have specified command status, which became effective on February 1, 1977. As such, the MAC commander could now deal directly with the Secretary of Defense and the Joint Chiefs of Staff and have a direct communication link with the National Command Center.

From the verbal orders in 1941 that created the Air Corps Ferry Command to the considered designation of MAC as one of the three specified commands, military airlift had come a long way. The latter part of that journey had been made easier by the research and development undertaken by the Air Force Systems Command and its predecessor organizations.

AIR FORCE RESEARCH AND DEVELOPMENT

Hap Arnold's beloved research and development torch was ignited by Theodor von Karman and then passed to the most appropriate hands, those of Bernard Schriever. Schriever's success with ICBM development and related space programs came because he was a brilliant manager who knew how to get along in the three communities he needed: military, scientific, and industrial. He infused these three sometimes antagonistic groups with a sense of national policy issues and the overall military strategy of the United States, and he saw to it that the interchange among the three elements was not confined to purely technical considerations. As a result, he elicited far more from the scientists and engineers involved than if he had attempted to limit them to their specialized fields. Further, he became a beacon for thousands of young scientists and engineers who valued enormous responsibility early in their careers more than higher salaries. All of these leadership concepts deserve review and emulation today, when top R&D leadership has been moved almost entirely away from the services that use the weapons and into the hands of politically appointed "temps" for whom a position in R&D is only a way station on a career path to higher office.

Schriever became Commander of the Air Research and Development Command (ARDC) in April 1959, and very shortly thereafter he established the Air Force Research Division (AFRD) under Brigadier General Benjamin G. Holzman. The AFRD was dedicated solely to basic research, which Schriever considered vital to the Air Force's overall research program.

ARDC's primary mission was the research and development of weapon systems. Its counterpart command, the Air Materiel Command (AMC), was responsible for the procurement of the weapon systems and their lifetime support in terms of maintenance, spares, and engineering changes. Naturally, ARDC and AMC had a long history of rivalry and overlapping functions that reached back to 1917 with arguments between the Engineering Division and

Materiel Division at McCook Field, and with mission conflicts with Fairfield Depot. Over the years, there were various changes in names and functions, and by the 1950s, coordination was enforced by having personnel from both commands man the weapon system project offices (WSPOs). Schriever found the duplication intolerable and suggested that ARDC acquire all of the Air Force's research and development activities, as well as AMC's procurement responsibilities. Despite stiff opposition from AMC and from the Chief of Staff, General White, Schriever persisted.

Time was on his side. Concerned about the Soviet advances in space, the new Kennedy administration offered Secretary of the Air Force Eugene M. Zuckert an incentive to consolidate ARDC and AMC. He was informed by Deputy Secretary of Defense Roswell L. Gilpatrick that if the "Air Force cleaned up its act" (i.e., got its ARDC/AMC conflict resolved), it would be awarded the coveted space mission. Zuckert accepted with alacrity, and General White now saw the merit in Schriever's proposal for ARDC to do some acquisition of its own—that of AMC's role in procurement.

With only two weeks' notice to the affected personnel, on April 1, 1961, ARDC and AMC were redesignated Air Force Systems Command (AFSC) and Air Force Logistics Command (AFLC), respectively. The timing was exquisite, for on April 12, the Soviet Union launched the 5-ton Vostok spacecraft into an orbital flight with Major Yuri Gagarin aboard. American pride twitched like a galvanized frog leg, and the Kennedy administration recognized that funds had to be allocated to recover the nation's position as the leading technological power. Vostok spacecraft would ominously pass over American cities with impunity hundreds of times in the future—at a distance far closer than any enemy ship or aircraft would have been allowed to pass.

Schriever, promoted to full general, assumed command of AFSC and launched it on a course that would be responsible for the remarkable flowering of research and development within the USAF and led directly to the creation of the advanced weaponry and intelligence-gathering systems that won the Persian Gulf War. AFSC also achieved a goal beyond its dreams or its plans by fielding weapon systems that convinced the Soviet Union that it had irrevocably lost the race for supremacy in military arms.

In the reorganization, Schriever suffered one disappointment. AFRD, devoted to basic research, was removed from his command and turned into the Office of Aerospace Research at Air Force Headquarters. But he could look with pride at his newly created divisions that reached across the spectrum of aerospace research, strengthening AFSC's scientific base. They included the Aeronautical Systems Division, Ballistic Systems Division, Space Systems Division, and Electronic Systems Division, all supported by special advisory groups, including civilian think tanks such as MITRE and RAND. Over the following three decades, these units would produce not only the weapon systems to defend the free world, but also many of the principal leaders of the USAF. With his new organization, Schriever had made it possible for individuals known for their engineering and scientific ability rather than their combat record to reach four-star rank, and the Air Force and the country was richer for it.

Secretary Zuckert and Schriever worked well together, perhaps in part because of their differing relationships with Secretary McNamara. The relationship of Zuckert and McNamara is a textbook example of bureaucratic incest. Zuckert had recommended McNamara to succeed him as Assistant Secretary of the Air Force (Management). McNamara had picked Zuckert to be Secretary of the Air Force. Each man obviously admired the other's capabilities. But Zuckert, who had been Stuart Symington's assistant secretary from 1947 to 1952, became intensely frustrated that McNamara's takeover of power had reduced the importance and influence of the office of Secretary of the Air Force. Often driven to the point of resignation, Zuckert could not conceal his antipathy for McNamara, who responded in the most annoying manner: not appearing to notice.

Schriever, who yielded to no man in his dislike of McNamara's methods, nonetheless had the quick wit and the credentials to get along with him on his own terms, i.e., to prove his point by quantitative analysis rather than by the intuitive foresight that was actually Schriever's forte.

It thus happened that Zuckert would claim that his greatest achievement as Secretary of the Air Force was directing Schriever to embark upon Project Forecast, a comprehensive study of long-range technologies for the USAF. Zuckert asked Schriever to determine, via Project Forecast, what the existing state of U.S. air power technology was, what discoveries might occur in the next five to ten years and what they would lead to, and what, in general, science had to offer to help improve the Air Force's ability to do its mission. In effect, Zuckert was asking of Schriever the same thing that Arnold had asked of von Karman: to take an educated look into the future for the benefit of the USAF. Once he had delegated the task, Zuckert then withdrew to let Schriever do things his own way.

With his customary energy and using his familiar network of contacts, Schriever established that Project Forecast would examine the pace of technological advance in the years from 1966 to 1975, relate this change to the planning activities of the Air Force, and see what effect it would have upon USAF weapons.

Ultimately, almost 500 people contributed to Project Forecast, from twenty-eight separate Air Force organizations, thirteen major government organizations including the Army, Navy, and Marines, forty-nine subordinate government agencies, twenty-six colleges and universities, seventy corporations, and ten nonprofit institutions. The most stellar names in government, industry, and academia were included in the panels, which covered six broad subjects, including technology, threat, policy and military considerations, capability, costing, and analysis, evaluation, and synthesis.

It took a man of Schriever's talent to keep a gargantuan committee of this size and scope within bounds. He did so with the assistance of then Major

General Charles H. Terhune, Jr., who saw to it that the panels emphasized technologies that had a direct usefulness to national security and a high chance of practical success at a reasonable cost and were, withal, a major advance over presently forecast systems.

The deliberation process on which projects should be nominated for inclusion in the final report went through months of winnowing, filtering, arguing, fighting, and inevitably horse-trading, before its delivery in March 1964.

In the end, a number of technologies met all four of the qualifications outlined above. These included opening significant new areas in aircraft and engine construction through the introduction of high-strength boron filaments and oxide-dispersion-strengthened metals. The new materials resulting from this technology made huge cargo transports and vertical takeoff aircraft not only possible but inevitable. A candidate that proved more elusive was the use of liquid hydrogen as a fuel for long-distance reconnaissance planes; the report was closer to the mark with the suggested use of high-pressure oxygen-hydrogen engines for reusable space-launch vehicles. Of the host of promising aerodynamic subjects, the most prominent were reduced-laminar-flow-control and variable-geometry wings.

Other technologies recommended for pursuit included improved nuclear weapons, some with lowered radiation, some with low yields to reduce collateral damage, and some with enhanced radiation, fission-fusion nuclear devices that operated against enemy personnel but did little heat or blast damage to equipment. The last, often termed the "neutron bomb," ran into particularly vehement opposition by groups that somehow saw it as more immoral to kill and not damage material than to kill and damage material.

Perhaps the most important of the recommendations was within the realm of guidance technology. New optical image-matching techniques were developed that promised almost 100 percent accuracy to air-to-ground missiles.

These technologies, sometimes diluted, found their way into weapon systems, from the Lockheed C-5 to the current crop of precision-guided munitions. As valuable as these weapon systems were, their intrinsic importance was vastly overshadowed by the impetus Schriever, AFSC, and the methods of Project Forecast had given to Air Force research and development. Over the next twenty years, from the divisions and laboratories of AFSC and its successor organization, a myriad number of projects would emerge. They would make significant contributions to the magnificent series of spy satellites that gave the United States a measure of intelligence advantage never before possessed by one state over another, one that completely overshadowed the importance of even Ultra and Magic in World War II. Classified until 1995, the Corona spy-satellite program will be discussed at greater length in a later chapter.

Some of the possibilities seen in Project Forecast were adversely affected by the drain of funds to the Vietnam War, but for the most part, its predictions came true. Perhaps more important, it was the vehicle through which Schriever transformed AFSC from a mere merger of two commands into something far larger. He rejuvenated Air Force science by making AFSC the institutional framework in which civilian science and military objectives could be combined. He saw to it that a balance was struck by fostering both basic and applied research and by connecting science to national policy and *practical* military requirements.

By 1962, AFSC had grown in size to employ 26,650 officers and airmen and 37,000 civilians, handling some 54,000 contracts annually, and with a budget of $7 billion at a time when a billion was regarded as real money. AFSC worked closely with the National Aeronautics and Space Administration, supplying the Atlas rockets used for the Project Mercury flights, one of which lifted the spacecraft *Friendship 7* to altitude for Marine Lieutenant Colonel John Glenn's three-orbit flight.

The scope of AFSC's work from its inception was breathtaking. In addition to all the standard feverish activity at the Edwards Air Force Base Flight Test Center, it was engaged in a joint USAF-NASA-Navy test program for the X-15 rocket plane, which Major Robert White flew to a speed of 4,093 mph. It was bringing advanced Atlas, Titan, and Minuteman ICBMs into operational status. The Discoverer series of satellites were being launched, applauded for their ultimate impact on the environment—their real task would remain classified for three more decades. And there was so much more, not all destined for ultimate fulfillment, including the X-20 Dyna-Soar, which anticipated in many ways the space shuttle; the North American RS-70 Mach 3 aircraft; a fantastic "Facet Eye" camera; the completion of the Semiautomatic Ground Environment (SAGE) air defense network; the design and construction of BMEWS, the ballistic missile early warning system; the SPADATS (Space Detection and Tracking System) for monitoring all objects in space from satellites to space junk; and events less worthy of headlines but nonetheless vitally important, such as the creation of the Aerospace Medical Division at Brooks AFB, Texas.

The very complexity of AFSC, the urgency of its mission, and the success Schriever had experienced with the technique in creating the ICBM fleet made him push for concurrent rather than sequential development of the various projects. Strong views against this were voiced at the top, including those of the Director of the Directorate of Defense Research and Engineering (DDR&E), Harold Brown, then the third most powerful man in DOD. Schriever prevailed, and for the most part, concurrency worked. Another element of Schriever's program that met with opposition was the encouragement of federally funded basic research by industry. This proved to be one of the most valuable assets of the AFSC R&D program, but ultimately would be killed because it was an easy political target. Firms could be accused (no doubt justly in some instances) of using government funds to develop products that they would profit from commercially. The result was a progressive reduction of such funding to the point that it jeopardized the research and development effort.

AFSC was an administrative handful. Schriever's initial attempts to control AFSC as he had controlled BMD while creating the ICBM fleet proved to be unworkable because of the increasing demands for reports and briefings from higher levels in DOD. The effect was to create dozens of informal layers of management intended to ensure that the briefings that went to the top were sufficiently inclusive. Schriever's answer was to persuade DOD that AFSC's review methods would be so rigorous that only the critical decisions would have to be made at the top.

The war in Vietnam put a budget ceiling on AFSC's activities for almost a decade and placed a more critical focus on immediate operational problems that had emerged in conflict. An example of this work was the development of the previously mentioned TISEO (target identification system electro-optical), a telescope that gave the pilots a chance to identify hostile aircraft visually at a greater distance. At the same time, the AFSC fostered the "Rivet Haste" modification on the F-4E, which introduced leading-edge slats to provide more maneuverability and a better platform for internally installed cannon. Although immediate requirements like these had priority, long-range work went on for aircraft such as the McDonnell Douglas F-15 and the North American/Rockwell B-1 as well as for missiles such as the AGM-69A SRAM (short-range attack missile) and Maverick AGM-65 television-guided air-to-air "smart weapon," of which more than 100,000 would ultimately be made. Also in the works was the first of what would become known as a "force-multiplying weapon," the AWACS (airborne warning and control system), which would become operational in 1977 as the Boeing E-3A.

With peacetime budget restrictions, risk in weapon system development was feared more than delay. The concurrency concept fell into disfavor, so that development and production were severed and a return was made to traditional "fly-before-buy" methods. Development programs were carefully measured at specified milestones, with special emphasis placed on any critical subsystem development that seemed risky. The delays caused by innumerable reviews added billions to the cost.

For several years after the conclusion of the Vietnam War, funding for AFSC remained relatively constant at about the $7 billion level. By 1982, inflation and expanded needs had pushed it to about $26 billion. While aircraft such as the General Dynamics F-16 Fighting Falcon, the McDonnell Douglas F-15 Eagle, and the resuscitated Rockwell B-1B Lancer and missiles including the LGM-118A Peacekeeper (MX) and advanced medium-range air-to-air missile (AMRAAM) were being developed or brought into use, a tremendous proportion of AFSC effort was devoted to command, control, communications, and intelligence (C^3I), intelligence-gathering satellites, and other smaller, less publicly spectacular programs.

It is a credit to AFSC leadership, from Schriever to the present, that personnel selected for the command remained at a very high level of competence even though the programs were becoming both increasingly specialized and attenuated. Gone were the days when a fleet of ballistic missiles could go from drawing board to launch in four or five years. By the very nature of their complexity and because of the micromanagement of Congress, important programs now required an excruciatingly long time to execute. Because they were usually highly classified, the scientists and engineers working on them were not only denied public recognition for their efforts but were not allowed to publish in the standard journals so that they might enjoy recognition by their peers.

As a single example of the depth and breadth of the work accomplished at AFSC, it is instructive to look at a summary listing of the major work of just one division in the mid-1980s. At the Electronic Systems Division at Hanscom Air Force Base, Massachusetts, no less than 122 *major* programs were underway, ranging from the Boeing E-4 airborne command post, the Milstar satellite communications program, and an upgrade program for the Royal Saudi C^3 system to microwave landing systems (MLS) for airfields. In addition, other programs too classified even to be acknowledged by a public project name were also in progress.

The same depth and breadth of effort were to be found at all the other divisions, as well as at the engineering and flight test centers. The combined results of these concerted efforts would be made manifest in the Persian Gulf War, where all the efforts leading to the use of space in warfare would bring about a decisive conclusion and validate all of AFSC's efforts for the preceding thirty years.

When AFSC and AFLC came into being in April 1961, some observers felt that the Logistic Command had received the short end of the stick. Politically, that might have been the case, but in practical terms, the importance of AFLC grew over the years to overshadow its former stature as AMC. One somewhat ironic reason for this was that the versatility of modern weapon systems—combined with the staggering cost of their replacement—kept aircraft and missiles in service for much longer periods. Unlike the 1950s, when ten new fighters, five new bombers, and four new ballistic missiles were introduced, new weapon systems were now a rare event. Instead, older systems such as the B-52 required AFLC to modify them continuously to keep them in service. Over the years the average age of first-line U.S. fighters, bombers, and missiles grew to twenty, then thirty years—and it keeps on growing. Faced with this unprecedented challenge of maintaining an aging fleet that had long since outlived the manufacturers of many of its components, AFLC managed to keep the USAF ready for combat.

Ultimately, the downsizing of the Air Force had its impact on AFLC and AFSC, which were merged on July 1, 1992, in the Air Force Materiel Command (AFMC). The name and mission changed but the heritage remained the same.

LOGISTICS: ALWAYS VITAL

The World War II organizations that preceded the Air Materiel Command (AMC) and AFLC had a host of names and responsibilities and succeeded in supplying the needs of the United States Army Air Forces on a scale never before achieved in history. The supply organizations and industry worked so well together that by the time of the Japanese surrender, there were huge surpluses of every kind of materiel, including as many as 35,000 first-line aircraft.

The immediate postwar task for AMC was somehow storing and disposing of this vast quantity of materiel without incurring congressional wrath for wasting funds, or generating accusations of competing with industry. All over the country, huge disposal sites were established, ranging from thousands of acres of western land dotted with endless rows of surplus aircraft down to city-center salvage yards disposing of precious metals. AMC accomplished the task even as it, like the rest of the USAAF, suffered a demobilization meltdown. It had just about completed the work when the Korean War began, and the airplanes it had so laboriously readied for storage would now be hauled out and returned to service. The rehabilitation of stored aircraft was an outstanding success and laid the foundation for today's Military Aerospace Maintenance and Regeneration Center at Davis Monthan Air Force Base in Tucson, Arizona. There aircraft and other materials are still stored for future use while an annual profit is turned on the materials sold.

The advent of the Cold War brought extreme pressures on the science of logistics as practiced in the past. The demand that huge forces of the most sophisticated kind be kept in constant readiness was without precedent in U.S. history. The Eisenhower administration's concept of massive retaliation implicitly recognized that the outcome of a war between the United States and the Soviet Union would be decided in the first ninety days. This new view of warfare—with its attendant ballistic missile systems—made the past logistics concepts largely derived from U.S. Army needs, obsolete. Fortunately, AMC had the capable leadership of General Edwin W. Rawlings, one of the first Air Corps graduates from the Harvard Graduate School of Business Administration, where he received a master's degree in business administration, cum laude, in 1939.

Rawlings was to AMC what LeMay was to SAC and Schriever to AFSC. He used his education and experience to place the command on a businesslike basis, adopting a "systems approach," as Schriever had with AFSC. Rawlings made AMC look at logistics as a whole, with a concern not only for acquiring and storing the needed parts and materials but also managing the inventories, improving communications with the field, minimizing and mechanizing record-keeping, making sure that the transport system was swift, and, impelled by the space age, ensuring that new quality-control procedures were applied to both new and old suppliers. Rawlings backed the creation of LOGAIR, an AMC airline that used contractors to fly high-priority cargo in a network that tied AMC depots together with all Air Force bases.

The old Air Corps had been small, and men like LeMay and Rawlings knew each other well on both business and personal levels. In 1954, LeMay told Rawlings that if war came, SAC bombers and tankers would need fast, flexible emergency maintenance service from the very first day. Within two years, Rawlings had established a unique reserve "minuteman" force of 4,000 civilian personnel divided into sixty mobile maintenance teams, ready to leave their assigned jobs on a few hours' notice and deploy to forward bases to repair SAC aircraft. The concept was later extended to ADC, TAC, and MATS.

Every military leader from Sun Tzu on has known the critical importance of military logistics. Yet in times of budgetary crisis, the accountants nonetheless invariably turn to logistic line items for savings. Where deferring procurement of a new aircraft is immediately obvious to all, particularly the Congressman from the district where the aircraft is manufactured, deferring the purchase of spare parts can remain almost unnoticed—until they are needed.

Similarly, the pace of urgently needed modifications can be slowed to save money without its being particularly evident to anyone except those in units flying the unmodified equipment. The result can be a gradual increase in accidents.

The neglect of logistics is felt in other ways: reductions in personnel, freezing of grade levels, and the politically driven use of outside contractors for work best done in-house. All of these methods have been seen in the past and will be seen increasingly in the future.

Having noted this, the fact remains that the logistics effort of the United States Air Force functions at a high level of competence, thanks to the energetic management provided AFLC/AFMC over the years by the top leaders who succeeded Rawlings. These included such stellar general officers as William F. "Bozo" McKee, Samuel E. Anderson, Mark E. Bradley, Thomas P. Gerrity, Bryce Poe II, Henry H. Vicellio, Jr. and many others.

The primary mission of all the commands devoted to logistics—AMC/AFLC/AFMC—has been to see that all Air Force combat units are properly equipped and ready for action at all times, anywhere in the world. This involves procurement of spares and equipment by the billions of dollars each year, along with supply, maintenance, and transportation functions. From being the "supply shack" operation of World War II, where canny sergeants rat-holed extra spare parts to keep their own squadrons' airplanes flying, AFLC capitalized on Rawlings's precepts and met the space age head-on with computers to manage inventories, new standards for test and calibration equipment, new manufacturing techniques in the repair facilities, and new methods of materials handling.

At its inception, AFLC was divided into nine air materiel areas (AMAs), huge facilities with as many as 20,000 employees each, which were allocated missions based on both functional tasks and geographic areas of responsibility. (In today's multicultural world, it is interesting to recall that in those early days, members of the San Antonio, Oklahoma City, and Sacramento AMAs proudly referred to their units as "the brown-eyed AMAs" to reflect the high proportion of Hispanic workers.) A combination of economies demanded by the war in Vietnam, improvements in computer control of inventory, and the quicker movement provided by jet aircraft permitted a reduction in the number of

AMAs to five—Oklahoma City, Oklahoma, Sacramento, California, San Antonio, Texas, Ogden, Utah, and Warner Robins, Georgia. The end of the Cold War saw further reductions, and ultimately the creation of the current AFMC.

The scope of logistics command work became bewilderingly complex. In 1980, a "typical" year in terms of its having no active combat, AFLC supported 9,000 aircraft and 1,000 strategic missiles. More than 1.5 million items were procured and managed, while 1,800 aircraft and 4,700 engines were overhauled. The demand for technical support varied from the very lowest level—a request for a particular type of sheet-metal screw—to the most esoteric, as in a requirement for inertial guidance systems. As many as 10 million individual requests for service were fulfilled. With assets valued in excess of $70 billion and a total stock fund of $12.7 billion, AFLC was by any standard one of the biggest businesses in the world.

Like the rest of the Air Force, AFLC responded well to pressure. At the time of the Cuban missile crisis, the command distinguished itself by shipping 168,000 tons of freight to Florida bases by air, rail, and convoys of trucks impressed into service from all Air Force bases east of the Mississippi. The effort was so professionally done that almost all the supplies were in place before all of the combat units had assembled.

The Vietnam War placed massive demands on AFLC. Maintenance efforts were now required 10,000 miles from home. Aircraft were literally coming unglued in the moist tropic air, as when the electrical component potting material self-destructed in the McDonnell Douglas F-4s. A new challenge was faced when the Air Force suddenly acquired large numbers of aircraft that had never before been in the inventory, such as the Douglas Skyraider, that had to be upgraded to Air Force standards and spare parts and manuals provided. The familiar task of pulling aircraft out of storage also remained; instead of refurbishing B-29s for Korea, AFLC began restoring B-26s and T-28s for Vietnam.

The command responded to the emergency with Project Bitterwine, a vast expansion system involving twenty-seven bases: ten in South Vietnam, eight in Thailand, three on Taiwan, two in the Philippines, two on Okinawa, and two in Japan. Like some gigantic amoeba splitting, AFLC fought the war by replicating stateside capabilities in bases overseas. As might be expected, forced expansion caused its own problems, and emergency situations were met by using new procedures with new acronyms, always the fastest guns out of bureaucratic holsters. Rapid Area Maintenance (RAM) teams were rushed to Vietnam to expedite the removal and repair of battle-damaged aircraft. Rapid Area Transportation Support (RATS, no less) teams were dispatched to expedite deliveries through water and aerial ports. Rapid Area Supply Support (RASS) teams were rushed to handle the excessive buildup of supplies at bases not equipped to handle it and ensure that accurate inventory and accounting records were kept. These acronyms combined to keep another crucial acronym in check: throughout the Vietnam War, such emergency AFLC support kept the infamous NORS (not operationally ready, supply) for units in the field at the same low level as stateside units. The rate dipped to 2.5 percent—the lowest yet experienced Air Force–wide—and by 1968 reflected new techniques like resupply by airlift, improved computers, a vastly expanded depot repair system, and top-notch management.

The war forced AFLC to improvise, as it did the entire Air Force. One of the most successful programs was the conversion of Korean-vintage Fairchild C-119 Flying Boxcars to gunships for ground support in Vietnam. No one who had ever flown the "ground-loving" C-119 would have imagined that it could have been a success as a ground strafer, but modified with jet pods, armor, and side-firing miniguns, it was a deadly addition to the in-country war.

Nothing affected AFLC's tightly scheduled facility like the sudden discovery of a fatigue problem in an operational aircraft. In the past, the AMAs had responded to similar problems in the B-47 and B-52 in peacetime, but discovery of a potentially catastrophic fatigue problem in the tail of the overworked KC-135s caused an immediate furor. AFLC responded with Project Pacer Fin, carried out at the Oklahoma City AMA, where sixteen KC-135s were put through a modification every day for forty-five days. The AMA worked around the clock and sent teams to Southeast Asia to repair those KC-135s whose mission requirements kept them on station there.

The cost of the Vietnam War caused Secretary McNamara to seek economies, and in 1964 he found some of these with a reduction in the number of air materiel areas to five, with the AMAs at Rome, New York, San Bernardino, California, Mobile, Alabama, and Middletown, Pennsylvania, being closed. One result was a massive migration of functions, along with the associated personnel and equipment to the remaining AMAs.

AFLC had concentrated the preponderance of its efforts on supporting the fighting in Vietnam; with the end of the war, budget cuts forced an immediate contraction. Yet the requirement for supporting nuclear deterrence remained, and in the post-Vietnam period, the command refocused its attention on war readiness. The fall-off in personnel strength (which has always been about 90 percent civilian) was precipitous; in 1957, AFLC had reached a peak of 225,000 persons, military and civilian. Even at the height of the Vietnam War, when the demands for service were both more numerous and more urgent, management decisions had allowed this number to decline to 134,000. It fell further to 90,000 by 1976.

Some of the reduction in force stemmed from increased centralization of common logistic items under the Defense Logistics Agency. The loss in personnel was offset to a degree by the modernization of depot facilities—some of which dated to 1920—improved material-handling equipment, increased use of private contractors, and, advanced data processing.

Like the rest of the Air Force, new life was breathed into AFLC activities with the Reagan administration's expanded defense spending. The modernization of strategic systems and the acquisition of new equipment such as the Rockwell B-1B, Northrop B-2, McDonnell Douglas C-17, and Lockheed F-117 were a challenge that also brought acquisition reform issues to the forefront. Congressional attention was whipped to a new intensity by media coverage of cost overruns, spare parts pricing, and questionable contractor overhead charges.

Congressional reaction was severe and comprehensive; a massive volume of reform legislation was passed, with the inevitable effect of magnifying the micromanagement aspects of congressional oversight. The President's Blue Ribbon Commission on Defense Management, led by industrialist David Packard, conducted a probing examination that will be explored in the next chapter. Further emphasis on curing the problem was implicit in the Goldwater-Nichols Department of Defense Reorganization Act of 1986.

The external reactions were matched by internal reforms designed to fulfill the multiple recommendations of the Packard Commission, as it came to be known. Amid this Sturm und Drang of reaction to vitally needed reforms, AFLC and AFSC nonetheless managed to work together, providing the USAF with the advanced weaponry it would use with such devastating effect in the Persian Gulf War.

AFLC, like AFSC, was vitally affected by the change in focus demanded by the Vietnam War, the subsequent drawdown, and the demands for improvements. The two commands were able to meet the challenge because the Air Force had been foresighted enough to provide its personnel with the necessary training and education. Thus, just as the aircrews were educated to use the most sophisticated computer-controlled weaponry, so were the ground support people prepared to use the modern computer tools just becoming available. The general success of AFSC and AFLC in meeting the twin crises of modern war and modern equipment was made possible by the contributions of the two "school" commands—the Air Training Command and the Air University.

INVESTMENT IN EDUCATION

Education has been the saving grace of the United States Air Force, in terms of capability and its immutable corollary, the retention of qualified personnel. Rarely able to compete in salaries even at the entry level, the Air Force has managed over the years to provide educational opportunities—technical, academic, and professional—that have not only induced people to join and to stay in, but produced the doctrine and the leadership that have made the USAF what it is today. A particular facet of education's contribution has been the creation of the highly professional noncommissioned officer corps, of which more later.

The three principal educating organizations in the history of the USAF have been the Air Training Command (ATC), the Air University Command (AU), and the United States Air Force Academy (USAFA). The ATC and the AU were combined—again—in 1993, to become the Air Training and Education Command. The Air Force Academy is a direct reporting unit (DRU), not connected to a major command and reporting directly to Headquarters, USAF.

THE AIR TRAINING COMMAND

Like every other command, ATC has had to face a roller-coaster ride in funding, manning, and bases over the years, but during times of emergencies, the officers guiding ATC were presented with a recurring problem. Just when more instructors were required to build up ATC to meet the demand for more graduates, the combat commands raided ATC units at all levels for experienced personnel—and these, invariably, were the instructors. As a result, ATC has always found itself gearing up for an emergency just when it was itself at minimum strength.

Traditionally, the Air Training Command handled flying training, the Air Force Reserve Officer Training Corps, Officers Candidate School, Officer Training School, the technical training of all the myriad Air Force job specialties, and special schools like the famous survival schools. ATC also had a number of additional missions. One of the most important of these, and directly allied to the basic function of ATC, was the USAF Recruiting Service. Another important unit was the Community College of the Air Force. Added to these were the San Antonio Joint Military Medical Command, a consolidation of Army and Air Force medical facilities, and many subsidiary functions like contracting, real property maintenance, and foreign military training affairs. Of all these fundamentally important tasks, the image of flying training—the "Taj Mahal" at Randolph Field, students crawling into aircraft, formation flyovers—always seemed to epitomize ATC.

PILOT TRAINING

The Air Training Command demonstrated its remarkable flexibility during World War II, when it raised pilot training from a few hundred per year in 1939 to 87,283 in 1944. The boom was made possible in large part by Hap Arnold's foresight in establishing Civilian Pilot Training (CPT) schools that bore the brunt of the training burden throughout the war. Training was cut drastically after the war, with fewer than 400 pilots graduating in 1947. When the Korean War broke out, nine new contract flying schools were established to take up the slack. During the three years of the Korean War, ATC graduated almost 12,000 pilots—about six times the normal peacetime rate. After Korea, and until the defense cutbacks of the 1990s, the number of graduates from pilot training varied from a peacetime high of 6,159 in 1955 to a low of 1,081 in 1979.

Pilot training methods varied over time in the phases of instruction and to reflect phenomenal advances in the trainers. After World War II, primary training was conducted in the ubiquitous North American T-6 Texan, while in advanced training students used the multiengine North American TB-25, an unarmed version of the plane in which Jimmy Doolittle raided Tokyo. Single-engine students flew the North American F-51 (until 1952), the Lockheed F-80, and later the Lockheed T-33.

Variations on this system continued after the war until 1959, when contract training was ended and specialized training for single- or multiengine aircraft was stopped. The advent of jet aircraft caused further changes; by 1961, Consolidated Pilot Training was begun, with preflight, primary, and basic courses combined and instruction provided in Cessna T-37s—the long-lived "Tweety-bird"—and the equally durable T-33s. The Northrop T-38 Talon, a supersonic trainer, was introduced the same year.

The T-37s and T-38s *still* form the backbone of today's much-reduced pilot training system. Preflight screening was long accomplished in the Cessna T-41 Mescalero (a version of the standard civilian Cessna 172), now replaced by the Slingsby T-3 Firefly, imported from Great Britain. The indomitable T-37, first flown on October 12, 1954, still provides training during the basic phase, although it will ultimately be replaced by the winner of the long-contested Joint Primary Aircraft Training System (JPATS), the Raytheon (Beech) Mk II, a variation of the Swiss Pilatus PC-9.

The sleek looks of the T-38 Talon belie the fact that it first flew on April 10, 1959, and for many years provided advanced training for all pilots. A requirement for more specialized training eventually combined with a need to extend the service life of the T-38, and in 1993, a new system, Specialized Undergraduate Pilot Training (SUPT), was put into effect. In the new system, pilots selected for reconnaissance, attack, or fighter aircraft will continue to train in the T-38, while those selected for tankers, bombers, or transport planes will fly the Raytheon (Beech) T-1 Jayhawk. The T-1 is a militarized version of the Beech 400A, itself a variant of the twin-jet Mitsubishi Mu-300 Diamond.

The one constant in pilot training has been the high standards maintained; Air Force pilots receive the finest flight training in the world. Physical qualifications are very strict, and applicants are screened carefully for their flying aptitude. Flying training students are constantly evaluated for their proficiency, a practice that continues after graduation and for as long as the pilot continues to fly.

OTHER AIRCREW TRAINING

The training programs for other than pilot aircrew members changed over time as aircraft and systems modernized. The traditional disciplines of navigator, bombardier, and radar observer converged, and a new requirement for electronic warfare officers emerged. After thirty-seven years, navigator-bombardier training at Mather AFB, California, ended in 1984, with the responsibility transferred to SAC's Combat Crew Training School at Castle AFB.

Undergraduate navigator training also experienced great variations in the numbers trained. A high point of almost 40,000 was reached in 1944, followed by a precipitous dropoff to only eighteen in 1947. Over the years, the number of navigators graduating ranged from 500 to 5,400, depending upon the needs of the service and the degree of the perceived emergency.

Electronic warfare officers (EWOs) became increasingly important as the years elapsed. Originally confined to bomber crews, EWOs became vital first to aircraft dedicated to electronic warfare, such as the Douglas RB-66, and then to fighters, when the McDonnell Douglas F-4 and the F-105F two-seat fighters appeared. The importance of EWOs grew in direct proportion to the increasing sophistication and strength of enemy defenses.

The requirement for training one of the great combat crew positions, that of gunner, ended in October, 1991 when the Strategic Air Command phased out the tail gunner on the B-52. Gunners had been the backbone of the giant formations of bombers during World War II and had scored additional successes in Korea and even Vietnam. But time and technology finally put an end to a great and honored tradition.

TECHNICAL TRAINING

The methods for providing technical training have varied widely over the years to meet contemporary budget and personnel situations. ATC always preferred to give complete training in residence at established technical training schools, no matter whether in avionics, cryptology, meteorology, supply, or any of the hundreds of other specialties that have been required. Training suffers from the same ratchet effect that airlift, logistics, and all the other disciplines do. In periods of peace, there is little money, and schools are allowed to wither and almost die. When an emergency looms, schools are rapidly expanded, even as instructors are siphoned off to take frontline positions in their specialties. Then, usually just about when the emergency ends, training reaches its peak, with adequate staffs, good recruits, and sufficient time. At the end of the emergency, training is drastically reduced, and the cycle begins again.

The retention problem has similar impact. When the civilian economy is booming, it usually makes economic sense for a first-term airman to leave the service and use the training he or she has received in civil employment. ATC has tried a number of methods to deal with this situation. In "first job" training, a young airman is given just enough theoretical information to enable him or her to learn the skills to perform the mission with extensive "on-the-job" training. New airmen, just out of basic training and perhaps a short course in the specialty, were assigned to mentors on the job who would teach them the task. Unfortunately, were usually too few technicians in the field already. When faced with the alternatives of getting their basic jobs done or training new airmen, the mentors most often chose to do the former.

The results of first job training were unsatisfactory. During 1992—designated the Year of Training by then Air Force Chief of Staff Merrill A. McPeak—an in-depth review of the training process was made. One result of this was the merging of Air Training Command and the Air University Command into the Air Education and Training Command in 1993. Another result was the emphasis placed on resident training and a turn away from on-the-job training.

PROFESSIONAL MILITARY EDUCATION (PME)

Because Professional Military Education (PME) is a prerequisite for advancement, service members understand that failure to be invited to a PME course often portends a career plateau. Competition is keen, and the slots are limited.

The lineage of the Air University can be traced back to February 10, 1921, and the founding of the Air Service School at Langley Field, Virginia. The school was upgraded the following year to the Air Services Field Officers School (i.e., for officers of the rank of major through colonel) and was given the mission of "preparing senior officers for higher Air Service command duty."

An important change occurred in 1926, when the school was redesignated the Air Corps Tactical School (ACTS) and opened to all officers. It then became a military think tank and a hotbed of ideas. It was at the ACTS where firebrands like Claire Chennault became advocates of Billy Mitchell's concept of air superiority, and where other brilliant young officers like future generals Haywood Hansell, Harold L. George, Muir S. Fairchild, Laurence Kuter, and others became proponents of the bomber.

The ACTS set the pattern for the future, for it elicited from its select student body the doctrines, tactics, and strategies of air war that were employed in World War II and subsequently. Its faculty and graduates included future USAF Chiefs of Staff Carl A. Spaatz, Hoyt S. Vandenberg, Nathan F. Twining, Thomas D. White, and Curtis E. LeMay—a very distinguished alumni group.

World War II brought about a suspension of the Tactical School; it was discontinued on October 9, 1942. A series of wartime applied tactical schools replaced it, but it was not until March 12, 1946, that, recognized as a major command, one of the replacement schools, the Army Air Forces School, was redesignated the Air University.

Just as Arnold had insisted on invigorating the research-and-development process with unfettered thinking, so did the backers of the new Air University insist that it break away from traditionalism and avoid the rigid doctrine and formalized instruction of typical military schools. The first Commander of the Air University, Major General Muir S Fairchild, insisted, "This is not a postwar school system—it is a prewar school."

The Air University was intended as a single military organization free to concentrate all of its intellect and energies on the leadership, strategies, concepts, and doctrines necessary for winning future wars. The AU has become famous for its ability to continuously revise course content to keep abreast of new technologies. The international scene is closely monitored with classes, lectures, and field trips tailored to keep tabs on events as they develop.

The staff of the AU is made up of the most highly qualified military and civilian instructors available; unlike most similar institutions, the AU encourages a rapid turnover of the faculty to facilitate fresh views. The turnover is possible because the best and brightest of students are tapped to become instructors at the appropriate point in their careers.

The Air University takes its position as the designated keeper of professional military education very seriously, and now has fourteen major subordinate units, including the world-renowned Air University Library. The heart of its PME program for commissioned officers rests on the Squadron Officers School (for company-grade officers), the Air Command and Staff College (for field-grade officers), and the Air War College (for senior officers). There are reciprocal arrangements for similar-level education in other services, including those of our allies, and assignment to these schools was and is hotly coveted.

As important as these schools were in teaching doctrine, and ultimately in synthesizing experience for the creation of doctrine, they nonetheless did not have the overall impact on the Air Force that another branch of the Air University did: the Air Force Institute of Technology.

AFIT: CREATING AN AIR FORCE OF PROFESSIONALS

The tremendous impact that the Air Force Institute of Technology has had stems from the large number of students it has graduated. More than 266,000 personnel, mostly from the Air Force but also from other DOD organizations, have attended AFIT in the last seven decades. Chief of Staff General Charles Gabriel once stated, "The AFIT of today is the Air Force of tomorrow."

The overall goal of AFIT is to maintain itself as a world-class institution of higher education in defense science, technology, and management. AFIT came into being because the large number of technically qualified officers needed by the Air Force could not be attracted because of the low pay and demanding conditions of military service. The alternative was to select qualified applicants who had already chosen the service and provide the education. The result has been the creation of a highly professional corps of career officers.

In essence, the Air Force needed officers with academic attainments to match those of their industrial counterparts, and to obtain them it grew its own. Most of the officers selected for AFIT programs are bright young company- and field-grade officers who avidly desire graduate degrees. It is not an easy choice, for AFIT training, either in residence or at a civilian institution, is demanding, with heavy course loads and an unbending requirement for excellent performance. A "gentleman's C" is a sure ticket to removal and assignment to some less than desirable job.

The mission of AFIT is to support the Air Force through graduate and professional education, research, and consultation. The institute performs two closely related services, graduate-level education and professional continuing education, both drawing on similar academic resources. The graduate-level education is provided to carefully selected officers and Air Force civilian employees to give them the educational background to understand their technological and cultural environments and to analyze and attempt to solve the Air Force's problems. Professional continuing education programs are designed to satisfy specific Air Force and Department of Defense needs for special applications.

The roots of AFIT go back to 1919 and two remarkable men, Colonel Thurman Bane, who could set up a laboratory on one day and fly the Air Service's first helicopter on the next, and First Lieutenant Edwin E. Aldrin, a gifted administrator. Aldrin was also an expert test pilot, and was the father of Apollo 11 astronaut Edwin E. "Buzz" Aldrin, Jr., the second man to step on the moon.

On Bane's instructions, Aldrin did the legwork to establish the Air School of Application at McCook Field, Dayton, Ohio, in 1919, and then became a member of the first graduating class in 1920. In 1926, this became the Air Corps Engineering School, and a year later it was moved to Wright Field. The pressures of war caused a suspension of school activities from shortly after Pearl Harbor until 1944.

In 1946, the Army Air Forces Institute of Technology was established as a part of the Air Materiel Command; this became the Air Force Institute of Technology in 1947. The primary emphasis remained on engineering, but maintenance, logistics, and procurement were also covered.

The Commander, Air University, was authorized by Congress in 1954 to confer degrees upon persons in the AFIT Resident College, and this marked the beginning of a period of rapid growth for the institute. The Resident College was later subdivided into the School of Engineering, School of Logistics, and School of Business, the last being transferred to civilian institutions in 1960. Full accreditation of AFIT programs at the doctoral level came in 1970.

The constantly changing environment of the modern Air Force has put a premium on training engineers in a broad spectrum of disciplines that range beyond the standard aeronautical, astronautical, computer, electrical, and nuclear engineering fields to esoteric areas like electro-optics, space operations, and the Air Force Test Pilot School program. Graduates of AFIT have made significant contributions to the nation's defense, including the original concept of the Joint Surveillance and Target Attack Radar System (Joint STARS or J-STARS) which was used with such deadly effect in the Persian Gulf War.

Perhaps the most unusual aspect of AFIT is its Civilian Institution Programs (CIP). Each year the CIP directs as many as 3,000 higher education students and 4,000 continuing education students in almost 400 school locations worldwide. The CIP budget of approximately $33 million per year represents about 50 percent of AFIT's entire budget. The CIP vastly expands AFIT's ability to provide schooling possibilities to Air Force personnel and offers advanced degrees in everything from engineering to medicine, dentistry, and law.

AFIT schooling comes with a quid pro quo: students who take advantage of it incur an obligation to remain in service for a specified number of years, depending upon the length of the schooling. The author looks back in gratitude to AFIT, which provided him with both a bachelor's and a master's degree.

OTHER MAJOR COMMANDS

Over the history of the modern USAF, there have been a number of important major commands, including the United States Air Forces in Europe (USAFE), Pacific Air Forces (PACAF), Air Force Communications Command (AFCC), Air Force Space Command (AFSPACECOM), and Air Force Special Operations Command (AFSOC). Each one merits a book of its own; unfortunately, space constraints prohibit telling each one's full story.

7

CHANGE AFTER VIETNAM

In the years since Vietnam, the USAF has steadily improved its weaponry, doctrine, and leadership, as well as the quality of its personnel and the quality of life they enjoy. The task has not been easy, for it is engaged in intense competition from industry for the kinds of financial and intellectual resources necessary to field a superior air force. The competition is particularly aggressive for qualified African-Americans, other minority personnel, and women, a situation that will be addressed later in the chapter. For much of the time, the Air Force faced a skeptical public, a hostile press, declining budgets, a long series of external threats, and the daunting and unfamiliar task of achieving dominance in space. As it did so, it retained its nuclear deterrent power, built up its conventional forces, and rallied its allies to new standards of proficiency and performance. It was a demanding, if ultimately rewarding, two decades.

The gradual rise of the Air Force to its present position of eminence can be attributed to many factors. Improvements in doctrine, leadership, and weapons will be addressed in the next chapter. Here we will consider three often overlooked but complementary phenomena that have been the foundation stones of the Air Force's success. The first of these is an enlightened leadership that quite literally inverted traditional management methods. The second is the creation of a superb enlisted corps, led by noncommissioned officers who rose to the challenge of a highly professional, technologically oriented service during a time of profound social turbulence. The NCOs were instrumental in seeing that the entire enlisted force was qualified to handle its far more sophisticated assignments. The third was the belated but welcome exploitation of a huge new pool of talent implicit in bringing women and minorities into the service on an equal opportunity basis. Together these three elements helped shape the Air Force into an instrument that national leaders could use with confidence. The last two phenomena were completely dependent in their execution upon the first, the complete turnaround in Air Force management methods.

THE MANAGEMENT REVOLUTION

General Curtis E. LeMay, the most influential Air Force leader of the postwar period, instilled the discipline and established the highly structured centralized management necessary to create the Strategic Air Command, which provided the nuclear shield for U.S. foreign policy. Secretary of Defense Robert McNamara capitalized on the Defense Reorganization Act of 1958 to lift LeMay's centralist methods to an extreme. His methods trickled downward, inhibiting subordinates at all levels from taking action on their own and creating vast paperwork bureaucracies where any idea had to be "coordinated" with many offices, any one of which could say no, but none of which could give a definitive yes. Decision-making on even minor issues could be made only at the highest levels—and that more often than not meant McNamara himself. As a result there was established a "numbers are everything" reporting mentality that stifled initiative. The momentum McNamara imparted to the process kept the Department of Defense on a centralist course for years after he left.

A far more important effect of this bureaucratic centralism was that it sapped integrity by making the report on a mission more important than the results of a mission. The ability to waffle—what is today called "spin control"—became for some a more important qualification for promotion than the ability to produce. A hierarchy developed in which many rose to the top because they understood the system and knew how to manipulate it; beneath them, too often overlooked for promotion, were the mission-oriented workers who carried out their tasks without regard to their own advancement.

One has only to look at the great ossified bureaucracies around the world to realize how difficult it is to reverse the course of rigid, centralized bureaucratic control. On rare occasions a turnaround occurs. The reversal is usually due to a number of factors and to a large number of people, but there is often a single person who becomes indentified with the change of direction. So it was with the Air Force. Identification of the man responsible emerged in the course of a number of in-depth interviews with four-star officers, including Generals David Jones, Larry Welch, Michael Dugan (all three former Chiefs of Staff), Robert Russ (COMTAC), and Michael Loh (COMTAC) and first Commander of the Air Combat Command). All were candid and forthcoming about their own careers in the Air Force and in assessing the contributions of others. Each one, quite independently, and usually in a different context, attributed the remarkable turnaround in Air Force management style, with its consequent increase in proficiency, efficiency, and improved quality of life, to one man: General Wilbur "Bill" Creech. Jones ranks him with Curtis LeMay as one of the two most influential men in his long Air Force experience.

A cynic might think that this was simply "good-old-boyism," praising a popular colleague. Nothing could be further from the truth, for Creech was personally unpopular with many people because of his demanding standards of performance. Those who did not come up to his standards certainly regarded him as being far too critical; some to this day somewhat ironically consider him to have been the spiritual successor to General LeMay.

However, the effect of Creech's work has been specifically acknowledged by a number of other sources, including, as we will see, by the Department of Defense. General Charles Horner, the air commander of the Persian Gulf War, has written at length about how, as a classic fighter pilot, he was at first vehemently opposed to everything Bill Creech stood for, from pressed uniforms and carefully combed hair to minimal drinking and less horseplay at the officers' club. Over time, however, he became not only a convert to Creech's methods but a self-described "Creech clone." Even though Creech had retired long before the war, Horner gives him "number one credit" for the victory of air power in the Gulf War.

Creech's methods worked management miracles in many organizations, creating, despite his admitted rigor, a climate of benign, forward-looking personnel relations throughout the Air Force that is still followed today. This climate—so foreign to anyone who had been dusted off in a McNamara briefing or frosted by a LeMay admonishment—fostered an inventive outlook and engendered an eagerness to create new approaches to old problems.

There were also other practitioners of proactive human relations. Prominent among these was the first man to rise from the ranks of combat crew member to become Commander in Chief of SAC, General Russell Dougherty, who was the complete antithesis of LeMay in personality, always genial and self-proclaimed as the "first nonhero to head SAC." Dougherty's towering intellect is well known in the service and in his retirement career as a prominent businessman and lawyer. As much as Dougherty admired LeMay, he recognized that the authoritarian leadership style had been carried too far. In one of his earliest talks to his commanders, Dougherty told them, "There's nothing in your job description or mine that requires either of us to be an unmitigated son of a bitch." (Dougherty is noted for his erudite and literate presentations to Congress, and his rare use of raw language was to drive home a point.)

When all is said and done, the Air Force that was driven to an absolute and crippling centralism by McNamara was turned around by Bill Creech, who found that many of his brother officers also recognized the problem and were more than willing to adopt his solutions.

"COMMAND MUST BE REASONED"

Many people, unaware of the extent of General Creech's contribution, will question the space devoted here to his career. However, the impact that he had upon the Air Force was both broad and deep, and it is worthwhile to see exactly how it came about.

A distinguished 1949 graduate of that grand old institution the Aviation Cadet program, Creech had a fast-moving career in fighters. In between his 103 combat missions over North Korea and 177 in Vietnam, Creech flew hundreds of aerial demonstration missions with the U.S. Air Force Thunderbirds and their USAFE equivalent, the Skyblazers.

Quality was the underlying principle of Creech's leadership. He created small teams of people who trusted each other and were genuinely concerned about each other's welfare. He added to this the ingredients of leadership and commitment operating from the bottom up. Time and again, he achieved levels of quality and productivity that appear miraculous when compared to the results of prior efforts.

The future General Creech was propelled into management on a larger scale commanding tactical fighter wings at Zweibrucken Air Base, Wiesbaden, Germany, and at Torrejon AB, Spain, during 1970 and 1971. It was at Zweibrücken that Creech's fanatic obsession with quality and productivity first received wide attention as, in nine short months, he turned a new wing and a derelict base into the best unit of its kind in Europe. He was sent immediately to the 401st TFW in Torrejon, which had flunked its last two war-fighting inspections. Two wing commanders in a row had been fired. In another nine-month whirlwind of activity, Creech once again proved that his principles worked.

General David Jones, Commander in Chief of U.S. Air Forces in Europe (CINCUSAFE), was impressed. Sensing a kindred soul, he nominated Creech to be his deputy for operations—a major general's slot. Jones had to go to Chief of Staff General John D. "Three Finger Jack" Ryan for permission. Ryan argued against placing a colonel (although Creech was on the promotion list for brigadier general) in a two-star position, but ultimately agreed. Now clearly identified as a comer, Creech received choice assignments as Vice Commander of the Aeronautical Systems Division and later as Commander of the Electronic Systems Division. He applied his management principles in both positions.

Driving himself hard at ESD, Creech, by now a three-star general, suffered a heart attack, which under Air Force policy meant automatic retirement. The Chief of Staff, David Jones, changed the policy. He allowed Creech time to recover, and then assigned him to command what had become the biggest challenge in the Air Force: the Tactical Air Command.

CREECH AT TAC

The magnitude of Creech's assignment was staggering. Promoted to be a four-star general on the day he took over command, he headed an organization that employed 65,000 members of the Air Force and a further 50,000 civilians in more than 150 separate installations around the world. He had a supersonic, nuclear-capable fleet of 3,800 aircraft maintained by a centralized bureaucracy that did not see itself as inefficient nor concern itself that its training sortie rate was dropping an average 8 percent per year, or that its pilots were opting to leave the service in droves after receiving more than $1 million in training. On any given day as many as 1,900 of his fleet of aircraft might have been out of commission. About 220 were certified "hangar queens," unable to fly for three weeks or more because of maintenance or spares deficiencies. A very high abort rate plagued the aircraft that attempted to launch.

Creech received authority to do a large-scale test of his concept of using decentralized, team-based systems. He set about changing TAC, beginning with his familiar tenet that well-trained, well-motivated teams of people, empowered to do their jobs and given a sense of proprietorship, were the most efficient. Creech told his commanders, "Gentlemen, command must be reasoned," and cautioned them against losing their tempers. He also told them that a leader's place was in the air, just as the place of the senior master sergeant on the flight line was in the trenches turning wrenches, not behind a coffee-and-doughnut-laden desk. Aircraft that previously were rotated through an anonymous centralized maintenance procedure now "belonged" to a dedicated crew chief, who had the privilege of painting his name on the side of the fuselage, just as the pilot did. Creech tells a story about one of his routine visits with the troops when he talked to the frontline workers about how things were going. A crew chief shook his hand and told him how much he liked the dedicated crew chief program. Creech asked him why he liked it, and the crew chief asked him a question in return: "General, when was the last time you washed a rental car?"

Some of Creech's attributes worked against him. He didn't have the stentorian voice of a stereotypical great leader. He was fussy in his appearance and demanded that his officers be similarly conscious of their uniforms. His enthusiasm, his leading by example, and most of all his results first commanded respect and then affection. But Creech was tough. If a pilot violated safety regulations, he would be disciplined without a voice being raised—but also without a moment's hesitation. When a hotshot pilot buzzed his hometown of Plano, Texas, Creech fired him. He was lobbied by the pilot's peers and by a petition from the people of Plano. Creech held firm: there was no recourse.

Under his direction, the TAC accident rate dropped from one every 13,000 hours to one every 50,000 hours. During the same interval that the accident rate dropped, the average aircraft sortie rate (flights per aircraft per month) rose from 11 to 21. In plain language this meant Creech had effectively almost doubled the number of aircraft under his command—with no additional procurement.

As hard as he was on safety violations, he was even more rigorous with any violation of personal integrity, which he treated with an instant, severe response. He discovered that a fast-rising brigadier general had lied to him about a safety incident. Creech called the errant BG and told him to resign immediately or face a general court-martial.

Many people—including, at the time, the future General Chuck Horner—thought that Creech went too far, that he was going to snuff out the raucous independence that had characterized fighter pilots since World War I. Creech maintained instead that he was channeling the pilots' energy into a new under-standing of their equipment. Pilots in World War I flew $10,000 Spads when the weather was good; his pilots were flying $30 million fighters in all weather. The difference was vast, and Creech knew that pilots had to conform to the new reality.

Despite a reduced budget, Creech had TAC substantially turned around within four years. With typical government myopia, the General Accounting Office was disturbed over the amount of money he was spending on improv-ing the appearance of his installations. The term "Creech brown" had become a joking reference to the number of buildings that had been painted in earth colors as he set about his "Proud Look" campaign, giving every base a sense of pride in its appearance.

In its lengthy investigation, the GAO did find that much money was being spent on "home improvement," but that it was primarily for materials; the air-men *and the officers* were doing most of the work in their off-time. The GAO also found that the new leadership style at TAC had increased productivity by 80 percent. The number of aircraft out for maintenance was cut by 75 percent. Later it was estimated that in a period of four years, Creech's methods had produced equipment availability and combat capability that would have cost more than $12 billion if purchased on the open market.

CREECH'S INFLUENCE ON THE PERSIAN GULF WAR

The profound change of course at TAC was not confined to managerial methods. Creech was a warrior who found TAC's tactics encumbered with what he called "go-low disease"—the perceived need to fly at minimum height to avoid enemy surface-to-air missiles. Creech argued that the buildup in en-emy antiaircraft artillery made the go-low approach dangerous, and that new methods were needed. He decreed that taking out SAMs was the first order of business—that the enemy defenses should be nullified and rolled back so that follow-on aircraft would have the flexibility to operate at high or low altitude in hostile territory, depending upon the nature of enemy defenses. Just as in Linebacker II, removal of the SAMs was essential.

The new tactics were drummed into TAC in practice in the rigorous Air Force Red Flag training at Nellis Air Force Base, and with the complementary training functions Creech had devised: Copper Flag for the air defense forces, Green Flag for electronic countermeasures, and Checkered Flag for rapid de-ployments.

The payoff for the realistic training came in the Persian Gulf War, where USAF casualties were remarkably low. Those coalition partners who stayed with the go-low philosophy suffered heavily as a result of their unwillingness to change tactics.

At the end of the Gulf War, the air commander, General Horner, had an opportunity to review the magnitude of the victory and realized that the USAF had succeeded because it had anticipated the nature of the threat, had trained its pilots to fight on the first day of the conflict, and had provided them with technology and tactics that saved their lives and kept their aircraft at an unprecedented level of operational readiness. He attributed this happy state of events to Bill Creech and three elements of his teachings. The first of these was the critical importance of decentralization in ensuring a maximum flexibility, responsiveness, and feeling of ownership. The second was the absolute necessity of getting leadership and commitment from everyone, regardless of rank or position. The third was the power of planning for quality in every action.

THE SPREAD OF TQM

TAC's turnaround did not go unnoticed; word spread rapidly through the Air Force, and the reins of centralized management were loosened. It was not only more efficient, it was a more pleasant way to work—the quality of life for the airmen improved because their job satisfaction increased.

Perhaps the most important aspect of Creech's methods was that they were exactly in tune with the changes going on in civilian attitudes toward work and the quality of life. Creech was hammering at the frontier-outpost mentality the Air Force had inherited from the Army and driving it into the mainstream of human relations management, at a time when this was absolutely critical. The old methods of the Air Force were not working and probably would not have worked in the post-1970s world. More important, perhaps, they would have left the Air Force unequipped to meet the challenges it would face.

Creech's methods were exactly counter to the ongoing centralist philosophy of the Department of Defense, which like some mythic giant octopus was absorbing as many functions of the services as possible. On December 1, 1982, Deputy Secretary of Defense Frank C. Carlucci signed a DOD memorandum calling for the study of a single defense agency that would "own and operate all DOD installations." All Air Force bases, Army forts, posts, and camps, and Navy stations would be consolidated under a single DOD manager.

Creech met the issue head on, inviting DOD representatives to visit TAC and see what decentralization had accomplished. Robert A. Stone was then Deputy Assistant Secretary of Defense, with the responsibility for all DOD installations worldwide, and was the man behind the memorandum Carlucci had signed. He believed in the efficiency of centralization and wished to reduce costs by bringing all the installations under one set of rules so that procurement, maintenance, repair, and other activities could be standardized.

Stone and his associates were well intentioned. They conducted a full-scale investigation of conditions in TAC and followed Creech's own methods by getting into the field and talking to the people who were doing the work. When they were finished, they reversed their position and asked Creech to help formulate a method for achieving the same success DOD-wide. In early 1984, only a little over a year after his own initial memo on centralizing installations had been signed, Secretary Stone created the DOD-wide Model Installations Program, or MIP. After two years of successful tests, on March 26, 1986, DOD issued another memorandum, "Defense-wide Application of the Model Installation Management Approach." The techniques to be used were those Creech had demonstrated in TAC.

Further objective corroboration of Creech's methods came with the President's Blue Ribbon Commission on Defense Management, headed by David Packard, who had made his fame and fortune running Hewlett-Packard in a maverick style. In his foreword, Packard applauded the new methods being used by DOD in managing its installations. It was truly a case where less was more: less management gave more results in installation cost, utilization, and appearance.

Creech began calling his approach Total Quality Management (TQM) in 1981; in 1988 the term was adopted officially by the Department of Defense, and its use was extended to vendors doing business with the military. (The full history of TQM, as well as an in-depth explanation of all its details, may be found in Creech's book *The Five Pillars of TQM: How to Make Total Quality Management Work for You*.)

Creech's influence might well not have been so widespread nor so long-lasting if it had not been institutionalized by the officers who followed him to command positions. Many future leaders who adopted Creech's management precepts were subsequently assigned to organizations outside of TAC. They not only employed his methods in their new positions, but, in turn, they indoctrinated their subordinates in Creech's management philosophy. Individuals often made minor changes, reflecting their own personalities; but for the most part, Creech's ideas became pervasive throughout the USAF.

General Creech died on August 26, 2003, at the age of seventy-six. In 2005, the Air Force renamed Indian Springs Air Force Auxiliary Air Field as Creech Air Force Base to honor him.

TQM IMPACT ON THE AIR FORCE'S QUALITY OF LIFE

General Creech's contributions to the Air Force's management revolution have been given special emphasis here because they were fundamental to two other massively important aspects of the Air Force's post-Vietnam turnaround.

First, the Air Force turned to the management methods espoused by Bill Creech at TAC with enthusiasm and relief. The new methods that brought about improvements in proficiency also happened to be attuned to the swift-moving changes in the lives of American civilians. The effect of TQM was to provide the Air Force with more efficiency while it provided Air Force people with a better life, one more nearly comparable in pay, benefits, and prestige to those of their civilian counterparts.

Second, this turnaround also coincided with the maturation of the enlisted forces as a significant managerial element. As we will see below, the enlisted ranks, guided by their noncommissioned officers, changed rapidly over the years and emerged from the trauma of the Vietnam War able to meet the challenges of the modern Air Force.

Before we review the evolution of the Air Force enlisted force from its early days to its present status, it might be salutary—and sobering—to get a snapshot of how the current quality of military life is gauged.

THE BASELINE OF MILITARY QUALITY OF LIFE

The Air Force, like the other services, traditionally placed more emphasis on equipment and readiness than on the quality of life of its members. It now faces a serious crisis in maintaining a quality of life competitive with that of the civilian population. An "all-volunteer force" will have no volunteers if there is not a reasonable parity in military and civil life.

In October 1995, the report of the Defense Science Board task force on "quality of life" was released. It reveals plainly that despite the best efforts of the services, the Congress has not kept its promises. The report emphasizes that preservation and improvement of the quality of life of members of the armed services is not a matter of kindness or generosity. It is instead directly related to the readiness of our military, to the retention of key personnel, and to morale.

Conceptually, five basic elements define the quality of life for military services. These include compensation, medical care, housing, personnel tempo, and community or family services. (Personnel tempo is defined as the amount of time a service person is forced to spend away from home compared to the amount of time spent at home. For purposes of the study, field exercises that keep the service person away overnight, even though conducted from a home base, are considered time away.)

As important as they are, compensation and medical care will be given only a brief mention here. They are the subjects of other task forces, whose reports were not completed at the time this was written. Military compensation has traditionally lagged well behind civilian compensation, particularly in times when funds are tight. Military medicine, once greatly admired and one of the chief inducements to make a career of the service, has fallen on even harder times than medicine in the civilian world, and the solemn promise of medical care for retirees has been monumentally diluted. Skyrocketing costs have reduced services everywhere, even as noncompetitive salaries drive the most competent professional medical people away from serving even a tour, much less a career, in the military. The inevitable result has been a lowering of standards. The waiting periods to obtain services are long and a variety of fiendish systems have been developed to make getting an appointment much more a matter of sheer dogged persistence than of medical need.

In the three areas covered by the report on quality of life, the basic findings run in sharp opposition to stated congressional intent and to the policies of current DOD leaders, both military and civilian. Secretary of Defense William J. Perry stated in his 1995 *Report to the President and the Congress,* "Readiness is associated most closely with the morale and espirit de corps of U.S. soldiers, sailors, airmen and Marines. These intangibles are maintained by ensuring the best possible quality of life for people in uniform and their families. Quality of life falls into three general categories: standard of living for Service members; demands made on personnel, especially time away from family; and other ways people are treated while in the Service."

Despite the good intentions of Congress and DOD, a long-term situation has evolved in which reduced budgets and increased commitments have been rationalized only by taking the difference out of the hides of service personnel. In effect, the policy of procuring hardware and deferring pay raises has been accepted as a way of doing business.

Here is a brief survey of the three quality-of-life areas covered by the report, along with some relevant quotations.

HOUSING

Excellent housing facilities and services shall be provided for all military members, their families and eligible civilians. Continual improvement in quality is a measure of excellence, and customers of housing services shall participate in their evaluation.

—Department of Defense Housing Management Manual,
September 1993

The Department of Defense owns or leases about 387,000 homes, which have an average age of thirty-three years. Deferred maintenance, repair, revitalization, and replacement have reached a total of $20 billion. In other words, Congress and the armed services have chosen not to spend the $20 billion known to be necessary to bring housing up to a decent standard. Sixty-four percent of these military homes are rated as "unsuitable" for a variety of reasons. Some 15 percent of military families live in private-sector homes also considered unsuitable. Thus, *79 percent* of military families are forced to live in housing unsuitable by civilian standards.

The situation is no better for bachelor housing. A $9 billion backlog in repairs and replacement has been amassed, and 62 percent of the 612,000 bachelor housing spaces are considered substandard.

The report notes that "the delivery system [of personnel housing] is so intrinsically flawed that it should be replaced with an entirely new system." The probability of this occurring is small, given that an entirely new system would require congressional funding at a time when deficit reduction is the political imperative.

PERSONNEL TEMPO

The drawdown has caused many Service members to question their long-term commitment and the prospects of a full career. The turbulence of consolidations and base closures has disrupted assignments and family life.

—Secretary of Defense William J. Perry,
November 1994

A direct correlation exists between family separations, spousal support for a military lifestyle, and retention rates. The armed services tend to define "personnel tempo" in a variety of ways. For example, the Navy only credits a unit—not an individual—for time away, and does so only when a deployment is underway for fifty-six days. The Marine Corps uses a ten-day period. Service members, however, define it on a realistic basis: the amount of time they are forced to spend away from home. The task force report also chose to view it in this manner, saying that its yardstick was "1 day away = 1 day away."

In the five years after 1989, the total strength of the Department of Defense decreased by 28 percent. Random samples of service personnel deployments indicate that the average serviceman or servicewoman will be away four times as often in 1995 as in 1989. In other words, as strength has declined, requirements for travel have increased. The results of the increase in the number of days spent away from home are many and varied, and none are good. They range from divorces to the loss of second jobs (a critical necessity for many enlisted personnel) to inability to compete for promotions. As an example of this last problem, technicians from the 429th Electronic Combat Squadron at Cannon Air Force Base, New Mexico, were kept so busy on deployments that they had no time to prepare for the highly competitive promotion exams. The fifty-five staff sergeants of the 429th were among the most highly qualified people in the service in their specialty, that of making sure that the complex electronic suites on their General Dynamic EF-111 Ravens were in perfect order. (The Ravens did a marvelous job in suppressing enemy electronics in the Gulf War.) Of the fifty-five, *not one* was selected for promotion to technical sergeant, because not one had time to prepare for the examinations. The impact on morale can be imagined.

In the USAF, people with certain skills, such as those required for AWACS or Fairchild A-10 aircraft, deploy as much as 75 percent of the time. These long-term deployments have an inevitable effect upon morale and family integrity; they also adversely affect training, as the tragic friendly-fire shootdown in Iraq of two Blackhawk helicopters by Air Force F-15s under the control of an AWACS demonstrated. While new managerial solutions have been applied to the problem of personnel tempo, the increase in demand for USAF participation in the last ten years has exacerbated the problem.

COMMUNITY AND FAMILY SERVICES

Military people stay in service because they like being part of something special. They won't stay long, however, if families aren't treated well.

—General John M. Shalikashvili, Chairman of the
Joint Chiefs of Staff May 1995

The mores and customs of members of the armed services mirror those of the civilian community. Women are well integrated into all the services, and the phenomenon of the single-parent military family—male and female—all but unknown a generation ago is becoming increasingly common. Almost 65 percent of military spouses work, and in far too many cases, both parents hold down two jobs to make ends meet. This creates a pressing requirement for child care facilities that are safe, affordable, convenient, and of high quality. (It is shocking to note that many service families, unable to sustain a two-parent-four-job employment blitz, are forced to resort to food stamps and other charities. A spouse fighting a war on another continent is unlikely to have his or her morale improved by knowing that only food stamps are keeping the family alive.)

Like housing, funding for child care has fallen steadily behind. Service members have approximately 1 million children under twelve years of age; about half of these are preschool children. The DOD provides child care at 346 locations with spaces for only about 155,000 children. There is a current waiting list of almost 144,000.

All of the services have done a great deal to help themselves, through the use of nonappropriated funds generated from such activities as the base theater and bowling alley and from user fees. However, resolution of the overriding problem of child care can come about only through congressional action and consistent, sustained funding.

The Defense Science Board came up with recommendations for most of the problem areas it cited; if carried out, they would go a long way to remedy the current shortfall in the quality of military life. Many of the recommendations are within the power of the individual services to act on, and undoubtedly they will do so. Unfortunately, many of the more critical recommendations require legislation and congressional funding. The probability of this happening is left to the reader's judgment.

In many areas, current Air Force practice was recommended by the Defense Science Board Report for emulation by other services. This was true of child care, housing, and, most particularly, education programs. The Community College of the Air Force, of which more later, was singled out for its exemplary work.

THE RISE OF THE PROFESSIONAL NONCOMMISSIONED OFFICER CORPS

The foregoing coverage of the DOD report on quality of life is useful as a background for understanding a significant managerial coup: the creation of what has been termed the most outstanding corps of noncommissioned officers of any service in the world—that of the United States Air Force.

The assessment of the quality of the noncommissioned corps comes from many sources. General Jones recalls that when he was Commander in Chief of USAFE, he was constantly asked by NATO officers how the USAF achieved such excellence in its NCO corps. General Robert Russ, former Commander of TAC, delights in telling the story of the first major exchange of visits between Soviet and U.S. leaders, initiated by the Chairman of the Joint Chiefs of Staff, Admiral William J. Crowe, Jr. One of the stops on the Soviet itinerary was a visit to Russ's headquarters at Langley AFB, Virginia, by a delegation that included Crowe's counterpart, Soviet Chief of Staff Marshal Sergei F. Akhromeyev.

In a debriefing prior to their departure from the United States, the Soviet leaders confided to the Admiral Crowe that the one thing that had impressed them most was not the advanced fighters, nor even the commissaries and post exchanges bursting with consumer goods. They had expected American leadership in these areas, and accepted it. But they were totally overwhelmed by the U.S. noncommissioned officer corps, which had no counterpart in the Soviet Union. In the Soviet Union and in its successor states, jobs typically held by noncommissioned officers in this country—crew chiefs, line chiefs, tank commanders—are always held by officers.

Russ attributes the Soviet assessment of the Air Force NCO corps in part to a fortuitous choice of the Langley Engine Shop in the tour he provided his guests. The visitors were amazed that this multimillion-dollar facility, conducting hundreds of millions of dollars' worth of engine repair each year, was run by a senior master sergeant. Their astonishment was doubled when they discovered that the senior master sergeant was a woman. The Soviets closely questioned the NCO shop manager, convinced that she was a ringer, a "Potemkin" leader. They were astounded when the shop manager not only demonstrated a sure knowledge of her job and of the complex procedures involved in it, but introduced them to the people working for her and explained their jobs as well.

The USAF's requirements for noncommissioned officers and enlisted personnel are somewhat different from those of other U.S. services, particularly the combat arms of the Army and the Marines, where a high turnover in personnel is desirable. It takes young, well-conditioned troops—trigger pullers—to engage in the rough and tumble of conflict, and it is desirable to build up a strong reserve of experienced personnel who can be recalled to duty if necessary. Thus it makes economic sense to bring them in, teach them the trade of ground war, and then replace them with younger troops at the end of their tours.

Air Force personnel are also required to maintain themselves in good physical condition, but other attributes are equally important when it comes to managing the sophisticated systems found in every Air Force discipline. Considerable experience is required for supervisory positions in the fields of jet engines, ICBMs, advanced electronic systems, and modern precision-guided munitions, to name just a few. As the Air Force, like other services, has drawn down, its senior noncommissioned officers have been hard pressed to manage their own assignments and to train new people at the same time. Therefore, retention of experienced people is one of their most important concerns. A first-term airman begins to earn his pay only when he approaches the end of his tour, so it is vital to retain that airman for another enlistment period—more than that if possible—to get a reasonable return on the training investment. A young airman whose name is painted on the side of a $30 million fighter might have civilian technicians—the famous "tech reps"—working under his or her direction who earn twice as much in salary. The demonstrated ability to handle responsibility of this magnitude makes young airmen highly desirable commodities in the civilian job market when the term of enlistment expires. One of the signs of a superior noncommissioned officer supervisor is his or her ability to convince young airmen that their long-term interests are better served by a career in the Air Force than by the plums being dangled by a civilian recruiter.

This would seem an impossible task, given the content of the previously cited Defense Science Board report on quality of life. A brief review of the history of the enlisted force will help put things in perspective, and show how certain fundamental factors, including simple patriotism, maintain their appeal.

BREAKING THE LOGJAM

The demobilization frenzy that had so debilitated the USAAF after World War II ended had its effect upon the enlisted force when the USAF came into being in 1947. The force was "old Army" and top-heavy with rank, for the people who elected to stay in were for the most part those whose seniority gave them an incentive to make a career of the service. Many of them were children of the depression whose motivation was, in the phrase of the time, "three hots and a cot," meaning three hot meals and a place to stay. This was no small matter for those who had come to maturity in the 1930s and, like everyone else, had no idea there would be a postwar economic boom of unprecedented magnitude and longevity.

Over the next several years, the logjam of high-ranking enlisted personnel was exacerbated by regulations that permitted officers who had been separated to enlist at the rank of master sergeant. Thus an outsider, already unhappy at his loss of commissioned status, could come in at the top of the enlisted ranks even though he might not possess the skills or the inclination to execute an NCO's duties with vigor. Another problem was inherent in the high ratio of officers to enlisted in the Air Force compared to other services. Officers were often assigned additional duties that normally could have been done by an NCO.

The advent of the Korean War brought about yet another difficulty. The Air Force again built up rapidly, and many reservists who were recalled elected to stay in service when the war ended in 1953. Many of these reservists, their lives interrupted for a second time, realized that they now had eight or ten years of active duty and that staying another ten or twelve to gain a retirement benefit made sense. This saddled the Air Force with what became known as the "Korean Hump"—an unbalanced rank structure with far too many senior NCOs and a shortage of enlisted personnel in the lower grades, E-1 through E-5.

The excess of senior noncommissioned officers, most of them still many years away from retirement, stagnated the promotion system, stifling the incentive of younger airmen. When promotions were made, the selection was done by local boards, and results were often skewed by the inevitable cronyism. The NCO corps was burdened with yet another problem, the rank of warrant officer, with which the Air Force never really came to grips in defining its place in the grade structure. Warrant officers were usually administrative specialists in a field, e.g., finance, supply, or medicine. They held positions that otherwise would have been held by NCOs and thus helped block the normal progression through the grades.

The promotion stagnation was partially alleviated in 1958 by the creation of two new "supergrades," E-8 for senior master sergeant and E-9 for chief master sergeant. This immediately offered promotion possibilities for master sergeants, and, in domino fashion, for lower ranks. The warrant officer problem was finessed nine months after the supergrades were introduced with a decision to make no further promotions to that rank. Normal attrition lowered their numbers until the last warrant officer was retired in the late 1970s. The problem of officers' performing additional duties in what might ordinarily have been NCO slots was alleviated over the years as the sophistication of Air Force weapon systems increased and there were more officer-level duties to be assigned.

The remainder of the promotion logjam was eventually broken up by two separate events. The first of these was under the impetus of the great friend of the enlisted man Congressman L. Mendel Rivers, a South Carolina Democrat. As chairman of the House Armed Services Committee, he held hearings that resulted in the creation of the Weighted Airman Promotion System (WAPS) in 1970. Under WAPS, promotion was contingent upon objective factors and used clearly defined, weighted criteria such as time in grade and test scores. These criteria were often within the control of the individual airman to manage—for example, if he or she made the effort to take training courses, it helped the score. It also removed the element of cronyism, although there were later charges that the tests had an inherent cultural bias that resulted in low scores by minorities. Work was done to redesign the tests to alleviate this problem.

The second major factor in the final breakup of the promotion logjam shows just how far the Air Force will reach in its search for a catchy acronym. In this case it was an "up-or-out" system for airmen known as TOPCAP, or Total Objective Plan Career Airmen Personnel. TOPCAP initiated a policy of forced attrition. In 1973, the high year of tenure for various ranks was established as follows: an E-4 (senior airman, the equivalent of sergeant) or an E-5 (staff sergeant) had one of twenty years; if not promoted by twenty years, it was necessary to retire. Time requirements for other grades were as follows: E-6 (tech sergeant), twenty-three years; E-7 (master sergeant), twenty-six years; E-8 (senior master sergeant), twenty-eight years; E-9 (chief master sergeant), thirty years.

There were factors adversely affecting enlistment other than promotion. Pay was the most obvious of these; military pay always lagged civilian pay, primarily because lawmakers always turned to a cap on military pay as a means of restraining the federal budget. Only when the gap grew so great that an exodus of officers and enlisted was eminent was some measure of "catch-up" provided.

Salary was only part of the compensation problem. Military people lost out on permanent-change-of-station (PCS) moves because the allowance was never sufficient to cover the expenses of uprooting a household and moving across the country or to another continent. One factor, never quantified in any military pay analysis, was that military people were unable to buy a home and live in it for years, thus building up equity. If they chose to buy, each time they moved they bought in at the current price at whatever interest rate they could obtain. (It was unusual for a military family to buy a home until the 1960s; it became more common during the 1970s and 1980s, but, with the advent of higher home prices and high interest rates, it became more difficult in the 1990s.) Many people at retirement were buying their first home in communities just as their new civilian neighbors, who had never had to move, were making their last mortgage payments.

Tours of temporary duty (TDY) away from home also generally cost money, and, in the case of those who had to make frequent trips, became a significant item in the family budget. TDY trips had another, undemocratic aspect. Officers' quarters were generally of acceptable quality, while for too long enlisted temporary quarters were rough open barracks.

Another factor that had far-reaching effects upon the services, and particularly upon the challenge faced by Air Force noncommissioned officers, was the establishment of the All-Volunteer Force by Congress in 1973. The Air Force and its predecessor organizations had always been volunteer, but for much of the time, the volunteers were motivated as much by the threat of the draft as by the desire to serve. After 1974, it could be presumed that enlisted personnel shared the long-term view of senior NCOs, i.e., that service in the Air Force was a full-time job that served the societal need of protecting the country. In addition, they viewed their job choice as desirable because of the specialized training given in sophisticated trades and the opportunity for gaining additional education.

Yet for the real achievers, another factor was fundamentally important: the Air Force offered the opportunity to obtain far greater responsibility at a much younger age than in civilian life. This is in part a function of the size of the Air Force and of the mobility of its personnel. Jobs turn over more swiftly, and a ready and waiting airman can take over responsibility as it becomes available.

It is difficult to overemphasize the importance of this most intangible of factors. The desire for responsibility is pervasive in the enlisted force, and the psychic return derived from the proper execution of that responsibility is a primary reason that many stay in service. The knowledge an airman has that the aircraft would not have been able to take off on time without his or her expertise is immensely rewarding; it is doubly so when it is acknowledged— usually by no more than a "thumbs-up" sign or a slap on the back—by the airman's peers and superiors. And this sense of reward from responsibilities well discharged is not confined to earth-shaking events. It comes in all jobs at all levels, from a beginning cook who creates a sparkling salad bar to an armorer loading missiles for a strike in the Persian Gulf War. The common denominator is that Air Force leaders at all levels see to it that accomplishments are recognized and appreciated.

The military retirement system was one aspect of service that a noncommissioned officer could use to induce enlisted personnel to remain on duty. The concept of retirement seems remote to a first-termer, but by the end of the second tour, it becomes a real selling point. In today's budget-cutting climate, the military retirement system (already subjected to considerable reductions) is under fire; if it should be significantly altered, it would have an adverse effect upon enlistment rates—and a catastrophic effect upon reenlistments.

As will be shown later in the chapter, the All-Volunteer Force concept helped in another important process: the widespread integration of women and minorities into the work force. Whole pools of previously ignored talent suddenly became available, and the smarter officer and noncommissioned officers were quick to take advantage of it.

CHIEF MASTER SERGEANT OF THE AIR FORCE

The change in the nature of the Air Force's noncommissioned officer corps was a joint product of the growing expertise required by all the systems coming into service and the increased responsibilities these systems demanded. In this context, the term "systems" includes not only weapon systems such as the Minuteman missile, but also the many accounting, inventory, personnel, and other systems that were becoming increasingly sophisticated as they became more automated. The day of the "old soldier", immortalized as Sergeant Bilko or Beetle Bailey's Sarge, a man who had performed the same set of duties for twenty years, was gone. Now new demands were placed upon the intellect and the energy of the noncommissioned officer corps every day. The pressures were welcomed by most, for it provided an opportunity to excel. The bell had tolled for the old-fashioned NCO who preferred to shuffle papers at a desk and wait for happy hour.

The Marines, always aware of the value of the noncommissioned officer, had established the position of Sergeant Major of the Marine Corps in 1957. The United States Army followed suit in July 1965, with the position of Sergeant Major of the Army. Talk of a similar position in the Air Force had been opposed at Headquarters, USAF, officially because there were already many channels of communication for the enlisted force. A more gripping reason was voiced unofficially by General John P. McConnell, Chief of Staff, who thought the position would be used as a means of circumventing channels.

As he did on so many occasions, Congressman Rivers stepped into the fray, proposing legislation that mandated a senior enlisted position in all four services. His bill was not passed, but his message was, and on October 24, 1966, General McConnell announced his decision to create the post of Chief Master Sergeant of the Air Force (CMSAF), stating bluntly, "The man selected to fill this job will be used as a representative of the airman force when and where this is appropriate and will serve as a sounding board for ideas and proposals affecting airman matters. It is not intended that he be in the chain of command or on the coordinating staff, but he will have unrestricted access to the Air Staff."

The position description was placed in slightly more cosmetic terms officially. The responsibilities of the CMSAF are "to advise and assist the Chief of Staff and the Secretary of the Air Force in matters concerning enlisted members of the Air Force." These matters were understood to include morale, training, welfare, pay and allowances, discipline, and promotion policies, among others. Traditional duties included representing the enlisted force at official social functions, serving on various enlisted welfare boards, and accompanying the Chief of Staff on visits to bases.

In the finest tradition of the noncommissioned officer corps, each CMSAF has interpreted this guidance broadly, expanding the functions and responsibilities to suit his management style. Far from being a backdoor means of circumventing normal channels, the office has become a major relief valve for complaints, often settling things before they ever become an issue, simply because the CMSAF has so much credibility with the enlisted force.

In hindsight, it is easy to see what the Air Force accomplished: it raised the hallowed role of the squadron or wing first sergeant (sometimes called the sergeant major, and always informally the top kick or the first soldier) to a new level, but with the same trappings of confidence and authority. All airmen and officers worth their salt know that, arriving on a base, the number one person to make a friend and confidant of is the first sergeant. No matter what the difficulty—a need for leave, a shortage of housing, a bully in the ranks—the first sergeant could solve the problem legally if possible, by other means if not. When the CMSAF position was created, it inevitably carried with it the cachet and power of the first sergeant in vastly magnified form.

On April 3, 1967, the first Chief Master Sergeant of the Air Force, Paul W. Airey, was sworn in. In 1944, Airey had been shot down and made a prisoner of war on his twentieth combat mission, flying out of Italy as a B-24 radio operator and waist gunner with the Fifteenth Air Force. He might be described as cool under fire; during his parachute descent he tore up his code papers, then lit up a cigarette as he enjoyed the view.

Prison life was tough; over the next year, his weight dropped from its normal 150 pounds to 100, but he survived to be liberated by British forces in May 1945. He remained in service, and in the course of his career, he served as NCOIC of communications at Naha AB, Okinawa, and there received the Legion of Merit for the anticorrosion, antifungus procedures he devised for electronic equipment. He then spent the next half of his career as a first sergeant.

Selected as CMSAF after an intense competition, Airey made an early personal decision that the official responsibilities with which he was charged were merely guidelines; each CMSAF should expand the office to suit his or her special talents. He also made sure that the CMSAF was seen as a spokesperson for the enlisted force, one who was neither a front for the Pentagon nor a lobby for malcontents.

He came to office at a bad time. The Air Force suffered its lowest retention rate in more than ten years in 1967. Airey attributed this not to the unpopularity of the Vietnam War, as most did, but to poor pay, too-numerous remote assignments, an inequitable promotion system, and excellent opportunities for civilian employment. He set about lobbying to change the system, principally by speaking his mind in all the many forums to which he was invited, including testifying before Congress.

Airey's approach has been followed by the eleven Chief Master Sergeants of the Air Force who have followed him. Each has adapted the role of CMSAF to suit his own talents. None has allowed the office to become the out-of-channels conduit that General McConnell feared. The Pentagon, like any large bureaucracy, has its own mores; one of the most important of these is never to allow someone you are working with to be blindsided. Early on, Airey made a practice of networking with the action officers in all the departments concerned with enlisted affairs, and if he knew of trouble brewing—a riot at one of the bases, a congressional investigation of an enlisted person's complaint, anything—Airey made sure the appropriate action officer knew about it before he told the Chief of Staff. If the Chief happened to call the action officer to inquire what was being done, the action officer could always say, "Right, General, we're working on it, and I'll have a report for you in the morning." Airey's practice has been followed religiously by succeeding CMSAFs, and the result has been an extremely cordial relationship with the Air Staff.

With the passage of time, the prestige and influence of the position of Chief Master Sergeant of the Air Force has grown. All the CMSAFs have been called upon to travel to bases with the Chief of Staff and the Secretary of Defense, who rely increasingly on their advice on enlisted matters. And the CMSAFs have changed with the times. CMSAF Airey began his career with B-24s and Morse code; he carried out his duties at the Pentagon with the aid of a telephone. The tenth Chief Master Sergeant of the Air Force, David Campanale, was not only on the road continuously, he was available on e-mail and had a direct line to his constituents. Campanale, an intense, focused personality with a quick smile and a grasp of all the major Air Force issues, had, like his predecessors, adapted the office of CMSAF to his own personal style. He believed strongly in the chain of command; if he was asked a question, he answered, but was not hesitant to tell the questioner that the information should have been available at squadron or wing level, as applicable. If he found the answers were not available where they should be, he swung into action, having identified a problem.

In 1970, the fabric of the noncommissioned officer corps was further strengthened by the establishment of the position of senior enlisted adviser to commanders. There are now about 230 senior enlisted adviser positions throughout the Air Force, filled by senior NCOs who channel information to and from the enlisted force to the unit commander. (Airey hates the term "senior enlisted adviser" as not being sufficiently military. He would prefer the position to be called "Chief Master Sergeant of the Air Mobility Command" or "Chief Master Sergeant of the 1st Tactical Fighter Wing.")

Initially, the senior enlisted adviser position was opposed by many ranking officers in the field who felt they were establishing a mole in their own office, one whose loyalties would run to the enlisted men below him and to the CMSAF above him. They were entirely wrong, for the senior enlisted advisers feel the same primary loyalty to the Air Force as a general officer does. The position quickly proved itself, and is now regarded as indispensable to the smooth operation of the service.

ENLISTED PROFESSIONAL MILITARY EDUCATION

The opportunity a recruit has for getting further education is one of the most important tools of a recruiter. The Air Force offers a wide spectrum of such opportunities, some general, and some, like the NCO academies, tailored to leadership requirements.

The first NCO academy was established by USAFE in 1950; it was followed by a SAC counterpart in West Drayton, England, in 1952. Other commands quickly followed SAC's example, and NCO academies have become an indispensable part of a noncommissioned officer's career path. In their early days, veteran NCOs sometimes went to great lengths to avoid attendance, feeling that there was something demeaning or at least unmilitary about going back to school midway in a career. No more; attendance at the NCO academies is hotly sought after, for it "adds weight to the WAPS (Weighted Airman Promotion System)." The NCO academies also conferred additional prestige to the senior NCOs who were named to be their commandants.

As the positive results of the command NCO academies became manifest, a decision was made in 1972 to create the Senior NCO Academy at Gunter AFB. The mission of the Senior NCO Academy was to prepare the top three grades in the enlisted force structure to handle the new challenges of the 1970s. The first class graduated on March 3, 1973; among the 120 graduates were three future Chief Master Sergeants of the Air Force, CMS Thomas N. Barnes, CMS James M. McCoy, and CMS Sam E. Parish.

The Senior NCO Academy conducts 280 hours of training for approximately 360 students per class. The typical student is a thirty-nine-year-old Senior Master Sergeant with nineteen years of service and three years of college credit. Graduation from the Senior NCO Academy is a requirement for promotion to Chief Master Sergeant.

Top NCOs are but part of the Air Force's concern for educating the enlisted force. It has always encouraged education, and in 1950, only three years after its formation, it established the Extension Course Institute, which offered correspondence courses that were especially useful to personnel on assignment in remote bases. The results of the ECI courses and attendance at courses taught on the bases were helpful, but more was needed if the enlisted corps was to have a higher representation of college-educated personnel to meet the sophisticated demands of modern technology.

THE COMMUNITY COLLEGE OF THE AIR FORCE

The most directly useful extension of professional military education for enlisted personnel is the Community College of the Air Force (CCAF). The college was created as a direct response to the twin phenomena of the noncommissioned officer corps taking over midlevel managerial positions from officers and the movement of those positions into the high-technology disciplines. (By 1972, more than 70 percent of Air Force jobs were considered to be in high-technology fields.)

To raise the standard of NCO education to the requisite level required a program that would be open to all members of the enlisted force and would provide them with a tangible benefit for participating. Air Force enlisted personnel found it difficult to bring their educational experience, both formal and informal, into a coherent pattern that related directly to their jobs. Civilian schools offered programs providing a wide range of academic course work, but few offered credentialing related to many Air Force specialties, such as munitions or missile maintenance. Air Force personnel also faced the common problems resulting from frequent relocations—credits would be lost in the transfer process, and course work would have to be duplicated to satisfy residency requirements.

After a number of intensive studies, Chief of Staff General John D. Ryan agreed to the establishment of the Community College of the Air Force, which was activated on April 1, 1972, at Randolph AFB, Texas. The CCAF would establish programs that would give credit for enlisted job training and experience and be a center for the accumulation of credit toward a degree.

At the time of its inception, the CCAF drew on the staffs and curriculum of seven major Air Force training schools: the five Air Force Schools of Applied Aerospace Sciences, the USAF School of Health Care, and the USAF Security Service School. The first major task of the CCAF was to implement a two-year program that would broaden the noncommissioned officer as a technician, manager, and citizen.

The Community College of the Air Force grew swiftly, affiliating with other technical and professional schools and becoming accredited on December 12, 1973. By 1980, the Commission on Colleges accredited the CCAF to award the Associate in Applied Science degree.

The expansion was not without problems, particularly that of maintaining a faculty with appropriate credentials. A major step forward was taken in 1994, when the college began registering other service instructors in the Instructor of Technology and Military Science degree program.

By July 1993, it had become the largest multicampus community college in the world, with affiliated schools in thirty states, the District of Columbia, and eight foreign locations and more than 6,700 faculty members. More than 275,000 Associate in Applied Science degrees have been awarded, and annual registration is about 380,000.

SIXTY YEARS OF PROGRESS

In sixty years, the enlisted force has been transformed from "old Army" into a space-age Air Force team. The process has been possible only because of the dedication, experience, and ability of a phenomenally devoted noncommissioned officer corps. The Air Force has created a class of first-rate executives who have direct access to the enlisted personnel and who lead them by example. There is a poignant patriotism involved in this, one that hearkens to the fundamental basis of this country. The noncommissioned officer corps is smart, and knows that it is underpaid and overworked. Yet the pride and satisfaction it derives from doing a difficult job in an excellent manner, *and being widely recognized for it*, keeps the ranks full.

SOCIAL FACTORS

Amid all these considerations of pay and promotion, the Air Force was rocked, as the nation was, by the events in Vietnam, and by the growth in racial problems. The USAF was caught off guard; it may not always have been perfect in its handling of such critical social questions as the integration of women and minorities into its ranks, but for the most part it had led the way. Early efforts at integration had been highly successful, more so than in any of the other services and certainly much better than in the country as a whole. Therefore the smoldering resentment of the African-American recruit of the 1960s and 1970s came as a surprise.

The path to integration of African-Americans into the Air Force may be best characterized as always superior to that of the civilian community, but uneven in execution. There is no little irony in the fact that integration ultimately came about as a direct result of enforced segregation during World War II. The famous Tuskegee airmen "experiment" resulted in the establishment of the segregated 99th Pursuit Squadron and subsequently the 332nd Fighter Group. In these were a body of skilled black airmen and ground personnel who proved their worth in combat against very difficult odds.

Many of the leaders of the U.S. Army Air Forces were not enthusiastic backers of integration, nor entirely approving of the performance of the 332d. General Arnold had insisted that black officers and airmen could only serve in segregated units because of the explosive social issues involved. Generals Spaatz, Eaker, and Vandenberg all commented that the relative difficulty and expense of sustaining segregated units in combat was not an efficient use of resources. What they failed to perceive was that the nature of the Army Air Forces as a service (as was subsequently true in the Air Force) made *segregation itself* the barrier to efficiency.

The barrier was not merely a moral one; it was also a practical one. African-Americans constituted about 7 percent of the Air Force enlisted force and about 0.6 percent of its officer corps. Given the disadvantages then implicit in African-American society in terms of education, cultural bias, and other factors, it was impossible to maintain segregated units with an equitable distribution of ranks and skills. The most prominent unit, the 332nd, could not be expanded upon mobilization, because of insufficient resources, nor could it be sustained with replacements if committed to combat. There were some specialties in which there were more qualified African-Americans than there were vacancies, and some specialties in which there were more vacancies than there were African-Americans to fill them.

Five men were most influential in the smooth Air Force preparation for integration. They include the Secretary of the Air Force, Stuart Symington, and his assistant, Eugene Zuckert, and Lieutenant General Idwal H. Edwards, Major General Richard E. Nugent, and Lieutenant Colonel Jack F. Marr.

Based on his business experience, Symington was convinced that integration made sense from both a moral and a cost-efficiency perspective. He was backed in his beliefs by Secretary of Defense James Forrestal and ably supported by Zuckert.

Under Edwards's prompting, a group under the leadership of General Nugent was appointed to study the problem of integration. It transpired that Edwards, Nugent, and Marr all saw the problem in the same light. Segregation was inefficient and could not be made efficient; integration would be efficient and solve the problem of the distribution of talent.

Marr, later described by Zuckert as "the indispensable man" in the integration process, wrote the plan that antedated Truman's order to integrate the service. Marr also stayed on to see his program adopted as a means of carrying out the President's order. His approach to the question of integration was entirely pragmatic: the available pool of talent in the African-American community could not be ignored, and it could not be used efficiently while the practice of segregation continued.

Under Edwards's recommendations, Air Force policy unequivocally endorsed President Truman's Executive Order 9981 and let it be known that ungrudging compliance was expected with both the spirit and the letter of the law. Perhaps the most important facet of Air Force policy was that it made local commanders fully aware of their responsibility for compliance. There could be no passing of the buck—the message was clear.

The result was a quick and uneventful transition of the Air Force from segregated to integrated status. The personnel of the 332nd, under the direction of its commander, then Colonel Benjamin O. Davis, Jr., were distributed to other Air Force units. This great American, the first black general officer in the USAF and the son of the first black general in the Army, had first led the 332nd to its height of success, and then led the fight for integration.

By June 1952, the personnel of the last all-Negro (in the term of the time) unit in the USAF had been distributed without comment throughout the Air Force. The Air Force had completed its integration process almost before the Army had begun. The Navy made only token efforts during the same period. (The Korean War spurred integration in the Army, when the demand for soldiers overcame the hidebound preference for segregation.)

The black community had watched Air Force performance closely. Indeed, one of the real watchdogs of the civil rights movement, the influential African-American-owned *Pittsburgh Courier*, ran an article on October 22, 1949, with the headline "The Job Is Done—Air Force Completes Integration." The story was a little premature, but is indicative of the appreciative view held at the time by the African-American community. A few years later, the *Courier* ran stories with headlines such as "Tan Fliers Over Korea" and highlighted not only then Captain Daniel "Chappie" James, on his way to becoming the first African-American four-star general, but also First Lieutenant Dayton W. "Rags" Ragland, a 336th Squadron F-86 pilot who shot down a MiG-15 on November 28, 1951.

The Air Force's efforts to enforce policy had mixed effect in communities near bases. The commander of Harlingen Air Force Base, located near Brownsville, Texas, received a letter from a local white church inviting all officers and enlisted personnel of the base, regardless of race or color, to attend any or all church services. This was a precedent-setting event; unhappily, it was not matched in all American communities.

Trouble was expected—and found—in the South. The Air Force might have integrated, but the South had not, and assignment to a Southern base could be torture for an African-American serviceman and his family. Travel was difficult. Common carriers like trains or buses were segregated. If a car was used, there could be problems just getting gasoline and oil, and there were few restaurants and fewer hotels where African-Americans were permitted to enter. If a trip had to be made by car, families soon learned to carry their food with them and, if possible, find accommodations at the home of a friend or acquaintance.

Life was bitterly divided: on base, the African-American was treated as a professional; off base, he and his family were subject to all the humiliations of local custom, including separate drinking and rest room facilities and the requirement to be servile with any white person.

The North was often not much better, particularly in rural areas where one might have thought the great traditions of freedom and independence held sway. The contrary was true; African-Americans were not well treated, and were excluded from service facilities in town even more rudely—and sometimes violently—than they were in the South. African-American airmen at bases in Maine, Michigan, Montana, North Dakota, South Dakota, and elsewhere complained to the National Association for the Advancement of Colored People and to their Congressmen. Pressure was applied on the Air Force to use its economic powers to force local townspeople to alter their attitudes.

The Eisenhower administration was curiously passive in its stance on civil rights, but President John F. Kennedy took the problem to heart, and on July 26, 1962, established the Committee on Equal Opportunity in the Armed Forces, headed by Gerhard A. Gesell. The Gesell Committee, as it was known, found that living conditions off base were intolerable for African-Americans and urged the services to take positive action to change the situation. The Air Force was nonplussed at every level. Local commanders knew that their jurisdiction ended at the boundaries of the base, and that they could jeopardize relations with the townspeople for all their personnel if they attempted to force integration off the base. Yet they also knew the Gesell Committee's analysis was correct. African-Americans were systematically denied housing and entrance to service facilities like barbershops. For dependents, the situation was usually even worse, for the question of integrated schools was perhaps the most inflammatory of all.

It was not until the passage of the Civil Rights Act of 1964 and the Voting Rights Act of 1965 that local base commanders could begin to make their influence felt. The Air Force revised and expanded its AFR 35-78 with the passage of the Civil Rights Act. It would not be until 1969, however, that commanders could instruct their personnel that they could not rent or lease property that was not available to all, regardless of race or color. The regulations had almost immediate effect in every area except the Deep South, where additional efforts were required to overcome the centuries-old viewpoints. In 1970, the regulation was reworded to read, "Commanders will impose off-limits sanctions against all segregating establishments . . . that discriminate against military personnel and their dependents."

It was social engineering, and in his 1968 book *The Essence of Security: Reflections on Office*, Secretary McNamara admitted using the military to attack what he termed "tormenting social problems." Military leaders resisted and resented his tactics at the time, and predictably, Southern politicians saw it as the beginning of a police state. McNamara's policies worked, however, and despite the war in Vietnam, the military continued to lead the country in its efforts at integration.

Perhaps because of this, the Air Force completely misread the trend on racial relations in the country and became complacent until the four-day race riot at Travis Air Force Base, California, May 21 to 24, 1971.

An accumulation of perceived slights precipitated the riot. There were complaints that punishments awarded for the same offense were different for whites and African-Americans and that equal-opportunity housing regulations were not being enforced. The use of the clenched fist salute was forbidden by the base commander. Entertainment facilities were not suited to the taste of young African-Americans. Everything came to a head when a fight broke out in a barracks over the volume of a record player.

After four days, 135 airmen were arrested, including twenty-five whites. Air Force embarrassment was extreme, for the event was played up nationally, and great emphasis was given to the fact that seventy civilian lawmen had to be brought in. There were more than thirty injuries, and one death—from a heart attack.

The Air Force was complacent no longer, nor was the Department of Defense. Education in race relations was made mandatory for all personnel, regardless of rank. Officers now had to be rated on their "Equal Opportunity Participation" on their efficiency reports.

In the years that followed, USAF efforts in the field of race relations have been as sensitive as possible. The linking of personal performance reports to efforts on the behalf of equal opportunity had an immediate salutary effect. Although there is no question that some individuals retain bigoted views, they keep them masked or risk immediate and forceful censure. The Air Force has not been 100 percent successful in its efforts to achieve equality in all aspects of service life. It has not been able to raise the percentage of black pilots significantly, despite extensive efforts to do so. It has not achieved the desired representation of minorities in higher grades, but continues to address the problem. One difficulty is competing for the services of highly qualified minorities in the recruiting process. A well-qualified, well-educated member of a minority will be offered civilian employment at a better salary and with greater benefits than the service can provide. A bright young member of a minority who has entered the service and done well is similarly an attractive potential employee for civilian firms. General Dougherty tells the story of a tremendous push made by Secretary of the Air Force Verne Orr to have more African-American and women generals. Dougherty and others tried to explain to him that minorities who were potential candidates for general were snapped up by industry with salaries and perquisites that the service could not begin to match. Orr would not buy it until he sat down and perused a huge pile of personnel records that proved that fast-track minorities were siphoned off by more attractive opportunities.

WOMEN IN THE AIR FORCE

In 1965, the total number of spaces authorized for women had fallen to 4,700, and these were confined primarily to the clerical, administrative, personnel, information, and medical fields. Some say that the decision to reduce the number of opportunities offered women stemmed from General LeMay, although others deny this. During the 1960s, women were not allowed to serve as flight attendants or in specialty positions in intelligence, weather, equipment maintenance, or control towers—despite the fact that they had served in all of these roles during World War II and into the 1950s.

The turnaround came when then Colonel Jeanne M. Holm was appointed Director, Women in the Air Force, in November 1965. She was to have an enormous influence on the role of women in the Air Force. Her distinguished career began as a Women's Auxiliary Army Corps (WAAC) enlistee in 1942. On July 16, 1971, she became the first woman in the Air Force to rise to the rank of brigadier general, and she was promoted to be a major general on June 1, 1973—again, the first woman to serve in that grade.

During General Holm's tenure as director, WAF strength more than doubled, assignment opportunities were greatly expanded, and uniforms were modernized. As a result of her many initiatives, women were soon allowed to enter almost all but combat positions. The most difficult hurdle, perhaps, was admission of women to flying school, which began in 1975.

The success of the integration of women into the service depended in every instance on the attitudes of the local commanders. Some difficulties occurred because there were a few dinosaurs who opposed the idea of women in the Air Force on principle—but not for long, as they soon discovered that another pool of unlimited talent had been opened. A more subtle problem area was male pride, which often was harder to overcome.

General Dougherty tells a story of some difficulties he encountered when introducing women into the Strategic Air Command's maintenance program. One of his key maintenance people was adamantly opposed to women's working on the flight line because he did not believe they were physically capable of doing the required work—or so he said. One morning he came in with what he thought was an iron-clad argument against their employment. The SAC maintenance manual called for changing alternators on a Boeing KC-135 tanker in one hour. Dougherty's maintenance man crowed that a woman just could not do the job, which required holding the heavy alternator while four bolts were removed, then replacing it with another, holding it in place, and inserting the bolts.

Dougherty asked, "How many electricians do you have on the base?" The man replied, "Probably a hundred and fifty." Dougherty said: "Get them all out on the ramp right now, and we'll see how many of the men can do it." As Dougherty expected, few of the men could do the job without assistance—and positive leadership resolved another impediment to using women in all jobs.

By 2005, women constituted almost 20 percent of the Air Force, with 13,000 officers and 60,000 enlisted. Over 99 percent of the career fields and positions in the Air Force are open to them. The few closed positions are in accordance with DOD policy, which excludes women from assignments to units whose primary mission is to engage in direct combat on the ground. No career fields are closed to women officers, but they are restricted from certain positions, including certain types of combat helicopters. The chronology included in this book documents the rise of women in the Air Force, but certain key events stand out, the most important of which was the establishment of the All-Volunteer Force in 1973. DOD policies followed which initiated a sex-neutral policy for spouses' entitlements and rescinded previous requirements to involuntarily separate women because of pregnancy and/or parenthood.

The apogee of the service revolution came with changes in the combat exclusion policy that allowed the first large-scale deployment of women to a combat zone in Operation Desert Shield. Of the total of 100,905 Air Force personnel deployed, 12.4 percent were women. The trend toward total equality culminated with a suspension of combat aircraft restriction on women on April 28, 1993; First Lieutenant Jeannie Flynn became the first Air Force woman pilot assigned to an operational fighter aircraft, the McDonnell Douglas F-15E. Lieutenant Flynn's achievement was complemented in 1995 by Lieutenant Colonel Eileen Collins, who became the first Air Force woman to serve as Space Shuttle pilot. Captain Nicole Malachowski was named to be the first woman pilot in the Air Force's world-famous Thunderbird squadron. She was a 1996 graduate of the Air Force Academy.

BUILDING ON THE FOUNDATION

The management revolution took full advantage of the potential of the capabilities of its enlisted personnel and exploited the new reserves of talent made available by the full integration of women and minorities into every aspect of Air Force life. This provided a solid foundation for the Air Force's renaissance after Vietnam, when new leaders, new doctrine, and new technology came together over a twenty-year period to produce a service worthy of a superpower.

8

Leadership, Doctrine, and Technology After Vietnam

*O*ver the centuries, the armed forces of combatant nations have rebounded from defeat in very different ways. Following the defeat of its armada by England's emerging (but at that time inferior) naval might in 1588, the Spanish navy lapsed into decay. The French, humiliated in the 1871 Franco-Prussian War, so thirsted for revenge that they leaped with élan into the maw of World War I, where they were saved from another disaster by their allies. The German army reacted to its loss in 1918 by viewing it merely as the conclusion of Round One. Two decades later, it began Round Two with confidence, made almost exactly the same mistakes, and lost again.

After enduring the embarrassing—and unnecessary—American defeat in Vietnam, the United States Air Force might have rebounded in any manner, from lapsing into a defeatist lethargy to thirsting militantly for revenge. Fortunately, a combination of good leadership from the top and the clear call for reform from below combined to set the Air Force on a different path, one that led first to space and then to victories in the Persian Gulf and the Cold War.

CONGRESSIONAL HURDLES

Just as the year 1972 was a vantage point from which to review the contributions of the major supporting commands, so it provides an excellent perspective for an examination of how the Air Force learned from and adapted to the experience of Vietnam, even as it maintained its awareness of contemporary changes in domestic and international politics. This learning process, still going on today, influences the USAF in matters of policy, leadership, and weapon systems.

The environment was totally different from the days of World War II, when the public, the media, and the Congress vociferously backed the armed forces. In the post-Vietnam period, the USAF had to conduct its fight on many fronts. Its primary mission, from which it never deviated, was the deterrence of aggression by the Communist powers. Secondary fronts arose from year to year—the *Pueblo* and *Mayaguez* incidents, Lebanon, Central America—that had to be dealt with. The third front was the most difficult, however—continually persuading the Congress first to approve and then to sustain procurement of necessary weapon systems. Getting the necessary approval for a needed weapon system had always been difficult; before World War II, for example, the Congress in its wisdom had preferred procuring more of the less capable Douglas B-18s, a twin-engine imitation of a bomber, rather than buying fewer of the far more capable four-engine Boeing B-17s. But in the post-Vietnam environment, every program was given a hard managerial scrub by teams of professional congressional staffers who often had far more experience in the business—and sometimes even in the particular weapon system—than the Air Force personnel presenting their cases. The staffers recommended to their committees which firms should win or lose in the tough competition among them for budget dollars. Congressmen had a natural penchant to support programs that provided jobs and dollars to their own constituencies, while either opposing procurements that went elsewhere or horse-trading support. It thus became a part of every procurement pitch to show exactly where the program dollars were to be spent and how many jobs would be generated in which congressional districts. If there were not sufficient dollars being spent in the right district, you can be sure that the contractor would make the necessary adjustments.

A further problem was more sensitive. There have been—and probably always will be—a number of Congressmen whose views make them automatically oppose any military procurement or R&D initiative, irrespective of its merit. To do so is entirely within their right, a part of the healthy democratic process of airing opposing views. However, a succession of elected officials, including Bella Abzug, Les Aspin, Edward Brooke, Ron Dellums, William Proxmire, and Patricia Schroeder, among others, so identified themselves with antimilitary views that their opposition became reflexive. Driving a desired procurement through these intractable congressional roadblocks takes patient, diligent effort, and consistently strong leadership.

LEADERSHIP AND CHANGE

In the two decades after the Vietnam War, the United States Air Force has had a succession of fifteen Secretaries of the Air Force and ten Chiefs of Staff, including those serving as "acting." Space prohibits a detailed examination of each of their contributions, but in general terms, the Air Force has been fortunate in its leaders, who seemed to have an uncanny adaptability to the requirements of the time, while always managing to keep the vital research-and-developments efforts going. One remarkable aspect of the leadership chain is the continuity of purpose in pursuing weapon systems through many different tenures of office and over long periods of time. (A list of all Air Force Secretaries and Chiefs of Staff may be found in Appendix 1.) Moreover, most of those leaders had the great good sense to foster the turbulent demand swirling in the ranks for improvements in the weapon systems.

As noted earlier, the Defense Reorganization Act of 1958 put greater power in the hands of the Secretary of Defense, power that was applied vigorously by Secretary McNamara and his successors. The effects of the 1958 reorganization were sometimes shocking. The service secretaries had been moved so far out of the decision-making loop by 1972 that the Secretary of the Navy was not informed about the mining of the harbors of Haiphong and other North Vietnamese ports, and Secretary of the Air Force Robert C. Seamans learned about the air raids on North Vietnam only when he saw them reported on television.

It was obvious that the mission of the Air Force Secretary had changed drastically. In Stuart Symington's day, the Secretary greatly influenced DOD policy; in later years the role was altered to smoothing and streamlining internal Air Force policies and actions. Yet, the loss of political power was partially compensated for by the extra time it provided for the Secretary to work within the service. Air Force success has depended upon technology realized under the leadership of its Secretaries who were also scientists, including Seamans, John L. McLucas, Hans Mark, Edward C. Aldridge, and Sheila E. Widnall, the first woman to serve as Air Force Secretary. All of them sustained the emphasis on research-and-development efforts exemplified by the vision of Hap Arnold, Theodor von Karman, and Bernard Schriever. This R&D effort proved itself in the striking advances of a wide range of weapons, intelligence-gathering systems, and war-gaming techniques that made the United States Air Force ever more powerful even as it exposed the shabby underpinnings of the former Soviet Union.

The American defense industry—one-half of the often-reviled "military-industrial complex"—participated fully in this process, benefiting in some instances from contracts for basic and applied research and in others developing (solely from its own resources) new ideas that were beneficial to the progress of the Air Force. Without the defense industry, the USAF would have been unable to pursue the fertile ideas its R&D had produced. An important interface in this relationship was the "think tanks" such as RAND, MITRE, and others, which were able to gather in a nucleus of scientists and thinkers, pay them at market rates rather than at fixed government salaries, and allow them to give dispassionate advice on research and development.

A LOOK BACK IN TIME

Although it is impossible to recount all of the many achievements and infrequent failures of the Air Force under its successive Secretaries and Chiefs of Staff, it is instructive to look at highlights to see how some of these leaders performed during the period from the middle of the Vietnam War until the end of the Reagan administration. This period reflects the roller-coaster fall and subsequent rise of the Air Force and carries it to the threshold of its greatest triumphs.

FIGHTING THE WAR: 1965–1969

The two top leaders in the Air Force during the period from 1965 to 1969 were not a good match. Secretary of the Air Force Harold Brown was a scientist, an intellectual who was definitely not a "people person." Brown was far more in the mold of Robert McNamara than in that of his Chief of Staff, General John P. McConnell. McConnell was hampered by having to fill General LeMay's shoes, and in his later years of service by an unfortunate drinking problem. Nonetheless, he presided over the buildup of the United States Air Force during the Vietnam War. One of the innovations of his tenure was the USAF modified "Total Force" concept of 1966, which brought U.S. Air Force Reserve and Air National Guard units into daily operations and was subsequently widely adopted by DOD. More important, he fostered the accelerated research and development that attempted to fix the deficiencies in USAF combat aircraft that the war in Vietnam exposed. Major decisions he and his successor as Chief of Staff, General John D. Ryan, made were driven by the Vietnam War, which under President Lyndon Johnson's guns-*and*-butter building of the Great Society was fought entirely from Air Force budgets that had not been increased to bear the additional expenses. McConnell had the bitter duty of reporting on his retirement that he left the Air Force with "less air power than when I became its Chief of Staff four and one-half years ago," one of the last comments that any Chief wishes to make.

DIFFICULT YEARS: 1969–1973

President Nixon selected the able Robert C. Seamans, Jr., as his Secretary of the Air Force. Seamans and his Chief of Staff, General Ryan, were an excellent match. Both men had pleasant personalities and an unruffled manner of doing business. Seamans was also a scientist and an intellectual, but unlike Harold Brown, he related well to people. Ryan, an experienced combat leader and former Commander in Chief of the Strategic Air Command, was also a man who demanded rigorous honesty of himself and all his colleagues. He displayed his rigor in 1972 when he relieved the Commander of the Seventh Air Force, General John D. Lavelle, for allegedly violating the rules of engagement and falsifying records to conceal the fact. (Many maintain that Lavelle had been implicitly encouraged by his superiors, including then Secretary of Defense Melvin Laird, to undertake the actions for which he was held accountable, and then, like a good soldier, had to take the fall.)

Seamans and Ryan had to prosecute the Vietnam War with an Air Force whose increasingly obsolete equipment was wearing out and that was running low on some key supplies of spare parts and munitions. Despite their necessary preoccupation with this task, they used their combination of intelligence, compassion, and energy to facilitate a whole series of important USAF projects, including aircraft such as the A-10, B-1, C-5, F-15, and F-111. And in a manner that was a fortunate characteristic of many of the top Air Force leaders, they also gave impetus to the sophisticated electronic warfare devices that would determine the outcome of future wars.

It was the astute Seamans who noted that the Soviet Union did not appear to be conducting its research and development and its production of weapon systems on the basis of what the United States was doing. He observed that instead it was increasing its military spending as its gross national product increased, without regard to the impact on the civilian economy. Further, it was carrying out extensive prototype programs for aircraft and missiles on a scale beyond anything the United States was attempting. He also noted an aspect of the Soviets' activity unusual from the American perspective: they operated at a sustained level of effort in their output of prototypes and production aircraft, and they almost never terminated a program, even when it overlapped another in capability. This was in sharp contrast to the United States, where programs were always vulnerable to termination regardless of their technical importance. The volume of production and the type and number of prototypes the Soviet Union was placing into production pointed to a desire to gain the very first-strike capability that the United States had abjured. The importance of this insight was overlooked until the Reagan administration.

REBUILDING FROM WITHIN: 1973–1976

The Nixon administration brought about remarkable changes in the Department of Defense under the leadership of two Secretaries, Melvin Laird (1969–1973) and James Schlesinger (1973–1975). Many consider Laird, a nine-term Congressman, to be one of the most outstanding Secretaries of Defense in history. He was even-tempered and used his political wiles to mend fences with the JCS even as he superintended the "Vietnamization" process, which phased out American participation in the war. He backed ending the draft and the creation of an All-Volunteer Force. To this end, he directed that the 1966 Air Force concept of using the Air National Guard and the Air Force Reserve in daily operations be followed by other services.

Schlesinger, his successor, followed up on this in 1973 with a call for full integration under the "Total Force" policy, in which members of the National Guard and Reserve were the initial and primary source for augmentation of active forces in emergencies requiring a rapid and substantial buildup. There were many who did not believe that these forces could be brought to meet regular service standards. They were wrong. Air National Guard and Air Force Reserve units so quickly achieved a readiness and proficiency equivalent to regular Air Force standards that General Ryan called their work a "proud chapter in Air Force history." They have since demonstrated their proficiency on a daily basis, in war and in peace. (The change in twenty years was dramatic.

When two future four-star generals, David Jones and Russ Dougherty, were young company-grade officers in a Louisville, Kentucky, airlift reserve unit, they had only single-engine, two-seat North American T-6 trainers to fly. Jones grins now and says, "You can imagine our airlift capability.")

Schlesinger was a much different sort of man from Laird. An academic, he was cold, pedantic, and without Laird's political insight. Yet he was wise enough to look for personable people he could work with—and who could work with Congress. He found one in genial John McLucas, who had succeeded Seamans as Secretary of the Air Force, and another as Air Force Chief of Staff in the hugely popular General George Brown. Brown's star was so ascendant that General Ryan, even before he formally took office, had informed Brown that he would succeed Ryan, and that he had better prepare himself. Despite his meteoric rise, Brown was an unpretentious man who insisted that he not be given the customary military honors upon arrival at a base.

His greatest personal triumph came during the Yom Kippur War in 1973, when on his own, without coordinating with the Secretary of Defense, he allocated two squadrons of McDonnell Douglas F-4s—then the standard USAF fighter—to Israel and began preparations for a massive airlift of munitions. Ironically, Brown would be pilloried as anti-Semitic in November of the following year after a speech at Duke University. An unfortunate choice of words led to calls for his dismissal from his new appointment as Chairman of the Joint Chiefs of Staff.

There is no question that Brown's candor with the press bordered on naiveté. He was caught up again by statements that could be interpreted as anti-Semitic and later in an interview that appeared to criticize the armed forces of the British Empire. Nevertheless, Schlesinger and his successor, Donald Rumsfield, always stood by their man, whose obviously sincere apologies managed to smooth things over. They both valued Brown's experience working with the office of the Secretary of Defense. He had earned it the hard way, working for Robert McNamara for two years and progressing from colonel to major general in the process. His rapport with Congress—something Schlesinger notably lacked—was outstanding even when under the pressure of his occasional controversial statements. In 1974, he was selected to become the first Air Force Chairman of the Joint Chiefs of Staff since Nathan Twining, some thirteen years before. Brown was picked not because it was the Air Force's turn, but because he was the best man available for the task.

Of all Brown's many accomplishments, one stands out as most important to Air Force navigators, bombardiers, and radar observers, for it was he who forced a change in the regulations that forbade command positions to anyone but pilots.

When Brown was selected as Chief of Staff, his two principal competitors had been Generals David C. Jones and John C. Meyer. Meyer, a fighter pilot, had scored twenty-four victories during World War II and two more in Korea. As CINCSAC, he had directed the eleven-day campaign against Hanoi and

Haiphong in December 1972. Jones had flown 300 combat hours over North Korea in B-29s. His competence had been noted early on by Curtis LeMay, and, like Brown, he had been given assignments that prepared him for the Air Force's top job, including command of an F-4 wing. It was said of Jones that he never left an assignment without having improved the organization he led.

Jones was selected to succeed Brown twice, first as Chief of Staff, and then in 1978 as Chairman of the Joint Chiefs. Of occasionally fiery temperament, Jones made large numbers of both friends and enemies as he approached his task with intellect and energy. He threw himself on the sword as an advocate of the B-70 long after Secretary McNamara ruled against it, believing that with the proper penetration aids, the B-70 could have operated against the Soviet Union at high speeds and altitudes and greatly compounded the Soviets' defensive problem. Later in his career, when President Carter canceled the B-1, Jones accepted the cancellation without resigning, as some had called upon him to do. Jones's response was that his resignation would not do any good, and, further, that the President had campaigned on the issue of canceling the B-1 and had been elected, and Jones was not going to oppose the will of the people.

Jones was an ardent advocate of reforming the Joint Chiefs of Staff and continued to campaign for reform after he retired. He was instrumental in the creation of the Goldwater-Nichols Act of 1986, which incorporated many of the changes he had recommended.

As Chief of Staff, Jones proved to be an innovator in hardware as well as in organization. He was a zealot for the Airborne Warning and Control System (AWACS) aircraft. With the head of the Defense Advanced Research Agency, William J. Perry (later Secretary of Defense under President Clinton), he had the Have Blue stealth program initiated as the first step toward what became the F-117 stealth fighter.

Like Brown, Jones worked well with Schlesinger, whom he carefully cultivated by adapting himself to the Secretary's special interests, of which ornithology was one. He tells an anecdote of the informal way they worked, meeting in civilian clothes on weekends, putting their feet up on the desk, and chatting. One morning Schlesinger asked him, "What would it take to get the Air Force to support a lightweight fighter?" (As we will see below, there was a great debate raging over high-cost fighters such as the McDonnell Douglas F-15E and low-cost fighters such as the proposed General Dynamics F-16 and Northrop F-17.) Jones knew that Schlesinger was not going to offer him any more funds or personnel, but he also knew that it was a chance to expand the force structure, a vitally important consideration, given the declining budget outlook. His reply was, "Four additional wings in the force structure." Schlesinger extended his hand, they shook, and the deal was done. Thus commenced the launch of what became the F-16 Fighting Falcon.

The one incident that Jones remembers with real distaste is the aborted attempt to rescue American hostages being held in Tehran in 1980. In a 1996 interview, Jones said that the biggest mistake in the operation was dividing up responsibility so that every service had a part to play, which resulted in the use of Navy helicopters and crews not experienced in rescue operations. He also faults the inability to rehearse the operation because of the absolute security requirements.

DIFFICULT YEARS: 1976–1980

When Naval Academy graduate Jimmy Carter was elected President, he came to office convinced that the Pentagon military should be administered by methods somewhere between those of Captain Bligh's handling of the lower elements on the *Bounty* and Admiral Hyman Rickover's treatment of aspirants to nuclear submarine command—in other words, treat them rough. In his election campaign he had promised to make the Pentagon more efficient, halt the B-1 program, and arrange an arms agreement with the Soviet Union.

Carter made little effort to conceal his contempt for the JCS. The JCS, in turn, did not trust Carter or his judgment, an attitude that quickly spread through the military. Carter shook the JCS to its roots with his inquiry as to how quickly the United States could get rid of its nuclear weapons, for he believed that as few as 200 atomic and thermonuclear devices would be a sufficiently large arsenal to deter the Soviet Union. This was essentially similar to the "finite deterrence" straw for which Navy officials had grasped in the 1959 roles and missions dispute, and was now identified as a "countervailing strategy."

The President had selected a soul mate, Harold Brown, to be his Secretary of Defense. Both men were engineers with a clinical approach to problems and, unfortunately, to people. Brown amplified the concept of countervailing strategy by stating that an "essential equivalence" with the Soviet Union would be maintained. He explained this to mean:

(a) That the Soviets would not be able to use nuclear forces to coerce other countries;

(b) That if the Soviets had an advantage in one area of armaments, the United States would have an offsetting advantage in another area;

(c) That the U.S. position would not be perceived as inferior to the Soviet position; and

(d) That nuclear stability would be maintained.

Carter's policy and Brown's explanation of it gave scant comfort to an Air Force bred on the LeMay tradition that it was to maintain an overwhelming nuclear superiority. The Navy was equally dismayed, now that it had become part of the strategic triad with the submarine-launched ballistic missiles (SLBMs). The Army was unhappy because it knew that neither it nor the NATO forces could withstand a conventional attack by the Soviet Union, given the latter's overwhelming superiority in troops, armor, artillery, and aircraft. Since 1991, when the Soviet Union seemed to dissolve like a cube of sugar in a cup of coffee, the sinister threat of its powerful forces has been discounted or forgotten. But during Carter's Presidency, the Soviets had an army of 1.8

million men, 50,000 tanks, 20,000 artillery pieces, and almost 5,000 tactical aircraft—all backing what seemed to be formidable Warsaw pact forces. The survival of NATO forces in Europe depended first upon the American nuclear shield, and second upon its tactical nuclear weapons. Carter's 200-weapon proposal dumbfounded the JCS.

In fact, the size of the nuclear deterrent force was not significantly reduced, but Carter did stop production of the B-1, announcing the substitution of the air-launched cruise missile (ALCM) as the preferred alternative. The ALCM had its origins in the AGM-86 SCAD (subsonic cruise armed decoy) and, at the time of Carter's announcement, was planned as a bargaining chip to be given away at the armament limitation talks.

When George Brown retired as Chairman of the JCS in June 1978, David Jones was named as his successor. (Brown died only five months later, a victim of cancer.) Jones's willingness to accept Carter's decision on the B-1 may have played a part in the appointment, but a more important factor was the President's appreciation of Jones's intelligence and the role he could play in the continuing disarmament negotiations with the Soviet Union. Carter then approved the selection of General Lew Allen, considered by many to be the most gifted scientist in the Air Force, to become the new Chief of Staff. Allen was a missile expert who would be crucial in fostering the development and ultimate acceptance (albeit on a far smaller scale than he had hoped) of the advanced MX missile. Carter would work with Allen and Jones to effect a deal: approval of the MX (later named the Peacekeeper) in exchange for military backing on the disarmament talks (SALT) in progress with the Soviet Union.

YEARS OF PLENTY: 1981–1985

President Ronald Reagan campaigned on the requirement to restore the United States' military prowess, and his new Secretary of Defense, Caspar Weinberger, saw to it that every promise was fulfilled. Curiously enough, relations between Weinberger and the JCS were not harmonious at first, even after he had demonstrated his skill at selling the massive budget increases to Congress. Nonetheless, in short order the B-1 program was reintroduced, the MX was approved, Army divisions were brought up to strength, and the Navy was authorized to build to a strength of 600 ships. The secret B-2 stealth program (which had been one of Carter's considerations in canceling the B-1) was still in the "black" (i.e., its budget was a highly classified secret) but received adequate funding. Military spending grew by more than $300 billion in the first four years of the Reagan administration, with about $75 billion earmarked for strategic modernization and another $75 billion for research and development. As welcome as the funds were, they caused concern in the JCS, which feared—correctly—that the funding would not be sustained and that expensive adjustments would have to be made later to programs already under way.

The Reagan program had what appeared to be a huge internal conflict, for it was embarking upon a massive modernization program that was intended to provide the capability to wage a protracted nuclear war and a limited war simultaneously, while it was at the same time negotiating with the Soviet Union for strategic arms reduction. General Jones was concerned that the Reagan team conducting the strategic arms limitation talks (START) might carry disarmament to an extreme position that would jeopardize national security. The Chairman, whose relations with Secretary Weinberger were distinctly uncordial, called on all his political skills to bring the dual tracks to convergence. He masterfully orchestrated a compromise that allowed the armament reduction to begin, but at a pace that did not place national security in danger or interfere with modernization of American strategic weapons.

While Jones and Weinberger sparred at the very highest levels on defense policy, the new Secretary of the Air Force, Verne Orr, worked with General Allen and his successor as Chief of Staff, General Charles A. Gabriel. Orr and Gabriel formed an excellent team that continued the modernization of the Air Force and its growth in capability. Both men were extremely personable and passionately concerned about the well-being of the Air Force and its personnel. Orr was not experienced in military aviation, but had full confidence in both Allen and Gabriel and supported them ably.

LEANER YEARS AGAIN: 1985–1988

Orr was succeeded briefly by Russell A. Rourke, who served only for four months and was then followed by one of the most popular men ever to become Secretary of the Air Force, Edward C. "Pete" Aldridge, Jr. Aldridge was a scientist in the mold of Seamans and McLucas, and his popularity did not impede him from being a coldly calculating deal-maker.

As will be shown, it would fall to Gabriel and his successor, General Larry Welch, to do some of the most farsighted—and in many ways most difficult—planning in Air Force history. They recognized far in advance of others, including other branches of the American armed forces, that the combination of a failing Soviet Union and the inevitable downward revision of military budgets meant that the Air Force would have to be reduced in size. They made a decision to sacrifice force structure for modernization and quality. The success of their efforts will be seen in the next chapter and are still apparent in today's air force.

IT'S A WONDER THAT ANYTHING GETS DONE: THE ADMINISTRATIVE MINEFIELD

In addition to the ordinary problems of running an Air Force alluded to above, there were many other obstacles of varying origins, as the following examples will show.

POLITICAL TURMOIL

Each President brings his own strategic agenda to his office. This agenda is often the product of campaign promises. These will usually be adhered to initially, no matter how out of tune they might be with reality, as in the cases of President Kennedy's "missile gap" and President Carter's campaign vow to cut $7 billion from the defense budget. To fulfill his agenda, the President will pack his cabinet and make changes in the Chairman of the JCS and in the service chiefs. The Chairman and the chiefs must then balance their loyalties to the good of the service with their sworn duty to the President.

The crucial Presidential cabinet choice is that of a Secretary of Defense who will mirror the President's interests. The power of the SecDef is so great that he can virtually dictate to the service secretaries and the members of the JCS, unless they are prepared to fight for their interests with rigorously researched arguments. This exercise of power goes far beyond the enunciation of policy. It frequently became the practice of successive Secretaries of Defense to go against the advice of their service secretaries and the Joint Chiefs of Staff and make their own decisions on the cancellation or acquisition of certain weapon systems. McNamara started the process with his cancellations of the B-70 and Skybolt, even as he ramrodded through the acquisition of the F-111 and the A-7. Such decisions virtually became a badge of office, as in the case of Secretary Melvin Laird with the Manned Orbiting Laboratory (MOL), Secretary Schlesinger with the lightweight fighter, Secretary Harold Brown with the B-1 and the ALCM, and Secretary Caspar Weinberger with a host of acquisitions.

ADAPTATION TO POLICY CHANGES

The amazing comeback of the Air Force occurred despite changing political climates. Credit must be given first to the great architect of nuclear air power, Curtis LeMay, and only then to the successive Secretaries of the Air Force and Chiefs of Staff who managed to keep their eyes on the technological ball and so create the modern Air Force.

Each succeeding administration coined new terms to reflect its spin on a national defense policy that at its heart remained unchanged. The names changed from "massive retaliation" to "assured destruction" to "flexible and appropriate response," which required an even greater buildup of nuclear power than did "massive retaliation." Later there came changes from "counterforce strategy" to "essential equivalence" to "countervailing strategy," yet all of these variations on a theme were feasible only because they were backed by the tremendous nuclear power that the Strategic Air Command's bombers and missiles represented. The latter were supplemented by the Polaris submarine-launched ballistic missile (SLBM), which completed the essential strategic triad upon which America's security is still based. No American strategy, no matter how it was named, would have been respected by the Soviet Union without first SAC's and then the triad's backing.

Rather than being administrative initiatives, the changes in nomenclature were actually mere reflections of each administration's degree of recognition of the growing strength of the Soviet Union's nuclear force. Massive retaliation was an appropriate strategy only so long as the United States enjoyed an over-whelming nuclear superiority. Soviet strength grew to match or exceed that of the United States in many areas, including "throw weight," the number, size, and sophistication of nuclear weapons at its disposal. Successive American ad-ministrations formulated ground rules for slowing down the arms race, sedu-lously avoiding the impression of attempting to achieve a first-strike capability and trying to negotiate arms limits initially, followed by reductions. As a part of the desire to slow the arms race (and to avoid spending the necessary funds), the United States had virtually no realistic civil defense and turned away from a ballistic missile defense system.

Unfortunately for American strategists—and the American people—the Soviet Union did not respond to these leads, as Secretary Seamans had noted. Just as the North Vietnamese government had done, it interpreted any Ameri-can concession as a sign of weakness. It signed agreements limiting certain classes of weapons only when it was hopelessly behind in the development of those weapons. It proceeded all the while doing exactly as it wished, building weapons with first-strike capability and supplementing these with a monu-mental civil defense effort that emphasized the survival of the Communist leadership. The Soviets' capability was immeasurably enhanced by an incredi-ble U.S. government decision in 1972 to sell them the previously restricted Centalign B machines for making the small ball bearings necessary for ex-tremely precise missile-guidance systems. The Soviet Union promptly bought 164 of the machines—twice as many as were in use in the United States—and by 1976 had increased the accuracy of their SS-18 and SS-19 ICBMs to the point that they could destroy an estimated 90 percent of our Minutemen in their silos on a first strike. These are the same missiles and technology that China sought for "commercial space launches."

TECHNOLOGICAL ATTENUATION

The development cycle of a weapon in the 1960s, from start to deploy-ment, was about five years. It is now about fifteen, with the result that new weapons systems inevitably must endure review by at least two and perhaps as many as four administrations. The Rockwell B-1 was born only after a twenty-year gestation period that included abortion by the Carter administration and resuscitation by President Reagan. The contract for the McDonnell Douglas C-17 was signed in August 1981; it entered service in late 1995.

The lengthened cycle meant that Air Force leaders had to determine the weapon systems they needed, then defend them through several changes of ad-ministration and as many as fifteen annual budgets before having the weapons in hand. When the inevitable changes in the programs occurred—increased costs, lengthened schedules, alterations in performance—they had to be de-fended anew. Congressional staffers are smart, competitive men and women, patriots who want the United States to have the best weapons at the most fa-vorable cost. However, it is a given that these staffers have the reelection of their member uppermost in their minds. Like gold miners, they pan weapon systems over their life cycles, looking for the glint of political capital. If something glitters it does not matter if it's fool's gold so long as announcing it will get the member favorable attention in the press. Staffers provided Congresswoman Patricia Schroeder the information for her long and hard campaign against the AWACs as a "billion-dollar boondoggle" (the alliterative term lingers through the years) without regard to the demonstrated need. Conversely, staffers gave Senator John Towers the information necessary to ensure that the General Dynamics F-111 was kept in production long after USAF requirements for it were filled. There are as many more examples as there are weapons systems, and, perhaps, as there are Congressmen.

Changes in Procedures

The very means of doing business has changed drastically over the years. General Larry Welch has commented that when General LeMay was Chief of Staff in the early 1960s, the Air Force did not have a line item for research and development—it had a lump sum that it could allocate as it saw fit. Now R&D items with budgets as low as $100,000 are analyzed separately by DOD and Congress. Further, if LeMay had a program he liked, he had only to confer with perhaps two people in the Department of Defense and four people in Congress—the heads of the four major military committees—to get it approved. Ten years later, General Jones had to deal with as many as five people in DOD and sixteen on the Hill. When General Welch became Chief of Staff in 1986, he found that to push a program forward, he had to deal with at least fifteen key people in DOD and hundreds in Congress.

The number is actually greater, for in the thirty-plus years since General LeMay's tenure, the number of congressional staffers has multiplied by a factor of thirty, and many of them have to be satisfied before a decision can be obtained. During the same period, the influence of the chairmen of the major military committees has declined. The power that Congressman Rivers or Carl Vinson once had to make a decision on the spot has been greatly eroded.

Technical Complexity

Technical complexity has grown almost as fast as political complexity, and the two are inevitably intertwined. Weapon systems requirements for stealth and precision have reached such exotic levels that successful achievement often depends upon breakthroughs in many sciences, such as metallurgy, electronics, and aerodynamics.

Despite the unpredictable path from conception to successful deployment in the field, weapon systems are scrutinized so that the program advocate has to declare from the outset exactly what a weapon system is going to do, how long it is going to take, and what it is going to cost—all information that is simply unavailable at the time.

This situation has a number of side effects, including a sharp reduction in the number of advocates who are willing to risk their careers to secure a needed weapon system when they know they must promise results that are problematic. Fear of failure makes managers tend to be less willing to make the great leap to the next level of technology, and they become subject to the lure of incrementalism, seeking minor advances in performance because they are safer—politically as well as careerwise—even though they might not meet the ultimate requirement.

In the quarter century after Vietnam, the Air Force overcame these and many other impediments in its march to a demonstration of pure air power in 1991.

THE RISE OF THE PHOENIX: FROM VIETNAM TO THE GULF WAR

Vietnam saw the unprecedented phenomenon of first the American media and then the general public turning against the military services for participating in war that was begun, controlled, lost, and ended by the civilian leadership. After 1972, it could well be argued that the Air Force's mascot should be the phoenix rather than the falcon, for it rose from the cold, dead ashes of the Vietnam War to its old place in the hearts of the American public. In the process, it increased its fighting power to an incredible degree with amazing new technologies—and the personnel to match.

LESSONS LEARNED IN COMBAT

The Air Force went from strength to strength in the years after Vietnam, the path made easier by the relatively few fundamental doctrinal changes. The technological focus of the Air Force was adhered to by all its leaders, despite the increasing difficulties imposed by centralized DOD control. The depth and breadth of the Air Force commitment to technology permitted it to endure severe force cutbacks, lean budgets, and a diminishment of its industrial base and still emerge with an air force worthy of the only superpower.

Although the Vietnam War was the low point in United States Air Force history, it nonetheless taught a series of powerful lessons about military operations and, most important, about military operations controlled by the politics of an election-oriented civilian leadership. Despite the speculation of the rabid *Seven Days in May* genre of fiction, the idea of the military taking over political direction as has occurred in other countries in similar situations was never attempted—and probably never contemplated.

Instead, working within the chaos of political change, the Air Force leadership came up with some hard rules that would be difficult to implement but were essential to the future:

1. Air superiority had to be earned immediately; gradualism was the path to disaster.
2. Air superiority, given the expected budget limitations, could be gained only by a huge investment in technology.

3. Significant components of that technology had to be revolutionary rather than evolutionary.

4. In the interim, much existing equipment had to be upgraded by modification rather than replaced.

5. The Vietnam concept of using large numbers of aircraft, each with a relatively low lethality, was wrong, as the Paul Doumer and Than Hoa bridges had conclusively proved.

6. A grinding approach to victory was not politically sustainable. If war came, it had to be fought immediately and won conclusively.

7. The enemy must never be given time to recover after the first attack. Sufficient aircraft and equipment must be available to do the job, and the civilian leaders must be persuaded to abstain from political gimmicks such as bombing halts.

Not listed as a lesson learned because it is beyond the Air Force's purview was the absolute requirement not to undertake any campaign not backed by the will of the American people.

These lessons, among others, became the operative philosophy for the post-Vietnam Air Force. Many additional subsets of lessons learned regarding the requirements for individual weapons, intelligence gathering and air transport of men and supplies supported these fundamentals, materializing in the weapon systems discussed below.

What Weapons Were Needed?

The Air Force emerged from the Vietnam War with the certain knowledge that it needed an air superiority fighter, a modern bomber capable of penetrating the Soviet Union, more aircraft with stealth characteristics, an improved ground attack aircraft, improved electronic countermeasures, precision-guided munitions, greatly improved theater airlift capability, and a modern, multiple-warhead intercontinental ballistic missile. All of these had to be tied together with an efficient C^3I system—command control, communications, and intelligence.

To obtain these weapons, the Air Force called upon the pool of experienced talent from the Vietnam War and the strong R&D base it had created with so much dedication over the years. These components provided the basis for America's technologically advanced industry to produce two new generations of weapon systems of unprecedented performance—and cost. These systems were overlapped in time with each other and with veterans of the past. Many books have been written about each individual weapon, and only a brief description is possible here.

First-Generation Weapons

Two key fighters were developed in the first generation. The McDonnell Douglas F-15 Eagle, covered in greater detail below, was the result of hot debate about what a new USAF fighter would be. The General Dynamics (now

Lockheed Martin) F-16 Fighting Falcon came as a joint by-product of the re-vitalized thinking on what a fighter should be and of the F-15's cost. The F-16 (also known as the Viper) was to be a lightweight fighter with less capability than the F-15, but inexpensive enough to be procured in large numbers. After much debate, it was decided that the Air Force was best served by a hi-lo mix—a smaller number of expensive F-15s as the "hi" portion, and a larger number of F-16s as the "lo" portion.

Both the F-15 and the F-16 proved to be superior to any other combat air-craft in their respective classes in the world, and they gave the USAF the ca-pability for establishing air superiority in any theater. They also gave impetus to the Soviet Union, which fielded remarkable competitors in the MiG-29 and Sukhoi Su-27. Other follow-on foreign competitors include the French Rafale and the Eurofighter 2000, the latter a product of a consortium that included the United Kingdom, Germany, Italy, and Spain.

The need for a close-air-support weapon prompted a competition which was won by the Fairchild A-10 Thunderbolt II—more commonly known as the Warthog for its uncomely appearance. The requirement called for an aircraft that had a large weapon-carrying capability, was uncomplicated and easy to maintain, required a minimum of ground facilities so that it could operate from forward bases, and could withstand extensive battle damage. A single-seat twin-jet aircraft, the A-10 can carry up to 16,000 pounds of ordnance and fea-tures a tank-killing 30mm GAU-8/A cannon. It proved itself beyond a shadow of a doubt in the Gulf War.

The Rockwell B-1B Lancer—familiarly known as the Bone to its crews—became the backbone of the long-range bomber fleet after an excruciatingly long development period that began in 1961 and was characterized by terrible acronyms as well as delays. The acronyms included SLAB (Subsonic Low Alti-tude Bomber), ERSA (Extended Range Strike Aircraft), LAMP (Low Altitude Manned Penetration), AMPSS (Advanced Manned Precision Strike System), and finally AMSA (Advanced Manned Strategic Aircraft—this had originally been the Advanced Strategic Manned Aircraft until it was noticed that ASMA was not ex-actly a warlike nickname). A total procurement quantity of 244 was planned for what had become the B-1A at the time of contract award in June 1970. The first flight took place on October 26, 1974. The B-1A's variable-geometry wings and blended wing/body configuration made it fast and long-ranged.

The development cost of the B-1A program approached $3 billion, creat-ing a demand for alternatives. These included a stretched General Dynamics FB-111G, a re-engined B-52, and a stand-off cruise missile launcher based on the Boeing 747.

None of these programs met the requirements as well as the B-1A—yet the B-1A was canceled by President Carter. The aircraft was revived by Presi-dent Reagan as the B-1B, which was similar in appearance to the B-1A but had a reduced radar signature and lacked a low-level supersonic capability. The

B-1B is heavier and can deliver nuclear or conventional weapons as well as stand-off missiles. A total of one hundred were delivered to the Air Force.

The Lancer has been controversial over its entire life, primarily because some of its complex electronic equipment did not perform to design specifications but also because of some well-publicized engine problems. The engine anomalies were par for the course for such an advanced aircraft and have been worked out over time. The electronic suite difficulties were in part overcome by superb crew training, but improvements are still planned. Despite this, the last Commander in Chief of the Strategic Air Command, General George L. Butler, stated unequivocally that the B-1B has a higher survivability factor in a penetration role against the Soviet Union without any electronic countermeasures (ECM) than the B-52H has with its ECM fully operational. A series of modifications to enhance the B-1B's conventional arms capability, including improved computers to support precision-guided munitions, makes its employment feasible well into the next century.

The Boeing E-3 Sentry is essentially a 707-320B airframe surmounted by a 30-foot-diameter rotating radome. The AWACS (airborne warning and control system) is capable of all-weather, long-range, high- or low-level surveillance of all air vehicles above all types of terrain. Its Westinghouse radar's look-down capability made it a tremendous advance over all previous airborne radar systems. Its initial task was to track the Soviet warplanes that would have accompanied an invasion of Western Europe. Since then, its role has been expanded to include antidrug patrols and enforcement of the "no-fly" zones imposed in Iraq and Bosnia. The AWACS is continually updated for new tasks, including electronic support measures to locate enemy stealth aircraft. The Sentry retains its capability as a force multiplier, enabling on-scene commanders to allocate forces with precision to the most urgent threats.

The General Dynamics F-111 was one of the most controversial warplanes in history. Deriving from Secretary McNamara's concept of the TFX, a multicapability fighter for use by both the Navy and the Air Force, it served well as a long-range strike fighter, and as the FB-111, as an interim strategic bomber for SAC. However, it reached its peak of utility with its introduction in 1981 as the EF-111A Tactical Jamming System. The high-performance, swing-wing F-111 airframe was combined with the most advanced, most powerful jamming system in the world, derived from the proven ALQ-99 equipment installed on the Grumman EA-6B. Grumman was selected as the contractor to combine the two elements into the EF-111A Raven. The Raven has the ability to loiter for up to four and one-half hours without refueling, yet, after sweeping its wings, it has the speed to keep up with the fastest Air Force fighters, as it demonstrated in the Persian Gulf War.

The temperature of the discussions over the several aircraft involved in the new generation did not rise above that of the fiery arguments over the single new missile, the long-debated MX. Eventually designated the LGM-118A, the Peacekeeper missile was unusual in that there was far more debate about its basing methods than there was about the missile itself. The Peacekeeper is a four-stage ICBM with as many as ten independently targetable reentry systems. It is far more accurate than the Minuteman III, carries more warheads, and has a greater range. SALT talks have placed limits on the number of warheads.

Because of its accuracy, it was conceded to become a prime target for Soviet missiles, and considerable thought was given to elaborate basing plans. One plan called for the missiles to be installed on railroad cars and shuttled around the country on a random basis, to make targeting difficult. Another called for basing the missiles closely together in a "dense pack," so that incoming Soviet missiles would interfere with each other in what became known as "fratricide." The most popular—and expensive—called for the creation (at a cost of $34 billion) of no less than 4,600 missile sites for 200 missiles. Like a pea in a shell game, each missile was to have been shuttled randomly from one spot to another to complicate Soviet target planning. Time and budget finally intervened, and a total of fifty Peacekeepers were finally installed in vacant Minuteman silos—an inadequate compromise in the minds of most.

SECOND-GENERATION WEAPONS

The very unorthodox, very advanced Lockheed F-117A Nighthawk stealth fighter led the second generation. The concept of very-low-observable, or stealth, technology has been with us since World War I, when camouflage and even Cellon, a transparent, celluloid-like covering, were used on aircraft to minimize their visibility. Being "invisible" to radar was another challenge, and was first demonstrated on the German Gotha Go-229 flying-wing jet fighter. USAF interest in the mid-1970s spurred development of prototypes to test how new shapes and materials might reduce a radar signature. One of these was the Lockheed "Have Blue" and another was Northrop's whalelike "Tacit Blue."

As a result of the Have Blue project the Air Force selected Lockheed to build five full-scale development models and fifteen production F-117As. Developed in Lockheed's famous Skunk Works under the direction of the brilliant Ben Rich, the new aircraft provided a study in contrasts. (The distinction between fighter, fighter-bomber, and bomber has become blurred as fighter-sized aircraft like the F-117A aegured bomber-like ranges, navigation, and bomb-dropping ability.)

The primary task of the Nighthawk is to penetrate enemy territory, unseen by active or passive defenses, and attack high-value targets with precision-guided munitions. A discussion of all the factors pertinent to the design of stealth aircraft is outside the scope of this book, but the F-117A's shape and materials either deflected or absorbed radar signals so that it was virtually invisible to radar. Its engines were hidden within the body so that its infrared signal was minimal. On a mission, it did not operate radar or other electronic equipment that could be picked up by enemy detectors. The shape and configuration

This CV-22 Osprey, its wings and rotors in a mid-position, fires countermeasures over a test range at Edwards Air Force Base, California. This tilt-rotor transport combines the maneuverability and lift of a helicopter with the speed of a fixed wing aircraft.

The Northrop Grumman RQ-4A Global Hawk, a high-endurance unmanned aerial vehicle (UAV), was deployed operationally during Operation Enduring Freedom in Afghanistan in 2001. It provides continuous all-weather, day/night wide area surveillance.

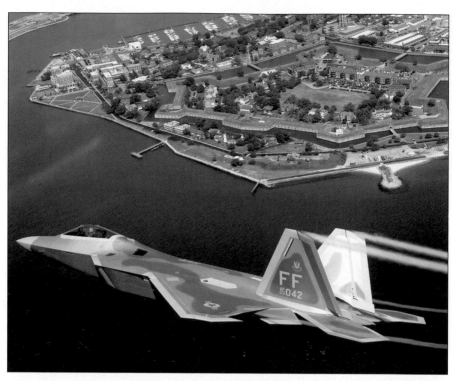

Lt. Col. James Hecker flies over historic Fort Monroe before he delivers the first operational Lockheed Martin F-22 Raptor to the 27th Fighter Squadron at Langley Air Force Base, Virginia.

The Lockheed Martin F-117A was introduced to battle in Panama, but proved to be the decisive difference in the Persian Gulf War when it operated unimpeded through the intense Baghdad antiaircraft defenses. *(Courtesy of Lockheed Martin)*

The first team in the Persian Gulf War. Two Lockheed Martin F-16 Fighting Falcons fly on the wings of a three-ship McDonnell Douglas F-15 flight. *(Courtesy of Eric Hehs)*

The most expensive warplane ever built—and worth every penny, according to fans—is the Northrop Grumman B-2, which combines stealth characteristics into a flying wing configuration. *(Courtesy of Northrop Grumman)*

The prototype Boeing YAL-1A Attack Airborne Laser aircraft will be the world's first operational air-borne high energy laser weapon system. The airframe is a modified 747. The laser weapon system is a Chemical-Oxygen Iodine Laser (COIL) which fires the laser through the large ball turret in the nose. Intended targets are enemy ballistic missiles in their early phase of flight.

The General Atomics Aeronautical Systems MQ-1 Predator is equipped with two Hellfire missiles and has already achieved significant success in combat. While cruise speed is only 138 mph, the Predator has an endurance of twenty-four hours.

General Bill Creech was always perfectly turned out, whether meeting with President Ronald Reagan or in a flight suit.
(Courtesy of W. L. Creech)

The Lockheed Martin A-10A Thunderbolt II has proven to be invaluable in providing close air support. Its armament includes a 3mm CAU-8 Gatling gun, plus up to 16,000 pounds of ordnance (about four times the amount routinely carried by a B-17 in WW II).

The U.S. Army's McDonnell Douglas AH-64 Apache is as lethal as it looks; a fearsome combat weapon, it devastated Iraqi radar stations to which it was guided by USAF Pave Low helicopters. *(Courtesy of Robert F. Dorr)*

The McDonnell Douglas C-17 had a long gestation period, but it will be the mainstay of the airlift fleet for almost the entire twenty-first century. *(Courtesy of McDonnell Douglas)*

With its great success in the Gulf War, the McDonnell Douglas F-15E proved just how much "stretch" there was in the basic design. Originally intended as an air superiority fighter, the improved electronics, navigation, bombing, and communication gear gave the Strike Eagle an entirely new capability. *(Courtesy of McDonnell Douglas)*

necessary to avoid detection by radar or infrared devices resulted in a wedgelike, multifaceted configuration so inherently unstable that it required a full-time fly-by-wire command augmentation system. No mechanical backup system was provided because no pilot could fly the aircraft without computer-controlled stability augmentation. The Air Force originally intended to purchase one hundred of the Nighthawks (also called "the Black Jet" by its crews) but budget limitations reduced this to only fifty-nine. As we shall see, the F-117A operated with a preternatural brilliance in the Gulf War.

Northrop's Tacit Blue was designed as a prototype for a reconnaissance aircraft with a side-looking radar array. Tacit Blue's existence was kept secret even longer than that of the Have Blue series, not being revealed until May 1996. The design was cancelled because its mission could be performed by other means, including unmanned vehicles. Another approach to stealth design was shown in the Northrop B-2 Spirit stealth bomber, which depended more upon composite materials and a smoother, blended shape than the angular faceting of the Nighthawk. About 80 percent of the B-2's airframe weight is made up of composites; titanium and aluminum are used for internal load-bearing structures.

The flying wing was a poignant if coincidental return to tradition for Northrop, which had pioneered the concept with the B-35 and B-49 bombers of the early postwar period. The great designer Jack Northrop lived long enough to learn that his long dream of a production of flying-wing aircraft would happen. Although it was remarkably similar in shape and planform to the flying wings of the 1940s, the primary reason for adopting the configuration was its low radar cross section and the convenience the thicker wing shape offered for shielding infrared emissions from engine operation.

The B-2 has been perhaps the most controversial aircraft to enter service in recent years, because of its high program cost. Initial production procurement was capped at twenty aircraft (plus refurbishing the prototype), although funding for an additional twenty aircraft has been proposed.

The McDonnell Douglas C-17A Globemaster III had almost as long a gestation period in all its prior forms as did the B-1. Driven by an urgent airlift requirement that was exacerbated by the wear and tear placed on C-5s and C-141s during the Gulf War, the C-17 became operational in January 1995. The unusual-looking cargo plane is designed for both inter- and intratheater airlift and can carry outsized cargo, including the M1A2 tank. A total of 120 production aircraft are currently planned, although this figure will undoubtedly grow over time.

The Boeing/Grumman E-8 Joint STARS (Joint Surveillance and Target Attack Radar System) consists of modifing "previously owned" Boeing 707-300 airframes with the radar and communication equipment necessary for ground surveillance, targeting, and battle management missions. The E-8 uses a Norden side-looking radar that is in a large fairing on the forward fuselage. The radar maintains surveillance over a wide area and is designed to find and track slow-moving ground targets. Fixed targets like tank parks, bridges, and airports are identified in a high-resolution synthetic aperture mode. The system can distinguish between trucks and tanks, so that weapons can be applied against the most valuable targets. J-STARS is in effect an AWACS for the ground war, directing attacks against enemy targets. Prototype E-8s were used with considerable success in the Gulf War.

The final aircraft weapon of this second generation is the most advanced fighter in the world, the Lockheed Martin F/A-22. The F/A-22 combines a highly maneuverable airframe at both subsonic and supersonic speeds with low-observable characteristics. Actual performance figures are still classified, but the aircraft is capable of supersonic cruising without the use of afterburner. The twin-engine F/A-22 has a very advanced weapon systems capable of engaging multiple targets. Budget limitations have forced a reduction in the total number of F/A-22s to be procured. Its stablemate, the Lockheed Martin F-35, is intended for the close-air-support role and is also facing a reduction in the number to be procured.

As outstanding as each of the weapon systems of these two closely linked generations is, their cumulative effectiveness would be zero without the network of space systems that provides them unparalleled intelligence, communication, navigation, and meteorological capability. This will be discussed below.

Their usefulness is similarly dependent upon the diverse electronic systems and advanced munitions that have simultaneously been developed for them. The time necessary for the creation of all these modern weapons was purchased by the continued employment of veterans such as the B-52, KC-135, and U-2 aircraft. The usefulness of these weapon systems, every one of which was decried by critics as wasteful folly at one time or another during their procurement cycles, seems unbounded by time or circumstance.

AGING GRACEFULLY

Military aircraft typically fly far fewer hours than their civilian counterparts and are built to different standards that reflect the possibility of harder usage (greater G-forces, for example) but fewer cycles of operation. For most of the history of aviation, military aircraft had an expected service life of perhaps five years in first-line operation, followed by another five in reserve or training applications. In the past thirty years, however, military aircraft life spans have increased dramatically, which is just as well, given that the time required to introduce a new weapon system has also increased.

The current champion combat senior citizen is the Boeing B-52, which first flew in 1952 and ended production in 1962. In 2005, the average age of the eighty-four B-52s still serving in the active inventory over 43 years. Some B-52s are projected to be in service until 2040, when the average age will be something like sixty-seven years, and when their pilots could conceivably be the great-great-grandchildren of the men who first flew them. (The mechanical equivalent would be for a 1929 Keystone bomber to be in frontline service in

2005.) The original average price of about $6 million per B-52 was greeted with outrage and astonishment. However, when one amortizes this over a period of many decades and several wars, it seems quite a bargain, something to keep in mind for later discussions.

The B-52 survived for a number of reasons. Its excellent performance has been increased by improved engines, and its large size has permitted it to carry a broad range of equipment and munitions for a wide variety of missions. Most important, almost from its inception it has been the beneficiary of continual modification, including rebuilding the wings and fuselage over the years.

The story is the same for the stalwart Boeing KC-135 tanker fleet, which has an average age of 43 years and now succors bombers, fighters, and transports with equal care. The KC-135 may well stay in service even longer than the B-52.

Both aircraft have been supplemented over the years. The B-52 had the sleek Convair B-58 as its companion for a time. The Hustler was a supersonic bomber but it could not adapt as well as the B-52 to changing conditions and was retired. The B-52 now shares the major duties of the bomber fleet with the Rockwell B-1B, a youngster averaging only 18 years in age. The KC-135 has been assisted by the addition of fifty-nine McDonnell Douglas KC-10s, averaging only about 21 years in age.

Age has had less effect on the transport aircraft than has the routine wear and tear of the demanding task of carrying huge loads to airfields of every description all over the world. The fleet of Lockheed C-141s averaged close to 40 years before the strain of their accelerated use forced them into retirement. Fortunately, a decision had already been made to purchase the Boeing C-17. The Lockheed C-5 fleet of eighty-one aircraft averages only 22 years, primarily because fifty of the fleet were built over a decade ago. The hardy Lockheed C-130s average 32 years and promise to be "hauling trash" for several more decades, in company with the highly advanced current model, the C-130J.

The fighter fleet is in no better shape. The F-15, still the first-line American air superiority fighter, averages about 18 years in age, while its companion, the Lockheed Martin F-16, averages about 14.

Among missiles, the 500 Minuteman IIIs that came into service in the early 1970s were supplemented by fifty Peacekeeper missiles, which are now retired. Like the B-52, the Minuteman has benefited from continual modification and updating. Also like the B-52, its purpose will have been perfectly served if it is never used to deliver nuclear weapons.

EVOLVING THE NEW WEAPONS

The program histories of each of these new weapons could fill a library. For the purposes of this book, it is worthwhile to examine how a representative example, the McDonnell Douglas F-15 Eagle, was brought into being.

Its predecessor as the USAF's standard air superiority fighter, the McDonnell Douglas F-4 Phantom II, had started life in 1953 as a proposed naval fleet defense fighter, armed solely with missiles. It evolved into such a highly capable aircraft that Secretary McNamara pressed for its adoption by the USAF in 1962. It became the premier American fighter in Vietnam, serving ably with the Air Force, Navy, and Marines and performing in every role, including air superiority, close air support, reconnaissance, and even as a "fast FAC," a swift-moving forward air control aircraft.

Yet combat revealed deficiencies in the F-4 that the Air Force wished to avoid in a follow-on fighter. Among these were a lack of visibility from the cockpit, the requirement for two men to operate the weapon systems, and the highly visible smoke trails left by the engines. The advantage of having two engines was diminished by the tendency for fire to propagate from one engine to the other. There were also ergonomic problems involving such things as the positioning of the control stick and armament switches. Maintenance was a headache—in some cases the ejection seat had to be removed to replace a radio.

The F-4 was originally armed only with air-to-air missiles, which proved to be a great disappointment. Expectations had been unrealistically high. Because guidance equipment was improved and missile speed was supersonic, their proponents assumed such a very high kill rate for them that guns were omitted from aircraft armament packages. The classic dogfight was considered impossible at supersonic speeds and thus a thing of the past. Ironically, the technology devoted to creating missiles had been aimed at destroying enemy bombers. When it turned out that most air fighting was conducted in dogfights between fighters at subsonic speeds, the missiles proved inadequate. The envelope within which the missiles could be used and the length of time required to prepare them for launch often exceeded practical limits in a dogfight. When elaborate rules of engagement were superimposed on these limitations, the kill ratio of American versus enemy aircraft fell to its lowest point in history.

The young fighter pilot veterans from Vietnam wanted a more agile aircraft that could outmaneuver the nimble MiG-21 as well as its inevitable successors, and they wanted it equipped both with a gun armament and better missiles. Two such young veterans tapped for their fighter experience were future Chief of Staff Larry Welch and future Commander of Air Combat Command Mike Loh.

In the late 1960s, Air Force Pentagon experts were attempting to create a computer simulation of dogfighting and air-to-air combat. Welch's task was to teach them how a fighter pilot makes decisions. After two months, it was obvious to Welch that he was never going to be able to teach a computer programmer about aerial attack. Instead, he took a quick course in FORTRAN programming and with another fighter pilot developed a computer program with variable parameters that was used in designing the F-15 Eagle and subsequent fighters.

Welch's work would augment studies begun in the mid-1960s under the direction of Air Force Secretary Eugene Zuckert to develop the F-X (fighter-experimental). The F-X was to be highly maneuverable even at the expense of sheer speed, and to have superior capabilities for air-to-air combat, ground attack, and all-weather operations. Curiously, the first design studies resulted in a behemoth 60,000-pound clone of the F-111, exactly opposite to then-current thinking as to what a fighter should be.

The situation was saved by an aggressive young Pentagon action officer, Major John R. Boyd, who led the effort to redirect F-X development to a highly maneuverable aircraft optimized for the air superiority role. Boyd was the outspoken advocate of the concept of energy maneuverability that had proved to be the key to the F-4's success in Vietnam. His precepts led to a demand for the F-X to have the capability to outturn and outaccelerate enemy fighters under all conditions encountered in a dogfight.

A massive paper competition under Secretary McNamara's Total Package Procurement (TPP) rules was held among three manufacturers, Fairchild, North American, and McDonnell Douglas. These rules were intensified by a "Demonstration Milestone Chart" that called for the winning design to meet twenty-four individual milestones, ranging from the preliminary design review through first flight, fatigue testing, complete flight testing, and the delivery of the first aircraft to TAC. Each of these milestones had to be met or further funding was delayed—the ultimate combination of contractual stick and carrot.

After 2.5 million man-hours of study by massive teams of USAF military and civilian experts, the McDonnell Douglas entrant was selected as winner by Secretary Seamans on December 23, 1969. Early the next year, another contract was signed with Pratt & Whitney to develop the F100-PW-100 engine to power the F-15 Eagle. (The cost of a TPP procurement to competing contractors was enormous, not only in the colossal dollar outlay, which could be near-ruinous to a smaller company like Fairchild, but in the opportunity costs of projects forgone because the resources were not available. The TPP experiment is a perfect example of bureaucratic bean-counting gone wild, in which the search for perfect numbers overlooks the cost to other elements of the economy and sacrifices the opportunity for progress in other vital areas.)

All of the contractors threw massive resources into the competition. McDonnell Douglas had conducted more than 10,000 hours of very expensive wind tunnel tests, along with extensive use of the then brand-new computer-aided-design equipment. In addition, McDonnell Douglas employed two cockpit air combat simulators to test its design against all possible competitors, with Air Force combat veterans flying the simulators and suggesting improvements in cockpit layout.

In the end, the F-15's world-beating maneuverability depended upon factors as old as the Wright Flyer: a low wing-loading (the relationship of wing area to gross weight) and a high power-to-weight ratio. It made its first flight on July 27, 1973, at McDonnell's St. Louis plant and entered service on November 14, 1974, at Luke AFB, Arizona. In late 1976, the 1st Tactical Fighter Wing at Langley was the first to be operationally combat-ready with the Eagle.

The F-15 encountered some airframe problems in its test program that, while troubling, were susceptible to relatively easy modification. Its engines, at that time the most complex ever used in a fighter, had both operational and maintenance problems that required a much longer time to resolve fully. Despite all this, the F-15 quickly proved that it was the fighter of the future, with a performance that eclipsed that of all existing aircraft of the Soviet Union. Highly maneuverable and equipped with a Hughes multimode pulse Doppler radar, it met all air superiority requirements. The Israeli Air Force would score notable successes with it in 1982, and it would again prove itself in the Persian Gulf War.

Such performance was expensive, and the initial estimate of F-15 unit cost of about $15 million rose through the years to a current $50 million—the result of modifications, inflation, and additional equipment. The F-15 also turned out to be larger than exponents of the lightweight fighter concept preferred. Members of the so-called "fighter Mafia," led by Major Boyd and his colleague Pierre Sprey, a systems analyst with OSD (Office of the Secretary of Defense), sought to raise interest in a smaller aircraft.

The idea of a lightweight fighter brought an instant adverse reaction from Air Force Headquarters. After having made a great effort to sell the F-15, the Air Staff did not wish to provide Congress with a "low-cost rationale" for canceling it. In an effective demonstration of how field-grade action officers can influence future events, Mike Loh saw to it that the nascent movement for a lightweight fighter was caught up by another wave of enthusiasm, that of Deputy Secretary of Defense David Packard's desire to select aircraft by competitive fly-offs. Loh became the program element monitor for what became called the air combat fighter and facilitated the competition between the Northrop YF-17 and the General Dynamics YF-16. Thanks to the presence of Air Force "young Turks" and timely intervention by the Secretary of Defense, it thus happened that the successful development of one fighter, the F-15, fostered rather than suppressed the development of another—the F-16.

Each weapon system has a similar history that runs from its initial concept in the minds of planners to its final destination at the smelting pots of the aircraft disposal unit at Davis Monthan AFB. In today's world of protracted development, strung-out production, and long service lives, the interval between concept and salvage may extend for sixty years, or more than three careers for the average airman—or legislator.

The F-15/F-16 saga shows that it takes genuine vision to conceive weapon systems that have to meet specific requirements while operating in an unusually demanding environment over many decades. The vision of a weapon system must be accompanied by diligent management, adroit politicking, and a steadfast adherence to doctrine, again through many changes of leadership and over a long period of time.

The full scope and depth of the task is rarely appreciated. The media, desperately seeking stories about a $600 toilet seat or $25,000 facsimile machines, would find nothing of interest in the magnitude of the problem of creating and fielding a single weapon system or in the fact that at any given moment the Air Force alone may have hundreds of major weapon systems (including munitions and electronic countermeasures) under development. Each of these might have literally tens of thousands of subcomponents, as well as additional thousands of items in support equipment, all of them requiring equivalent care. Every one of these elements has to be designed to criteria that have to be specified in enormous detail to obtain the desired performance, strength, longevity, and capability. (There is a trend away from this sort of specification where commercial items will serve. This is a valid concept when those components already exist, but in many cases they have to be invented for a specialized weapon system—and there is little commercial pressure for such exotic invention.)

Many of the weapon systems have to be designed in anticipation of threats and uses that are not yet fully defined. Personnel have to be trained to manufacture, deliver, use, and maintain them. Most important, American policy and American lives depend upon their being brought to fruition successfully and on time.

Are mistakes made? Yes. In the fabric of the huge society of military and civilian personnel involved in the programs, are there a few miscreants who will take advantage of the system? Yes. This was illustrated by the Darleen Druyun scandal of 2004, which compromised the procurement of badly needed tankers. (Druyun pled guilty to conspiring with the Boeing Company to defraud the federal government in regard to the acquisition of tanker aircraft.) Does the vast scope of the endeavor inevitably yield stories that are "newsworthy" for their signs of human failure? Of course. But on the whole, the entire apparatus is carried out with a degree of honesty and success of which any organization, industry or government, could be proud.

As we will see in the next chapter, the Air Force met all of its challenges with skill and panache in so quickly creating the two generations of war-making equipment it knew it required. At the same time, it created the simultaneous advances in the military use of space upon which the proper functioning of these weapon systems demanded.

THE EXPLOITATION OF SPACE

Although there has not yet been a significant groundswell to change the name of the junior service to the United States Aerospace Force, the resources devoted to space and the benefits derived from its exploitation make the change almost certain in the future. The tidal movement toward this new name may be charted in the organizational development that follows.

The initial approach of the Air Force to space was both curious and coincidental, and, not surprisingly, it followed traditional organizational patterns, with responsibilities divided among AFSC, ADC, AFLC, SAC, and the Defense Communications Agency (DCA). The need to create ICBMs put the Air Force in the space business before it could formulate extensive plans for the exploitation of space. That the propulsion and guidance elements of the ICBM could be used for the insertion of satellites into orbit was immediately apparent, but it took time to grasp the full range of possibilities offered by extensive systems of satellites. Further, the technological infrastructure necessary to develop and use satellites as intelligence, communications, navigation, and meteorological networks had yet to be developed. As will be seen, when the technology became available, the USAF created a space superiority that corresponded to older concepts of air superiority.

The earliest concepts on the USAF in space were intuitive. That great thinker General Thomas D. White saw space as a continuum of our atmosphere. He realized that the United States could not dominate all of space. In his view the United States should establish a space superiority to control space activities, just as it had established air superiority to control enemy air actions.

The full economic implications even of this limited strategy were not understood. It soon became obvious that the Soviet Union intended to use space as an extension of its military power. The only certain way to prevent this was a preemptive strike by ICBMs, an act abhorrent to U.S. policy and national character. So it happened that the great Cold War competition between the United States and the Soviet Union went boldly—but blindly—toward a new frontier.

HESITATION IN SPACE POLICY

The Eisenhower and Kennedy administrations both wished to avoid the expense and hazard of an arms race in space. Eisenhower introduced the concept of "freedom of space," which ensured overflight, and saw to it that the first sentence in the 1958 law creating the National Aeronautics and Space Administration (NASA) confirmed this desire: "The Congress declares that it is the policy of the United States that activities in space should be devoted to peaceful purposes for the benefit of mankind." But peaceful purposes embraced space defense support missions such as reconnaissance and communications. (Eisenhower must be given great credit for his courage and foresight in authorizing the Corona program detailed below.)

The law went on to delegate to DOD the responsibility for the use of space for these defensive systems. The Air Force had earlier been given similar responsibilities, a situation confirmed by DOD Directive 5160.22 of September 1, 1970, which provided that "the Air Force will have the responsibility of development, production, and deployment of space systems for warning and surveillance of enemy nuclear capabilities, and all launch vehicles, including launch and orbital support operations."

President Kennedy chose the Apollo program as a prestigious but nonthreatening means to demonstrate American technological superiority in a way that would not spur an arms race. The Apollo program was wildly successful and restored U.S. international prestige, but it also reinforced the determination of the Soviet Union to seize for military purposes what inevitably was called the "high ground of space." Soviet launch activity surpassed that of the United States in 1971, because Soviet satellites proved less reliable and possessed shorter lives in orbit. By 1973 the Soviets were putting almost five times as many satellites and other space objects into orbit as the Americans. The Soviet Union had no inhibitions about testing antisatellite weapons, conducting tests throughout the late 1970s and early 1980s. Its system proved to have limited capability, in altitude and azimuth.

Unfortunately, the American reaction to the well-trumpeted Soviet space activity was exactly parallel to its handling of events in Vietnam. There American policy was artfully crafted to induce certain positive reactions from the enemy—which never came. The same was true in space. The U.S. thirst for accommodation with the Soviet Union led to a stifling of technological progress in an attempt to avoid destabilizing the quest for arms reduction. The Soviet Union did not reciprocate. Despite this general trend, U.S. satellites became much more reliable than their Soviet counterparts, operating for ten years and more.

In the McNamara era, the accommodation was often accomplished internally by a political variation of "bait and switch" sales tactics, with the switch here coming before the bait. McNamara insisted that military space efforts be consistent with NASA's programs, and then used NASA programs as a reason for canceling Air Force space projects. The first of these was the Boeing X-20 Dyna-Soar, a rocket-boosted vehicle with a single pilot that was to orbit the earth, then return to land at a designated spot. Conceptually a forerunner of the space shuttle, it traced its engineering roots back to German World War II "skip-bomber" concepts originated by Dr. Eugen Saenger.

The Dyna-Soar had been a key element in the Air Force program for developing a manned military patrol capability in space. The hypersonic X-20 was to have tested controlled, maneuverable reentry from orbital flight beginning with its first launch in 1962—the same year in which future Senator John Glenn made his successful orbital flight in the Mercury spacecraft *Friendship 7*. Without detracting an iota from the remarkable work of the Mercury program, the X-20 would have been a far more advanced vehicle than the Mercury spacecraft, and at the very least would have removed the "pilots as monkey" odium from those early orbital flights.

Convinced that the Mercury-Gemini-Apollo programs could accomplish the Dyna-Soar's mission, McNamara canceled the X-20 on December 10, 1963. That was the switch. His simultaneous announcement of the development of a Manned Orbiting Laboratory (MOL) was the bait to allay congressional and service resentment over the cancellation of Dyna-Soar. As things worked out, the only net gain from the Dyna-Soar was some of the experimental work done on the project that transferred to both the X-15 and Space Shuttle programs.

This shell game of program switches became part of what was known as "paralysis by analysis" within the DOD. Any program not on McNamara's priority list would be so bombarded by questions, demands for briefings, requests for alternatives, and other inquiries that the program officers would be forced to reduce activities on the program just to meet the demands for information. Then, if the programs still progressed, a cancellation was made with the offer of a new program in a similar field. It was a technique that worked well for McNamara and some of his successors, if not for the services or the taxpayers.

The Manned Orbiting Laboratory was a small space station intended to perform a reconnaissance function. In his last year in office, McNamara began to reduce funds for the MOL program, which had a predicted launch date of 1971. As time progressed, the combination of the success of unmanned satellites, funding demands of the Vietnam War, and the Apollo program proved to be too much. Ironically, it fell to an ardent supporter of the MOL, Secretary of Defense Melvin Laird, to cancel the program in June 1969—after $1.37 billion had been spent.

Some of the Air Force's early efforts in space were reactive. The demand for an antidote to a Soviet threat gleefully wielded by Khrushchev, the orbital bomb, led in 1963 to a hurry-up call for an antisatellite capability. The Space Systems Division of AFSC responded in just over a year with Program 437, using an earth-based Thor rocket positioned on a Pacific island to intercept satellites. The guidance systems could place the antisatellite missile within 5 miles of an enemy target satellite, well inside the lethal radius of the missile's nuclear warhead. The system remained operational for ten years before it was ordered deactivated.

In 1961, the Air Force also developed plans for a precursor of President Reagan's Strategic Defense Initiative—"Star Wars"—in the form of a satellite system capable of infrared homing on and destruction of enemy rockets in their initial boost phase. This ballistic missile boost intercept weapon (inevitably known as Bambi) received short shrift from McNamara as being too technically demanding for consideration. Yet it was exactly the sort of revolutionary leap forward that the technology required, and its cancellation illustrates the danger to research and development when practical applications have absolute priority over basic research. McNamara instead preferred to create what he called "technological building blocks" that could be used as the basis for a variety of programs—a space version of what he termed "commonality" in aircraft like the TFX.

Despite these setbacks vitiating announced Air Force space programs, work proceeded on a series of other defense-support space programs: missile defense alarm, communications, and meteorological, navigational, and reconnaissance satellites. These highly classified programs would ultimately be supremely successful and demonstrate that space was not a mission in and of itself, but rather a medium in which to exercise the Air Force's traditional role of controlling the combat arena. As that work progressed, there was an increased call for the creation of a space doctrine and the establishment of a space organization. The call would be heeded at successively more important levels.

The first steps toward what became a separate space organization were sponsored by then Colonel Robert T. "Tom" Marsh. In a non-turf-conscious manner that was to be typical of his leadership, Marsh decided there was a requirement for a separate directorate of space within AFSC, and in 1965 he put together a briefing for General McConnell, who approved the idea. Later, as a four-star general commanding AFSC, Marsh again gave up turf by advocating the creation of a separate space command.

By 1977, General Jones, as Chief of Staff, issued a letter on "Air Force Space Policy" in which he included the development of weapon systems and the conduct of military operations in space as among the Air Force's prime responsibilities. The following year, President Carter issued Presidential Directive 37, which, given his generally anti-Pentagon attitude, lent a surprisingly strong Presidential backing to the military aspects of space.

The chill fog of the Cold War obscured the rationale behind the increasing rate of Soviet satellite launches—almost a hundred in 1981. The replacement rate was in fact dictated by the shorter lives of the Soviet satellites, but it could also be inferred that the Soviets were planning frequent replacement as would have to be done in the event of a space war. Soviet professional military journals, normally the most opaque of reading material, openly discussed the military uses of space as a given.

The intensification of Soviet efforts coincided with the increasing U.S. dependence on space systems and the perceived need to take advantage of the imminent arrival of the Space Shuttle. After 1981, with its program of rearmament and the rapid growth of satellite systems discussed below, the Reagan administration set the stage for Air Force Chief of Staff General Lew Allen's June 21, 1982, announcement of the formation of Air Force Space Command (AFSPACECOM), effective September 1, 1983.

Allen crafted AFSPACECOM so that its commander would also be CINCNORAD and CINCADC and have as his deputy the commander of AFSC. This conglomeration of acronyms meant simply that Space Command would have within its jurisdiction the air and missile defense of North America and easy access to the research-and-development potential of AFSC. Allen also selected a whirlwind of energy as the first SPACECMD commander, pugnacious General James V. Hartinger, a former football star and prototypical fighter pilot. Hartinger was given the mission of managing and operating all assigned space assets, including controlling operational spacecraft and managing DOD space shuttle flights. In addition, he had to centralize and coordinate the management of the blossoming series of highly classified space programs.

In an unselfish move, Secretary of the Air Force Orr and Chief of Staff General Gabriel began to advocate a unified command for space, declaring that no single military organization had authority over military space systems. The need for a clear chain of command from the National Command Authority also called for a unified command. The Army agreed, but the Navy demurred, setting up its own space command and opposing a unified command. It was to no avail, for a new unified U.S. Space Command (USSPACECOM) was established by the DOD on November 23, 1984. It was time, for the lower limits of space were becoming crowded with a galaxy of ever more sophisticated satellites.

At this writing the organization of Air Force Space Command is in flux. Air Force Space Command organizes, trains, and equips the forces that are provided to U.S. Strategic Command (STRATCOM). It also operates the following space-related activities: missile-warning radars, sensors, and satellites; national space launch facilities; worldwide space surveillance radars and optical systems; worldwide environmental systems; position, navigation, and timing systems. It provides command and control for DOD satellites as well as missile warning to the North American Aerospace Defense Command (NORAD), United States Northern Command (NORTHCOM), and STRATCOM.

All of these are part of its mission to deliver war-winning expeditionary capabilities to the warfighter through technology, acquisition, support, and sustainment.

All of the services had other subcomponents devoted to space, but the most advanced, best funded, and most varied were managed by the organizations that came under the umbrella of the Air Force Materiel Command. The number of current space programs managed by AFMC numbers in the hundreds, and ranges from such stalwarts as the Atlas launch vehicle program, in which refurbished Atlas E boosters lift satellites into orbit, to airborne lasers to destroy enemy vehicles in their boost phase, to determining the effects of energy from the sun on Air Force satellites. The array of programs is impressive, not least because each one has survived wire-brush scrubbings by internal Air Force teams, the DOD, and Congress.

TOWARD THE UNKNOWN

All of the American expenditure on all space programs may well have been more than repaid by a single intelligence-gathering program. It began on March 17, 1955—two and one-half years before Sputnik would set the United States upon its scientific ear—with a requirement issued for a strategic reconnaissance satellite. A program called Corona, operated by Air Force and CIA teams working in the National Reconnaissance Office, would tear open many of the secrets of the Soviet Union and China as it mapped 510 million nautical square miles of Communist territory, revealed that there was no missile gap, identified each type and counted each individual submarine, bomber, and fighter, monitored missile site activities in both countries, and provided data in depth on heretofore inaccessible subjects such as atomic weapons storage sites. The scope of this intelligence windfall was unparalleled, being orders of magnitude more important than a combination of the Ultra and Magic intelligence coups of World War II. With admittedly much greater effort, the United States could still have fought and won World War II without either Ultra or Magic. Corona won the greater victory of providing the vital information to win the Cold War while helping to avoid a nuclear conflict.

Masquerading under the code name Discoverer, a cover program operated by the Air Force, Corona was kept absolutely secret for twenty-three years after it had passed out of use. The entire story was unveiled in February

1995, when Vice President Al Gore officially lifted the security classification from the nation's first spy satellite project. The Corona program had operated between 1960 and 1972 and yielded more than 800,000 satellite images.

The CIA provided the project director, Richard Bissel, as well as the funding and the tight security planning necessary to maintain secrecy. Bissell was special assistant for planning and coordination for CIA Director Allen W. Dulles, and was already involved in the equally secret Lockheed U-2 program. Later he would be a prime mover in the Lockheed A-12/SR-71 triumphs. His deputy was the veteran Air Force test pilot Major General Osmond J. Ritland, who had flown more than 200 different aircraft, including British, German, and Japanese combat planes. Their opposite number at the prime contractor, Lockheed, was James W. Plummer, who ran a streamlined management organization patterned after Kelly Johnson's famous Skunk Works.

The heart of the Corona package was the Itek camera used for most of the 130 launches. Resolution early on was about 40 feet; this was improved over the years to less than 10 feet, so that items as small as individual automobiles could be seen and evaluated.

The program lasted from 1956 through 1972 and cost a bargain $850 million. The going was not easy. The first launch, on January 21, 1959, was designated Discoverer 0 and was a failure. The second test was a technical success, although it had been flown without a camera. The next eleven launches were failures, and the pressure on Plummer, Bissell, and Ritland to deliver was extreme. Gary Powers had been shot down in his U-2 on May 1, 1960, and all Soviet overflights were suspended, eliminating the CIA's principal reconnaissance intelligence-gathering mechanism.

Working with a swiftness unfettered by bureaucratic oversight, sometimes making significant contract changes verbally, often over the telephone, the Lockheed and Air Force teams were able to achieve success precisely in the nick of time. On August 10, 1960, Discoverer XIII lifted off for the first completely successful mission. Subsequent Discoverer flights were almost uniformly successful, and aerial recovery of the capsule by specially equipped Fairchild JC-119 and Lockheed NC-130H aircraft became, if not routine, at least predictable. The Corona system provided about 500 times more coverage than the U-2 had been able to do, and without any risk to a pilot.

Although the Corona project was a brilliant success, security prevented the proper acclaim being given to the determined leadership, which had persevered through the bleak months of failure.

There were also many scientific spin-offs from the program. The reentry data gathered from the Corona capsules had direct bearing on the Apollo spacecraft reentry design and materials. The experience was also a key factor in strengthening Lockheed's Space and Missiles Division, which went on to great success in the Polaris, Poseidon, and Trident missile programs, among others. Most important of all, in the great Cold War poker game, Corona served as the mirror in which the United States could read the Soviet Union's cards.

In terms of the involvement of genius, expenditure of resources, resolution of problems, and delivery of results, the Corona program was a paradigm for the ensuing major U.S. space programs, each of which has its own spectacular history. These programs were naturally more sophisticated and faced even greater uncertainties and difficulties than did Corona. Unfortunately for this book, the story of these problems and their resolution is for the most part still classified. The author can only describe what the systems are and what their import is.

MAKING THE UNKNOWN KNOWN

The USAF shares the responsibility for using space to advantage with dozens of other governmental and civilian agencies. The government agencies include the CIA, NASA, the National Reconnaissance Office (NRO), the National Oceanic and Atmospheric Administration (NOAA), and the National Security Agency (NSA). Civilian agencies include INTELSAT, run by the 131-member International Telecommunications Satellite Organization, and Landsat, a privately operated system for monitoring Earth's landmasses and oceans.

Space has provided the venue for a number of military functions. These are supported by a galaxy of satellite systems and their concomitant ground facilities. A brief description of the functions and some of their operating systems follows.

NAVIGATION

The military services in general and the Air Force in particular have reaped enormous benefits from the navigational accuracy conferred by the Navstar Global Positioning System (GPS), operated by the 50th Space Wing located at Falcon Air Force Base, Colorado. A constellation of twenty-four orbiting satellites sends signals to earth to furnish military and civil users with extremely accurate three-dimensional location information, i.e., latitude, longitude, and altitude, plus velocity and time. The service is available twenty-four hours a day, providing continuous real-time information to an unlimited number of users under all weather conditions.

The GPS satellites orbit the earth every twelve hours, emitting continuous navigation signals so accurate that time can be figured to within a millionth of a second, velocity within a fraction of a mile per hour, and location to less than 30 feet. Because of military concern that a potential enemy could make use of GPS accuracy to pinpoint valuable targets like Minuteman silos, the accuracy permitted civilian users was raised to about 300 feet. Demand forced a change in this practice to "differentially corrected" GPS signals, but the military reserved the right to distort the signal for civilian use in times of emergency.

The GPS system achieved fame in the Gulf War when it was used by everything from stealth bombers to B-52s flying the longest combat missions in history to special operation forces navigating in the desert. GPS was essential in establishing precise reference points upon which inertial navigation systems could be calibrated. The demand was so great that many military personnel purchased their own small, hand-held GPS units from civilian sources, and British pilots flew their Jaguar fighter-bombers with civilian hand-held versions attached with Velcro to the instrument panels. They were especially valuable during the periods of bad weather.

GPS was also effective in enhancing the accuracy of other attack equipment like the LANTIRN (low-altitude navigation and targeting infrared for night) system. F-16s integrating GPS and LANTIRN were used in the ground attack role against mobile launchers. Other F-16s used GPS for fast-forward air control work, passing back precise coordinates of targets rather than having to lob in a target-marking rocket. A great attraction of the system is that the information is continuously beamed and received passively—there are no emissions to alert the enemy and give away positions.

More modern equipment includes the AN/AAQ-28 LITENING Advanced Airborne Targeting and Navigation Pod and the Joint Direct Attack Munitions (JDAMS), both of which are playing an increasingly important role in precision aerial warfare, and both of which use GPS.

Use of the GPS system is spreading rapidly, and there is a movement toward having GPS positioning become an essential part of a primary instrument landing system for commercial aircraft.

The first Navstar GPS satellite was launched on February 22, 1978. At the time of the Gulf War, sixteen satellites were available for use. There are currently twenty-four GPS satellites in service, including three spares, each with a design life of seven and one-half years. The GPS satellites are 5 feet wide and have a 38-foot-long solar array. They weigh 2,570 pounds. They are deployed in a near-circular orbit at an altitude of 12,532 miles. Solar panels generate up to 2,900 watts to provide power. The Delta II expendable launch vehicle is used to launch the GPS satellites in orbit. Rockwell International provided the first two series of satellites, and Lockheed-Martin provided the third.

MISSILE DEFENSE ALARM

The need to track potential incoming missiles has existed since the earliest days of Soviet ICBM success. The concern about attack from the former Soviet Union has diminished, but ballistic and cruise missile technology has proliferated, and the need to track missile firings is as great as ever.

An ideal missile defense would detect launches and destroy the missiles during their boost phase, so that the warheads would fall back on the launcher's territory. Such defenses have been rejected as impractical or too expensive in the past, but the migration of technology to Third World countries makes the concept increasingly attractive. New approaches to this tactic are being developed.

Terminal-area destruction of missiles was demonstrated by the Patriot system during the Gulf War. Initially viewed as an almost 100 percent success story, later analysis indicated that there were drawbacks in the system, of which the random fall of the intercepted missile and warhead on friendly territory was the greatest. Detection of missiles has been more successful. The Defense Support Program (DSP) was first deployed in the 1970s and has since furnished an uninterrupted early warning capability. It is designed to detect and track all ICBMs, IRBMs, space launches, and nuclear detonations, as well as small tactical missiles. The DSP system worked well tracking Scud missile launches during the Gulf War. In 1995 a new method of processing data called ALERT (attack and launch early reporting to theater) was developed. The new method provides improved warning of attack by short-range missiles against U.S. and allied forces overseas.

TRW is the primary contractor for the DSP satellite, which operates in a geosynchronous orbit at about 22,000 miles altitude. Each satellite weighs 5,200 pounds and is 14 feet in diameter. It is 32.89 feet long with its solar arrays deployed. The solar arrays generate 1,250 watts of power. Twenty-one of a projected total of twenty-four DSP satellites have been launched. A space-based infrared system (SBIRS) has endured some development problems and when perfected will replace the DSP system.

The 21st Space Wing, located at Peterson AFB, Colorado, operates DSP satellites and reports warning information to the North American Aerospace Defense Command (NORAD) and U.S. Space Command. Both organizations have early warning centers within the fabled Cheyenne Mountain complex at Colorado Springs. Command and control of the DSP satellites is provided by the 50th Space Wing.

COMMUNICATIONS

The first radio message from an aircraft was transmitted in 1910; since then the importance of communications has grown until it is now the most important element in the command, control, and communications arena. Without it, neither command nor control could exist. A number of satellite systems are used to facilitate military traffic for everyone from the national command authorities down to the squadron level. These include the Defense Satellite Communications System (DSCS), the Military Strategic and Tactical Relay (Milstar) system, and the Air Force Satellite Communications (AFSATCOM) system. The Fleet Satellite Communications system is used both by the Air Force and the Navy.

The DSCS is used for high-priority communications between defense leaders and battlefield commanders and to transmit space operations and early warning data to various users.

Air Force Space Command operates thirteen Phase III DSCS satellites at an orbit of more than 22,000 miles. Each satellite uses six super-high-frequency transponder channels capable of providing secure voice and high-rate data communications. DSCS III also carries a single-channel transponder for disseminating emergency action and force direction messages to nuclear-capable forces.

The DSCS III satellite weighs 2,716 pounds and is 6 feet long, 6 feet high, and 7 feet wide. It has a 38-foot span with the solar arrays deployed. The solar arrays generate an average of 1,500 watts.

Members of the 50th Space Wing's 3rd Operations Squadron at Schriever Air Force Base, Colorado, provide satellite bus command and control for all DSCS satellites.

As we shall see, the ability to change satellite positions paid dividends during the Gulf War, when a DSCS II satellite was repositioned in geosynchronous orbit above the Indian Ocean, creating excellent communications in an area where they previously had been inadequate.

The Milstar satellite communications system is a joint service system that provides secure, jam-resistant worldwide communications for high-priority military users. It links command authorities with ships, submarines, aircraft, and ground stations.

It consists of five satellites in geosynchronous orbits at 22,500 miles above the earth. Lockheed Martin is the primary contractor for the huge, 51-foot-long satellites, which are 116 feet wide with solar panels extended fully. Each one weighs about 10,000 pounds. The solar arrays generate 8,000 watts of power.

The Milstar system is composed of three segments: space (the satellites), terminal (the users), and mission control. The 4th Space Operations Squadron at Schriever AFB, Colorado, is the frontline organization that provides real-time satellite platform control and communications payload management.

METEOROLOGY

The unsung heroes of the space age satellites are those of the Defense Meteorological Satellite Program (DMSP), which for more than forty years have been providing continuous real-time information on global weather conditions.

Two DMSP satellites are in polar orbits at all times, at about 500 miles altitude, and survey the entire earth four times a day with both visual and infrared imagery. The data is gathered by tracking stations and analyzed at military weather centers that can determine the location and severity of thunderstorms, hurricanes, and typhoons. The DMSP satellite weighs 2,545 pounds, is 11 feet, 6 inches high, 4 feet wide, and a little over 20 feet long. Built by Lockheed Martin and launched by a Titan II vehicle, it has a solar array capable of generating 1,250 watts.

RECONNAISSANCE AND SURVEILLANCE

It is exceedingly difficult to keep a single project, such as the U-2 or the F-117A, in the "black" for an extended period of time. Yet the work of an entire government agency, the National Reconnaissance Office (NRO), was successfully concealed for many years even as it opened enemy skies to constant surveillance. The progenitor of the NRO was first formed by President Eisenhower on August 25, 1960, and was redesignated the National Reconnaissance

Office in 1961. The Director of the NRO reports directly to the Secretary of Defense.

The term "reconnaissance" in NRO's title is a euphemism for its basic mission of spying from the sky. The satellite cameras do either reconnaissance, i.e., looking for specific information, or surveillance, i.e., a continual search of general areas for items of interest, as when it is tasked to determine compliance with arms treaties.

The principal mission of spying is augmented by ferret satellites monitoring radar and other electronic emissions, weather satellites, and special top-secret communication links. The constant use of data provided by the NRO satellites, circling the earth twenty-four hours per day, permits an assessment of the degree of military threat in foreign countries at all times and inhibits the use of camouflage and decoys as misinformation. The satellites also yield data useful in diplomatic negotiations. In the Bosnian conflict, for example, the NRO satellites were used to take photos of the notorious concentration camps and mass burial grounds.

The NRO acts essentially as a middleman, receiving instructions as to areas of interest by a committee of the Director of Central Intelligence. NRO does not make its own analysis of its data, leaving this to the using agency. The USAF has used this information for decades to plan and execute both strategic and tactical missions.

The success of the Corona series of satellites already referred to only hints at the extent of surveillance done since 1972. NRO surveillance, besides removing the missile gap myth, pinpointed secret Soviet activities from new ICBMs to the building of a phased-array radar that violated the ABM (antiballistic missile) treaty. NRO satellites were able to identify Chinese CSS-2 IRBMs being deployed in the Saudi Arabian desert, a clear indication of Beijing's expanding interest in arms exportation.

Each advance in computer technology enhances the capabilities of the satellites, which have grown steadily swifter in their processing and transmission of data so that today that they can be extended to nonmilitary uses, including monitoring the environment.

CONVERGENCE

The final decades of the twentieth century saw disparate trends within the USAF converge, against the odds, into a finely honed war-winning mechanism. One trend, maintained resolutely since its invocation by Hap Arnold, was the inexorable demand for improved technology. A second trend, reinforcing the first, was the spread of a culture of enlightened management that obtained the very best results from both the officer and enlisted forces. The third, opposing, trend was the tremendous drawdown after the heady first four years of the Reagan administration, when funds were virtually unlimited. But this drawdown was unlike the catastrophic demobilization after World War II. The severe reduction of almost 40 percent in numbers of aircraft and personnel was carefully managed, so that the resulting "base force" remained extremely capable.

Amid these merging, if partially opposing, trends, the USAF conducted successful operations in smaller emergencies such as Libya, Grenada, Panama, and Bosnia. In addition, it fought a successful war on a much larger scale in the Persian Gulf. The USAF's demonstration of unequaled military prowess in that conflict was demoralizing to the Soviet Union and contributed in large part to its breakup, and to the successful conclusion of the long and bitter Cold War. In the next chapter, these events will be covered to lay the foundation for a look at the Air Force of the twenty-first century.

9
≋

VICTORY: SPRINGBOARD TO THE FUTURE

*T*he United States Air Force began the twentieth century with confidence. For most of its existence, it had been fortunate to have charismatic leaders who were also hardheaded managers able to anticipate changing world conditions. These leaders assessed the genuine and truly awesome threat of the Soviet Union's nuclear striking power and contained it. In doing so, they simultaneously contained adventures by the Communist ground forces.

Mistakes were made in the process. The numbers of bombers, missiles, and warheads fielded over the years have been called excessive—yet they had the desired effect, and a lesser number might not have. It must be remembered that the Air Force was forced to operate under the most difficult policy conceivable, one that was without historic precedent. Its forces had to be able to accept a massive nuclear first strike from the enemy and *then* go on to win the subsequent conflict decisively. Thus, a large quantity of Air Force delivery vehicles and weapons was necessary to ensure that a war-winning force survived. It happened that this was a message the Soviet leaders could understand, one of the few "signals" sent by the United States that was properly interpreted. The massive armament investment also contributed to the breakdown of the Soviet economy as its leadership attempted—in vain—to meet all threats.

The United States probably did invest too much in its air defense efforts over the period by imputing more striking power to Soviet long-range aviation than it possessed. This was the legacy of World War II, when the Battle of Britain and the Eighth Air Force's hard fight over Germany demonstrated the necessity of an extensive air defense system. However, U.S. expenditures on air defense, massive as they were, never approached those of the Soviet Union in either absolute or relative terms.

An argument was made at the time, and could also be made today, that the money could have been better spent. The Air Force often did not have adequate conventional forces, and for the entire period of its existence has been near the breaking point in terms of required air transport capacity. Yet on balance, the task of defending the United States was fulfilled brilliantly, especially if one compares it to the corresponding efforts made by the leaders of Soviet Union, whose mistakes contributed significantly to the ultimate demise of their system.

The success of USAF planning efforts also must be considered in the light of the wildly gyrating budget. The Air Force's budget in 1980 was about $39 billion. It began rising in 1981 to a total of $46 billion and peaked in 1985 at $99.4 billion.

The subsequent fall was significant, the budget slumping to $71.2 billion in 1996. The war on terror has seen the budget climb to a peak of $125,245 million in 2003 (all figures in current dollars).

Abstract terms like constant dollars and gross domestic product sometimes obscure rather than illuminate the effect of budget reductions, which are felt in the field by the realities of unavailable spare parts, reduced flying time, and shortage of personnel. In 1987 the Air Force had 7,245 active-duty aircraft; by 2004, this number had been reduced 40 percent to 4,584, about the same number with which the Luftwaffe began World War II. Active-duty personnel have dropped at about the same rate, from 607,000 in 1987 to 359,700 in 2005. Air National Guard (ANG) and Air Force Reserve (AFR) totals have declined during the same period from 263,000 to 181,000. However, ANG and AFR integration into "Total Force" operations has significantly increased Air Force capability over the same time period.

Although the threat of attack from the former Soviet Union has all but disappeared the number of missions assigned to the Air Force has increased—from preparation for smaller conflicts to peace-keeping to nation-building to narcotics control. No one has yet grappled with the frightening notion of the states of the former Soviet Union and the Republic of China selling nuclear weapons to any country, association, or individual with the hard cash to buy them. Opinions from think tanks that no rogue nation would have the capability to use an intercontinental missile against a U.S. city until about 2010 are not reassuring. If that estimate is wrong, the United States might lose New York City, Washington, or Los Angeles in an instant.

The declining Air Force budgets have been further strained by the combination of the rising costs of a volunteer force and the upward spiral in the price of weaponry. The leaders coped by creating a smaller, leaner, modernized striking force backed by advanced technology and first-rate personnel. In addition, clear representations have been made to the Department of Defense, Congress, and the Executive Office on just how far air power's shortened blanket will stretch in terms of conducting a certain number of wars, large and small.

The Air Force was able to use the welcome but temporary aberration of greatly increased budgets in the early 1980s to increase force readiness through greater expenditures on the often neglected operation and maintenance funds, creating forward operating bases and increasing training. Then when the reductions appeared inevitable, Generals Gabriel and Welch planned the phased drawdown that traded manpower and older equipment for smaller, more capable forces.

Through all the budget gyrations, the import of Hap Arnold's emphasis on research and development had done far more than create equipment beyond even his imagination. It had created a culture that looked to technology not for a few magic bullets aimed at specific threats, but instead as a renewable source of power to harness for the entire spectrum of offensive and defensive needs. This culture would draw on the lessons learned in Vietnam to create new doctrine. To use the new equipment and the new doctrine effectively, the Air Force, a champion on the comeback trail from Vietnam, would require several tune-up fights before reclaiming its title before the world in the Persian Gulf War.

EVERYTHING COMES TOGETHER

The 1990s saw the convergence of the major elements previously highlighted. A broad research-and-development program provided Air Force leaders with a portfolio of modern weapons. This would have been far less useful had it not been for the remarkable change in the tenor of Air Force leadership from centralized and authoritarian to decentralized and participative. This change on the one hand encouraged the growth of a superb corps of noncommissioned officers and enlisted personnel who could fulfill the new responsibilities required by the advanced technology, and on the other created within the officer corps a sense of identity and purpose that permitted the full exploitation of the new technology.

A striking example of this phenomenon of total managerial revolution can be found in a publication that came from the very top, the white paper *Global Reach—Global Power*, issued in June 1990 by the very effective Secretary of the Air Force Donald R. Rice. In it, Rice, a former president of the RAND Corporation, asserted in measured terms that land-based air power now projected the national image of power and prestige that sea power had done in times past. Rice's paper did not denigrate the Navy or the Army, but instead spoke of the increased requirement for joint operations. But his message was unmistakable. Air power was now endowed with an unprecedented technical capability, one orders of magnitude greater than sea power at its most effective.

Rice pointed out that modern technology provided even cargo carriers with jet-age speed; air refueling enabled fighters, bombers, and cargo planes to fly nonstop to theaters thousands of miles away. The combination of speed and range gave any mixture of these forces a unique flexibility. New space-based navigation and communication equipment enabled Air Force units to use the new precision-guided munitions with an unprecedented accuracy so that the long-touted "surgical strike" became feasible. The combination of these characteristics gave an incredible lethality to air power without the brutal overkill of a nuclear weapon. *Global Reach—Global Power* did not eschew nuclear power—the umbrella of deterrence fashioned by Curtis LeMay was still required. But it did clearly state that air power could now reach anywhere in the world within hours to intervene decisively in any emergency. In essence, Rice maintained that the Air Force could now project dominating power to any point on the globe in *hours,* rather than in the days, weeks, or months that sea power still required. In most emergencies, time and strength are equally critical; x amount of strength in the first hours of a crisis can be worth 10x five days later.

Global Reach—Global Power was a cool, dispassionate representation of contemporary reality. The Air Force, still maintaining the always necessary nuclear shield, already possessed the capacity to execute any other mission that national policy demanded.

In the military services as in business, it is not enough to have new ideas; those ideas must also be adopted and put into use. The change in leadership style that would eventually permit placing the ideas of Rice and his colleagues into practice was also reflected in the way emergencies were met during the 1980s. Even as the Air Force was testing procedures and trying to acclimate to the changing methods of warfare required by the new technologies it had developed, a series of crises had to be mastered.

FIRST TESTS OF THE NEW AIR FORCE

The unfortunately named Operation Urgent Fury, the much-mocked invasion of Grenada on October 25, 1983, has generated such derision through the years that it has become a stock ingredient for stand-up comics. Often cited as a massive overkill, the steroid sledgehammer descending upon a malnourished fly, the intervention was more than justified to protect American lives. To the airmen participating, the antiaircraft fire was real and deadly: almost a dozen helicopters were shot down, and considerable damage was done to the USAF transports that brought in troops and cargo. The greatest outcome of the operation for the Air Force was the realization that ten years after Vietnam, command and control functions still had kinks that needed to be worked out.

The kinks showed up again in a more complex endeavor, the retaliatory strike against terrorists nurtured by the Libyan government. It was the most intense combat operation by the Air Force since 1972. Stephen Decatur had asserted newly found American strength in his attack on Tripoli in 1804; the United States Air Force would reassert that strength in Tripoli in 1986 with Operation El Dorado Canyon.

Disappointing political circumstances dictated that the mission be executed with old equipment, aging General Dynamics F-111F Aardvarks stationed in Great Britain. As had happened so often in the past, the superb skills of the aircrews managed to overcome planning deficiencies and daunting defenses. With sheer courage and superb flying skills, a handful of crews were able to bring off a marginal military success that had excellent political consequences.

STRIKE AGAINST QADDAFI

The persistent terrorist campaign by radical Muslim factions had reached a new height with simultaneous attacks on airports in Vienna and Rome on December 27, 1985. There was an almost immediate flurry of speculation in the media that the United States intended to respond with a chastening military attack on Libya. The inflammatory oratory of dictator Colonel Muammar Qaddafi and the verified existence of terrorist training camps made Libya a reasonable target, even though other nations were almost certainly involved. The situation was brought to a boil by terrorist attacks on a Berlin dance hall and a TWA airliner in April 1986, both of which took American lives.

Planning for a retaliatory raid had begun in January. The assignment to strike selected targets in the Tripoli area on the night of April 14–15, 1986, was given to the 48th Tactical Fighter Wing at Lakenheath, England, which was equipped with General Dynamics F-111 Aardvarks. Six Navy McDonnell Douglas F/A-18 Hornets, six Vought A-7E Corsair IIs, and fourteen Grumman A-6E Intruders from the carriers *America* and *Coral Sea* were to support the USAF mission, while other naval forces engaged in attacks on separate targets in eastern Libya. Most of the crew members going on the mission had never been in combat.

Secretary of the Navy John Lehman (who would have preferred to conduct the operation solely with naval assets) later wrote that only three targets behind the entire iron curtain had more sophisticated air defenses than those of Libya. More than 3,000 Soviet technicians directed the operation of the integrated defense system, which included 500 aircraft, a massive number of surface-to-air missiles, antiaircraft guns, and radar.

It is still galling to record that after the contributions of the United States to Western Europe (combat in World Wars I and II, the Marshall Plan, the containment of the Soviet Union, and the stiffening of NATO, to cite just a few), some of America's longtime allies capitulated to the fear of terrorism. France refused overflight permission (even as it "talked tough" to Qaddafi), while Spain, Italy, and Greece refused the use of their bases. In consequence, the mission had to follow a meandering over-water route around the coastlines of France and Spain. To attack the hazardous target, the crews were forced to spend fourteen hours flying 6,300 miles and refueling in flight eight times. Normal F-111F missions ran two to four hours. The cockpit of the F-111 is fairly commodious, but does not permit a short walk to stretch the legs.

The attack force was followed by many civil and military radars throughout its flight, but only the Italian radar operators went so far as to report the F-111s' passage, alerting counterparts in Malta. The Maltese government promptly warned Libya, removing any remaining vestige of surprise.

The F-111s, most fifteen years old, were tasked with the longest fighter combat mission ever flown in terms of time and distance. Recognizing that the mission duration would put a severe strain on engines and the aging electronic systems, twenty-four F-111s were launched, beginning at 17:36 Zulu (Greenwich Mean Time) on April 14. The aircraft were assessed en route, and the eighteen considered in the best condition were assigned to the attack. The F-111Fs were provided electronic countermeasures support by three of the superb EF-111A Ravens (aka "Spark Vark"), the most effective and sophisticated aircraft of its type.

The unified force hit Libya simultaneously early in the morning of the 15th in a thirteen-minute attack that destroyed five selected targets, including missile sites airfields and other installations. The heavy defensive fire lit up the Libyan sky, a preview of Baghdad a few years later. Two crew members were killed when one F-111 (70-2389) was lost for reasons unknown.

The attack came perilously close to being a failure. Rigorous rules of engagement, designed to minimize civilian casualties, ended up preventing six aircraft from even hitting their targets. Eight other F-111s did not strike their targets because of equipment difficulties. Fortunately, four F-111s hit the targets as planned, giving a preview of the future by using their AN/AVQ-26 Pave Tack infrared and laser targeting system to destroy a line of Ilyushin Il-76 transports, valuable four-engine aircraft corresponding to the Lockheed C-141. Some bombs struck within 50 feet of Qaddafi's residence, including the tent he used as Arab-style living quarters. Being on the receiving, rather than the giving, end of bombs severely shook the Colonel's composure.

If the military results had not been all that was desired, the political fallout proved to be exceptionally worthwhile. Even as the European media discounted the efficacy of the raid, member nations of the European Economic Community began taking much more forceful political and economic action against Libya. The Soviet Union confined itself to the usual propaganda attacks, but did nothing, and even Arab nations gave only pro forma expressions of support. Most important, not only did Libyan terrorist activity go into an immediate decline, that of Syria and Iraq fell off as well. The attack had effectively isolated Qaddafi, which was more than had been hoped for by the mission planners.

OPERATION JUST CAUSE

The changing nature of the world was reflected in the deployment of USAF assets to combat the drug traffic originating in Central and South America. Panama, crucial to U.S. defense needs since the opening of the canal, had become a center for drug traffic under the cruel dictatorship of General Manuel Noriega, a prototypical Central American strongman who had seized power from the legitimate government. A long series of incidents, including the assassination of a U.S.. Marine lieutenant, and the abduction and beating of a Navy lieutenant and his wife prompted President George Bush to order the armed services to intervene so that Noriega—once a protégé of the United States—could be removed from power.

Highly classified preparations for Operation Just Cause, the 1990 invasion of Panama, had been going on for some time. In many ways, the operation would be a test of concepts emerging from the Goldwater-Nichols bill of 1986, which gave theater commanders enhanced control over the forces of all services. Both the planning and the execution would be important dress rehearsals for the team that would just over a year later conduct the Persian Gulf War. Bush had selected a former Congressman, Richard Cheney, to be his Secretary of Defense and teamed him with General Colin Powell, the first black Chairman of the Joint Chiefs of Staff. The Air Force Chief of Staff, General Welch, had anticipated the downsizing of the Air Force and crafted its modernization; now he would see how his efforts would pay off.

As the leaders of Operation Just Cause started their work together, an interesting interpersonal drama began that was to have sudden, explosive impact a year later during Operation Desert Shield, the defensive buildup in Saudi Arabia. A testy Secretary of Defense Cheney, new to his office, was apparently determined to let everyone know who was running the DOD. At his first press conference as Secretary of Defense, Cheney gave General Welch a stern public scolding—before having discussed the matter with him—for his discussions with various Congressmen on the mix of Midgetman (a small, single-warhead ICBM) and MX missiles. The attack was both unprecedented and unwarranted, for Welch had a strong reputation for being loyal and cooperative and had advised DOD officials of his perfectly routine conversations on the Hill. No egotist and a good soldier, Welch retained his composure over the incident, realizing that any protest would be harmful to Air Force interests. Why Cheney acted in such a manner was a subject of debate, but the incident was widely interpreted as a signal of his determination to be regarded as the number one person in DOD. Some even thought that he intended it as a touch of the lash, a calculated insult to the military. Of this, more later.

The real villain in Panama, Noriega, was no stranger to hubris or narcotics. He proclaimed that Panama "is declared to be in a state of war" with the United States on December 15, 1989, and applauded the murderous assaults on U.S. military personnel stationed in the Canal Zone.

Operational secrecy was lost, thanks to a cable television broadcast reporting that Lockheed C-141s destined for Panama were taking off from Pope Air Force Base with the 82nd Airborne Division. Nonetheless, the USAF struck just after midnight on December 20, 1989, leading off with a brand-new weapon. Six Lockheed F-117A Nighthawks flew from their secret home base at Tonopah, Nevada, to Panama and back, refueling in flight six times.

It was not recognized at the time, but their assault heralded an entirely new era in warfare, that of stealth aircraft. Two of the F-117s dropped their 2,000-pound bombs at Rio Hato, deliberately striking a large open field next to the barracks of the Panamanian Defense Forces (PDF). The Nighthawks' mission was not to kill Panamanian soldiers, but instead to "stun, disorient, and confuse" them.

(This nicety of judgment was both a forecast of future policy and a reflection of public opinion, which wants wars won with minimal damage to personnel on either side.) Major Gregory Feest of the 37th Tactical Fighter Wing dropped the first bomb—as he would just one year later over Baghdad.

In the first wave of the attack, the Military Airlift Command flew 111 missions, seventy-seven of them by C-141s, twenty-two by C-130s, and eleven by C-5s. A second wave of thirteen C-5s and forty-four C-141s brought reinforcements. The MAC aircraft were refueled by seventeen KC-135As and six KC-10As from SAC. Ten regular Air Force wings participated, as did thirteen Air National Guard and four Air Force Reserve units—the Total Force in action. Seven participating 1st Special Operations Wing Lockheed AC-130H Spectre gunships used their 105mm howitzers and 40mm cannon with devastating effect on PDF vehicles.

Despite the 40 percent drawdown in strength, the modernization program of General Gabriel and Welch had borne fruit, for the F-117As worked with other highly specialized Special Operations unit aircraft to achieve a shock effect. Operation Just Cause used five Sikorsky MH-53J Pave Low helicopters, equipped with nose-mounted FLIR (forward-looking infrared), terrain-following and terrain-avoidance radar, and sophisticated electronic countermeasures equipment. Four Sikorsky MH-60 Pave Hawk helicopters, designed for bringing in or taking out personnel on a clandestine basis, and equipped like the Pave Low aircraft, were also employed, backed up by three Combat Talon I Lockheed MC-130Es. The latter are complex, Rambo-style aircraft, designed to battle their way in and out to conduct in-flight refueling of Special Operations helicopters at night or in adverse weather conditions. Well equipped with electronic countermeasures, the MC-130Es can deliver cargo or the biggest conventional weapon in the U.S. arsenal, the 15,000-pound BLU-82 propane bomb. More conventional HC-130 tankers and AC-130A gunships were also used.

The object of the endeavor, Manuel Noriega, initially gave the invading force the slip by the time-honored but totally unexpected expedient of gaining sanctuary. There was no little irony in Noriega's seeking asylum in the Vatican's diplomatic mission in Panama City, given that Noriega's perhaps limited religious beliefs ran more to the witchcraft of Brazilian Santeria. Noriega surrendered on January 3, 1990, and was flown to Florida in a MC-130E to be arraigned.

With Operation Just Cause, the USAF and its sister services had moved a long way toward a unified conventional war-fighting capability. At the same time, they found a new problem related to modern warfare. NBC had asked permission to land one aircraft at Howard Air Force Base.

The Air Force agreed, anticipating that a Learjet or similar executive aircraft would arrive with half a dozen newspeople. Instead, a chartered Lockheed L-1011 landed, carrying hundreds of media representatives of all sorts, all of whom wanted instant gratification in the form of inside stories. If the Nighthawks were the precursors of the air war in Baghdad, then surely NBC's L-1011 was the precursor of the news carnival in the Persian Gulf.

THE CALM BEFORE THE STORM

Almost surreal political events punctuated the last months of 1989 and the first months of 1990 as the fearsome Soviet Union suddenly began to crumple. Headlines almost weekly depicted some new movement of the Eastern bloc towards democracy, capitalism, and freedom. Hungary declared itself a free republic in October 1989. On November 9, the Berlin Wall was opened and hundreds of thousands of East Germans crossed to the West. On December 1, Presidents George Bush and Mikhail Gorbachev met on stormy seas off Malta to discuss arms limitations and the virtual end of the Cold War. The odious Romanian President Nicolae Ceausescu and his wife were killed on Christmas Day, and a reform government of sorts was established. In valiant Czechoslovakia, where Moscow's domination had been fiercely resented and resisted through the years, Communist leadership was replaced on December 29. The playwright and opposition leader Václav Havel was elected president. Lithuania declared its independence on March 11, even as the two Germanys began the process of reunification, which ultimately took place at midnight on October 2, 1990.

The USAF continued to hone its new capabilities as it reassigned resources to accommodate these welcome international developments. The Air Force Special Operations Command (AFSOC) came into being on May 22, 1990, with a proud history reaching back through predecessor organizations to clandestine operations in World War II in both Europe and Asia and to full-scale participation in Operation Just Cause. SAC began turning its FB-111s over to TAC, where they were redesignated F-111Gs. General Michael J. Dugan crowned a magnificent Air Force career when he became Chief of Staff on July 1. On July 12, the last of the fifty-nine Lockheed F-117As was delivered. By July 24, world conditions were deemed to have eased to such an extent that the Strategic Air Command ended its "Looking Glass" operations after more than twenty-nine years of continuous airborne alert. More than 275,000 hours had been flown without an accident, and best of all, without any requirement for action, for Looking Glass was the post-attack command and control system (PACCS) designed to take effect if the SAC command post at Offutt Air Force Base, Nebraska, was destroyed in a nuclear attack. Boeing EC-135s conducted the mission, while Boeing E-4s (modified 747s) acted as national emergency airborne command post (NECAP) to provide an airborne link between the National Command Authority and the U.S. military forces. The quiet cessation of this mission seemed to speak volumes for the prospects for peace.

Things changed abruptly when Iraq invaded Kuwait and menaced Saudi Arabia.

OPERATION DESERT SHIELD

Many books have been written about the Persian Gulf War, which proved to be in all respects the perfect arena in which the lean, modernized, highly technological Air Force could demonstrate true Global Reach and Global Power.

Air Force and DOD leaders had made the possibility of war in the Persian Gulf the subject of many exercises. Most of these were aimed at a potential drive by the Soviet Union through Iran to secure Persian Gulf oil, but the possibility of other nations intervening was also considered. As the Soviet threat declined, the new Chairman of the Joint Chiefs of Staff, General Colin Powell, encouraged planning against a possible attack by Iraq on Saudi Arabia, via Kuwait. (When war came, Powell would be very hesitant to conduct operations and actively sought a solution by sanctions or other means that did not involve open warfare.)

The continuing American desire to be able to respond to "limited" emergencies such as an attack on Saudi Arabia had a long history reaching back to 1962, when Strike Command was formed at MacDill Air Force Base, in response to President Kennedy's desire to have more conventional capability. The organization matured over time into the U.S. Readiness Command in 1972, and in 1980 into the Rapid Deployment Joint Task Force. In 1983, U.S. Central Command (CENTCOM) was formed, with the responsibility for the Middle East, Southeast Asia, and Northeast Africa.

CENTCOM was essentially a planning agency with no forces and a staff of about 700 from all four services. In the event of an emergency, the services would be tasked for forces, as required. A man whose father was better known to the American public, General H. Norman Schwarzkopf, was Commander in Chief of Central Command (CINCCENT). (His father had hosted the popular radio show *Gangbusters*.) Lieutenant General Charles A. Horner commanded U.S. Central Command Air Forces (CENTAF). He also commanded the Tactical Air Command's Ninth Air Force at Shaw Air Force Base, South Carolina.

CENTCOM planners had developed Operations Plan 1002-90, which called for a massive deployment of force to Saudi Arabia. In anticipation of this and other requirements, the USAF had prepositioned more than $1 billion of war supplies, from fuel to bombs, in Oman, Diego Garcia, Guam, and on ships in the Indian Ocean. Saudi Arabia had cooperated in regal fashion, building magnificent airfields and other installations for use in an emergency.

As tensions increased, CENTCOM ran the exercise Ivory Justice, in which U.S. tankers would refuel fighters from the United Arab Emirates—a bit of airborne gunboat diplomacy that Iraq's dictator, Saddam Hussein, duly noted but discounted. The Iraqi leader evidently assumed that the United States was still locked in its Vietnam syndrome and would not go to war. He was confident that if it did, the U.S. public could not endure the thousands of casualties that his battle-tested army would inflict upon American forces. In some respects, Hussein had adopted the tactics of Adolf Hitler, who was willing to commit his forces to defensive battles in which the enemy possessed decisive air superiority, counting on his tough troops to hang on and exact a high toll of casualties from the enemy in the process. And despite his oft-demonstrated military incompetence, Hussein must have at least realized that he would not have air superiority. He apparently believed that if the Americans indeed decided to fight, they would do so as they had done in Vietnam. He expected his tough, experienced land armies would inflict enough casualties that the American public would rebel. He was explicit in this in a conversation with the American ambassador, April Glaspie, in which he commented that the United States was a society that "could not accept ten thousand dead in one battle." He knew that *he* could bear his army's casualties without a qualm from his underground bunkers.

IRAQI IRE

When Iraq concluded its debilitating eight-year war against Iran in August 1988, it had expended more than $40 billion and 100,000 lives. One might have thought it would have welcomed a period of peace. Instead, Hussein intended to recoup his country's fortune by the same means by which he had lost it: war. On June 17, 1990, he threatened neighboring Kuwait and the United Arab Emirates, restating his past demands that their oil production be reduced (to raise oil prices) and that they not only forgive Iraq's war debt, but compensate it for the cost of the war against Iran. He also included some Texas-style accusations that the Kuwaitis had stolen $2.4 billion worth of oil by slant drilling and demanded compensation. On June 27, the Organization of Petroleum Exporting Countries (OPEC) agreed to raise the price of oil to $27 a barrel. By June 30, the Kuwaitis had agreed to reduce their oil production, and Hussein assured the leaders of other Arab states that he did not intend to invade Kuwait. Yet Hussein's ultimate objective was obvious; if he invaded Kuwait and then Saudi Arabia, Iraq would control more than 50 percent of the world's supply of crude oil.

At 0100 local time on August 2, 1990, experienced Iraqi units swarmed across the Kuwait border and Iraqi helicopters attacked Kuwait City. The Iraqi Air Force had about 1,000 aircraft; its 550 combat aircraft included MiG-29s, one of the finest fighters in the world. It had opened its war on Iran in the Israeli style with a strike on airfields, but was dilatory in Kuwait, not going into action until 0500 A.M. Several Iraqi helicopters fell to the pitifully few Kuwaiti Mirage interceptors available. It was a harbinger of things to come: the Iraqi Air Force was not spoiling for a fight.

U.S. REACTION

Of all the potential trouble spots in the world, none was more directly a matter of national interest than the oil fields of the Middle East. President Bush was determined to fight rather than relinquish U.S. interests there. The question was how to do it quickly and effectively. With Secretary Cheney and General Powell, he listened to a briefing from Generals Schwarzkopf and Horner on August 4, which detailed a plan based on previous planning exercises and which would become Desert Shield. Horner intended to pack Saudi Arabia with sufficient air power to deter Hussein, while Schwarzkopf followed up with 250,000 troops. The air power reinforcement could begin immediately upon approval and reach the minimum necessary to give Hussein pause within a week, but it would take at least three months to bring in a quarter million troops. Bush agreed to the plan and sent the four men to Saudi Arabia to meet with King Abdul Aziz ibn Fahd in Riyadh.

The king was in a desperate position. He risked alienating the Arab world by allowing American military men—and worse, American military women— into his country. But he risked his country if he did not, for Hussein already had more than ten divisions in Kuwait, some 200,000 troops, and was massing them on the Saudi border. King Fahd gave approval on August 6 for American aid. Horner remained in Riyadh, bringing in Brigadier General Buster C. Glosson to be director of campaign plans.

At 0900 Greenwich Mean Time, August 8, 1990, a Lockheed C-141B flown by an Air Force Reserve crew from the 459th Military Airlift Wing, Andrews Air Force Base, Maryland, landed, carrying airlift control elements (ALCE). It was followed within a few hours by forty-eight McDonnell Douglas F-15C and D aircraft from the 1st TFW from Langley Air Force Base, Virginia. The great World War II ace David Schilling, the pioneer of transoceanic flights by fighters, must have been looking down with approval on the longest operational fighter deployment in history. The Eagles were serviced six or seven times by tankers on the flight, which took from fourteen to seventeen hours, depending upon the wind. No major mechanical failures were reported after landing, and a little more than a day later the crews were standing combat alert.

The forty-eight Eagles and the competent but small Royal Saudi Arabian Air Force were all that stood between Saddam Hussein and the conquest of Saudi Arabia. Iraqi tanks could cover the 200-mile distance from the Kuwait border to Dhahran in less than a day. At that moment in time, Saddam Hussein still possessed the sixth-largest air force and the fourth-largest army in the world. His forces included 5,530 main battle tanks, 7,500 armored vehicles, 3,500 pieces of artillery, and 1,800 surface-to-surface missiles. The Iraqi air defense was formidable, with as many as 17,000 surface-to-air missiles and about 10,000 antiaircraft guns linked with high-tech equipment. It was indeed a threat to be reckoned with.

OPERATIONAL PLANNING

General Schwarzkopf had called the Air Force Chief of Staff, General Dugan, and requested a plan he could implement if the President wanted "to take some intiative" in December. Thus even while talks were being conducted with King Fahd, operational planning was underway at the Pentagon. There the Directorate of Plans' "Checkmate" team (so called for the war gaming facility created by General David Jones when he was Chief of Staff) initiated planning the air war against Iraq. The group began a very satisfying around-the-clock effort that produced a plan called "Instant Thunder," the name being an implicit repudiation of the failed policy of graduated response that characterized Operation Rolling Thunder in Vietnam. Instant Thunder called for a massive strike at the crucial Iraqi areas—their "centers of gravity"—such as command and control, communications, oil production and distribution, the Iraqi Air Force, Scud missiles and their launchers, and the nuclear, chemical, and biological warfare plants. Every planning effort was made to avoid civilian casualties; when targets were located within highly populated areas, care was taken to assign precision-guided munitions for their destruction.

In brief, the aim of the air campaign was to expel the Iraqi field army from Kuwait. Conceptually, the first task was to prevent any Iraqi disruption of allied air operations. The second was to destroy the Iraqi offensive air threat. Then the target, the Iraqi field army, was to be isolated and reduced by attrition. Finally, the air forces would support the allied ground-force operations.

The U.S. military operation in Saudi Arabia was given the official name Desert Shield, implying the intent to defend Saudi Arabia from harm. The Checkmate unit's plan was passed swiftly up the chain of command, receiving approval from Chief of Staff General Dugan and then from General Schwarzkopf.

Horner was not impressed by the plan, feeling that it lacked the almost infinite number of details that would be required for the massive air attack contemplated. He tasked General Glosson to create an operational plan to include many more targets to be attacked over a much longer period of time. Horner considered that plan to be too optimistic as well, in that projected a collapse of Saddam Hussein's regime after a relatively brief air campaign.

AN EXPLOSION DURING THE BUILDUP

Forces poured into Saudi Arabia at an amazing rate. By September 11, the USAF had 398 fighters in place; by January 17, when Desert Storm started, there were 652. By that date there was an "aluminum bridge" of C-5s and C-141s, which were landing every seven minutes at Dhahran. They, and for the first time in history Civil Reserve Air Fleet (CRAF) aircraft, had flown in more than 125,000 personnel and almost 400 tons of cargo. The CRAF drew on sixteen civilian air lines to provide eighteen long-range passenger aircraft and twenty-one cargo aircraft, with their crews. The military airlift to the Persian

Gulf would eclipse all previous airlifts by any standard of measure, reaching the rate of 17 million ton-miles per day, exactly ten times the peak rate of the Berlin Airlift. The total tonnage of the Berlin Airlift was exceeded in the first twenty-two days.

An unheralded but revolutionary change in warfare occurred when the Air Force Space Command positioned satellites of the Defense Satellite Communications System to establish communication links for Desert Shield. With this and the Navstar Global Positioning System, space now cloaked the battlefield with an enveloping power that transcended Iraqi understanding.

There was bathos in Saddam Hussein's response. To this massive display of air and space power, he turned his army to doing what it did best: building field fortifications in the desert sands. Massive defensive positions were constructed along Kuwait's border, even as Kuwait was formally annexed as another province of Iraq. If instead Hussein had flung his army across the border during the first week in August, he could easily have overrun Saudi Arabia and made the response of the United States and the coalition it was forming immensely more difficult. It would not have changed the ultimate outcome, but the war would have been protracted, perhaps going on for years, and infinitely more costly, for he certainly would have destroyed the Saudi Arabian oil fields as he did those of Kuwait.

Fortunately, Hussein hesitated. In the next forty-five days, another 750 American aircraft arrived in Saudi Arabia, even as the ground forces began to build up. Desert Shield, the protection of Saudi Arabia from Hussein, was well underway.

THE BLOWUP

On September 16, 1990, the *Washington Post* printed a headline article quoting Air Force Chief of Staff General Dugan: "The Joint Chiefs of Staff have concluded that U.S. military air power—including a massive bombing campaign against Baghdad that specifically targets Iraqi President Saddam Hussein—is the only effective option to force Iraqi forces from Kuwait if war erupts, according to Air Force Chief of Staff Gen. Michael J. Dugan."

The article went on at great length as Dugan commented on potential targets, and the efficient, decisive results he expected from American air power. He had been quoted correctly, and in the presence of other reporters from the *Los Angeles Times, Aviation Week,* and *Space Technology,* but the inferences drawn from the quotes were often incorrect.

The remarks infuriated Colin Powell and Richard Cheney, who interpreted them as putting too much emphasis on the Air Force and not being in the spirit of joint operations that so characterized Desert Shield. It was unfortunate, for Dugan not only believed in working with the other services, he had directed the Air Force planners to make sure that all planning was done in total cooperation with the Army, Navy, and Marines. Ironically, it was the very ease of manner, articulate presentation, and friendly, outgoing attitude that had carried him to the pinnacle of his profession that now tripped him up. He spoke about Air Force doctrine and what he thought about air power, the reporters were knowledgeable and understood him, but out of the context of his sincere appreciation for joint operations, his words were subject to harsh interpretation.

After having cleared the matter with President Bush, Secretary Cheney summoned General Dugan to his office on the morning of September 17 and demanded his resignation. Dugan understood that protest would hurt the Air Force and accepted the dismissal stoically, a good soldier to the end. It was a bitterly disappointing moment, for Dugan loved the Air Force and had labored to bring it to the peak of proficiency that he knew it now possessed. Then at the moment of truth in the Persian Gulf, he was no longer a player. In a press conference that afternoon with Dugan's written resignation in his hands, Cheney ticked off his reasons for the "firing," including "lack of judgment" and "demeaning the contributions of other services."

The Air Force was badly shaken by the popular Dugan's dismissal, coming as it did after the rough handling Cheney had given Welch the previous year. Dugan's discharge also coincided with increasing speculation on the potential value of air power in Iraq, especially since Hussein was assiduously digging in his armor and dispersing his aircraft in shelters hardened to sustain a nuclear blast. The whole unfortunate affair was given a poignant grace note by Dugan's farewell letter to the service he loved.

If there was one bright spot in the episode, it was the manner in which the Air Force responded, working cheerfully under Mike Loh as Acting Chief of Staff and then marching smartly off in a new direction under the leadership of General Merrill A. "Tony" McPeak when he took over on November 1, 1990. McPeak had been the alternative candidate when Dugan was selected, so it was naturally assumed that he would be summoned to fill the gap. He was well known to and appreciated by Secretary Rice.

McPeak made his presence known immediately. At a meeting with President Bush and the other Chiefs of Staff at Camp David, most of the other Chiefs were somewhat general in their presentations. McPeak spoke out forthrightly on the tactics to be adopted, what the target list should comprise, what the phases of the air campaign should be, and how long each phase could take. He warned that as many as 200 aircraft would be lost, and estimated that a hundred pilots would be saved by ejection. Of these hundred, fifty would be recovered and fifty would be captured, and he warned Bush that the American public might see captured American aircrews being paraded through the streets of Baghdad. He also warned that there would be collateral damage and civilian casualties. McPeak was conservative in his estimates of how long the actions would take and deliberately overestimated the number of probable losses, to make sure that he was presenting a rigorous picture. At the end of his remarks he promised, "In any case, at the end of thirty days you can kick off the ground campaign and it will be a piece of cake because we will have

done our job." Bush later asked his National Security Adviser, Brent Scowcroft, "Does McPeak know what he is talking about?" and Scowcroft gave an
enthusiastic affirmative reply.

Early in January 1991, McPeak went out to the Gulf and flew six F-15 sorties, including one sixty-ship mission that used the Italian Tornado aircraft as
aggressors in a simulated combat role. He came back impressed with the state
of readiness, and was asked by President Bush to lunch at the White House,
where the other guests were Scowcroft and Secretary of Defense Cheney.
Colin Powell, the Chairman of the Joint Chiefs of Staff, who reportedly was
seeking another thirty-day delay before initiating combat, was absent. When
asked by the President what he thought, McPeak responded, "The only real
mistake would be another postponement: these guys are ready to go." Bush indicated his agreement.

During those turbulent times, the effort to provide the necessary tactical
air forces in the Gulf War was sustained by General Robert Russ, the TAC
Commander, who took a huge workload off General Horner's shoulders and
saw to it that forces, supplies, and personnel were funneled into the Gulf at an
increasing rate. McPeak attributes a great deal of the success of the war to
Russ's preplanning and to the actions he took to achieve Total Force participation by ensuring that Air National Guard and Air Force Reserve forces were
used. Russ specifically asked that the 157th TFS of the 169th TFG of the South
Carolina Air National Guard be sent because it had done so well in the weapons
meets. The entire Air Force wanted to get into the fight, and those regular
units that had to remain in the States were discomfited that Guard and Reserve units were sent. Russ also decided to take only two squadrons from each
wing, so that the home base had a squadron in place to continue training and
have replacement forces available. The net result took a great load off the personnel system and, in typical Russ fashion, caused the well-deserved plaudits
to be spread around equitably.

PRESIDENT BUSH: CAUTION, RESOLUTION, THEN SELF-RESTRAINT

President Bush managed the buildup of Desert Shield with caution, loosely
weaving a coalition and securing a "hands-off" agreement from President
Mikhail Gorbachev of the Soviet Union. He worked carefully with the United
Nations to obtain the Security Council resolutions that first condemned Iraq's
invasion of Kuwait on August 2 and then progressively placed more pressure
through sanctions and embargoes. Finally on November 29, 1990, he received
the authorization of UN members to use "all means necessary" to force Iraq to
withdraw from Kuwait. By taking his time and avoiding extreme actions, Bush
was able to bring a coalition of thirty-eight nations, including several Arab
states, into an alliance against Iraq. Israel had to remain out of the coalition,
and had to be restrained from taking unilateral action against Saddam's provo-
cations, which would include Scud attacks.

With his cautious approach, Bush, a warrior himself in his youth, also had the self-restraint to leave the prosecution of the war to his military subordinates. The war would not be run from Tuesday luncheon meetings of the sort made infamous by Lyndon Johnson and Robert McNamara, but instead prosecuted by professionals in whom he had confidence.

DESERT SHIELD BECOMES DESERT STORM

Under the leadership of General Glosson, the initial Checkmate planning concept was converted into an extraordinarily detailed operational plan. The original Checkmate paper had envisaged a forced withdrawal of Iraqi forces from Kuwait, but General Powell, who was recognized now as a man determined to avoid open warfare if at all possible, demanded changes that would instead freeze them in place, for ultimate destruction by ground forces.

The growth in the number of aircraft available to the coalition forces, acting as a true unified command, made an expansion of the target list possible. A classical air operation planning system developed in which Schwarzkopf, as CINCCENT, would provide commander's guidance to the CENTAF staff. Each day, they in turn prepared a huge air tasking order (ATO) detailing which aircraft attacked which targets. Routes in and out, refueling, armament, radio frequencies, and all of the other information necessary to execute an air strike were spelled out. It was a mammoth undertaking, one that defined the nature of the unified command and confirmed that the war was not going to be fought like the war in Vietnam. The CENTAF Commander, General Horner, was also the Joint Forces Air Component Commander (JFACC) and thus had operational control over every aircraft that flew. The ATO went out from his headquarters to every element of the coalition forces. (The armed services had come a long way from the days of runners and hand-cranked field phones. Using today's technology, the ATO was distributed electronically by the Air Force Computer-Aided Management System (CAFMS) to most units. However, the lack of an interface with the U.S. Navy computer system required that a floppy disk containing the electronic data be delivered each day.) The degree of control conferred an unstated but implicit control over any possible response by Israel, for without access to the coalition's IFF (identification friend or foe) settings, not even the Israeli Air Force could penetrate the combat area except at prohibitive risk.

During the period of the buildup, the coalition air forces continuously flew missions that simulated the coming offensive. Fighter missions were launched, tankers flew the refueling tracks, electronic countermeasures aircraft conducted operations, AWACS aircraft executed their control functions. Feint after feint was made toward the Kuwaiti and Iraqi borders, testing the enemy integrated air defense systems—and the will of the enemy pilots. By January 15, 1991, the buildup had reached mammoth proportions. The Air Force had 1,133 aircraft in place, including 652 fighters, eighty-seven other combat aircraft (bombers, reconnaissance, electronic warfare, and special operations), and 394 support aircraft (AWACS, tankers, and theater airlift). Navy and coalition aircraft raised the total to 2,614, of which 1,838 were fighters and attack aircraft. Having previously secured the backing of both houses of Congress concurring with the UN resolution that set a date of January 15, 1991, for withdrawal from Kuwait by Iraq, on January 15 President Bush signed the National Security Directive authorizing military action, accepting that all attempts at negotiations with Hussein had failed.

By this time, the air war plan had been elaborated and was divided into four phases. The first one, scheduled to take eight days, called for the establishment of air superiority, the destruction of Iraqi Scuds and any nuclear, biological, or chemical warfare capability, and the disruption of command and control. In the subsequent four-day period, all Iraqi air defenses in the Kuwaiti theater of operations were to be suppressed. This accomplished, the coalition forces would enter phase three, in which emphasis would be shifted to attacking Hussein's field army in Kuwait. When all this was completed, it was expected that the principal emphasis would be concentrated on air support of ground operations until the war was over. Things would work out rather better and far more swiftly than planned.

THE ATTACK BEGINS

Crack elements of the joint Air Force/Army/Navy team began the air war in the early morning of January 17 with three totally different forms of attack: Special Operations helicopters, stealth fighters, and cruise missiles.

The first of the three elements, Task Force Normandy, crossed the border at 2:20 A.M., a full forty minutes in advance of H hour, 3:00 A.M. local Baghdad time. It consisted of two pairs of USAF Sikorsky MH-53J Pave Low helicopters from the 1st Special Operations Wing, each pair providing the navigation and timing for four Army McDonnell Douglas AH-64 Apache attack helicopters of the 101st Aviation Brigade. Their targets were two Iraqi air defense radar sites located 13 and 18 miles behind the border.

The helicopters struck the two targets simultaneously at 0238. The Apache's Hellfire missiles and 70mm rockets were supplemented by the 30mm cannon fire of their M230 chain guns to destroy the radar sites and create a hole in the Iraqi air defense systems. This was the beginning of the payoff for the years of research and development. The MH-53J helicopters were themselves magnificent machines, with a range of 450 miles and a top speed of 196 mph. But in their Pave Low III configuration, they became corsairs of the night, their crews using special night-vision goggles and FLIR (forward-looking infrared). The Pave Low helicopters could maintain exact track of their course by means of their navigation systems, which included the Global Positioning System and inertial guidance and Doppler radar. Their electronic countermeasures suite and their terrain-following and terrain-avoidance radar enabled them to avoid detection by flying close to the surface, sometimes ducking down into the gullies that laced through the desert.

The helicopters gave new and valid meaning to the term "surgical strike." In their swift, slashing attack, they began firing from a range of about 2 miles, aiming first at the radar site's electrical generators, then at the communications facilities, and then at the radar equipment itself. An avalanche of twenty-seven Hellfire laser-guided missiles and nearly one hundred rockets, supplemented by thousands of rounds of 30mm cannon fire, pulverized the installations.

The second element of this overwhelming initial assault was the Lockheed Martin F-117A stealth fighters, the "Black Jets" of the 37th TFW. Led by Lieutenant Colonel Ralph Getchell, CO of the 415 TFS, ten of the F-117As had departed their remote base at Khamis Mushait (jokingly called Tonopah East because of its resemblance to the stateside base), conducted night in-flight refueling with the customary radio silence, then separated to attack targets deep within Iraq, flying different altitudes and courses. It was a daunting experience; of the sixty-five pilots available to the 37th TFW, only four had ever been in combat before.

The F-117As epitomized the value and the risk of Air Force research and development over the years. Much was riding on their success—and until the war came, no one could say for certain that they would function as designed. And, despite the famous Skunk Work's having produced the F-117As under budget and ahead of schedule, external budget pressures reduced procurement of the Nighthawk to only fifty-nine aircraft.

The Nighthawks were extraordinarily sophisticated aircraft. Despite their awkward, angular appearance, they were pleasant to fly. They had a top speed of 646 mph and an unrefueled mission radius of 656 miles. Packed inside the compact wedge-shaped airframe were the exotic systems that made the aircraft effective. Depending on an inertial guidance system like that of the B-52 to get to the target area, the F-117 pilot then used his FLIR and DLIR (downward-looking infrared) equipment to locate the exact target and track to it. ("Exact" here had a new meaning in bombing terms. Where during World War II "exact" might have meant a square-block area, or in Vietnam a target the size of a bridge, it now meant a specific window or door of a specific building. So demanding were the new standards that if a precision-guided missile did not hit the specified window exactly, and only hit the same building, it was counted as a miss.) The infrared/laser turret next designated the target, and the system released the bomb to home unerringly to the precise spot designated by the laser. The 2,000-pound GBU-10s and the GBU-27 Paveway III laser-guided weapons amazed television audiences around the world as they watched the bombs speed unerringly to their targets. (Subsequent evaluations, including those of the Government Accounting Office, raised questions about the accuracy of precision-guided munitions and their cost relative to conventional weapons. The USAF position is unequivocally that precision-guided weapons are the weapons of the future.)

The importance of the years of research effort at AFSC and the quality control efforts of AFLC were put to the acid test on this first black night of

Desert Storm. Any failure of any part of the weapon system—airframe, engines, radar avoidance, navigation, laser, infrared, or bombs—would have negated the F-117s value. A failure in pilot training, maintenance, or mission planning would have had the same effect. If one multiplies the number of parts in each of the components times the number of components in each of the systems times the chance that any one of them might fail, the absolute probability of failure seems very high. Yet there were no failures. The F-117s blasted their targets in a bewildering display of technological superiority, the first notice of their presence being the explosion of their bombs. The stealth fighters took out hardened facilities in Baghdad and in the far reaches of Iraq. Headquarters, airfields, communication and control centers, and air defense sites all fell victim, gutting Iraq's capability to fight back. The invulnerable F-117As pressed their attack hard for the next forty-three days of combat. They were never touched by bullet or SAM, and as far as can be determined, were never even tracked by Iraqi radar. As Lieutenant Colonel Getchell later remarked, his men had been in the right airplane at the right time. The F-117A's success was due to a long train of people, places, and events that led back to the Lockheed Skunk Works, to billions of dollars spent on research and development, and ultimately to Hap Arnold himself.

The third element of this early-morning surprise for Hussein was the cruise missiles, sea- and air-launched. The air-launched missiles came from seven Boeing B-52Gs of the Eighth Air Force. Departing from Barksdale AFB, Louisiana, at 6:36 A.M. on January 16, the Buffs carried AGM-86C cruise missiles armed with a 1,000-pound conventional warhead. They arrived at their launch points after a tiring fifteen-hour flight that included many refuelings. The B-52s, originally designed to fly singly and at high altitude to drop nuclear weapons, now fired thirty-five missiles, thirty-one of which hit their high-value targets. After the launch, the B-52s flew back home against strong headwinds, landing back at Barksdale after a record-setting thirty-five-hour flight. A brilliant representation of Rice's dictum of Global Reach—Global Power, the mission was but a part of an even greater, more elaborate plan that was just unfolding.

The Buff's ALCMs were supplemented by fifty-four TLAMs (Tomahawk land attack missiles) fired from the Navy's battleships *Wisconsin* and *Missouri*, veterans of four wars and now on their last hurrah, and the missile cruisers USS *San Jacinto* and USS *Bunker Hill*.

The combination of these three initial elements—Task Force Normandy, F-117As, and the ALCM and TLAM cruise missiles—opened the way for an avalanche of 650 coalition aircraft, including 400 strike aircraft. They poured through the hole torn in the Iraqi air defense system, their sophisticated mix of weapon systems expanding the gap. Fifty-three of the old faithful F-111s from the 48th TFW hit major airfields, hardened aircraft shelters, and chemical-weapon storage areas with laser-guided missiles. They also used the Durandel runway-cratering bombs to help keep the Iraqi Air Force on the ground, which was where, as things developed, it preferred to be. Nineteen *LANTIRN*-equipped F-15E Strike Eagles went after missile sites and Scud launchers. The Royal Air Force sent Tornadoes in low to hit three airfields, using JP-233 airfield-denial weapons. The RAF would do brilliant work in destroying bridges all through the war, part of the scheme that crippled Iraqi transportation.

Curiously, an unarmed EF-111A Raven electronic counterwarfare plane inflicted the first air-to-air loss on the Iraqi Air Force. Attacked by an Iraqi Dassault Mirage fighter firing missiles, Captain Jim Denton of the 42nd ECS took evasive action in a rolling, diving turn. The Mirage pilot tried to follow, but was observed by Captain Brent Brandon, the Raven's EWO, to fly into the ground, blowing up. Shortly thereafter, Captain Steven "Tater" Tate of the 71st Fighter Wing, flying an F-15C, got a more conventional kill, shooting down a Mirage with a Sparrow missile from a 15-mile distance. (The Sparrow, much improved since Vietnam, would prove to be the primary air-to-air weapon of the war, scoring thirty-one victories—two by the Royal Saudi Arabian Air Force—compared to ten by the heat-seeking AIM 9L Sidewinder.) With the AWACS aircraft able to provide positive identification (the most important element lacking in Vietnam), the AIM-7s could be launched from beyond visual range; consequently, most combats were conducted with Sparrows rather than Sidewinders.

In the course of the war, fourteen USAF, fifteen Navy and Marine, and six Army aircraft would be lost in combat for a total of thirty-five. Eight coalition forces aircraft were lost in combat, six of them Tornado low-level attack planes of the RAF. During the period from August 29, 1990, to March 31, 1991, the United States had thirty-seven noncombat losses, including ten from the USAF.

Nothing was more in keeping with this strange, almost bizarre war than the fact that American television commentators were broadcasting from Baghdad during the height of the attack. The American public could see the skies over Baghdad filled with an apparently impenetrable curtain of antiaircraft fire, unaware that U.S. aircraft were flying through the impressive display without any hits being scored. In one of the most incredible real-time confirmations of success in military history, U.S. leaders in the Pentagon and in Riyadh were watching the CNN broadcast from Baghdad when the station went off the air, a victim of a cruise missile.

Never before had there been anything to compare to this attack in the history of air warfare. Pearl Harbor pales in comparison beside it, as does the Nazi May 1940 blitzkrieg. Even the brilliant eleven-day attack on Hanoi and Haiphong in 1972 did not achieve the paralyzing results of this tidal wave of air power. In the first twenty-four hours of the Desert Storm air war, the coalition forces had established air superiority, decapitated Hussein's excellent—until then—command and control system, shut down Iraq's electrical production, and seriously reduced the effectiveness of the many SAM sites and antiaircraft batteries.

The very effectiveness of the attack forced a change in the war plan, for the substantial destruction of the important strategic centers and the integrated defense system took place within the first thirty-six hours. Everything, including the attacks on the field army, was now accelerated. The Iraqi Air Force had been effectively removed from the war; it would be seen again only in isolated sorties, or in the ignominious decampment of 120 aircraft to its former enemy, Iran. After the war, Iran somewhat predictably confiscated the aircraft rather than return them to Saddam.

Despite all this, hard fighting was still to come. Even though now being fired in barrage patterns rather than under radar control, Iraqi antiaircraft batteries put up a tremendous curtain of fire through which coalition aircraft had to fly. The battle-hardened Iraqi soldiers were tough and, not yet having felt the full fury of the coalition air attack, were prepared to inflict heavy casualties in the ground war.

And at the bottom of it all, Hussein remained as before, stubborn and brutal. He demonstrated the latter quality on January 17, when he fired the first two Scuds at Israel. Early the next morning, a fusillade of seven Scuds aimed at Tel Aviv followed. Fortunately, only ten people were injured in this opening act of terrorist warfare. The same morning also saw the first Scud fired at Saudi Arabia. It was destroyed when hit at 17,000 feet altitude by a Patriot missile, the very first antimissile missile to be used in combat.

Hussein's purpose was clear: if Israel entered the war, the Arab members of George Bush's coalition would be put in an untenable position and would have to withdraw. Immoral as Hussein's attack was, it was a practical tactical move, for it immediately diverted coalition resources for a campaign-long, 2,500-sortie assault against Scuds and their transporter-erector-launchers, launch sites, storage areas, and production facilities. Two "Scud boxes" were established, one in the west to strike missiles targeted on Israel, and one in the south, where they were targeted against Saudi Arabia. Within these boxes, Scuds were hunted mercilessly by strike aircraft, with at least two A-10s attacking each area twenty-four hours a day.

Reminiscent more of the German V-2 of World War II than of the later ICBMS, the SS-1 Scud was a crude weapon of 1957 vintage, designed and built in the Soviet Union and widely exported. Both Iran and Iraq used it indiscriminately against each other's cities during their eight-year war, each side sorry only that they lacked nuclear warheads. About 38 feet long, with a diameter of 33.5 inches, the Iraqi Scud B had a weight of 14,000 pounds and a range of 175 miles with a 2,205-pound warhead. The MAZ-537 TEL (transporter, erector, and launcher) made the missile highly mobile. With an inertial guidance system that provided only a limited accuracy of about 1.5 miles, the Scud was in fact an indiscriminate "area bomber," as much a terror weapon as a car bomb. Fortunately, the Scud did not hit a vital vulnerable target like a munitions dump or a hospital. If it were fitted with chemical, nuclear, or biological weapons, however, it could be extremely dangerous. The Iraqis had created two improved versions of the weapon, the Al-Husan and the Al-Abbas. By reducing warhead size and substituting fuel, the Iraqis were able to increase the range to about 400 miles.

The coalition intelligence on Iraq had generally been deficient, and nowhere more than in its underestimation of Hussein's Scud capability, where it was off by a factor of five. The Iraqis possessed more than 225 launchers and were as skilled in concealing them as they were in their use. After a launch from a previously surveyed site, the crews could be on their way in just a few minutes to a designated shelter area to hide. The most effective solution against these mobile targets was the patrolling McDonnell Douglas F-15E Strike Eagles. Their long range and high speed could make the best use of their LANTIRN and synthetic aperture radar equipment.

Ultimately it proved impossible to suppress the Scud launches completely, even though a large share of the coalition's resources was devoted to the task. A breakthrough in space-related warfare occurred when the Defense Satellite Communication System satellites, designed for use against ICBMs, proved to be most effective in detecting Scud launches. Another brand-new system, the Boeing/Grumman E-8 Joint STARS aircraft, was still in its experimental test program. Flown by USAF crews and linked to the Army battlefield forces, its radar peered as much as 100 miles into enemy territory, alternating between the Doppler and the synthetic aperture mode to detect transporters on the move or standing still. Lockheed U-2Rs and TR-1As, which flew reconnaissance missions monitoring Iraq before, during, and after the conflict, were also used to detect Scud launches. Immediately after launch, target information was provided to patrolling F-15s as to the TEL's whereabouts, and a warning was dispatched to Patriot missile sites. The Patriot search radar would acquire and track the Scud; as it approached, a salvo of Patriots (usually two) would be launched, guided by the ground-based radar. Sometimes, the Scuds broke up as they reentered the earth's atmosphere, making their tracking and interception more difficult. The Patriot's radar would pick up the incoming missile—or some of its components—guiding the Patriot to a point where its detonation would blow up the Scud. In the first flush of Patriot kills, the general sense was that the Scuds had been mastered. Later analysis showed that although the Patriots had been extraordinarily effective, they did not measure up to the initial impression. The intercepted Scuds still descended upon Israeli or Saudi Arabian territory, the warhead sometimes still intact, and debris from the rest of the missile and the Patriot itself caused damage. (It must be remembered that the Patriots rushed to defend Israel served to keep Israel out of the war during the most crucial period of the Scud assault. The Patriot's political effect might well have been greater than its military effect.)

By chance, one Scud was fired to arrive during the time the appropriate Patriot missile battery was standing down for routine maintenance. The warhead hit a barracks near Dhahran, killing twenty-eight American soldiers and wounding ninety-eight others. In the course of the war, the Scuds killed a total of forty-two people and wounded 450. More casualties were prevented when a group of twenty Scuds being prepared for a mass launch on Israel were discovered and destroyed on February 27.

The air onslaught against the Scuds was effective; fifty had been fired by January 27, but only forty-three more were fired during the remaining thirty-three days of the war. Despite the mammoth effort, the possibility remains to this day that the Scud could be mated with a nuclear, biological, or chemical warhead and do incalculable damage.

With air superiority decisively established, the Iraqi electrical system and C^{3}I (command, control, communications, and intelligence) destroyed, most radar sites incapacitated by SEAD (suppression of enemy air defenses) teams, and the Scud threat muted if not eradicated, more attention was turned to the destruction of the powerful Iraqi army, hunkered down behind their anachronistic desert defenses. The devout Iraqi soldiers believed that death in battle entitled them to an immediate ascent to paradise. Many would make the trip; many more, however, would endure a long hell on earth as coalition aircraft systematically reduced Iraq's mighty army to an inchoate force of starving, frightened soldiers, eager to surrender to anything from a roving Special Operations armored car to an unmanned remotely piloted vehicle.

MORE INTENSE ATTACKS: GREATER PLANNING PROBLEMS

The massive preparation for Desert Storm yielded results beyond anyone's imagination. One of the results was a huge discontinuity in the planning effort that was supposed to be translated each day into an Air Tasking Order (ATO) to assign the next day's missions. The quick success was welcome, even if it was a case of too much too soon. The forced compression of the four planned bombing phases, the diversion of assets to hunt for Scuds, and unexpectedly bad weather over the target and on the refueling tracks combined with inadequate resources for bomb damage assessment (BDA) to make planning after January 18 very difficult.

Air Force estimates of bomb damage were considered by the national intelligence community to be too high, although they were in fact conservative. Ground force commanders tended to believe the more pessimistic reports, and in a manner that exactly echoed their complaints in World War II, Korea, and Vietnam, demanded more air-to-ground strikes in their own individual areas. Fortunately, General Schwarzkopf took a broader view and relied on the reports furnished him by Horner and Glosson, which in turn were based on the hardest intelligence available: strike camera photos confirmed by the personal observations of the aircrews.

The combination of complicating factors plus the worst weather in fourteen years made the preparation of the ATO so lengthy that mission commanders did not have adequate time to study it or to accommodate the inevitable changes. One result of the changed circumstances was an increase in mission cancellations—456 on January 19 and 431 on the following day. Another was the loss of two F-16s on a mission into the heart of Baghdad that proved the Iraqis could still use their antiaircraft artillery effectively in barrage fire and fire their SAMs ballistically even with their C³I impaired.

Yet the situation soon stabilized, and General Horner was able to direct a well-disciplined, systematic round-the-clock air campaign that would destroy most of the remaining important targets in Iraq, while simultaneously waging a war of savage attrition against Hussein's now immobile army. The B-52s performed the same role they had in Vietnam, dropping huge quantities of weapons upon troop positions, while the strike aircraft, particularly the F-16s and A-10s, sought out and destroyed tanks and artillery.

VERSATILE FORCES = BLOODY ATTRITION

Hussein expressed his frustration in a series of random acts of terror, beginning on January 25, with the destruction of Kuwait's main supertanker loading pier, allowing millions of gallons of crude oil to be dumped into the Persian Gulf and creating an ecological disaster. A precision air strike by F-111Fs, using 2,000-pound GBU-15 (V)-1/B electro-optical bombs, against the pumping station and oil manifolds managed to stop the flow.

No act of vandalism by Hussein could shield his forces from the ferocious aerial attacks. The years of research and development, test and modification had borne fruit; when combined with the sophisticated new munitions, every one of the several types of USAF aircraft proved potent in battle.

The Fairchild A-10, officially named the Thunderbolt II, but invariably called the Warthog because of its starkly functional appearance, became the favorite of the Army troops, for it did a great deal of its work "up close and personal." A-10s fired almost 5,000 Maverick missiles and claimed 4,200 successes—1,000 tanks, 2,000 vehicles, and 1,200 pieces of artillery.

The Lockheed (formerly General Dynamics) F-16 Fighting Falcon was the most numerous USAF fighter in the theater, with 249 on hand. In almost 13,500 sorties, the F-16s attacked every sort of target, from SAM sites to nuclear plants. Ground crews rallied around the aircraft to maintain an unprecedented 92.5 percent in-commission rate, a full five percentage points better than the peacetime average.

The overall effectiveness of some F-16 operations was blunted by a lack of advanced equipment. Some F-16s had the LANTIRN navigation pod but not the targeting pod, and were forced to rely primarily on the standard 2,000-pound Mark 84 "dumb" (i.e., not precision-guided) bomb. Antiaircraft fire forced the F-16s to drop from higher altitudes, degrading their accuracy.

Working in conjunction with the experimental J-STARS aircraft, the multimission Lockheed (also formerly General Dynamics) F-111s and the deadly accurate Lockheed F-117s turned their attention to hardened aircraft shelters, using laser-guided 2,000-pound bombs. Designed to withstand a nuclear attack, the strongly built shelters served instead to contain the force of the exploding bombs, reducing aircraft and equipment inside to a jumbled mass of rubble. These attacks finished off the Iraqi Air Force: the coalition forces had gone beyond air superiority to air supremacy. As clear confirmation of this, by the second week of the war, KC-135 tankers were permitted to fly into Iraq to refuel the F-117s.

With the hardened shelters destroyed, the F-111s and F-15E Strike Eagles now turned their attention to "tank plinking," a term so chillingly apt that even American tankers disliked it. The J-STARS aircraft were very effective in locating targets and the strike fighters would use up their ammunition before their fuel, saving time and tankers. Iraqi armored vehicles were dug in deeply in the sand but were still betrayed by their heat signature to the infrared sensors of the FLIR pods. The USAF fighters used precision-guided munitions— "smart bombs"—to destroy them at the rate of 100 to 200 per night, "plinking" them like tin cans at a dump site. Iraqi tank crews that formerly took shelter inside their tanks now made sure they bunked down elsewhere. The task was admittedly made more difficult by the Iraqis' extensive use of sophisticated decoys, which mimicked the infrared and radar signatures of the real weapon and required some careful cross-checking by the surveillance aircraft.

The 48th TFW, commanded by Colonel Thomas Lennon, became adept at killing tanks at night. Flying in flights of two or four aircraft, the F-111Fs would patrol the "kill boxes"—60 by 30-mile areas. Each aircraft carried four GBU-12 laser-guided bombs; the 500-pound units had been considered too light to kill tanks, but, dropped from medium altitudes and guided with precision, they were deadly. In just twenty-three days, the F-111s flew 664 successful sorties, taking out tanks, trucks, artillery—anything that radiated enough heat for the Pave Tack infrared pods to pick up. In economic terms, it was a profitable exchange, for the GBU-12 cost about $10,000 versus the open-market price of $1.5 million for a T-72 tank. The F-15Es were equally successful. On one mission, they batted 1.000—two Strike Eagles, each carrying eight GBU-12s, destroyed sixteen armored vehicles.

The ALCM launch by the Barksdale Buffs has already been described. Less well known is the almost simultaneous attack thirteen other of the venerable Boeing B-52Gs made in their first-ever low-level combat mission, a task that had been practiced for years. About 4:00 A.M. Baghdad time, the B-52s, in flights of twos and threes, raced across the Iraqi desert at only a few hundred feet altitude, guided by terrain-avoidance radar, to strike Iraqi air bases with CBU-89 Gator mines and 1,000-pound runway-busting bombs.

The bulk of B-52 activity came later, when more than eighty Buffs were on call for combat duty, flying out of bases in England, Spain, Diego Garcia, Egypt, and Saudi Arabia. Flying in cells of three aircraft, Vietnam-style, they bombed targets every three hours. Each aircraft could carry fifty-one bombs, twenty-seven internally and twelve on each wing pylon, almost always 750-pound M-117s. The cell, flying at altitude, would drop 153 bombs in a swath a mile and a half long and a mile wide, an ear-popping, sinus-shattering symphony of disaster. Because the first warning to the Iraqi troops would be the bombs exploding, the psychological effect was enormous. For twenty-four hours a day, the Iraqi troops were conscious that the next second might see them dead. One Iraqi battalion commander surrendered not because his unit had been under the B-52s' bombs, but because he had seen the devastation wreaked upon another formation exposed to them. The B-52Gs flew 1,624 missions, dropping 25,700 tons of bombs. Perhaps the most amazing facet of the B-52s' performance, and a tribute to the enlisted personnel who maintained them, was their 81 percent in-commission rate—two percentage points better than the peacetime average, and an incredible achievement for a system so large and so old.

FINISHING TOUCHES

As Hussein felt his grip on Kuwait slipping, he began setting fire to the Kuwaiti oil fields, the smoke layering the sky over the battlefield and reducing visibility to 3 miles. As he assessed the effect of this wanton destruction upon the progress of the war, General Horner also made a personal estimate of how future sorties would be allocated. He believed that a total of about 450 sorties would eliminate any remaining Iraqi airfields, complete the destruction of current stocks of Scuds and their production facilities, and eliminate residual electricity and petroleum production. Another 200 sorties would be required to destroy munitions factories, munitions storage sites, and similar military support facilities.

Thus, the major remaining task was the destruction of Iraqi ground forces, especially the elite Republican Guard. He estimated that to do this would require between 17,500 and 20,000 sorties, perhaps even more, depending upon the ability of the Iraqis to resist the unremitting assault.

While the world waited for "G day"—the beginning of the ground attack that Hussein said he wanted to happen more than anyone—Horner accelerated the air assault, doing far more than "prepare the battlefield." By the unremitting ferocity of its attack, the coalition air forces deprived the Iraqi army of its will to fight. With communications destroyed, reinforcement impossible, food and water scarce, and their major strengths, artillery and tanks, being plinked into oblivion day by day, morale fell rapidly. Desertions began, and units were quick to surrender. By the time G day arrived on February 24, 1991, many Iraqi units were at or below the 50 percent point in their nominal combat strength; when morale was factored in, they were even less effective.

By G day, the coalition had 2,790 fixed-wing aircraft in the theater, of which almost 2,000 were "shooters," i.e., strike aircraft. When the ground war was launched, this force intensified close support efforts, with sorties reaching a peak of 3,500 per day on February 27, the day before the fighting stopped. A final example of the close relationship of R&D to the war effort occurred on the same night. The Iraqis had been foresighted in constructing bunkers that would have made Hitler proud, sunk deep in the ground and heavily covered in reinforced concrete. One such command bunker was located at the Al Taji airfield, just north of Baghdad.

Working with Texas Instruments and the Lockheed Missile and Space Company, a quick-reaction Air Force team at Eglin Air Force Base created a bomb tailored to reach deep into the earth after Iraqi leaders. The BLU-113 penetration warhead had been developed in only seventeen days to meet the requirement. Designated the GBU-28-B, the bomb casing of these bunker-busters was machined from 8-inch artillery tubes, filled with 650 pounds of molten tritonal explosive, and fitted with a hardened steel nose cone. Thirty bombs were built, none with any authorizing paperwork. The Paveway III guidance system was modified by Texas Instruments for the new 4,700-pound bombs, which were special-delivered to the F-111Fs of the 48th TFW. Dropped from altitude at supersonic speed to increase their kinetic energy, the bombs had the capability to penetrate 100 feet of earth or 20 feet of concrete. The damage they did sent a clear message to Hussein that there was no longer anywhere to hide.

Did air power win the Gulf War by itself? The answer is no, for the Army, Navy, and Marines rendered glorious service. But the fact remains that the most extensive and successful preparation of the battlefield in history had been accomplished by air power. The Iraqi army, the sword that Hussein hoped the U.S. Army would throw itself upon, was hammered into a shattered mass, incapable of fighting effectively and highly susceptible to surrender. Saddam Hussein himself, his communications reduced to the Revolutionary War standard of runners hand-carrying messages, may never have known the state to which his army had fallen.

The success of the one-hundred-hour ground campaign can be attributed to the most effective air campaign in history. If the 111,000-plus sorties of the air campaign had not been planned, flown, and executed as they had been, the ground campaign might have been a hundred days or a hundred weeks, and, instead of being brought to a climax by the great "Hail Mary" flanking movement, might have required a bloody novena of costly frontal battles.

REFLECTIONS ON LESSONS LEARNED

The Air Force had applied much of what it had learned at such bitter experience in Vietnam with satisfying results. Using a new and benign management attitude that permitted leaders to elicit the very best from all personnel, officer and enlisted, it had applied its funds wisely among the varied needs for new equipment, modification of older equipment, training, and the prepositioning of assets. It had adhered to the doctrine it had developed of suppressing and rolling back enemy air defenses to gain air superiority. Reliance was no longer placed upon sheer weight of ordnance and volume of sorties but instead upon the accurate placement of that ordnance. In a curious fashion, the pendulum of war had swung away from mass destruction to pinpoint elimination of critical nodes; instead of the doleful mutual deterrence concept of measuring casualties inflicted (and received) in the millions, the point was now to inflict minimum casualties and sustain none, if possible.

To the surprise of no one who had monitored the process, the "Total Force" concept had once again been validated, for of the 54,706 USAF personnel in the theater of war, 12,098 were Air National Guard or Air Force Reserve personnel—more than 22 percent. Nor were these "support troops." Instead, the thirty-seven Guard and twenty-eight Reserve units flew C-130s, KC-135s, A-10s, RF-4Cs, HH-3Es, F-16s, C-141s, and C-5s, all first-line equipment, flying and fighting in the front line. Their excellent performance would point the way to the future of the Air Force, for with declining budgets, the only possible way for the service to retain its striking capability was to transfer even more responsibility to Guard and Reserve units.

Space-age warfare was introduced with the remarkable success of the Global Positioning System, which at the time had sixteen satellites in place, five short of the number needed for complete worldwide coverage and eight short of the ultimate number planned. One of the sixteen suffered a failure that was vital to providing the three-dimensional coverage necessary in desert warfare. Air Force personnel at Falcon AFB's Air Force Space Command developed software that stabilized the satellite, placed it in the right attitude, and made it useful.

The two Defense Satellite Communications System (DSCS) satellites employed in the complex, spaced-based communications network were soon overloaded. Once again AFSPACECOM members stepped in, this time to execute a historic first in space warfare. A DSCS II satellite was being held in reserve over the Pacific, in a stationary orbit 22,300 miles above the earth. Space Command specialists commanded it to start its motor and scooted it westward for several days to a fixed point over the Indian Ocean, where it solved the communications overload. The satellites were exploited by more than thirty ground satellite terminals. These were the basis for an elaborate communications network that integrated the torrential volume of communications flowing over the varied equipment of eight countries. Astute preliminary planning, carried out over a long time span, had resulted in the interoperability of the communications equipment of the forces of the United States, Great Britain, and France.

DOMINATING THE BATTLEFIELD

The post-Vietnam Air Force has moved entirely away from the concept of contests to win air superiority in the mode of the Battle of Britain or the later battles of the Eighth Air Force against Germany. It now wants to win battles over the enemy heartland with such overwhelming superiority that there are few if any USAF casualties, while the enemy is completely subdued. Excellent aircraft and intensive training are but a part of the new strategy. There is now an overriding requirement for an electronic supremacy of the battlefield that provides U.S. forces with complete information on enemy strength and intentions while denying the enemy intelligence not only about U.S. forces but even about his own forces. The first demonstration of the efficacy of this concept was provided by the Persian Gulf War.

One of best-known illustrations of this new method of fighting was the remarkable success of the airborne command and control aircraft. These airplanes, the subject of bitter opposition during the period of their procurement because of their great expense, proved to be invaluable. The hard-working Lockheed EC-130E airborne battlefield command and control center (ABCCC) controlled air-to-ground attacks. More familiar because of its huge signature rotating radar dome, the Boeing E-3B Sentry airborne warning and control system (AWACS) controlled the masses of airborne aircraft, including tankers and reconnaissance and strike aircraft. The two experimental E-8 Joint STARS aircraft ferreted out ground targets such as Scud launchers and monitored traffic flow. No battle before had ever been fought with such superb command-and-control facilities—no future battle should ever be fought without this capability.

One of the less well known but vital tasks, the collection and analysis of electronic emissions, was handled with panache by the Boeing RC-135 Rivet Joint aircraft. Using the flight deck crew, electronic warfare officers (EWOs), and airborne intelligence technicians (AITs), the Rivet Joint team used its Elint (electronic intelligence) capability for three main tasks: (1) to provide indications about the location and intention of enemy forces; (2) to broadcast a variety of voice communications, especially combat advisory broadcasts and imminent threat warnings, which warned of SAM launches, assisted in search and rescue, and even helped aircraft on air defense suppression missions; and (3) to operate the data and voice links to ground-based air defenses, providing target information on incoming aircraft or missiles.

Elint is characterized by long hours of work on station, and patient analysis of enemy transmissions, punctuated by brief moments of urgency when the vital—often life-or-death—information is transmitted to the appropriate receiver. Three Rivet Joint aircraft of the 55th Strategic Reconnaissance Wing (SRW) were in the theater, providing twenty-four-hour coverage. The Iraqis were well disciplined in their use of electronic equipment before the start of Desert Storm, but the RC-135s were able to ferret out the locations of most of the communication centers.

The RC-135s were complemented by the versatile Lockheed EC-130H Compass Call aircraft, which jammed enemy transmissions with deadly effect. Few Iraqi radios were on the air for more than a few moments until the Compass Call aircraft electronically obliterated their transmissions.

LEADERS WHO WERE WINNERS

The war was started by the ego of one man, Saddam Hussein, but it was ended by a group of men who suppressed their egos for the common good. A great deal of credit must be given to the inner circle of U.S. leaders directing the war who permitted the new command structure envisioned by the Goldwater-Nichols Act to work as intended. The commander, General Schwarzkopf, was allowed operational control of the war without interference.

These U.S. leaders were no strangers to ego—one does not become President, Secretary of Defense, Chairman of the Joint Chiefs, or "even" a three- or four-star general without having acquired a considerable sense of self-worth along the way. Yet every one of them—Bush, Cheney, Powell, Schwarzkopf, McPeak, and Horner, to name only the most obvious—subordinated their egos to the decision-making process. This is not to say that there were not flare-ups, and in the case of one or two individuals, even tantrums. There were, but they did not adversely affect the prosecution of the war. It will be remembered that Secretary Cheney's track record with Generals Welch and Dugan did not indicate any partiality for the Air Force. Therefore, there was considerable comfort and satisfaction from Cheney's oft-quoted postwar comment "The air campaign was decisive."

HARD NUMBERS

The cold statistics of the war give dimension to the role the Air Force played. It flew 59 percent of the total of 109,876 sorties flown and dropped 74 percent of the total U.S. bomb tonnage of 88,500 tons and 90 percent of the U.S. precision-bomb tonnage of 6,520 tons. A formidable enemy force had been shattered so that the ground forces could execute their task with maximum speed and minimum risk.

Yet there were still lessons to be learned. A total of 210,800 gravity "dumb" bombs had been dropped, compared to only 15,500 units of precision-guided munitions. With a declining force structure, it was obvious that this 13.6-to-1 ratio was uneconomic and had to be reversed in future conflicts. Intelligence-gathering prior to, during, and even after the war was inadequate, particularly in regard to bomb damage assessment. As mentioned previously, an equipment mismatch prevented the electronic transmission of the Air Tasking Order to Navy units, an anomaly solved easily enough subsequently. And, despite the brilliant efforts made by MAC and CRAF aircraft, there were simply not enough airlift aircraft of the correct capacity available to meet the need. Had the McDonnell Douglas C-17 been available in quantity, the task would have been much easier. (The number of ships available for sealift was even more deficient, but that's another story.)

Two shortcomings were most evident. The first achieved wide notoriety—casualties due to friendly fire. The public that applauded the technology that put bombs in ventilation shafts was appalled that equivalent means had not been developed to prevent killing our own troops. The second serious deficiency was less obvious, except to those unfortunate enough to suffer from it: the failure to provide the quantity of high-technology search-and-rescue equipment necessary for a campaign of the magnitude of the Persian Gulf War. Search and rescue is both a moral and a morale issue. The level of funding required to maintain a first-class search-and-rescue capability compared to the total funds required to field a modern air force is small—therefore no decisions not to have adequate capability based on economy can be justified. Yet when funds are reduced functions such as search and rescue inevitably get cut.

Curiously enough, the greatest effect of the victory in the Gulf War may be in the minds of the American public. General Loh put it in sporting terms, saying that the Gulf War created a new standard in which the U.S. must win quickly, decisively, with overwhelming advantage and few casualties. It must, in short, prevail "by 99 to 1, not 55 to 54 in double overtime."

THE AFTERMATH OF VICTORY: CONTINUED REDUCTION AND TOTAL REORGANIZATION

The U.S. Air Force covered itself with glory during Desert Shield and Desert Storm, and most of its units returned in high spirits to welcoming crowds in the United States. Having done so, it continued the pell-mell process of downsizing and reorganizing to a degree unprecedented for a victorious force. Normally after great victories, armed forces tend to stay the same for years, content that they've solved the problem of warfare. But even before the parades and welcoming parties were over, USAF leaders persisted in the process that would transform the steady, ten-year drawdown into a new, better, more effective service.

In the past, the Air Force had brought leaders to the fore who were appropriate to the challenge. General McPeak came to his position as Chief of Staff with a vision for a change in the structure of the Air Force that he says would have been appropriate whether the Air Force was going to build up, draw down, or remain static. McPeak, no stranger to controversy before or after his accession to the top USAF job, had the drive and the personality to handle the dynamics of a revolutionary restructuring of the service he loved. Evidence of his success was found in the quick agreement he obtained on his plan for restructuring the Air Force from Secretary Rice. In a series of late-evening briefings, McPeak convinced Rice of the soundness of his plan, and he obtained essential agreement by Christmas 1990.

McPeak continued the trend of a fighter pilot as Chief of Staff that began with Charles Gabriel. An ROTC graduate, McPeak flew as a solo demonstration pilot with the famed Thunderbirds and accumulated 269 combat missions as a North American F-100D attack pilot and as a high-speed "Misty" forward air controller (one of the most demanding, hazardous jobs in the war) in Vietnam. His twenty-six assignments, which included twenty-four changes of station over thirty-three years, constituted textbook preparation to become Chief of Staff, for he did everything from flying as a grunt instructor pilot to commanding, in turn, a fighter wing, a numbered air force (the Twelfth), and, finally, PACAF. McPeak has flown over 6,500 hours in more than fifty types of aircraft and achieved combat-mission-ready status in the F-4, F-15, F-16, F-100, F-104, and F-111. He continued flying as Chief of Staff and earned adverse publicity for doing so—four-star generals were not supposed to be flying single-seat fighters. McPeak stoutly maintained that it was the best way to keep his finger on the pulse of the Air Force—and his hand on the stick of a fighter. President Bush, a pilot himself, went out of his way to praise McPeak for flying. McPeak wanted his numbered air force commanders to fly, and posted the monthly hours they flew on a chart.

During his career, McPeak never lost sight of the value of the Air Force as an entity—he did not see it as a fighter pilot's Air Force, but as an organization that functioned well because *all* of its components—cooks, mechanics, air policemen, medics, pilots—were making vital contributions to the best of their abilities. He appreciated technology as well as any of his predecessors, but understood that without realistic, Red Flag-style training such as General Robert Dixon had instigated, the value of technology was diluted.

The decline in strength that Generals Gabriel and Welch had planned for had now materialized. For the ten-year period from 1986 to 1995, the total obligational authority of the budget had declined 34 percent; the active personnel strength declined 27 percent, the total number of aircraft (including Guard and Reserve) declined 20 percent, and base installations declined 24 percent. There was no avoiding the budget cuts; the question was how to manage them. McPeak's view was the same as that put forth by Gabriel and Welch: if the change occurred solely because the Congress reduced funds, outside influences would control the effect of the cuts, but if the Air Force anticipated the problems and took the necessary steps, the Air Force could control the way it was restructured.

Another factor looming in addition to the tremendous reduction in force under way was the superimposition of unified commands like the U.S. Transportation Command (USTRANSCOM), U.S. Strategic Command (USSTRATCOM), and U.S. Space Command (USSPACECOM). These commands, intended to coordinate all four military services in accomplishing unified missions, imposed requirements for staffing and equipment that the corresponding USAF commands could not afford to duplicate.

McPeak thus understood that a dissolution of some of the old commands and their reorganization into fewer, leaner, new commands was essential—but, he maintains, not as a result of the end of the Cold War nor even as a result of the downsizing, but simply because they needed to be more focused on Air Force operations and the warrior concept. An essential part of his approach was to reduce the overhead, particularly in headquarters. To achieve this, organizations such as Air Divisions, which had historic precedent but whose function had been overtaken by time, were to be eliminated. In a similar way, "staff creep," the inevitable growth of staff at each level of headquarters, had to be eliminated.

McPeak envisioned an "objective Air Force," not the Air Force that existed, and not an Air Force that would ever exist, but one that would represent an ideal to strive far. As subsets of this concept, he talked about "objective numbered air forces, objective wings, and objective squadrons." As a result of his long experience as a commander of every size of unit, from squadron to major air command, McPeak knew how he wanted to reorganize from the start, and he was determined to initiate the process in the first six months that he was Chairman. His goal was to focus the USAF on its principal function, operations, and to ensure that it was instilled with the warrior concept: the Air Force was a fighting outfit, not just pilots and doctors and mathematicians all in the same uniform. It was there to fight.

Throughout the process, McPeak cautioned that when reductions led to mergers of organizations to achieve economies of scale, the real power of the organization was often centralized and moved up an echelon. His view was that of General Creech—organizations had to be decentralized to empower people on the line to do their jobs better. The aim of the restructuring was to flatten organizational charts, reducing the levels of command but at the same time clarifying the roles and responsibilities of support functions.

Realizing that examples were more important than words, McPeak saw to it that the Air Staff was reduced 21 percent, although he now dismisses this as more of a paper change than reality. Yet the biggest hurdle, emotionally and organizationally, was still to come: the disestablishment of the proud Strategic, Tactical, and Military Airlift Commands and the subsequent reorganization of their functions into new commands. Other commands were also affected, and for their members there was the same sense of uncertainty and nostalgia. But for the public at large the loss of SAC, TAC, and MAC was almost sacrilegious.

McPeak also agreed with a concept that had been inherent in the Air Force since its creation, that the division of air power into strategic and tactical elements was a mistake. He often quoted General LeMay, who had proposed in 1957 to combine SAC and TAC into a single Air Offensive Command. LeMay had said, "Whether we choose to recognize it or not, SAC and TAC are bedfellows. . . . They must deter together through their ability to defeat enemy air power together."

The plan put together under McPeak's leadership was draconian. For many years there had been thirteen major air commands, seven operational and six support. The former included the Strategic Air Command, the Tactical Air Command, the Military Airlift Command, U.S. Air Forces in Europe, Pacific Air Forces, the Air Force Space Command, and the Air Force Special Operations Command. The support commands were the Air Force Systems Command, the Air Force Logistics Command, the Air University, the Air Training Command, the Electronics Security Command, and the Air Force Communications Command.

In the reorganization process, the Air Force Space Command and the Air Force Special Operations Command retained their organizational identities. The first major cut came with the changeover of the Communications Command to three field operating agencies, with a reduction in personnel of more than 40,000. The Electronics Security Command was consolidated and replaced by the new Air Force Intelligence Command, which in turn was redesignated the Air Intelligence Agency on October 1, 1993.

Two of the best known of the support commands, AFSC and AFLC, were combined into the Air Force Materiel Command (AFMC). In addition to the pressing overall need to reduce personnel and consolidate functions, one of the principal reasons for combining the organizations was a fundamental change in the way the Air Force did business. In the rush toward centralization, the Department of Defense has essentially taken over procurement, with an assistant secretary for acquisition making the decisions on all major hardware purchases—in essence, doing work that AFSC formerly did. The services do not like it, and most senior Air Force officers do not consider it judicious to have the decisions on weapons to be used by the force to be made by temporary political appointees. But it is currently a fact of life, and made the creation of AFMC logical.

Even though the Air University and the Air Training Command had often been interrelated in the past, their combination into the Air Education and Training Command was traumatic. McPeak wanted to bring the Air Force Academy under the AETC umbrella, but recognized that it would require too much lobbying effort on his part to do so.

The disestablishment of all the commands involved an awareness of the sensitivities involved. A great deal of preparation went into the process, including soliciting ideas for the changes from the affected commands SAC, TAC, and MAC themselves.

Elements of each unit were combined to form the Air Combat Command (ACC) and the Air Mobility Command (AMC). In the initial planning, ACC received the fighters, bombers, reconnaissance aircraft, intercontinental ballistic missiles, some tacitcal airlift, some tankers, and the C³I functions. AMC received the strategic airlift, most tactical airlift, some tankers, aeromedical evacuation, and search and rescue. These allocations would change over time; the most important change was to reverse what McPeak calls his greatest mistake in the process, the allocation of ICBM forces to ACC instead of to Space Command. This was rectified about a year later.

Air Mobility Command was given the mission of worldwide strategic deployment. Air Combat Command was given the duty of providing reinforcement forces to the overseas command and is itself able to conduct independent, integrated air operations.

One of the most curious aspects of the great consolidation of these three premier commands was the delicacy with which it was handled and the courtesy with which members of one command treated their opposite numbers in another. In many ways this ran counter to the usual rough-and-ready humor of flying units, where joking insults can usually be expected to fly between bomber and fighter proponents. This was different, and they knew it. Understanding the situation full well, McPeak had to find a way to impose his clearly defined vision of the new Air Force structure and still permit members of the old commands to feel that they were participating in the decisions as to what the new commands would be called, what their missions would be, who would get which assets, what the new identifying insignias would be, and so on. Mc-Peak was extremely sensitive to the heritage issues, and made his staff work hard to ensure that the numbers of the most famous units were preserved, along with their heraldry. After a considerable amount of often heated debate on this subject, the numbers and heraldry of most of the most famous units were saved.

Yet the net effect of a massive change entailing the transformation of three proud and distinguished commands into two new organizations is not measured merely by insignia and thoughtfulness. At every level it is a visceral challenge to humans whose jobs, reputations, promotions, and futures are on the line. It reached out to involve the families, not only because of the stress felt by each service member but because moves were going to be necessary, houses would have to be sold and schools changed, again. This came about during a period when the Air Force, for the first time, was suffering such severe reductions in manning strength that even dedicated members who had done an excellent job were being forced out. Given the commitment necessary to volunteer for service life, and given the lack of similar outside institutions, the uncertainty that racks military members during a downsizing of this magnitude is as great as that which grips the personnel of a civilian company like AT&T or IBM. Many had stayed in the service because it was a stable organization, with a predictable career path and an adequate retirement package if one performed well. Now all that was gone, as well-qualified people with outstanding records suddenly found themselves forced to leave.

General Mike Loh was COMTAC at the time, and he recalls the process with a rueful pride, recognizing all the hazards that were attendant to the merger and all the efforts made to make things go smoothly. A "graduate" of the Creech school of quality consciousness, Loh would be the first Commander of the Air Combat Command, where he took the concept of TQM a step further to create a climate of what he called "ACC quality" in every aspect of the organization. But there were many pitfalls before ACC came into existence.

Loh notes that no corporation had ever gone through a simultaneous downsizing and restructuring of the magnitude undertaken by SAC and TAC, which were so different by the nature of their missions, SAC was a designated command i.e. with a defined mission under the control of the National Command Authority, very centralized, and still focused on the role of nuclear deterrence. TAC was far more decentralized, in part as a result of the influence of Generals Creech, O'Malley, and Russ, and in part because it was TAC's mission to furnish forces to theater commands for operations.

And there was a genuine rivalry. Just thirty years before, as noted earlier, General Walter Sweeney had come from SAC to "Sacimsize" TAC; now, to mix metaphors, the shoe was on the other foot, and it seemed that TAC would be ascendant. It was now a fighter pilot air force, from the Chief of Staff, McPeak, to the first Commander of ACC, Loh. Even General George L. Butler, CINC-SAC, and the last commander of that great organization, had started his career as a fighter pilot, flying F-4s in Vietnam.

McPeak had warned Loh that if too many resources were placed in Air Combat Command, the result would be "Air Combat Command being Snow White, and the seven dwarfs being the other major commands." Both men were determined to come up with a better way to provide balance, including putting the tankers into the Air Mobility Command.

The ACC mission seemed to call for the acquisition of the search-and-rescue forces. The stateside C-130 units were also shifted to Air Combat Command, on the basis that ACC had to provide that capacity to the theater commands in the event of emergencies. Loh knew that most C-130 units, including regular Air Force, Air National Guard, and Air Force Reserve, had always felt that they were "second-class citizens" in MAC, because their aircraft lacked the glamour and the capability of the C-141s and C-5s. He conscientiously went about the task of emphasizing how important the C-130s were to Air Combat Command because of theater operations. The result was that C-130 unit morale actually rose because of the transfer. Further, Loh took the additional step of seeing to it that people from the C-130 force (and the rescue force) were given positions of greater responsibility throughout ACC.

He adopted the same tactics with the 20th Air Force, which operated the ICBMs under the leadership of Lieutenant General Arlen Dirk Jameson. In SAC, the "missiliers" had always felt that they were not given the same recognition as warriors as were the bomber crews. Loh changed that, pointing out that the 20th Air Force could put its weapons on its targets faster than any other unit in the Air Force. The 20th Air Force was transferred to Space Command, with its ICBMs remaining a component of the U.S. Strategic Command for ICBM forces.

PATCHING UP THE REORGANIZATION

Big events are sometimes best understood in the small vignettes that accompany them. In the case of the gigantic reorganization of SAC, TAC, and MAC—household names for forty-six years—into the new AMC and ACC, a great deal of heat and energy went into the design of the patches the members would wear.

The patch became a symbol of who had won and who had lost the organizational wars. Loh recounts that he had solicited his new command for designs and submitted eight of them to Air Force Headquarters for consideration. McPeak rejected them all, insisting that the ACC combat patch be a duplicate of the old TAC patch, with the words "Air Combat Command" substituted for the words "Tactical Air Command."

The selection was a red flag to former members of SAC. Former SAC members were not mollified that the old SAC patch was slightly modified to become the U.S. Strategic Command patch. The MAC patch soldiered on as the AMC patch. Fortunately, as the new commands began to function, matters like patches and positions were forgotten in the drive to achieve the required proficiency with the new structure.

On June 1, 1992, General McPeak made a whirlwind tour, delivering addresses at Langley Air Force Base, Virginia, Scott Air Force Base, Illinois, and Offutt Air Force Base, Nebraska. His speeches symbolized the activation of Air Combat Command, Air Mobility Command, and U.S. Strategic Command, even as their predecessors, SAC, TAC, and MAC, were deactivated.

In each talk, McPeak made the same points. He congratulated the members and their predecessors for their professionalism and for their great victory in the Cold War. He pointed out how each of the previous commands had done brilliant work in its past endeavors under the command of the greatest names in the Air Force pantheon of heroes. He emphasized that they now carried their heritage on to a new mission in changes that were dictated by events, technology, and the passage of time. He reminded them that the past of each of the deactivated commands had been glorious, and predicted that the future for each of the new commands would be demanding—but equally glorious.

In these and other speeches, McPeak emphasized that the changes should not be regarded as a paring-down of the old Air Force, but instead as the building of a brand-new Air Force from the ground up to meet the challenges of the next century.

The great changes in commands were accompanied by further changes down the administrative hierarchy. The great numbered Air Forces were restructured so that they were no longer management headquarters but tactical echelons, with their commanders wearing flight suits (or fatigues) to work each day. Staffs were reduced by 50 percent. The function of a numbered air force commander was changed from a commanding general to that of an inspector general, checking on each of the bases in his unit to test their wartime capability.

Further down the chain of command, wings were restructured so that one man (usually a general officer) would run the base and the wing. Composite units were created so that fighters and tankers (or fighters and airlifters) would no longer be in separate wings but part of the same unit.

McPeak's concept of the composite wing met with philosophical opposition, particularly after an F-16 collided in midair with a C-130 at Pope Air Force Base, North Carolina. However, the level of resistance on composite wings was nothing compared to the furor created when he introduced a new Air Force uniform that seemed to many to be too similar to the Navy or airline uniforms. McPeak intended the new uniform to symbolize the new Air Force that had emerged from the restructuring: streamlined, clean, and without encumbrances. He was certain that his successors would begin adding insignia to it, and insists that it was well received by the enlisted force, which constitutes some 80 percent of the Air Force, as well as by half of the remaining 20 percent, the officers. Others perceived it as a defamation of the blue suit that had served so well so long.

Nonetheless, McPeak was the man of the hour when the Air Force was at its point of greatest change, and he had the personality, the drive, and the confidence of his superiors to push through the changes he believed in and had crafted. But as General Dougherty has pointed out, after any period of great change, a period of stability is always required, and two people emerged to foster this transition. One was the first female Secretary of the Air Force, Sheila E. Widnall, whose performance has impressed everyone in the service and out. The second was the new Chief of Staff, General Ronald R. Fogleman, the first airlift commander to accede to the post. Dr. Widnall had a distinguished academic career at the Massachusetts Institute of Technology and is internationally known for her work in fluid dynamics, particularly in the areas of aircraft turbulence and the vortices created by helicopter rotors. Like McPeak, Fogleman had flown F-100s and operated as a Fast FAC in Vietnam. After he had served as an F-15 demonstration pilot in many air shows, Fogleman's career path eventually took him to tanker and airlift aircraft. He became Commander in Chief of U.S. Transport Command and Commander of Air Force Air Mobility Command.

Fogleman put his stamp on the Air Force immediately with some very slight but very wise decisions involving compromises on the uniform controversy. Fogleman authorized the U.S. insignia to be worn again, and allowed officers to shift their insignia from the sleeve to the shoulder. He was thus able to satisfy most people with little effort and little expense. It was symbolic for the most part: some of the troops had spoken and Fogleman responded sensibly to their cry.

In the intervening months, Fogleman has made a series of statements on the future of the United States Air Force in light of national requirements, budget realities, and a series of studies prepared both within the Air Force and by the Joint Chiefs of Staff. He has called for the synergistic combination of all the services' capabilities, from the Air Force's stealth aircraft to the Navy's carriers to Army and Marine combat units, to provide deterrence to would-be aggressors of any type.

Fogleman has noted that the United States must transition from its past strategy of annihilation and attrition warfare to a concept which leverages our military capabilities by applying what he terms an "asymmetric force" strategy.

Demonstrated in part during the Persian Gulf War, an asymmetric strategy directly attacks enemy strategic and tactical centers of gravity—targets already defined by commanders in chief while developing war plans for their theater of operations. These centers of gravity include the enemy's leadership elite, command and control, internal security mechanisms, war production capabilities, and its armed forces. They comprise the enemy's ability to wage war effectively.

THE UNITED STATES AIR FORCE WON THE COLD WAR

The collapse of the Soviet Union and the end of the Cold War has been attributed to many causes, ranging from the wisdom of Ronald Reagan to the reasonableness of Mikhail Gorbachev to the incredible growth in the economies of the western democratic nations. There is an element of truth in each of these claims, but the fact is that the United States Air Force *won* the Cold War by its unstinting efforts over the entire period from 1945 through 1991.

The USAF was helped by the other U.S. armed services, by U.S. diplomatic efforts, and by the assistance of allied nations. But the bottom line is that the United States Air Force contained Soviet expansion from the very beginning. And it was the United States Air Force which built up such a powerful force that the leaders of the Soviet Union never dared exercise their policy of a first strike. The first two legs of what became known as the strategic triad were the bombers and the missiles of the United States Air Force. The third leg was the very important and very effective U.S. Navy submarine-launched ballistic missile force.

The importance of the United States Air Force in the Cold War must be considered in the light of the times when it was most needed. The first great need was met by the superb performance of the USAF in breaking the 1948–49 Soviet blockade of Berlin with a valiant airlift. There followed a period of five years when Stalin's actions were ever more unpredictable, and when massive Soviet ground forces could have overrun Europe in a matter of weeks. They did not do so, because their homeland was threatened by the growing power of the Strategic Air Command, under General Curtis E. LeMay. Unlike Hitler in Berlin, or the rigid Japanese leaders in Tokyo, who allowed their countrymen to die by the thousands to gain a few more weeks or days of life for themselves, the Soviet leaders knew that they, too, would die in an American atomic onslaught. It was sobering to them.

Next there was a period when the strength of the Soviet intercontinental ballistic missile force grew at a rapid if unknown rate, and the United States responded with its own ICBM program, led by the master of missiles, General Bernard A. Schriever. The Air Force once again provided the balance of power necessary to make the concept of Mutually Assured Destruction valid. The Soviet leaders knew that no matter how large their first strike might be, the United States Air Force had the resilience and the capacity to launch a devastating second strike that would destroy the USSR.

Over time the number of nuclear weapons possessed by both sides in the Cold War grew to incredible portions. As the years passed, in instance after instance, the Soviet Union demonstrated its quest for power, its brutality and its implacable enmity to the United States. It backed North Korea in the Korean War, crushed revolts in Hungary and Czechoslovakia, and supported anti-U.S. "wars of national liberation" everywhere.

During the same period it revealed an amazing scientific and military prowess, first with Sputnik, in October 1957, then with a succession of space and missile firsts that seemed to prove to the world that the USSR was approaching the capabilities of the United States. Khrushchev and his successors used these undeniable triumphs to go further. They promised the citizens of the Soviet Union that the poverty and squalor of their lives would be alleviated by rising to new standards of living, higher than that enjoyed by the United States.

A number of incidents, including the destruction of the Lockheed U-2 flown by Gary Powers, seemed to increase the hubris of the Soviet Union, reaching a peak in the Cuban missile crisis of 1962. There the danger of nuclear war came closer than anyone, including the Communist leaders in the Kremlin, knew. It has only recently been known that the individual Soviet submarine commanders possessed their own authority to use their nuclear weapons if their judgment (or lack of it) dictated. The decision as to whether there would be an international exchange of nuclear weapons at catastrophic cost to the world had been allowed to devolve to the hard-pressed captains of Soviet submarines. One of them, being suppressed by U.S. warships, came within a hairbreadth of deciding to use his missiles.

All during this period, the United States Air Force maintained itself on a war-ready status, able to launch retaliatory attacks within minutes of notice of a Soviet first strike. This capability was made possible by extreme measures in which as many as 50 percent of the bombers of the Strategic Air Command were on fifteen-minute ground alert. Other bombers were on airborne alert, fully armed, and able to penetrate Soviet defenses within minutes of being ordered to do so. "Looking Glass" aircraft, able to execute and control the national war plan if required, were aloft every minute of every day for twenty-nine years.

Soviet Premier Nikita Khrushchev later admitted that the thought of Boeing B-52 bombers orbiting on the borders of his country was crucial to his admitting defeat during the Cuban Missile Crisis.

All of these achievements came at the cost of many Air Force members' lives and at the sacrifice of time with their families. Sadly, far too many families were dissolved in the divorce court as a result. The sacrificial effort was made not for income, for there were enormous pay inequities at the time, but for patriotic reasons, the servicemen knowing that they were serving a noble cause.

This sense of nobility and service extended to all elements of the Air Force as it changed over time. It fought a long and distasteful war in Vietnam, suffering the opprobrium of the public but all the while maintaining the essential nuclear deterrent. At the same time it made enormous advances in social equality within the Air Force so that problems of race, ethnicity, and gender became minor. It carried out humanitarian efforts unprecedented in the history of the world, much less in the history of any armed service.

There were many factors in the final, rather pathetic, dissolution of the Soviet Union. But there was one primary cause for the victory in the Cold War, and that was the United States Air Force.

10

UNREMITTING WARFARE

The victory in the Gulf War and the subsequent reorganization of the United States Air Force was preceded by the dissolution of the Soviet Union. Incredibly, almost no one predicted this, the most important political event during the last fifty years of the twentieth century.

Called with uncanny accuracy the "evil empire" by President Ronald Reagan, the Soviet Union drove itself into the ground with inefficiency and corruption. No longer able to compete militarily with the United States, totally incapable of supplying even the basic consumer needs of its population, and beset with demands for independence by many of the states it had brought within its influence, the Soviet Union expired relatively quietly, its final demise announced in a poignant speech by Mikhail Gorbachev on December 25, 1991. Addressing the nation for the last time as president of a nonexistent USSR, Gorbachev noted that his country had possessed everything in terms of natural riches, as well as enjoying enormous human resources. He went on to say that, despite this abundance, the command-bureaucratic system had followed a doomed ideology and thus suffocated Soviet society.

All attempts at reform had been defeated, but the repressed society's demand for freedom had at last liberated it, politically and spiritually. The former president announced that their country now lived in a new world, with the Cold War ended and economic freedom restored. He ended on a hopeful note, saying that he was convinced that "our nations," the member states of the former Soviet Union, would each enjoy a prosperous and democratic society.

Many consider it a miracle that the Soviet Union ended so quietly, going out with a whimper instead of a nuclear bang. The disappearance of the Soviet Union and the Cold War made it seem certain that the United States would be able to relax its military vigilance and harvest an enormous peace dividend in anticipation of decades free from strife.

More important, perhaps, the collapse of the Soviet Union signaled the total inability of the intelligence community, political pundits, media, or anyone else to predict the future with any degree of accuracy. As an illustration, as late as 1990, many respected American writers were of the opinion that the Soviet Union had turned the economic corner and would soon surpass the United States not only militarily, but in its standard of living.

This worldwide inability to forecast the future emphasizes the vital importance of placing capable people in the highest military positions. These must be individuals who can look ten to twenty years in the future, not to predict a single event, but to prepare for a wide range of eventualities that are breathtaking in their scope. This preparation includes the delicate process of selecting high-risk, extraordinarily expensive weapon systems two or three decades in advance of their need. It is also necessary for subsequent leaders to nurture the development of those weapon systems even as they are making their own forecasts for the future, guiding them through Congress, maneuvering them through budgetary shortfalls, and creating the necessary personnel, logistic, and training infrastructure to support them when they are at last operational. A perfect example of this will be described later in the chapter—the initiation, growth, and maturation of the Air Expeditionary Force concept under three successive USAF Chiefs of Staff.

Unfortunately, the collapse of the Soviet Union did not begin an era of peace and harmony. Instead, there were many wars. Estimates vary with the definition of war, but at all times there were as many as fifty to one hundred major conflicts raging in all parts of the globe. The United States and its partners in NATO and SEATO became the principal responders to the requests made to the United Nations for both military security and compassionate relief.

Domestically, new social programs and a need to keep the federal budget within limits dictated that the United States Air Force would be reduced in size in the years to come, even as the unceasing demands for its services increased. Fortunately, the engrained Air Force ethic was to respond to these demands with fortitude and skill. Over the years, as ever greater requirements were placed upon the regular Air Force, the Air National Guard and the Air Force Reserve, capability has increased and morale has risen. This was in spite of a general and understandable concern about the increasing demands of the operational tempo, which required more and more people to return to overseas assignment more rapidly than ever in the past.

Nonetheless, there existed an inverse ratio between excessive demands and greater capability and excellent morale that is difficult to explain in ordinary terms. The only plausible explanation is that it derives from the intrinsic psychic satisfaction enjoyed by each individual understanding just how important his or her Air Force job is.

On Guard over Iraq

On February 27, 1991, President George H. W. Bush ordered what many considered to be a premature if not ill-advised unilateral cease-fire to Operation Desert Storm, one hundred hours after the ground offensive started. To many military observers the "one-hundred-hour cease-fire" sounded more like a sound bite than military strategy, particularly as it meant that Saddam Hussein remained in power.

The Allies and the obviously chagrined Iraqi military leaders met to discuss terms for the formal cease-fire to end the Gulf War. The Iraqis agreed to abide by all United Nations resolutions, and on March 4 the first Allied prisoners of war were released. An official cease-fire was signed on April 6. Some 532,000 Americans served in Operation Desert Storm. Initial reports indicated that there were 100,000 Iraqis killed and another 300,000 wounded. There were 147 U.S. battle deaths, 145 nonbattle deaths, and 467 wounded in action.

The Allies dictated that Iraqi fixed-wing aircraft could not fly in defined northern and southern regions of Iraq, in order to prevent that country from rebuilding its air power and violating UN resolutions. When the Iraqis requested that helicopters be allowed for communication purposes, the U.S. Central Command chief, General Norman Schwarzkopf agreed. An opening was thus woven into the agreement by which Iraq could use attack helicopters against its own populace.

Five days prior to the cease-fire, the United States had begun Operation Provide Comfort, an airlift that delivered 12,400 tons of food supplies, medical aid, and fuel to more than a million Kurdish refugees in the mountains along the Turkish-Iraqi border. The Kurds huddled in makeshift camps, often with little more than plastic sheets to protect them from the elements in the mountains. Long the victims of Saddam Hussein, they had attempted an uprising in the aftermath of the 1991 Gulf War. When Iraqi attack helicopters attempted to suppress the Kurdish revolt, a "no-fly" zone was established north of the 36th parallel on April 10, 1991, to provide a safe haven for the Kurds. This was the beginning of Operation Northern Watch. Ultimately, more than 700 USAF aircraft flew in support of Operation Northern Watch and its complementary effort, Operation Southern Watch, all organized under the Combined Forces Air Component Commander, then Lieutenant General T. Michael "Buzz" Moseley. Moseley's command comprised eighteen air and space expeditionary wings located in eleven countries throughout the region, and was supported by another 3,000 USAF airmen at eleven bases in six countries in Europe. Several thousand more airmen operated in smaller units at several other locations throughout the region. Many of these expeditionary wings and bases were created to support specific operations and existed only as long as the requirement did. Only a small percentage of these airmen were on permanent-duty assignments in the region; the vast majority were temporarily deployed in support of operations. They came from bases all over the world to fill expeditionary tasks, and when operations no longer required their presence in theater, they returned to their home stations. In addition to USAF forces based in and around the region, the air and expeditionary wings were supported by USAF long-range strike, airlift, and tanker forces based in the continental United States, as well as overarching USAF space capabilities that provided continuous, precise, and essential weather forecasting, communications, navigation, and surveillance for all coalition forces in the region.

Iraqi repression—including air attacks—of the Shiite "marsh Arabs" of southern Iraq gave birth to the requirement for Operation Southern Watch. As of August 27, 1992, Iraq could not fly military aircraft below the 32nd parallel. British and French forces joined the USAF in this effort. Over a decade, the larger Southern Watch area logged more sorties than the Northern area.

It should be noted that Iraq never acknowledged the authority of the allied nations to impose such controls, and over the next ten years would fire more than one thousand missiles at allied aircraft. Fortunately, the coalition did not lose a single aircraft to these efforts.

The seriousness of U.S. intent was revealed on December 27, 1992, when an Iraqi MiG-25 pilot locked his radar on to a USAF F-16. The MiG was shot down, and the Iraqis responded by moving more antiaircraft batteries into the no-fly zones.

The worst event in the lifetime of these two noble ventures came on April 14, 1994. Through a combined failure in communication and technique, two USAF F-15s shot down two U.S. Army Black Hawk helicopters, killing all twenty-six aboard.

As innocuous as their titles sounded, the two operations would become vital elements for the next three presidential administrations as they contained the still restive military instincts of Saddam Hussein. Besides protecting the Kurds in the north and the Shiites in the south, they also ensured the protection of Kuwait, enforced UN economic sanctions against Iraq (to the extent possible, given the UN's endemic corruption) and solidified the coalition of nations who sought to contain Iraq.

There were other, less obvious results. The twin campaigns consumed enormous resources—approximately $12 billion—and piled more flying time on Air Force aircraft than had been scheduled. Before they were ended, the two operations originated more than ten times the number of peacekeeping sorties than had been flown in the entire Gulf War. General John P. Jumper observed that it was not yet certain that operating in the two zones would actually wear the aircraft out faster, because most of the missions did not involve the violent maneuvering which so rapidly ages aircraft.

The patrols were flown with live ammunition, and an average of about seventy strikes were made per year; but for the most part, the pilots felt that they were accumulating flying time without getting the severe training that future air-to-air combat would require.

The two campaigns also exposed how strained Air Force resources had become, and pointed the way to a reorganization which would economize on noncombat assets so as to maximize the warfighting force. On the positive side, the campaigns provided enormous intelligence about Iraqi defenses, which, despite their relatively straitened circumstances, continued to evolve. When war began again in 2003, there were few Iraqi targets not covered precisely in a target folder. Coalition forces became intimately familiar with Iraqi geography, both visually and as represented on radar screens. The United States and other countries in the Gulf region were able to standardize procedures and air traffic control and permit ongoing joint exercises. Alternative basing options became available over time, freeing the United States from its high degree of dependence upon Saudi Arabian facilities. In addition, there were many opportunities to try new concepts and equipment in the no-fly zones with a great deal of safety. The Predator unmanned aerial vehicle (UAV) was tested and the Combined Air Operations Center (CAOC) gained in efficiency over the intervening years.

Saddam was nothing if not relentlessly hostile to his own people, and authorized a ground attack against the Kurds north of the 36th parallel. In response, the United States launched Operation Desert Strike, which was targeted against surface-to-air missile sites and command-and-control sites in southern Iraq. After fruitless debate about the creation of a third no-fly zone in western Iraq, the same end was accomplished by moving the southern border northward to the 33rd parallel, which ran just south of Baghdad. France objected, and restricted its pilots to flights below the 32nd parallel.

David A. Deptula, then a major general, commanded Northern Watch from April 1998 to October 1999 and helped introduce new rules that permitted coalition forces to respond to hostile acts in a fashion which limited Iraqi capacity for aggression. The coalition aircraft generally showed restraint, having been threatened or fired on thousands of times, while retaliating only about 500 times—on the average about three to five times per month.

There were exceptions to this, such as Operation Desert Fox, which took place over four days in December 1998. Its mission was to attack locales where Iraq was suspected of deploying, making, or hiding weapons of mass destruction. Other targets included airfields, air defenses, communication nodes, and oil fields believed to be illegally exporting oil. After Desert Fox, the rules of engagement were changed to permit strikes on any Iraqi capacity to threaten the coalition.

As tensions mounted later in the decade, both Northern Watch and Southern Watch activities would increase to the extent that they could reasonably be regarded as a preliminary bombing campaign for Operation Iraqi Freedom in 2003.

Operations Northern Watch and Southern Watch had yet another legacy, to be dealt with later in this chapter, and that was in shaping an expeditionary, rather than a garrison, mind-set within the Air Force. This came about as the two operations required a constant deployment of a mixed force that flew as a composite group, not unlike those that make up a typical Red Flag training exercise. The composite group's assets included Lockheed Martin F-16CJ and F-16CGs to suppress enemy air defenses, Boeing F-15Cs for air superiority, Boeing F-15Es for defense suppression, Northrop Grumman EA-6Bs for electronic jamming, KC-135s for air refueling, Lockheed Martin EP-3s or Boeing RC-135s for signal intelligence, Boeing E-3B/Cs for an Airborne Warning and Control system, Sikorsky HH-60G helicopters for combat search and rescue, Lockheed Martin MC/HC-130C tankers to refuel the helicopters, Beech C-12 Hurons for theater transport, Royal Air Force SEPECAT Jaguars for reconnaissance, and Vickers VC-10K tankers for air refueling both the Jaguars and the EA-6Bs. Other weapons were integrated into the process, including a U.S. Army Patriot missile battery.

The statistics on Operations Northern and Southern Watch, and subsequent operations clearly illustrate how valid the concept of Total Force had become. Of the aircraft deployed, 65 percent were active-duty Air Force (AF), 28 percent Air National Guard (ANG), and 7 percent Air Force Reserve (AFR). The Combat Search and Rescue force—one of the most respected and vital in the theater—was 49 percent ANG, 28 percent AFR, and 23 percent AF. The ANG dominated Lockheed Martin A-10 operations with 69 percent; the Air Force had 31 percent. In contrast, the Lockheed Martin F-16C force was 83 percent AF and 17 percent ANG. Among tankers, the Air Force had 61 percent of the KC-10 force versus 39 percent for the AFR, while 68 percent of the KC-135s were Air Force, 29 percent ANG, and 3 percent AFR.

The Total Force also demonstrated the capability of the USAF's professional personnel. Experts in many disciplines—civil engineering, communications, force protection, fuel, munitions, services, and medicine—could set up expeditionary bases from scratch to conduct combat operations from the middle of nowhere.

When Operation Iraqi Freedom began in 2003, Operations Northern Watch and Southern Watch had been extant for 4,365 days and 3,857 days, respectively. These two operations, almost ignored by the public and the media, were conducted for many time longer than World War I (584 days), World War II (2,174 days), Korea (1,125 days), and Southeast Asia (2,909 days).

Yet even as Operation Northern Watch and Operation Southern Watch were in the process of becoming routine, further demands were made on Air Force resources.

BALKAN TROUBLE SPOTS ERUPT

Despite the ongoing costs of restraining Saddam Hussein, Air Force budget reductions continued unabated, causing problems in retaining personnel, obtaining new equipment, and continuing research and development. In constant Fiscal Year 2000 dollars, the Air Force budget declined from $116,476 million in 1990 to $79,100 million in 1996—a 30 percent drop. But even as assets were decreasing, challenges were increasing.

At about 9:50 P.M. on the evening of June 25, 1996, a sewage truck filled with explosives exploded with the force of ten tons of TNT against the concrete structure of the Khobar Towers in Dhahran, Saudi Arabia and killed nineteen Americans. The largest terrorist device ever used against the United States to that time pointed the way to the long, intractable future of terrorism. Inevitably, it also precipitated a series of investigations to find out who had caused the explosion, and if any blame was to attach to those responsible for the safety of those in the Khobar Towers. President William Jefferson Clinton pronounced, "We will pursue this. Those who did this must not go unpunished."

Unfortunately, the administration was more successful in defining blame for the person responsible for the defense of the Khobar Towers than in punishing the perpetrators. Although two separate investigations revealed that the commander of the 4404th Wing (Provisional), Brigadier General Terryl J. Schwailer, had taken "reasonable and prudent" action to protect the force, Secretary of Defense William Cohen decided to punish Schwailer by blocking his imminent promotion to major general.

The following week, Schwailer resigned from the Air Force. Air Force Chief of Staff Fogleman had assured Schwailer of his support if the investigations determined that Schwailer had not been at fault. The time line is indistinct, but it now seems that that Fogleman had submitted his resignation prior to Cohen's final decision on blocking Schwailer's promotion. Fogleman later stated that his resignation had been an "early retirement." His comments indicated that he was generally disillusioned about politics in the Pentagon and had lost confidence in the leadership he was supposed to be following. Despite this, his resignation or retirement was widely interpreted as a courageous gesture of protest against the prevailing antimilitary sentiment in the Clinton administration and that it expressed the strong resentment Fogleman felt about Schwailer unjustly being made an administration scapegoat.

General Ralph E. Eberhart served as acting Chief of Staff until General Michael E. Ryan assumed the position on October 6, 1997. Ryan, whose father, General John D. "Jack" Ryan, was the seventh Chief of Staff of the Air Force, brought with him extensive experience as a fighter pilot and commander. His experience included commanding U.S. and NATO forces in Operation Deliberate Force, the first of two conflicts in the post–Gulf War period where the United States and its allies relied principally on air and space capabilities to achieve policy goals.

OPERATION DELIBERATE FORCE

The conflict in Bosnia arose from the decline and breakup of the Yugoslav Federation during the 1980s. After Croatia, Slovenia, Bosnia, and Macedonia pulled away, there was left a rump state, the Federal Republic of Yugoslovia, comprising Serbia (with Kosovo as an autonomous province) and Montenegro. It was ruled by President Slobodan Milošević and Serbia was the dominant power.

Under Serbia's influence, Serbian minority groups created conflicts in the "breakaway" states, either to force the majority to remain under Serbian rule, or to carve out Serbian enclaves that could then become part of Greater Serbia. Vicious battles were fought, and ethnic cleansing (a euphemism as brutal as "final solution") became a political practice.

The scale of the killing was such that international organizations felt compelled to intervene in Bosnia, which had been recognized as an independent nation in 1992. In June of that year, United Nations peacekeeping forces were sent to Sarajevo airport to protect the flow of food supplies into the country.

The April 12, 1993, UN Security Resolution 815 decreed Operation Deny Flight to further protect UN peacekeepers. Intervention air planning continued for three years. It was frustrated by the general inability to determine exactly who was right and who was wrong in the bitter internecine conflict, given the atrocities perpetrated by both sides. It became aimed primarily at peacemaking, with the intent to force a cease-fire and create a political settlement.

However, there was a divisive gulf between those exercising NATO's airpower and the will of the United Nations, which put extensive restrictions on air attack, reminiscent of the rules of engagement used during the Vietnam War. The air campaign was designed not to defeat an army, but rather selectively to destroy military capability so as to create a relative parity between the opposing forces. Fortunately, the American experience in the 1991 Gulf War had conditioned its airmen to think in terms of creating an environment suitable for subsequent surface operations (as distinct from the "air-land battle" concept of simultaneous operations) and to contribute directly to the achievement of national policy.

Things began to accelerate when NATO jets attacked Serbian ammunition depots around the city of Pale in May 1995. The Serbs responded by taking 370 UN peacekeepers hostage and chaining some to potential targets. Shortly thereafter, the Soviet-style air defense system shot down Captain Scott O'-Grady's F-16 in a demonstration of its capability. A brilliant combat search-and-rescue operation extracted O'Grady, but the Serbs made much of the shootdown. In July, Serbian forces continued to push forward, taking the key city of Srebrenica and violating the Sarajevo safe area.

The immediate cause of Operation Deliberate Force was the explosion of a mortar shell in the Mrkale market in Sarejevo on August 28, 1995. It killed 38 civilians and injured 85 others. The television coverage of the bloody incident made NATO action imperative. Admiral Leighton W. Smith, Jr., Commander of NATO's Southern Region, and his UN counterpart, French Lieutenant General Bernard Janvier, decided it was time to launch the long planned and long delayed air campaign.

Ryan later laughingly remarked that he learned how the combined air operation center (CAOC) functions the hard way during Deliberate Force. He called additional aircraft from the Air Force, the Navy, and the Marines into the Aviano Air Base and tasked the carrier USS *Theodore Roosevelt* to launch the initial strikes.

The attacks began at 0140 on August 30, against a triangular area that stretched in from the coast about 150 miles on each side. For twelve out of the next twenty-two days, aircraft and tactical land-attack missiles (TLAM) rained down on the Serbs. By September 14, the Serb Republic accepted the terms of the United Nations, and the campaign closed on September 20.

It was a magnificent combined effort, with aircraft from all three U.S. services joining those of France, Germany, Great Britain, Greece, Holland, Spain, and Turkey in the assault. Some 3,515 aircraft sorties were flown and 1,026 weapons were dropped. There were 338 aiming points within 48 target complexes. These targets included the heavy weapons, command-and-control bunkers, ammunition depots, and other high-value targets. A parallel NATO air campaign, called Operation Dead Eye, was directed against the Serbian air defense network.

Later analysis revealed that the Serbian leaders had difficulty at first in assessing the effect of the bombing. When they finally understood what was happening, they were so impressed by the precision involved that it influenced their decision to negotiate for peace terms.

Compared to the Gulf War, Deliberate Force was a minor effort in terms of the number of sorties, but it demonstrated a major improvement by the Air Force in the application of airpower. Only about two percent of the air weapons of the Gulf War were precision-guided munitions, while in Deliberate Force, about 69 percent of the ordnance was precision guided. The effect of this was particularly noticeable in bomb damage assessment (BDA) photos. In all previous history, BDA photos typically showed a great many bombs hitting around a target, doing tremendous collateral damage. In the photos from Deliberate Force, it was usually only the target itself that showed bomb damage. Less than two weapons per target had been used in Deliberate Force, and they hit 97 percent of the targets assigned.

A principal reason for this was that General Ryan had viewed every bomb as a "political bomb" and exercised very close control over every sortie. He personally authorized the type of weapon to be used and the desired mean points of impact (DMPI) for each target to avoid collateral damage. He ordered that a minimum number of bridges be struck, and then only at night, to avoid civilian casualties. He was later quoted as saying, "If anything went wrong, well, they had one person to hang, not a sergeant working in the photo shop saying 'That looks like a good one to me.'"

Ryan believed that precision-guided munitions were the key to any future combat. He also felt that the U.S. and NATO forces had been fortunate in having almost three years to build up to Deliberate Force, as well as having a history of some forty years of training experience in joint operations. All of this combined experience would soon be put to use again in the same Balkan theater of operations.

It might be concluded that if the Gulf War was the first campaign-level application of modern aerospace capabilities to achieve foreign policy goals, then Operation Deliberate Force was the second. The Bosnian-Serb acceptance of the Dayton Accords was an airpower success that would have gladdened the heart of Billy Mitchell.

OPERATION ALLIED FORCE — MARCH 24 TO JUNE 9, 1999

The necessity for further intervention in the Balkans became apparent in early 1998 when large-scale fighting in Kosovo resulted in the displacement of some 300,000 people. With winter threatening a humanitarian crisis, a cease-fire was agreed upon in October, enabling the refugees to find admittedly primitive shelter.

There was no decrease in violence, however, and conditions grew worse by January 1999. The Yugoslav delegation refused to agree to conditions of a peace conference held at Rambouillet, near Paris, which broke up on March 19. With the collapse of the negotiations, Milošević initiated Operation Horseshoe, with which he intended to cleanse Kosovo of its ethnic Albanian population in seven days or less, before effective NATO forces could be brought into action. He is believed to have hoped that the United Nations rules of engagement would preclude extensive air strikes and that there would be dissension in the ranks of NATO leadership.

Operation Allied Force was the NATO contingency response designed to enforce compliance with the United Nations Security Council Resolution 1199 of September 23, 1998. This resolution was intended to promote regional stability, cooperation, and security. It demanded an immediate cessation of hostilities and maintenance of a cease-fire in Kosovo. Among many other things, it also called for the leadership of both the Federal Republic of Yugoslavia and the Kosovo Albanian leadership to improve the humanitarian situation, and for a political dialogue to begin that would resolve the situation.

Control of the operation was delegated to the Commander, Allied Air Forces Southern Europe (COMAIRSOUTH), while conduct of the day-to-day missions was delegated to the Commander of the 5th Allied Tactical Air Force at Vicenza, Italy.

The American component of Operation Allied Force was Operation Noble Anvil. Bombing was delayed for 96 hours to allow time for negotiations between U.S. ambassador Richard Holbrooke and President Milošević to produce results. When they did not, NATO forces began air operations over the Federal Republic of Yugoslavia on March 24, 1999, at 1900 GMT.

The operation's goals included stopping military action, ending the repressive actions in Kosovo, forcing the Serbian military forces to withdraw, and forcing the Serbian authorities to conform to the requirements of Resolution 1199.

The USAF suffered a blow on March 27 when a Lockheed Martin F-117A Nighthawk was shot down. The pilot was recovered in a bold operation, but

Serb propaganda made much of the incident. The action is still sensitive, but unofficial sources state that there was inadequate jamming support because rigid security requirements (colloquially, the "Green Door") precluded the communication of the stealth aircraft's jamming requirements.

There were further, more important difficulties. The Supreme Allied Commander in Europe, General Wesley K. Clark, was skeptical about the role of airpower and was concerned that the Serbs might be able to carry out their ethnic cleansing of the Muslim population before airpower could be applied to stop them. As a result of this skepticism, he submitted a list of only 51 targets for air attack, and these were primarily Serbian field forces. He did believe in the effectiveness of Army Apache helicopter to back up his preferred plan for ground operations in Kosovo. Ironically, this plan, which ultimately would have required some 200,000 American and European ground soldiers, was opposed by almost everyone, including his own Army Chief of Staff, General Dennis J. Reimer.

As a direct result of Clark's actions, the NATO air war against the Serbs got off to a very slow start. It was not until six weeks had elapsed that NATO was able to boost the number of targets bombed in Serbia and Kosovo. As the tempo picked up, the number of strike aircraft involved doubled, from 120 to 400, while the total number of aircraft increased to 700.

During the initial part of the operation, NATO aircraft focused on Phase I targets, which included the Soviet-inspired Yugoslav Integrated Air Defense System (IADS), which had 16 MiG-29 Fulcrum fighters on strength. Other attacks were directed against Serbian command and control (C^2) and heavy weapons in Kosovo.

Besides the Boeing B-52, the USAF supplied its most advanced aircraft, including the Boeing B-1B and the Northrop Grumman B-2A bombers. The Boeing F-15E and Lockheed Martin F-117A conducted the deep-interdiction missions, while Lockheed Martin F-16C and F-16CJ aircraft executed air-to-ground strikes. The attacks were orchestrated by support aircraft that included the Boeing E-3 AWACS, Northrop Grumman E-8 Joint STARS, and Lockheed Martin C-130s. Boeing KC-135 and KC-10 tankers were, as always, indispensable. The Navy entered the fight with Boeing F/A-18s, Northrop Grumman F-14s, and Boeing AV-8As. Attack aircraft from twelve other countries also joined in.

When Milošević did not give in, NATO moved on to Phase II targets on April 3. Serbian military forces south of the 44th parallel were targeted. On May 1 NATO began attacks on electrical power grids, refineries, and lines of communication in Serbia, hoping to destroy the morale of the Serb population. Sorties now were being flown at a rate 60 percent greater than in the first three weeks of the war.

By May 5, more than 15,000 sorties had been flown, including 5,000 bombing attacks. Once again, NATO efforts were governed by stringent rules of engagement. More than 90 percent of the precision-guided munitions used were either the Joint Direct Attack Munition (JDAM) or laser-guided bombs. The commander of Air Combat Command, General Richard E. Hawley, had earlier announced that the USAF use of precision-guided munitions was so great that it was now facing a severe shortage.

Much of Serbia's heavy equipment was destroyed, as were almost all of its MiG-29s. Special U.S. weapons blacked out the Serbian power plants, severely disrupting the Serb command-and-control system. The nation's oil refineries and rail lines had been heavily hit. All the bridges across the Danube had been destroyed or made impassable for vehicles.

NATO forces were embarrassed on May 7 when missiles struck the Chinese embassy in Belgrade, killing three people. Profuse apologies were made, and the accident was attributed to "incorrect information."

The precision of the continuous bombing efforts helped force the Yugoslav military authorities to agree to NATO demands on June 9. Subsequently, many analysts have stated that Operation Allied Force was the first time in military history that airpower achieved victory in a conflict on its own. It was certainly true that Operation Allied Force played a prominent role in ending ethnic cleansing in Kosovo, returning over 840,000 refugees to their homes and restoring a semblance of peace to the troubled province.

However, there were other key events to be considered, including the threat of a NATO ground offensive and the Kosovo Liberation Army offensive against Yugoslav forces on May 26. That offensive failed, but it forced the Yugoslav army to react and thus exposed them to air attack.

The Allied Force air campaign unfortunately illustrated a deficiency on the part of the campaign commanders in the use of aerospace capabilities. As Major General Charles Link has noted, "One does not easily eject forces from Kosovo by bombing in Serbia." Unfortunately, the political authorities failed to announce publicly their intention to prevent the adversary from ever being capable of transgressing in the future. Most egregious was the policy pronouncement that there was a firm intention not to use ground forces. This permitted the Serbs to understand that their efforts could be focused on aerospace capabilities alone and that the allies had placed a limit on their commitment to the conflict.

THE EXPEDITIONARY AIR FORCE/AIR EXPEDITIONARY FORCE (EAF/AEF)

The unfortunate combination of a declining budget and vastly increased responsibilities led USAF leaders to reorganize the force once again, to make the maximum use of limited resources. The objective of the Air Force had clearly shifted from the containment policy of the Cold War to the global engagement of a wide variety of emerging threats. The United States no longer had a large forward presence in the world, where it could operate from bases with large infrastructures. Instead, it had to respond in all areas of the world and operate from bare bases with very limited infrastructures.

The extent of the global engagement might be inferred from the fact that during the 1990s there were almost 1,000 deployments to points in Europe, South America, Africa, the Middle East, and Asia, including more than 500 humanitarian missions to the former Soviet Union alone. An early example occurred on November 19, 1997, with Operation Phoenix Scorpion I. When Iraq remained intransigent over the United Nation's inspection of weapons of mass destruction sites, the United States deployed additional forces to Southwest Asia and the Indian Ocean. The Air Mobility Command established an air bridge over the Pacific to support the deployment of Boeing B-52s and KC-10s to Diego Garcia. Another airlift over the Atlantic by AMC provided airlift and refueling support to deliver an Air Expeditionary Force with Boeing F-15s, Lockheed Martin F-16s, Boeing B-1Bs, and KC-135s to Shaikh-Isa, Bahrain. A partial Patriot battery and Lockheed Martin F-117As were delivered to Al Jabar, Kuwait. In this instance, the sudden arrival of forces caused the Iraqis to back down and acquiesce in the inspections by UN observers—at least for a time.

Retired Chief of Staff Fogleman recalled that one of the big issues in the Air Force at the time was the operations tempo. This was accentuated by the requirement to deploy troops to the Middle East in support of the no-fly concept. Fogleman observed that after the Gulf War had been over for several years, the other services had been able to bring their personnel home—but the Air Force had not. It was faced with the requirement for daily support of Operations Northern and Southern Watch, and this was particularly demanding upon its low density–high value assets.

Fogleman recalled his days in the Tactical Air Command, when there was a "quick reaction force" concept in which the first squadron of a wing could deploy in 24 hours; the second would go in 48, and the third in 72. In discussions with Central Command, Fogleman asserted that with the available contemporary theater and national intelligence assets, Central Command would always have at least 48 hours warning of a potential problem. By stating that he could return forces to the theater within 24 to 36 hours, he obtained agreement to bring more Air Force resources home from the Middle East.

Fogleman's ability to move toward an AEF posture was facilitated by the fact that General Joseph Ralston, who was also well founded in the ideas that Bill Creech had preached at TAC, was about to become commander of Air Combat Command. Fogleman called Ralston in and emphasized to him the new concept. In the past, all deployments had been based on previously prepared war plans. Now the Air Force was going to adopt an expeditionary mind-set.

The Air Expeditionary Force concept was fleshed out by the deployment of four experimental Air Expeditionary Task Forces to Bahrain, Jordan, and Qatar between 1995 and 1997. General Fogleman directed that doctrine be developed for the idea in 1996. It was released in October 1998, as Air Force Doctrine Document 2, *Organization and Employment of Aerospace Power.*

At the time, one of the major advantages cited for the new concept of an

Air Expeditionary Force was that it would stabilize the life of Air Force personnel, guaranteeing them more time at home and a reasonable rotation cycle overseas.

General Michael P. Ryan, eighteenth Chief of Staff of the USAF, has stated that there were many other considerations involved in the original planning for the AEF. Ryan recalls that in order to meet the many demands being placed on the Air Force, it was necessary, when another trouble spot emerged, to select an available force and throw it at the problem. This had to be done without regard to rotation cycles, training requirements, or the intensity of the operations tempo.

There was a second, less obvious problem. With the end of the Cold War, there was no longer a competent and reliable enemy whose strength was a basis upon which to compute the necessary USAF force structure. An attempt was made to finesse this situation by postulating a force necessary to fight two wars simultaneously in different parts of the world.

The old concept that force structure was to be based upon sets of war plans put the Air Force in a difficult position, in that there was no way to determine how large a force structure was required.

This was compounded by outside events. The Central Command began to ask for specific combinations of aircraft, rather than for a certain capability. That is to say, instead of asking for the forces to support a certain action, they would instead demand so many F-15s, so many tankers, so many jamming aircraft, and so on, to meet their needs. This contrasted to the manner in which the other services responded to such requests. When asked for assistance, instead of individual weapon packages, the Army might respond with a brigade, the Navy with a carrier battle group, or the Marines with a Marine Expeditionary Force—or in each case, the portions of these units that they thought appropriate.

The Air Force needed to create a structure so that it could avoid piecemeal demands and instead respond with an Air Expeditionary Force.

Ryan states that three main reasons emerged for the Air Expeditionary Force. The first was to have a new structure to meet the appetite of Central Command commanders for Air Force support, the second was to avoid running certain indispensable units into the ground, while the third was to have a plan on which to base future force structures.

When Ryan became Chief in 1997, he called in Brigadier General Charles F. "Chuck" Wald and Colonel Robert M. Worley and told them that he wanted to turn the Air Force into a rotational force, using the Air Expeditionary Force as a construct. A year was spent going through variations of the plan, and on March 5, 1999, the change was announced, citing the implementation of what was called the Expeditionary Aerospace Force (EAF) concept. The operational date for the change was given as January 1, 2000.

The initial planning for the creation of the Air Expeditionary Force was quite simple. The planners totaled the number of individual types of aircraft—

fighters, bombers, transports, specialty types, etc.—currently in the Air Force, and then divided that number by ten. This gave the Air Force people a means to anticipate when they—and their equipment—would be deployed, so that they could prepare on both a personal and a professional basis. Perhaps the biggest saving grace of the concept was giving Air Force personnel a good idea of when they would come home, again for both personal and professional reasons.

The new organization solved many problems. It permitted the formation of habitual relationships among the organizations in each AEF. They could work together continually, as they did during Red Flag exercises. Thus when they were sent off, for example, to the Operation Southern Watch no-fly zone, they were already accustomed to each other. It also provided a method of meeting the CENTCOM's appetite for forces by allowing the Air Force to announce that it could sustain two AEFs on line for any length of time—but no more. This paralleled the Navy's method of operation for its carrier battle groups.

It was of course understood that if a major event took place, then the Air Force would generate all of its forces to meet the emergency.

Initial planning indicated that an AEF would rotate about every fifteen months, and have a 90-day vulnerability period for this rotation. This was soon raised to 120 days. A notional AEF would have about 175 aircraft composed of 75 forward deployed and 100 on call. The forward deployed units would consist of fighters, bombers, tankers, and airlift, with the high demand low density assets called on as required. About 12,000 people support each AEF, which includes C⁴/ISR and combat search-and-rescue elements.

OTHER MAJOR AIR FORCE EVENTS

The long and eventful period from the Gulf War to the ghastly acceleration of terrorist events in 2001 was not filled solely with acts of war. Instead the rich community and family life of the Air Force flourished as its personnel were engaged in thousands of humanitarian missions that reached to every corner of the globe, without regard to former conflicts or current political events. These humanitarian efforts ranged from Tashkent to La Paz and from Beijing to North Dakota. These important activities were complemented by political events that included the historic visit of President William Jefferson Clinton to China in June 1998 and significant milestones of the end of the Cold War, such as the destruction of the last Minuteman II missile silo.

There were also technological advances that included the introduction of new equipment, such as the Boeing C-32, the Raytheon T-6A Texan II, Lockheed Martin C-130J, Bell Boeing CV-22 Osprey, Lockheed Martin X-35, and the Northrop Grumman RQ-4 Global Hawk.

There was scientific progress everywhere, ranging from the resupply of scientists at McMurdo Station to the airlift of Keiko, the Orca killer whale, to the delivery of the NASA satellite for tropical rainfall measuring in the first joint American and Japanese space effort. Nor was nostalgia neglected, as celebrations were held for the twenty-fifth anniversary of the return of the prisoners of war from North Vietnam and the fiftieth anniversary of Brigadier General Charles "Chuck" Yeager's epic breaking of the sound barrier.

The United States Air Force had always been in the lead in the integration of minorities, and this six-decade-long commitment was exemplified by three merely representative events concerning women. The first of these occurred on March 21, 1997, when Lieutenant Colonel Marcelyn A. Atwood became the first woman to command a training squadron, and the first USAF officer to command a Navy squadron at Pensacola, Florida. On February 24, 2001, Lieutenant Colonel Stayce Harris became the first African-American woman to command USAF flying squadron, the 729th Airlift Squadron at March Air Reserve Base, California. Then, on July 23, 1999, Colonel Eileen M. Collins, who had been the first woman pilot of a space shuttle in 1995, became the first woman commander of a space shuttle mission on a flight of the *Columbia*.

But ordinary life in the Air Force, and everywhere else, was brought to a sorrowful end by the final, long-delayed, recognition that a global war of terror had been declared upon the United States and its partners in the free world.

11

TERROR AND RESPONSE

*T*he enormity of the absolute failure of American intelligence efforts in the years prior to September 11, 2001, is so great that it not only fosters but requires criticism by hindsight. The information-gathering apparatus of the United States was emasculated by a combination of the proprietary concerns of the individual agencies and the endemic national malaise of political correctness. The latter had managed to work its way into the legal framework of intelligence gathering. These two factors combined with others (e.g., emphasis on technical rather than human intelligence gathering and the failure to train sufficient people in the language and culture of the Muslim world) to inhibit the proper response to the murderous actions and the numerous declarations of war by the militant Islamic fundamentalists. To a notorious degree, they still do.

Historians often remind us that Adolf Hitler's frightful actions were made known to the world in his virtually unreadable book *Mein Kampf*. Hitler did his very best to carry out his deliberate and terrible plans for genocide and world conquest, and was thwarted not because nations took heed of his predictions, but because he overstepped himself.

There is no excuse for the failure to take action against more than three decades of Islamic fundamentalist terrorism during which articulate Muslim spokesmen clearly and precisely broadcast their evil intent. They did so in actions and in far more understandable verbiage than Hitler's. Further, their goals were analyzed explicitly for an uncaring public in timely books such as Yossef Bodansky's classic *Bin Laden: The Man Who Declared War on America* (1999).

A more complete list of terrorist acts may be found in Appendix Three, but the following selections show the successful pattern of assault and intimidation.

On March 2, 1973, Cleo A. Noel, the U.S. ambassador to Sudan was assassinated at the Saudi Arabian embassy in Khartoum; there was no retribution. On October 23, 1983, 242 Americans were killed in Beirut in the bombing of the Marine barracks; the American presence was withdrawn. In February 26, 1993, a truck bomb was used in the first assault on the World Trade Center in New York City. Six persons were killed and a thousand were injured, but no effective retaliation followed. On October 3, 1993, Task Force Ranger raided the Olympic Hotel in Mogadishu to begin a seventeen-hour battle in which eighteen U.S. soldiers were killed and eighty-four were wounded. The bodies of the dead American soldiers were dragged through the streets of Mogadishu. In response, President Clinton withdrew U.S. troops from Somalia. When the terrorists attacked the U.S. embassies in Kenya and Tanzania in August 1998, attempts at retaliation were feeble. Cruise missiles were launched against ill-defined and sometimes incorrect targets, with no perceived results, and with at least the suspicion that their launch was motivated by domestic political reasons.

The radical Islamic movement thus repeatedly demonstrated to the world—and most importantly, to the Muslim community—that you *could* attack the United States and kill its citizens without fear of an effective response.

Even more tragic, there were literally thousands of other related occurrences, less obvious to the public, but of vital importance to the intelligence community, which were ignored.

All of this led inevitably to the terrible events of September 11, 2001, which will be only briefly recounted here. At 8:45 A.M. American Airlines Flight 11 was flown into the north tower of the World Trade Center. Eighteen minutes later, United Airlines Flight 175 was flown into the south tower. At 9:43 A.M., American Airlines Flight 77 was flown into the Pentagon. At 10:10 A.M. United Airlines Flight 93 crashed into a field southeast of Pittsburgh.

All four airliners were hijacked by a team consisting of nineteen Arab men, fifteen of them from Saudi Arabia. The fourth aircraft was probably intended for the White House, but crashed after the passengers rebelled against their captors.

These attacks killed 3,025 Americans and foreign nationals—more than had died at Pearl Harbor. President George W. Bush indicated that Osama bin Laden was the primary suspect and announced that the United States was in a state of war with international terrorism.

In vivid contrast to the feeble efforts of previous administrations, President Bush began energetic military and administrative actions against the enemy, the first of which was Operation Noble Eagle. President Bush authorized a partial mobilization of the reserves on September 15, 2001, for homeland defense and civil support missions. About 35,000 airmen were designated to serve in Operation Noble Eagle—the overall plan to protect North America from air attack.

In its initial response, the Air National Guard kept two fighters flying over

New York and four over Washington twenty-four hours a day. Random patrols were also flown over other major cities.

The Air National Guard became the backbone of homeland air defense, and in the first year flew 75 percent of the combat air patrol missions, while executing 62 percent of the refueling sorties. The ANG maintained almost 100 percent of the alert bases, which began to operate around the clock. There was an immediate requirement to update and modernize many of the alert base facilities to meet the surge in demand.

Within fifteen months, more than 13,000 homeland defense flights had been flown from twenty-six bases. The effort has been arduous for the personnel. In 1958, there were thousands of fighters on alert; by early 2001, this number had dropped to twenty. After September 11, the total increased to more than one hundred.

By 2006, the Air Force was conducting Operation Noble Eagle with an average of forty-five to sixty fighters, two or three AWACS, a dozen tankers, and the command-and-control network.

In addition to Operation Noble Eagle, the administration froze the terrorist's financial assets and broke up their intricate fund-raising networks. As this was going on, American troops were deployed overseas to attack the terrorists in their most formidable stronghold, Afghanistan.

Many said that the downfall of the Soviet Union began in December 1979, when it launched a war against Afghanistan that would end ten years later in utter failure. In 2001 there was concern and even predictions that the United States would become bogged down in a similar way.

Official spokesmen made it clear from the beginning that the United States was engaged in a war on terrorism and not on Muslims in general. From the start, every effort was made to avoid giving offense. The initial name for the campaign was Operation Infinite Justice, but for fear of offending the Muslim community who attribute infinite justice only to Allah, the name was changed to Operation Enduring Freedom (OEF). The military objectives of OEF were announced to the country by President Bush during his address to a joint session of Congress on September 20, 2001.

On October 7, when the first attacks of OEF were launched, Bush addressed the United States, stating the objectives of the war to be the capture of Al Qaeda leaders and the destruction of their numerous training camps in Afghanistan. There the Taliban government had instituted a new dark age of repression that benefited only its leaders, and Al Qaeda.

In a supplementary address on the same day, Secretary of Defense Donald H. Rumsfeld added that OEF would see the destruction of the Taliban military so that a change in regime would be possible. He also promised to deliver humanitarian supplies to the Afghan people. The British government defined similar goals for its own military actions in what was termed Operation Veritas.

General Tommy R. Franks, commander of U.S. Central Command, directed Operation Enduring Freedom, which by 2002 had a coalition of sixty-eight nations contributing to it. Even before the opening hostilities, Franks ordered reconnaissance flights by Lockheed Martin U-2 s to gather targeting information to supplement that acquired by satellites.

The initial attacks began on October 7, 2001.

American aircraft, including the Boeing B-1B, Northrop Grumman B-2A, Boeing B-52, Northrop Grumman F-14, and Boeing F/A-18 aircraft launched attacks in concert with Tomahawk cruise missiles launched from U.S. and Royal Navy ships and submarines. The primary targets were warning radars, surface-to-air missile sites, antiaircraft batteries, armored vehicles, troop concentrations, and runways.

Lieutenant General Charles F. "Chuck" Wald, the Joint Forces Air Component Commander (JFACC), estimated that a substantial portion of the integrated air defense elements were eliminated in the first fifteen minutes of the attack. All fifty aircraft of the Taliban air force were destroyed. The bombing was aided by small units of U.S. and British Special Operations Forces that collected information and designated aim points for the attacking aircraft. Among their equipment, the Special Forces, some on horseback, carrying laptops, used binoculars equipped with laser range finders. This enabled them to determine specific geocoordinates for the targets they were observing. These coordinates were then relayed via "burst" signals to bombers orbiting high above. The bombers entered the coordinates in their computers, and a Joint Direct Attack Munition (JDAM) was dropped on the target within minutes.

The stealthy B-2As flew directly from Whiteman AFB, Missouri, to targets in Afghanistan on grueling missions lasting 44 hours. They landed at Diego Garcia and left the engines running while a new crew came aboard for the flight back to Whiteman.

Diego Garcia was home to eight B-1Bs and ten B-52H bombers that flew an average of four missions per day per type. (The small size of the force—eighteen aircraft—hints at the tremendous capability that C⁴ISR and PGMs endow them with.) They dropped leaflet dispensers, JDAMs, and Mk 82 500-pound "dumb" bombs. The B-52 also introduced the new Wind-Corrected Munitions Dispenser (WCMD) to combat. The WCMD has a tail kit attached to the bomb, which has an inertial guidance system to adjust for launch transients and wind effects.

The Air Force effort was supplemented by four U.S. Navy carriers (later reduced to two) that operated Boeing F/A-18s and Northrop Grumman F-14s on fighter-bomber missions that averaged five to seven hours. Air Force tankers kept them refueled en route. When basing became feasible, Marine AV-8B Harriers were also used.

Afghanistan was a tremendous challenge. The war zone was remote, and there were no convenient bases from which to operate. The American forces depended upon Navy carriers and the long legs of Air Force aircraft for their success.

By the time organized Taliban resistance crumbled in mid-December, the Air Force had flown 6,800 sorties, of which about 3,400 were by tankers. USAF airplanes dropped 6,100 tons of ordnance, about 75 percent of the total dropped in the campaign. Almost three-quarters of the bombs dropped were precision guided, and their use severely depleted the JDAM inventory.

The precision weapons had a devastating effect upon the morale of the Taliban soldiery, which had become inured to the ineffective air-to-ground efforts of the Soviet Air Force. The deadly accuracy of the precision-guided munitions introduced them to an entirely new kind of warfare with which they could not cope. In addition, the successful strikes impressed members of the Northern Alliance, leading them to cooperate with American forces. On a different scale, Lockheed Martin "B-130s" (C-130s) dropped the 15,000-pound Daisy Cutter bomb at Tora Bora against cave and tunnel targets.

All operations were facilitated by the Combined Air Operations Center (CAOC) operating at the Prince Sultan Air Base in Saudi Arabia. The CAOC's access to high-bandwidth commercial satellites enabled it to serve as a collection point for information from all intelligence sources. It also permitted bombers to launch and receive their target information en route, allowing attacks on "pop-up" targets.

The Chief of Staff, General John A. Jumper, emphasized that the time required for finding, identifying, tracking, and attacking a target—the "kill chain"—was greatly reduced during OEF. Jumper emphasized that the goal was to get the kill chain so refined that an emerging target could be hit in "single digit minutes." (There were some instances when an emergency target was hit in less than a minute after it was found.) One lesson learned was that all USAF "shooter" aircraft needed to be equipped with digital data-sharing systems, so that the time required to pass target coordinates and verify them by voice could be eliminated.

Another function of the CAOC was to direct the "streaming video" from Predator UAVs directly into some aircraft. The capability enabled Lockheed Martin AC-130 gunships to lay down precise fire on targets with a minimum of collateral damage.

When OEF began, the Taliban controlled 80 percent of Afghanistan and the anti-Taliban forces were in retreat everywhere. Afghanistan was a terrorist state, a sanctuary for Al Qaeda members.

The early elimination of the Taliban air defenses system enabled tankers to fly closer to the bomber and fighter routes, permitting an increase in sortie rate. Electronic warfare and signals reconnaissance aircraft then jammed enemy cell phones and located commanders and Al Qaeda fighters by triangulation.

Not all air operations were lethal. Messages were broadcast to the Afghan people, confirming that the United States was fighting to liberate them from the Taliban. It backed up the psychology with Boeing C-17A missions which dropped two million Humanitarian Daily Rations to the Afghan people.

C-17As also distinguished themselves by flying heavy equipment into Afghanistan, which permitted the Marines to establish forward operating base (FOB) "Rhino." The big, expensive C-17As flew out of unimproved desert strips in great clouds of dust, just as their brochures had promised so many years before when they were touted as the "pipeline to the frontline." The C-17As inserted troops at night, under fire. They used maximum-descent tactical approaches and then spent an absolute minimum of time on the ground to off-load. Among their vital services was one that company brochures had not foreseen, dropping oats to the horses that Special Force detachments used as they ranged around the country, inspiring the anti-Taliban forces.

On December 22, only seventy-eight days after the beginning of operations, General Franks witnessed the inauguration of an interim, anti-Taliban government in the capital city, Kabul. In less than three months, Operation Enduring Freedom stretched the Air Expeditionary Force concept very thin in some specialty areas, including force-protection troops and combat search-and-rescue forces.

The war in Afghanistan was won by an incredible combination of high-technology devices and a primitive style of warfare. The new technology reduced the targeting cycle time for Tomahawk missiles down to 19 minutes, compared to the 101 minutes it had required during Operation Allied Force.

Yet for the Air Force there was an annoying and persistent irony in the undeniable victory. In Operation Allied Force airpower was recognized as the key to victory. In Operation Enduring Freedom, the results of network-centric warfare and effects-based operations led to even more powerful results for airpower—which nonetheless was perceived as playing a supporting role to ground operations. The Air Force seemed to be caught in a psychological–public relations trap of its own making. It now appeared that the more successful its high-tech efforts were, the more they were taken for granted.

Operation Enduring Freedom demonstrated a new level of brilliance in joint operations by the armed services of the United States. In startling contrast to the decade-long debacle of the Soviet Union in Afghanistan, the American and coalition forces quickly obliterated the oppressive Taliban regime. Yet OEF was but a hint of the almost unbelievable proficiency and ferocity with which the United States would lead another coalition in Operation Iraqi Freedom (OIF)—and parenthetically, once again be somewhat overlooked for reasons of excellence.

BACKGROUND TO THE WAR

The United Nations knew that Iraq had used weapons of mass destruction in the past. It also knew that if Iraq still possessed such weapons, it would use them in the future, either internally against Iraqi dissidents or against Israel. There was genuine concern that Iraq would succeed in acquiring nuclear weapons. If they used them or supplied them to terrorists, the United States might suffer a disaster of unimaginable consequence. Consequently, President Bush determined to remove Saddam Hussein's cruel and repressive regime if the Iraqi dictator did not conform to the resolutions laid down by the United Nations.

Bush's goals were opposed in the United Nations by a number of countries, including Russia, France, and Germany, despite the fact that the latter three had all voted for UN Resolution 1441 (see Appendix Four). Each of those countries had its own rationale for wishing to keep Saddam Hussein in power. Their motivation was later explicitly impugned by the scandals involved in the notorious Oil for Food program of the United Nations.

Adequate justification for military intervention had been provided by previous Iraqi violations of UN resolutions. Iraqi forces had fired at British and U.S. aircraft on a regular basis as they enforced the no-fly zones. President Bush therefore determined to make Saddam Hussein live up to the rigid requirements of UN Resolution 1441, despite opposition from nominally friendly countries.

The first step came on March 17, with President Bush's unique public ultimatum to Saddam Hussein, which called for him to step down and leave Iraq, with his two sons, within forty-eight hours. Bush promised that failure to comply with this ultimatum would lead to U.S. intervention.

There was a general exodus of foreign nationals as embassies closed and visiting workers left. All UN personnel, including those still searching for Iraqi weapons of mass destruction, were ordered to leave.

Apparently buoyed by a conviction that his hired agents in France, Germany, and Russia would prevent the United States from proceeding, Saddam Hussein rejected Bush's ultimatum. Since Desert Storm, Saddam had been shedding his traditional secularism, and traded on this by threatening a worldwide jihad against the United States.

Everyone supposed that the initial air attacks on Iraq would be of such a catastrophic nature that those of Desert Storm would seem mild in comparison. As the clock ticked down toward the deadline, the buzz words "Shock and Awe," meaning a tremendous demonstration of unrestrained aerial attack, gained prominence.

During the same countdown, the media and many others—including the Army Chief of Staff—were concerned that the number of coalition troops was far too small. Everywhere the cry for "more boots on the ground" was heard. Specters of the United States in Vietnam and the Soviet Union in Afghanistan were invoked, despite the latest demonstration of U.S. proficiency in that impoverished country. Among the most mindless claims were those that war on Iraq would provoke retaliation from al Qaeda—as if any provocation was needed.

Pessimistic pundits were thrown another bone of contention on March 2, when the Turkish Parliament refused access to the U.S. 4th Infantry Division

(Mechanized). The division was diverted south to Kuwait. This immediately brought speculation from some military commentators that the entire war plan was endangered, for they believed that at least 60,000 troops had to invade Iraq from the north to assure victory.

Saddam had a total of about 400,000 soldiers available to him on the eve of Operation Iraqi Freedom, far less than the almost 1,000,000 men he mustered at one time. Most of these were conscripts, used in seventeen regular army divisions for internal security, as they were considered untrustworthy and a threat to the regime. His main strength was his six Republican Guard divisions, which received the best equipment and training along with special treatment and privileges.

Still, on paper the Iraqi army was formidable, with as many as 2,000 tanks, including perhaps 700 T-72s. It also had nearly 4,000 light tanks and armored personnel carriers and 2,000 heavy self-propelled artillery pieces. Most of these would be transformed by American airpower from weapons of war to scrap iron.

The Iraqi integrated air defense had been well tested throughout Operations Northern and Southern Watch. Iraq had far fewer surface-to-air missiles than in 1991, but still disposed of about 1,000 of the smaller man portable SA-7/14/16s and 400 of the larger SA-2, SA-3, SA-6, SA-8, and Roland varieties. There were, in addition, 6,000 antiaircraft guns.

General Franks had a force of 230,000. About 190,000 of these were from the United States and consisted of Army infantry and airborne troops along with the Marines and Special Forces. The United Kingdom provided 42,000 troops, including Royal Marines and paratroopers. These were joined by 2,000 Australians, including Special Forces troops, and a very small but welcome number of Polish forces. In the next three weeks Franks's total force grew to 340,000.

Perhaps the most important of the ground forces were the Special Forces troops. These Special Forces had operated clandestinely from well before the war, ranging all over the country. A force approaching 10,000 personnel from the Army, Navy, Air Force, the United Kingdom, Australia, and Poland was more than 50 percent larger than the 6,000 used in Afghanistan and very much larger than the number used in the 1991 Gulf War. Oddly enough, the diplomatic efforts of the Special Forces may have transcended their military triumphs, for they worked on an intimate, informal basis with tribal leaders. Perhaps as a consequence, the brilliant and ever dangerous campaigns of the Special Forces resulted in light casualties.

Their positive effect on political, military, economic, and psychological goals was immensely important, so much so that both U.S. and British forces will in the future contain a higher percentage of Special Forces. But raising the number of Special Forces from an army is easier to plan than to do. So few personnel possess physical and mental standards required by Special Forces that you have to cast a wider enlistment net—a difficult task in a world continually at war.

It was in part the prehostilities efficacy of the Special Forces that confused Iraq and facilitated Franks's decision to go in early with ground forces. Doing so avoided having to use the much heralded shock-and-awe tactics and reduced collateral damage to a level previously unknown in warfare. It allowed the full employment of effects-based operations, while perhaps mitigating Arab resentment of the United States for invading an Arab country.

It must be said that there is a small but growing belief among some analysts that forgoing shock-and-awe tactics and being excessively concerned about collateral damage may have been a mistake, given the psychology of the Arab nations. This school of thought maintains that it would have been better—more merciful, even—to have rained down destruction on Iraq to stun it into submission and to give the Arab world warning of what could happen in the future if they opposed the United States. Opponents of this idea say it smacks of gunboat diplomacy, but it may be that in this clash of cultures and, indeed, of civilizations, that gunboat diplomacy is a surer way of winning hearts and minds than furnishing Meals Ready to Eat.

General Franks commanded one of the most experienced combinations of armed forces in history. The services of the United States, along with many of its coalition partners, had been engaged together in conflict for years, and were staffed with experienced veterans comfortable with working with other services. The techniques of command and control introduced in the 1991 Gulf War and polished in the subsequent years, were now at a peak of proficiency. Airpower was still the key, and the Joint Forces Air Component Commander was then Lieutenant General T. Michael "Buzz" Moseley. As JFACC, he brought to bear the most modern, most effective, and ironically, most humanitarian assemblage of airpower in history.

Moseley, a veteran fighter pilot and commander of 9th Air Force, had an enormously powerful force at his disposal. Its 1,801 total aircraft included 735 fighters; 51 bombers; 268 tankers; 46 command and control (C^2) airplanes; 118 intelligence, surveillance, and reconnaissance (ISR) airplanes; 72 special operations and rescue craft; 73 Special Operations Force craft; and 126 airlifters.

This diverse and flexible force benefited from having an experienced and efficient CAOC. (It was called Combined Aerospace after integrated satellite information became part of the mix.) The CAOC provided current operational assessments to almost every level of command, so that everyone, regardless of service or country of origin, had precisely the same information at the same time. Originally based at the Prince Sultan Air Base in Dhahran, the CAOC was moved to a $40,000,000 facility at the Al Udeid Air Base near Doha, Qatar.

At the CAOC, enormous amounts of ISR data were gathered by satellites and aircraft. General Moseley was also charged with the responsibility of integrating all Department of Defense space activities for all services, and also all intelligence agencies.

Through refined "reachback" techniques, the combat forces received ISR data directly from databases in the United States. Using the Internet model, chat rooms were established, allowing the expert ISR analysts to communicate directly with troops in the field. The analysis was performed by personnel working at two distributed ground systems (DGS) that operated independently. DGS 1 was at Langley Air Force Base, Virginia, while DGS was at Beale Air Force Base, California. (The two units were later combined into the 408th Intelligence Wing.)

Using laptops (carried on everything from a horse to a jet), deployed forces received information that had been uploaded to satellites. During the early part of OIF, the system would provide more than 30,000 intelligence reports that identified more than 1,000 targets.

The CAOC (which also controlled the Marine Air Wing) functioned expertly, reflecting a new era in the warfighting capability of the United States. Colonel Anthony C. Cain has described this in the *Air & Space Power Journal* as "predictive battlespace awareness (PBA)," stating that in OIF there has been "a conceptual and organizational transformation that redefines how we think about employing air and space power to achieve national strategic and operational goals. Using air and space power's PBA, we can alter the character, pace, depth, and scope of operations to suit our timing and objectives." Cain goes on to remark that "operations in Iraq reveal a *seamless* integration of air and space power's capabilities with those of other components. . . . The result is a complex, integrated, and synchronized campaign that allows coalition commanders to outthink *and* outfight enemy commanders and their forces."

PBA, using the mass of analyzed intelligence data, permits the United States to develop a strategy based on an accurate assessment of the enemy's capabilities, force structure, and probable plans. With this ability to predict events, the scope and method of operations can be precisely tailored to achieve objectives with a minimum of collateral damage. Further, the same shared intelligence ensures that joint and combined operations are carried out smoothly. The success of these tactics was maximized in operation against formal forces; they are less successful against insurgency operations.

Thus shock and awe was translated from a thundering campaign that leveled cities to one that artfully shaped the battlefield. The hand-in-glove cooperation of air and surface power began the war with an amazing demonstration of flexibility. General Tommy Franks surprised the world by attacking on the ground (G-day) in advance of attacking via the air (A-day).

The brilliant opening move saw the heavy 3rd Infantry Division (Mechanized) race more than 200 miles in a short time. The 101st Airborne Division (Airborne Assault) provided additional reach and striking power with many Apache helicopters. The 1st Marine Expeditionary Force launched deep into the desert from the beaches, while United Kingdom forces operated primarily in the south. The union of Army and Marine ground units under one land component commander was unique in history and sets the pattern for the future.

A brilliant decision had been made to "embed" the media with the fighting forces, and they became the lens through which the world viewed the war. Unfortunately for the Air Force, it is far easier to embed the media with fast-moving tanks or infantry than in the cockpit of a fighter or bomber. The result was that on television and in the press, the war seemed to be almost totally a ground war. Moseley's forces were in fact heavily engaged from the start, reaching a sortie rate of 1,000 on the second day of the war and thereafter sustaining or exceeding that rate.

Problems soon arose concerning flank defense and supplies of water, food, ammunition, and fuel. So swift was the attack that the United States was already inside the Iraqi decision cycle, spreading confusion and anticipating reactions. Airpower was used to cover the flanks as the 3rd Infantry Division peeled off individual battalions to stop Iraqi forces from intervening. They were given close air support by F/A-18 Hornets, AV-8B Harriers, and A-10 Warthogs.

The smooth integration of aerospace capabilities was demonstrated on March 24, when hundred-mile-an-hour sandstorms obscured the battlefield. This brought operations of both land vehicles and helicopters to a virtual halt, but the combination of aerial sensors and precision-guided munitions (PGMs) made it possible for Boeing B-1B bombers of the 405th Air Expeditionary Wing to continue their operations. The "Bone," as it is called, came into its own once again, demonstrating its capability by dropping as many as twenty-four Global Positioning System/inertially guided JDAMs on a single mission.

Despite the weather, fourteen hundred strike missions were flown, with the major targets remaining the same: Republican Guard divisions, key command-and-control facilities, and surface-to-surface missile facilities. The venerable Boeing B-52 continued in operations from its deployed base.

Even during the sandstorm, any moving Iraqi vehicle was immediately identified by the Joint STARS Moving Target Indicator (MTI) systems and its geographic coordinates determined by GPS. Once identified and located, it was killed by precision-guided munitions from a coalition aircraft. In a brilliant example of "effects-based operations" the coalition did not resort to blowing up bridges, but instead destroyed troops on the move.

F^2T^2EA: FIND, FIX, TRACK, TARGET, ENGAGE, ASSESS

The war in Iraq saw a refinement of the target cycle concept that General Jumper termed "Find, Fix, Track, Target, Engage, Assess (F^2T^2EA)," and is more colorfully known as the "kill chain." The armed services of the United States have the ability to find targets anywhere in the world, fix them in a database, track them as necessary, designate them as a target, engage them with the appropriate weapon at the appropriate time, then assess the damage done to the target—all in a very compressed period of time. The goal became giving the services a similar but even swifter capacity to do the same to targets that emerge suddenly from hiding.

Three elements of the equation, F^2 and T (find, fix, and track), have been operational for years, and are provided by the following systems: Boeing E-3C airborne warning and control system (AWACS); Northrop Grumman E-8C Joint Surveillance Target Attack Radar System (Joint STARS); Lockheed Martin EP-3; Beech RC-12 Guardrail; Boeing RC-135 Rivet Joint; General Atomics Aeronautical Systems RQ-1A Predator UAV; Northrop Grumman RQ-4A Global Hawk; and Lockheed Martin U-2.

Once found, the target has to be "fixed," that is, its location must be determined with precision, most often using GPS satellites. Then aircraft with precision-guided munitions attack the target. After the attack, bomb damage assessment is done. This represents airpower influencing history in its purest form.

In Operation Iraqi Freedom, the use of PGMs became almost standard, for during the ten years after Desert Storm, a tremendous effort was made by all the services to adapt them to all of their aircraft. Even the Northrop Grumman F-14 Tomcat, after decades reveling in its fleet defense role, was given a new mission as the "Bombcat," using PGMs.

Increased reliance was placed on the JDAM, which was both accurate and relatively inexpensive. Several other PGMs were used, including the Maverick, HARM, AGM-154 joint standoff weapon (JSOW), the wind-corrected munitions dispenser, and several varieties of laser-guided bombs.

Moseley was especially pleased with the work of naval and Marine aviation in Operation Iraqi Freedom. A case in point was the sterling performance of the Boeing AV8-B Harriers, which took off from the carrier, flew a mission, went to a forward operating base to rearm and refuel, flew another mission, and then returned to the carrier. This doubled the availability of the aircraft.

The Harrier's capability was enhanced by the use of the Litening II ER targeting pod. The video downlink of the ISR version of the Litening pod enabled forward air controllers (FACs) to see what the Harrier pilots saw and talk them to the targets.

The Navy's claim for the effectiveness of carrier task groups was fully supported by the performance of the Boeing F/A-18s, which did very well with the new Super Hornets functioning in both tanker and fighter roles. They were joined by Northrop Grumman F-14s and a host of support aircraft.

Contrary to most expectations, the coalition forces avoided slugging it out on the ground, preferring to rely on airpower to provide close air support, on a rapid-reaction basis. FACs operated both on the ground and in the air. A wide variety of aircraft, including Boeing F-15Es and F/A-18s, Lockheed Martin F-16s and A-10s and Northrop Grumman F-14Bs provided round-the-clock cover in and over Baghdad, all armed with precision-guided munitions. Iraq saw a changing face in close air support, as it was no longer defined exclusively by fighter bombers dropping down close to the ground to deliver weapons. Instead, huge bombers, operating at high altitude, delivered weapons with uncanny precision on individual trucks, tanks, and bunkers. The bombers had some great advantages over the fighter-bombers, including much longer loiter times and far more on-board computer capability.

Iraq had anticipated the increased use of PGMs and attempted to counter them with electro-optical and GPS jammers supplied by Russia. One raid was described by Major General Victor "Gene" Renuart as "ironic." In it, the Iraqis were using six GPS-jamming units to circumvent the JDAMs, but all were destroyed by JDAM bombs using GPS/INS guidance.

It is useful here to compare past capabilities with current ones to gain insight into the true nature of B-1B warfare. During World War II, a Boeing B-17 equipped with the Norden bomb sight had a 3,300-foot circular error probable (CEP). This meant that, to be absolutely certain of destroying a target, the attacking force had to drop 9,000 bombs, from 1,000 bombers, putting 10,000 airmen at risk. In Korea, a Republic F-84F had a 1,000-foot CEP. To guarantee the destruction of a target, a total of 1,100 bombs had to be dropped from 550 fighters, putting 550 airmen at risk. During the Vietnam War, the McDonnell Douglas F-4 had a 400-foot CEP. To guarantee the destruction of a target, 30 fighters had to drop 176 bombs—and put 60 airmen at risk. Stealth and precision weapons vastly improved things in Desert Storm, in which the Lockheed Martin F-117A had a 10-foot CEP. It could kill two targets with two bombs from one fighter and put only one airman at risk. The current Boeing B-1B using the JDAM with its CEP of 20 feet, can destroy 24 targets with 24 bombs from one bomber . . . putting the single crew of four airmen at risk.

The Iraqis adopted an attrition strategy, hoping to inflict enough casualties on American forces that the people of the United States would rebel against the war. But under the benevolent umbrella of airpower, the ground forces moved forward quickly, Army and Marine forces working hand in hand. Airpower enabled the coalition forces to kill Iraqis at a distance and prevented them from responding with any lethal effect.

On April 3, the fifteenth day of the war, the sortie rate increased to 1,900, with 850 of these being strike sorties using PGMs. Coalition air forces continued to perform two key functions. First came the suppression of enemy mobile forces and the destruction of massed Republican Guard troops as they assembled. Then it was possible to provide cover to the flanks and the rear, allowing coalition ground forces to advance, knowing they would not be ambushed or cut off. The coalition advances were not unlike the island-hopping strategy of the United States in World War II, when thousands of Japanese soldiers were bypassed on Pacific island fortresses. But where the Japanese were marooned by water, the Iraqi concentrations were marooned by the threat of airpower pounding any movement.

On April 5, this kill-and-cover technique allowed the 3rd Infantry Division to send twenty-six M1A1 Abrams tanks and ten Bradley fighting vehicles on a twenty-five-mile run through the streets of south and central Baghdad, amazing not only the Iraqis but the world. Early reports indicated that the Thunder Run killed two to three thousand Iraqis who had attempted to interfere with the raid.

The armored spearhead, which paraded within two miles of Saddam's famous bunker and onto his favorite military parade ground, burst the bubble of media and public discontent over the slow progress of the war, demonstrating that with absolute air and information dominance, coalition forces could move when and where they wished to move, even in the heart of Saddam's citadel, Baghdad.

Despite the small number of Boeing B-1Bs deployed to take part in Operation Iraqi Freedom, they did a disproportionate share of the bomb dropping. In the course of the campaign, B-1Bs flew 432 sorties and dropped 2,250 tons of bombs, almost all precision guided.

On the nineteenth day of the war, April 7, a B-1B bomber named *Seek and Destroy* demonstrated just how flexible and responsive bombing had become. *Seek and Destroy* and its crew of four belonged to the 34th Bomb Squadron, 38th Bomb Wing, at Ellsworth Air Force Base. The wing was then commanded by Colonel James Kowalski, who was serving as Commander of the 405th AEW.

A Boeing E-3 AWACs aircraft notified the B-1B of an emerging target. Vetted information indicated that Saddam Hussein and his two sons were meeting at the al-Saa restaurant in the al-Mansur section of western Baghdad.

Aircraft Commander Chris Wachter, pilot Captain Sloan Hollis, Weapons System Operator (Offensive) Lieutenant Colonel Fred Swan, and Weapon System Operator (Defensive) Lieutenant Joe Runci knew that they had just twelve minutes to complete all their respective tasks and drop the four bombs they had been directed to use.

Their tasks included locating the exact target, probing enemy air defenses, confirming all decisions with AWACS controllers, arming and setting the fuse on the specified weapons (two instant and two delayed JDAMs), and finally placing the exact coordinates of the target in the bombs' guidance mechanism. (The JDAM receives constant updating from the aircraft avionic systems while it is in the bomb bay; once dropped, the inertial guidance system takes over, with periodic updates from GPS.)

Seek and Destroy was flying at 30,000 feet and at 500 knots when, just twelve minutes after receipt of their orders, the first two GBU-31 hard target penetrator JDAMs were released by the bombing system. Three seconds later, two standard GBU-31 JDAMs were released.

The bombs hit the target exactly as desired—but as would be the case so often with Saddam Hussein, he had moved on. *Seek and Destroy* moved on as well, hitting seventeen other targets in two other target groups, separated by two hundred miles.

Though there was natural disappointment that Saddam Hussein had escaped, the mission demonstrated an incredibly quick reaction time to an emerging target. It indicated clearly that bombing platforms with long loiter times and precision-strike capability were the instruments of a new phase of air warfare, one that General David E. Deptula has described as that of "mass precision."

Lieutenant General Moseley (who was later to become a four-star general and Air Force Chief of Staff) issued the first data compilation. By April 25, 2003, the coalition was shown to have flown almost 50,000 fixed-wing missions, of which the USAF flew roughly 30,000, the Navy 9,500, the Marines 5,500, and the Royal Air Force and Royal Australian Air Forces about 5,000. Of the missions, some 36 percent were strike sorties, 20 percent tanker missions, and 25 percent airlift missions, with 4 percent miscellaneous.

Of the 28,820 munitions dropped, 19,060 were precision guided, an amazing 66 percent. The aerial refueling figures showed that 417,137,233 gallons of fuel were off-loaded by tankers of all the services. Another 195,753,818 gallons of jet fuel were provided by land- and ship-based sources.

The surprise launching of the ground war validated General Franks's decision to strike with the relatively small number of forces in the theater prior to a heavy air attack. Yet it was the utter air and space dominance that permitted Frank's plan to work. The coalition's overwhelming C^4 ISR capabilities enhanced the tactical surprise, the speed, and the high quality of coalition troops. Sadly, in most popular analyses of the campaign, the vital synthesis of air and space elements in the latest application of airpower has been largely overlooked, while surface tactics that were used in World War II by the Desert Rats of the British Eighth Army have been extolled.

STRANGE CONCLUSIONS ON OPERATION IRAQI FREEDOM

Air power had an undeniably important impact on the conduct of the war, for it exercised a killing power that exceeded in lethality anything known before. It also allowed ground forces to operate without regard to a logistics train or the threat of flanking movements. A totally new phase of warfare opened in which, on some occasions, airpower would flush enemy ground forces out into the open to be finished off by surface forces, and others in which the movement of surface forces forced the enemy to react—and then be finished off by airpower.

Despite this, and only partially as a result of the embedded media, airpower emerged from the war perceived as a support arm for surface forces, rather than as the decisive air and space instrument that made every one of the ground successes possible. There are some who say that the feeling of being in a support role is beginning to infiltrate the Air Force itself, and if true, this is a trend that must immediately be halted and reversed.

It is not stretching the truth to say that the United States Air Force has become a victim of its own successes since the first Gulf War. The USAF has performed in a succession of conflicts with such excellence that it has begun to make its mission seem too easy. The other services have come to expect the Air Force to provide freely of its tanker force, space resources, intelligence-gathering capability, and firepower upon demand and without regard to the effect that demand has upon the Air Force budget. Unfortunately, the same services forget the importance of the Air Force contribution when it comes to the annual budget discussions. This may be a human enough response, but it is one which must be faced squarely in the future, perhaps by requiring some sort of industrial funding in which the other services pay for the use of Air Force resources.

The following chapter will contain a general survey of lessons learned in the USAF in the last ten years. But there were a number of immediately obvious lessons learned in Operation Iraqi Freedom, and unfortunately most of them are laden with the seemingly unbearable burden of recapitalization, the buzzword for buying vitally needed new equipment.

The value of the bomber was demonstrated over and over. The performance of both the B-2A and the B-1B surprised many, as did the emergence of the B-52 as a close-support aircraft. The effect of the bomber in swift, long-range power projection cannot be overestimated, and the combination of long loiter time, C^4 ISR, and PGM offers commanders an amazing capability for waging network-centric warfare. With such qualities of persistence, the bomber also lessens the critical demand for tankers. Yet the cost of bringing a new bomber into the fleet will be staggering. The probability of its occurring in the next twenty years is almost nonexistent, and the probability of its occurring in the next forty years is very low.

In a similar way, the space satellite systems operated by the Air Force worked almost flawlessly for *all of the services*, providing intelligence, navigation, communications, and meteorology during Operation Iraqi Freedom. Yet these systems are aging and must eventually be replaced at staggering, budget-busting expense.

The performance of American fighters was without reproach, but all Air Force fighters are aging, and their potential replacement by the Lockheed Martin F-22A and F-35 is by no means certain. Critics of the two new fighters contend that the decades-old Boeing F-15 and Lockheed Martin F-16 are still world-beaters, but do not take into consideration the potential for a country like China to develop superior fighters in the near future. Given China's booming economy, and the relations it has established with the top Russian design bureaus, it is almost a certainty that fighters even more advanced than the Sukhoi Su 35 will be built, and in quantity.

The emerging capability of unmanned systems is another cause for alarm in terms of the inadequacy of resources. Part of the ubiquitous "low-density–high-demand" contingent, the American fleet of unmanned aerial vehicles (UAVs; now, more commonly, unmanned aerial systems, UASs) has seen one success after another. UASs were particularly effective in Operation Allied Force, in which they were used to examine Serbian air defenses, carry out electronic intelligence missions, and monitor ethnic cleansing. They also served to act as airborne communication relays, jam enemy communications, and even perform an often overlooked task, bomb damage assessment. Yet despite their success, they are still inadequately funded.

The ante was raised in Operation Enduring Freedom when the amazing long-range Northrop Grumman RQ4 Global Hawk debuted. The smaller, more familiar General Atomic MQ-1 Predator used Hellfire missiles against high-value targets. The concept of persistent—that is to say, universal and unending—surveillance by UAS is certain to see the demand for their services increase. The goal now is to extend the striking range of armed UAVs to match the range of their surveillance counterparts. Yet the production quantity on order of UAS is deplorably low, and is not apt to rise at any great rate in the near future.

And perhaps no element of the Air Force is more in need of rejuvenation than the tankers. The Boeing KC-135s are as old as the B-52s and are used much more intensively. Reliance on tankers continues to increase—the tanker-to-total-sortie ratio of Operation Iraqi Freedom was twice that of Desert Storm. The solution to the tanker problem seemed almost at hand at one point, but blew up in the face of an acquisition scandal. There is recurrent interest in the problem, but the scandal has so tainted the program that its swift prosecution is improbable. And, of course, its capital cost will be as inhibiting for it as for any other major program.

The USAF—and all of the American armed services—must adapt to meet the new requirements of combat. Although it must maintain a capability to engage in a major war against a large nation-state, it must also be capable of dealing with lesser conflicts in rather more primitive countries, such as Afghanistan and Iraq. To do so, it has streamlined itself, adopted a new organization, and emphasized equipment using the latest technology. The U.S. Army has followed suit, if to a lesser extent.

The changes in the Air Force and the Army lend themselves to quick victories against an established military. However, those same changes are inimical to conducting what has been called "stability operations"; i.e., the maintenance of peace in the country after the war has been won. Forces designed to use high technology to win quick wars are not the same as those needed to patrol the back alleys of forlorn cities.

In the end, all of these challenges—new bombers, tankers, fighters, satellites, peacekeeping troops—depend upon an adequate budget for their solution. The question is whether or not the American public is prepared to pay for its own defense, and perhaps even more important, to personally participate in that defense.

It might be pertinent here to make another comparison to World War II. At that time, a nation of about 132,000,000 put 16,000,000 into uniform—or 12.8 percent. Fighting against Germany and Japan, neither one of which could actually do harm to the continental United States, America devoted 45 percent of its gross domestic product to defense. Now there are perhaps 310 million

Americans, and the armed forces are capped at about 2.5 million, including active duty, National Guard, and Reserve forces. That is only 0.8 percent of the population, which spends only about 4.0 percent of its gross domestic product on defense.

From these figures, our nation's enemies can hope that the United States does not have the will to defend itself. They must be proven wrong if our nation is to survive.

12

ASSESSING PROBLEMS AND
FACING CHALLENGES

*B*efore surveying the present status and future of the United States Air Force, it is instructive to look to the past to gain a perspective on the vision and the flexibility with which it has operated. All through its existence, the Air Force has conformed to the policies established by its civilian leadership. Successive Air Force leaders have undertaken the immediate defense of the nation while simultaneously planning for the almost infinite variety of combat scenarios that might emerge ten, twenty, or even more years in the future.

This placed inordinate demands upon both the military and civilian leaders of the Air Force, who each year had to balance the budget needed to solve pressing problems against the huge expenditures required to conduct the necessary research and development for the future. All too often the budget difficulties were partially resolved by postponing badly needed personnel benefits, deferring recapitalization, and maintaining equipment long after its useful life should have expired.

Solutions for future problems often had their roots in the past, as, for example, did tanker aircraft. First demonstrated in 1924, in-flight refueling languished until the postwar period when the short range of jet-powered bombers made the technique imperative. After much experimentation with its flying boom refueling system, Boeing provided first the piston-engine KC-97 and then the KC-135. The latter made in-flight refueling a standard, if never routine, operating practice.

The Stratotanker—as no one called it—entered service on June 18, 1957. Its ability to multiply forces became apparent quickly, and KC-135s were soon refueling fighters and transports. Not even its most ardent advocates at Boeing and the Strategic Air Command could ever have imagined that the aerial refueling tankers would still be the most important single weapons system in the Air Force's arsenal in the twenty-first century. Virtually no important operations, including those of the Army, Navy, and Marines, can take place without it.

Fortunately, the requirement to purchase large numbers of tankers came at the beginning of the Cold War, when the Air Force began an unparalleled peacetime expansion. Both the public and the Congress recognized the Communist threat and provided funding for radar warning systems, large fleets of bombers and fighters, and several generations of intercontinental ballistic missile systems.

A very few people, with then Major General Bernard A. "Bennie" Schriever first among them, were foresighted enough to see that the science and technology involved in the ICBM were the basis for a revolution in space warfare. The vast, expensive, and technologically demanding systems of satellites were introduced to provide quantum leaps in communication, intelligence gathering, meteorology, and navigation. Hotly debated at the time, these satellite systems are now crucial to all American military operations.

Unfortunately, the attitude of both the public and the Congress changed as the Cold War came to be perceived as a stalemate. The unpopular Vietnam War fostered opposition to defense spending. From the mid-1960s on it became increasingly difficult to obtain funding for new programs. This was particularly true of advanced programs whose details, and thus their rationale for existence, were so secret that they were considered "black" programs. Politicians, their fingers always to the wind, sensed the change in public opinion, and the more self-seeking were quick to voice opposition to programs from which results would not be obtained until many years in the future. A telling and typical example is the Boeing E-3 Sentry, an early "C-Cubed" (C^3: Command, Communication, and Control) aircraft.

Called AWACS, for Airborne Warning and Control Systems, the E-3 was essentially a modified Boeing 707 with a radar housed in a thirty-foot rotating dome. Intended to do exactly what its name implied, provide an airborne platform to warn of an approaching enemy and to control American assets, the AWACS was roundly derided as a waste of money. Critics claimed that it was, in a phrase popular at the time, a "billion-dollar blunder." The April 13, 1974, issue of *The New Republic* called the AWACS' mission "a complete phony" and described it as another ploy designed to keep hard-earned taxpayer dollars flowing to bloated defense contractors scamming the government. The article was titled "AWACS: The Plane That Would Not Die" and sneered at it as a "mushroom with elephantiasis."

The magazine had no understanding of the mission, the technology, or the requirement for the AWACS. Instead, it was exercising its routine knee-jerk reaction to military spending. Yet the media fueled opposition to the AWACS in Congress, where Colorado Representative Patricia A. Schroeder and Missouri Senator Thomas F. Eagleton teamed up to kill it. Fortunately, a congressional committee reviewed the system, and concluded that it would work and was worth funding.

The first AWACS aircraft was delivered on March 24, 1977, and has since proved to be the embodiment of the "high-demand, low-density" weapon. Had the critics investigated properly, they would have found that the AWACS' research, development, and production costs benefited many other weapon systems. Further, they would have perceived that it was a force multiplier that allowed a handful of aircraft to do the work of hundreds. An astute observer might have noted that the AWACS was infinitely more versatile and capable than the Chain Home defense system that had saved Great Britain during World War II, or the DEW Line upon which U.S. defenses had previously depended. It could not only provide warning of incoming threats, it could manage the assets available to meet those threats. Further, it was useful in tactical applications, serving as an airborne command-and-control center.

France, Saudi Arabia, the United Kingdom, and other members of the NATO alliance bought sufficient aircraft to increase the worldwide AWACS fleet to sixty-six aircraft. Further versions have been purchased of an AWACS based on the Boeing 767 airframe. The AWACS has earned its pay in conflicts around the world, including those in Panama, Grenada, Haiti, Kuwait, Kosovo, Serbia, Afghanistan, and Iraq. It was invaluable in both Gulf Wars.

But perhaps the most telling note on the AWACS and its short supply was that after September 11, 2001, no fewer than five NATO E-3 aircraft were deployed to the United States to monitor commercial traffic—a mission never foreseen by any planner, but one which captures the uncertainty of weapon system planning—and the hazards of budgetary shortfalls for necessary equipment.

Weapon systems still receive the same media wire-brush attention accorded the AWACS, and these include the Northrop Grumman E-8 Joint STARS (Surveillance Target Attack Radar System), the Lockheed Martin F-22A Raptor and F-35 Joint Strike Fighter, the Bell Boeing CV-22 Osprey, and a wide array of unmanned combat aerial systems (UCAS). All of these weapon systems must achieve their technological breakthroughs under continuous, critical political scrutiny.

Such adverse opposition by the media is often effective and is aided by a budgeting system that calls for the annual defense of programs, no matter where they are in their development cycle. Congressional staffs have mushroomed in size, and their interest is often less in the needs of national defense than in how a proposed weapon system affects the prosperity of the political base of the representative or senator for whom they work.

One additional aspect of this almost obsessive media and congressional review is that a single failure in testing can now doom a system. In the past, it was understood that experimental weapons were just that. When failures occurred they were analyzed and fixed for the production article. The incomparable

Corona program, the first of the intelligence-gathering satellites, suffered twelve straight failures before it had a successful flight. Now, however, a test failure is often seized upon as a reason for program cancellation.

All of these difficulties make it apparent that Air Force leaders have a doubly demanding task. They must have a vision of what will be required for the defense of the nation for many years into the future. At the same time, they must be proficient in the political gamesmanship necessary to shepherd the ideas of their predecessors through all the hazards into operational use. In effect, Air Force leaders must divide their attention and their efforts among three main issues:

1. Preserving the efficiency of legacy (a deftly spun euphemism for obsolete) equipment far beyond any reasonable expectation of their service life. At this moment, it translates into keeping forty-year-old bombers, tankers, and helicopters and thirty-year-old fighters effective, often by brilliant improvisation.

2. Defending the still-emerging ideas of their predecessors—even if they might not agree with them.

3. Conducting the research and development of advanced weapon systems that will be effective against a broad spectrum of future threats.

Weapon selection and procurement is but one aspect of the demands placed on the Air Force leaders. They must also routinely deal with doctrine, strategy, tactics, organization, training, and performance. What follows next is a synthesis of the thinking of many present and past Air Force leaders on these key subjects.

DOCTRINE

Current opinions on Air Force doctrine vary from flat assertions that "the Air Force has no doctrine" to well-thought-out expositions of what is perceived as official Air Force doctrine.

The confusion is of long standing. On September 21, 1961, General Curtis E. LeMay, then Chief of Staff, noted that "a very knowledgeable reporter stated recently that in the early 1950s, he felt he knew what the Air Force stood for, but today he doesn't. His statement puzzled me. It also alarmed me because understanding our doctrine and concepts is basic and important to our very existence."

LeMay's alarm was understandable, since Brigadier General William Mitchell had referred to "our doctrine of aviation" in 1920. In 1943, the seminal Army Field Manual 100-20 acknowledged that "land power and airpower are co-equal and interdependent forces; neither is an auxiliary of the other."

Air Force doctrinal manuals have had a long and checkered history. The very first one came in 1921 in the form of a manual intended for instruction at the Air Service Field Officers School. The manual called for developing a doctrine which would underlay all the principles of employment of the Air Force.

But it was not until April 1, 1953, that General Hoyt S. Vandenberg approved Air Force Manual 1-2, commenting that "the dynamic and constant changes in new weapons make periodic review of this doctrine necessary." AFM 1-2 was revised in April 1955 and still reflects lessons learned in World War II.

By 1959 responsibility for doctrine had shifted from the Air University at Maxwell Air Force Base to the Air Staff in the Pentagon, and the version published that year reflected the contemporary opinion that "the best preparation for a limited war is proper preparation for a general war." Participation in the ground-bound constrained conflict in Southeast Asia was only two years away when this was written.

Even the bitter experience of Vietnam did not change things. Like Korea, Vietnam was written off as a doctrinal anomaly. The focus of Air Force doctrine was set on nuclear deterrence. For years, what passed for doctrine was a collection of proven or "best" practices, instead of a leading strategy. It evolved continuously, striving to keep up with technology and, ideally, to remain just a little in the forefront. Unfortunately, the acceleration of technology in recent years made this difficult in theory and almost impossible in practice, for the very process of creating and approving documents defining doctrine takes longer than the development and employment of some new systems.

By the 1980s a struggle for control of Air Force doctrine had developed between the Air Staff and the Air University (AU). The balance began to shift when the Center for Aerospace Doctrine, Research, and Education (CADRE) was officially opened on January 3, 1983, by Lieutenant General Charles G. Cleveland, just after he assumed command of the Air University. The consolidation of many agencies already in existence at the AU, CADRE was designed to be an air-division-level organization dedicated to improving the AU's contribution to the development of ideas, concept, and doctrine.

Yet confusion on doctrinal issues continued, so much so that Dr. James A. Mowbray commented in the Winter 1995 *Airpower Journal* that USAF doctrine suffered from four essential problems:

1. The Air Force had neglected airpower theory as a basis for its mission or purpose. This neglect impaired the ability of the Air Force to write sound doctrine, particularly operational doctrine.
2. The Air Force needed an established and institutionalized process for the development and transmission of basic and operational-level doctrine.
3. The Air Force feared to commit itself in doctrine to more than that which in fact it could deliver.
4. The Air Force was obsessed with winning budget battles for hardware without the underpinning of airpower theory.

The continuing struggle to establish doctrine was resolved in August 1996, when General Fogleman established the Air Force Doctrine Center (AFDC), reporting directly to him as Chief of Staff. It was Fogleman's desire to highlight the growing importance of joint war fighting, and to establish a single clear voice for matters of doctrine.

Major General Ronald E. Keys established the new center at Maxwell Air Force Base and assumed command in February 1997. The right man for the job, Keys had long maintained a "little red book" in which ideas on doctrine were entered, and this subsequently became the basis for the development of the first manuals from AFDC. (Another important entry in Keys's book was the modern concept of an operational air-control center [OAC]).

AFDC provided a renewed focus and commitment to doctrine, with emphasis on teaching it at the lower levels of command. In the foreword to the current version of Air Force Basic Doctrine (AFDD-1), published in November 2003, General Jumper wrote, "As airmen, we have not properly understood or consistently applied our air and space doctrine. As great operators we have preferred to improvise over sound, repeatable principles."

AFDD-1 takes an evenhanded approach, applying to both global and local levels of conflict. It maintains the importance of air and space power (note: not aerospace power) and states that the 1991 Gulf War proved that airpower can be a maneuver element in its own right, coequal with land and maritime power. It underscores the cardinal characteristics of airpower—speed, range, and firepower, which can turn the opening phase of a conflict into the decisive phase. Yet it emphasizes the importance of obtaining the desired national objectives no matter which forces are involved.

Some recent developments will ultimately find their way into later editions of AFDD-1. Lieutenant General David E. Deptula cites two important changes. The first of these is the efficiency and the effect of the Air Expeditionary Force (AEF). He notes that "the AEF construct is how we prepare and then deploy our forces on a rotational basis to ensure the right force is sent to the right place at the right time. It also helps ensure we don't overstretch our people or our equipment by providing the appropriate reconstitution time."

General Deptula also notes that the second important doctrinal addition is the establishment of the Warfighting Headquarters (WFHQ). Under the WFHQ concept, the Unified Combatant Commanders will be provided with a dedicated capability designed to support the Joint Force Commander across the full spectrum of military operations.

While there is still some speculation as to the direction the WFHQ will take and the importance it will assume, it is a major factor in current planning.

The WFHQ had as its original impetus the need for an organization to exist with sufficient time and resources to plan and execute air and space operations at the operational level of war. It is purposely insulated from organizing, training, and equipping forces. These functions are handled by the major commands. Warfighting headquarters are intended to plan and execute air and space power more efficiently. The warfighting headquarters will have a command section, and air and space air operations center, and sufficient staff to interface with the combatant commander's staff, the Air Staff, other services, and Air Force major commands. It will be positioned above the wing level, and eventually supplant some numbered air forces.

The effort to define doctrine is ongoing. And just as agreement on and postulation of doctrine is difficult, so is it difficult to define strategy for the United States Air Force.

STRATEGY

The strategy of the United States Air Force has changed over the years as the service struggled to adapt to changing national policies on defense. It should be noted that warfighting is now recognized as a joint responsibility and that the Air Force's role is to organize, train, and equip air and space forces, not to fight them. This was not always the case—and some wish it were not the case now, for the Air Force role in joint operations is too often not given the recognition due it.

In 1947 the perceived nuclear capability of the new United States Air Force was seen as an offset to the preponderance of military power enjoyed by the Soviet Union.

This strategy was consistent with the philosophy of winning wars by aerial bombardment that had been the heart and soul of the Air Corps and the US-AAF, and was adhered to during World War II, Korea, Vietnam, and the Cold War.

The goal of nuclear deterrence—sometimes termed "counterforce"—was the destruction of enemy military capability, rather than its civilian population. As the capability of the Soviet Union grew, the destruction of Soviet nuclear capability assumed first priority.

By 1963 the strength of the Soviet Union's nuclear forces was such that the United States was forced to adopt a policy termed "countervalue," which became known as "assured destruction." In this strategy, the United States, after absorbing a nuclear strike, would still have enough retaliatory power to destroy the population of the Soviet Union. In time, the term became "mutual assured destruction" which gave the uncannily appropriate acronym MAD to the idea.

As the years passed, the strength of the Soviet Union grew both in terms of numbers and quality of offensive and defensive missile systems. The United States allowed the Soviet Union to gain a critical first-strike capability that was deterred only by the Soviet fear of the Strategic Air Command and the growing threat of submarine-launched ballistic missiles. Increasingly impoverished, the Soviet Union in 1972 agreed to the idea of strategic arms limitation, with President Richard Nixon and Premier Leonid Brezhnev signing the SALT I treaty in 1972.

In 1974 Secretary of Defense James R. Schlesinger warned that the Soviet Union had developed a very dangerous first-strike capability. It was becoming evident that the Soviets were planning on fighting, winning, and surviving a nuclear war. Recognition of this led first to a strategic concept called "essential equivalence" between U.S. and Soviet forces—the world turned upside down from earlier eras. This led to yet another strategy, that of "limited nuclear options," which forswore massive retaliation for the destruction of military capability.

Next on the flip-flop list of national defense policies came the "countervailing strategy" of the Carter administration, which still acknowledged mutual assured destruction as a policy, but offered other options as well.

A seemingly jocular but truly incisive remark by President Ronald Reagan set the tenor for the next approach to strategy. When asked in 1983 what his policy on the Soviet Union was, he replied that some might think it simplistic, but that it was "We win, they lose."

Reagan appointed a bipartisan committee headed by Brent Scowcroft to examine the requirements of a strategic force. The commission concluded that the Soviet Union possessed sufficient ICBMs of high reliability, accuracy, and warhead yield to destroy all of the ICBM silos of the United States, and still have an adequate supply of weapons remaining. This was contrary to both the policy and the capability of the United States, which could not initiate a first strike and take out the Soviet ICBM force.

The security of the United States was maintained by its strategic triad of land-based ICBMs, bombers, and submarine-launched ballistic missiles. They provided a combination that offset the numerical and qualitative preponderance of Soviet ICBMs and ICBM launching silos.

Reagan abandoned the concept of strategic parity and began a huge rearmament program. He also began the Strategic Defense Initiative (SDI), which was taken seriously by the leaders of the Soviet Union despite widespread opposition and derision as "Star Wars" within the United States.

The SDI was ultimately canceled but it had already served its purpose, for the ultimate objective of Reagan's policy was shrouded by the defense buildup. His real intent was to spend the Soviet Union into bankruptcy as they struggled to catch up. He succeeded brilliantly. The Soviet Union's belief in the possibility of the SDI, its concern over burgeoning U.S. military strength, and its ever deteriorating economy resulted in the fulfillment of Reagan's "We win, they lose" idea.

Technical advances permitted refining the Star Wars concept. The old idea of its being a defensive anti-ballistic missile shield was scrapped. Instead, a new system of warfare was developed that combined satellites, intelligence-gathering equipment, and new operational techniques such as effects-based operations and network-centric warfare. The latter requires the identification of critical target sets and the most vital nodes of those sets. The new ideas were demonstrated in the 1991 Gulf War, and were perhaps the final push that sent the Soviet Union tumbling into Reagan's "ash bin of history."

Many leading thinkers in the Air Force believe that both doctrine and strategy were optimized in the motto *Global Reach, Global Power* enunciated by the Secretary of the Air Force Donald B. Rice in 1990. Under the *Global*

Reach, Global Power concept the unique characteristics of the Air Force—speed, range, flexibility, precision, and lethality—were the vital underpinnings of national security. The Air Force contributed mightily to national defense with the following capabilities:

1. Continuing to deter any nuclear conflict, with its bombers and missiles constituting two legs of the strategic triad.
2. Projecting power around the world, and being able to swiftly and effectively conduct theater operations on a sustained basis.
3. Having the rapid global mobility inherent in an ample fleet of airlift aircraft and tankers to permit sustained operations by all the services.
4. Continuing to control "the high ground" with its space and command-and-control systems.
5. Building the influence of the United States by strengthening its security partners, and building relationships through training, combined operations, and compassionate relief missions.

The basic concept of *Global Reach, Global Power* remains inherent in the ability of the United States Air Force to bring a variety of forces rapidly to bear on challenges across the spectrum of contingencies in a way that no other service can match. The Air Force's global approach to strategy has been enhanced by its ability to hit carefully selected targets in adverse weather conditions using GPS equipment. The USAF can now base more of its capability within the United States and still be sure that it can deter aggression, defeat enemies, or rapidly provide compassionate relief to places in need.

As noted earlier, this very capability has had an unfortunate budgetary byproduct. The very fact that the USAF can do these things with a fraction of the resources required by the other service components to do anything similar means that it is often taken for granted in times of need but ignored during budget debates. The example of air refueling is but one example. Every mission—whether combat or compassionate, whether conducted by the Army, Navy, or Marines—requires a robust Air Force refueling capability. Yet there is no support among those services during budgetary debates on the reequipping of the air-refueling fleet.

Another difficulty arises for the USAF from the current national philosophy of inflicting minimum collateral damage. The success of the Air Force in demonstrating its skill with precision weapons has led to a public expectation of clinically clean wars, untrammeled by casualties. As a result, a single error in execution can have enormous political consequences that may negate otherwise sterling work.

In retrospect, the strategy of the United States Air Force was for decades inextricably linked to a changing series of national policies which made massive nuclear deterrence the inevitable last resort. Fortunately, the incredible technological advances made in the last twenty years have given the USAF such enormous capability that national policies can be much more flexible, even as a nuclear deterrent is maintained against possible eventualities. However, no matter how advanced the USAF's technological leap may be made, it is almost worthless unless there are concomitant advances in training and tactics.

TRAINING AND TACTICS

The important advances in Air Force equipment (discussed below) have imposed tremendous requirements on both training and tactics.

The computer is not only the heart and soul of the advances in equipment, it is also at the center of those in training and tactics. Nowhere is this better demonstrated than in the easy translation from Distributed Mission Training (DMT) to Distributed Mission Operations (DMO).

The concept of DMT has gained influence within the Air Force since 1998. In DMT, simulated combat, cyberspace collaboration, and realistic virtual representations provide aerial warfighters with team training from remote locations while still interacting in real time in a synthetic battlespace. Using high-speed communication equipment, simulators are linked together so that pilots at distant locations can engage in amazingly realistic combat.

The transition to DMO comes readily, with the simulator training proving not only valuable, but in some instances being psychically more demanding than real combat. Being able to link geographically separated units in air operations saves time, effort, and resources, and is equally valuable to real-life experiences. For example, F-15 pilots in simulators at Elmendorf Air Force Base interact with AWACS controllers in a simulator at Tinker Air Force Base, with the same training results that an actual mission would have conferred.

The next step in both DMT and DMO is to introduce the processes at the interagency level, so that each of the services understands the value of the others' capabilities. Within the Air Force, there has been a large-scale removal of compartmentalized security restrictions (colloquially, "opening green doors"), particularly in such major exercises as Red Flag. Doing the same thing on an interservice basis will be more difficult, but if accomplished, will make mission and campaign planning much more effective because of the increased knowledge of the real capabilities resident in the joint forces.

DMT and DMO add to the proficiency gained in joint and coalition operations where there are large air force exercises. The proficiency is enhanced by a better integration of air and space assets, greater use of night vision devices (DVD) and the avoidance of low-level tactics by using medium-altitude drops with precision-guided munitions (PGM). The use of PGMs by the B-52 has introduced a new concept of close air support, one that is not accepted by all, but which promises to give the Buff and the Bone an even longer lease on their already incredibly long lives as first-line equipment.

But advances in computers are not the only way to improve training. Retired Chief of Staff General Merrill A. McPeak has noted two rather basic but fundamentally important contributions to training. The first of these was placing combat crew training in the Air Education and Training Command (AETC), thus making it a departmental responsibility to provide mission-ready aircrews to the using commands. The second was the realignment of and strengthening of enlisted training, making it more formal and rigorous. This was accompanied by the reduction of the number of occupational specialties, broadening the usefulness of individual airmen.

Changes in training are more easily translated into changes in tactics because the Air Force has been continuously engaged in combat at varying levels of intensity since 1991. But changes in political attitudes have come to affect tactical considerations as they formerly affected strategic considerations.

Perhaps most onerous of these, particularly from a budgeting standpoint, is that air superiority has changed in the minds of many planners from being a campaign objective to being a "given" planning assumption. This places the Air Force in an invidious position, for there is a constant expectation that it will furnish adequate assets on demand, without any regard to the harm done to training, morale, and efficiency by an excessively high operations tempo.

Tactics are also constrained by a new circumstance which was previously never much considered in planning a war. The importance of avoiding collateral damage has gone from a mitigation issue to an overriding planning input. Every isolated tactical event now has major strategic implications for the efforts of a joint force if a mistake is made.

The very nature of the Air Force and its weaponry has caused most of the changes in tactics. It must now plan for two extremes. On the one hand, it must be able to conduct operations in high-density threat areas over hardened targets. On the other, it must be able to conduct persistent operations over soft, mobile target sets. These extremes must be met with the same set of increasingly complex weapons and aircraft systems. Fortunately, these systems are versatile and permit mission planners to envision the number of strikes per sortie instead of sorties per strike. Yet more targets per sortie poses a difficult problem in battling an insurgency, where the location and identity of the targets are uncertain. It helps that targets can now be attacked in all weather conditions at night with precision-guided munitions.

Precision-guided munitions, introduced in Vietnam, but continuously refined since, are but a part of the broad array of modernized legacy and new equipment that forms the Air Force arsenal today.

EQUIPMENT

The tightening of the sensor-to-shooter loop has been mentioned previously. It lies at the heart of the Air Force's challenge to meet the demands of the future against an enemy who has resorted to improvised explosive devices as his primary weapon. The new combat question about equipment is not whether a fighter is the fastest in the world, or a bomber the most efficient, but whether modern technology, with its stealth, precision, and expense can defeat the primitive tactics of terrorists using cell phones to blow up wedding parties.

The USAF's ultimate goal in the war against individual terror bombers must be a satellite or aircraft (especially Unmanned Aerial Systems) with the capability to detect and kill individual homicide bombers and improvised explosive device fabricators at the moment they gather up their explosive cache. This sounds unachievable, but anything short of this will mean a continuation of the present era of random terror attacks.

The array of USAF equipment currently employed both as a nuclear deterrent and in the war on terror is both extensive and remarkable. While some of it is brand-new, an even greater proportion is "legacy equipment" updated with new capabilities, as the following examples show.

1. UNMANNED AERIAL SYSTEMS (UAS)

Formerly called Unmanned Aerial Vehicles (UAVs), the UAS perform multiple functions. These include sensor operations, communications relay, and serving as an attack platform. These broad capabilities allow UAS to provide real-time information on a global basis. The General Atomics Aeronautical Systems MQ-1 Predator is a medium-altitude, long-endurance, remotely piloted aircraft. Predators armed with Hellfire missiles have reduced the killchain time to a minimum. The MQ-1's primary mission is interdiction and conducting armed reconnaissance against critical, perishable targets. When the MQ-1 is not actively pursuing its primary mission, it acts as the Joint Forces Air Component Commander–owned theater asset for reconnaissance, surveillance, and target acquisition in support of the Joint Forces commander.

The Northrop Grumman RQ-4 Global Hawk's high-altitude, long-range, and endurance capability provides near-real-time, high-resolution intelligence, surveillance, and reconnaissance imagery. The Global Hawk provided Air Force and joint warfighting commanders with tens of thousands of images in support of Operation Enduring Freedom.

Many observers believe that UAS and Unmanned Combat Aerial Systems (UCAS) technology is in its infancy and will soon make significant advances. When the UAS achieve their full potential, they will be employed on a spectrum of missions in which their enhanced endurance, persistence, range, and low radar signature will be invaluable, and are not achievable with conventional inhabited craft. UAS will keep aircrew members out of harm's way. The future will probably see formations of UAS being controlled by a single pilot based thousands of miles away.

There are also indications that the success of some prototype UCAS (e.g., variations of the Boeing X-45A) has been such that they have been withdrawn from public view and placed in "black" programs for further development.

2. PRECISION-GUIDED MUNITIONS (PGM)

Precision-guided munitions, particularly the joint direct attack munition (JDAM) take away the enemy's ability to use bad weather as a sanctuary. They are a boon to ground forces because they confer real potency on long-loiter-time platforms like the B-52. In Operation Iraqi Freedom, the Air Force's capability to bomb during heavy sandstorms gave ground forces close air support when ground maneuver was stalled by bad weather. They are also, compared to other advances, relatively inexpensive to develop and manufacture.

3. ADVANCES IN NET-CENTRIC CAPABILITIES

The development of net-centric capabilities permits a common operating picture and a distributed situational awareness. Much written about and much discussed, Net-Centric Warfare (NCW) requires a change in thinking about mission planning. A new approach must be made to organization, particularly the interrelationship among all echelons and at all levels of warfare. With these new concepts, modern equipment enables networked units and individual platforms to function together in ways that were impossible before. One of the main characteristics is the ability of forces, widely dispersed geographically, to have a common awareness of the battlespace. NCW links people, platforms, sensors, and weapons into networked forces that are faster and more synchronized in their efforts, and are thus able to achieve the effects of a massed force with far fewer numbers. Another important characteristic is the effect of information moving down through the ranks, which allows decisions to be made at a lower level and enhances the responsibilities of junior officers and noncommissioned officers.

When such network capabilities are combined with GPS and communication satellites, Blue Force Tracking (friendly force tracking) is vastly improved and the chances of fratricide much reduced. During the major combat phase of Operation Iraqi Freedom in March and April of 2003, it was clear that the change in tactics had reduced fratricide. Only 11 percent of the 115 U.S. battle deaths in OIF were caused by fratricide. In Operation Desert Storm, 35 of the 148 U.S. battle deaths (24 percent) were so caused. Net-centric capability has also improved Combat Search and Rescue (CSAR).

4. COUNTER-SPACE CAPABILITIES

During Operation Iraqi Freedom U.S. forces used space systems for communication, navigation, weather forecasting, targeting, and intelligence-surveillance reconnaissance functions. The satellites greatly reduced the number of personnel required in the theater because they allowed reachback capability to the databases of experts in the United States.

Potential enemies know that the United States depends upon its satellite systems. Therefore the United States must be able to defend these systems from enemy attacks, and also be able to prevent a potential enemy from developing comparable systems. There is concern that a rogue state such as North Korea or Iran, knowing that the loss of our satellite systems would severely handicap, if not negate, our principal weapons, might elect to attack them.

There are at least three systems under development for defending our satellite systems. They include the space-based surveillance system (SBSS) which is an orbiting telescope able to watch objects in space with greater clarity than ever before. The second system is the orbital deep-space imager (ODSI). Like the SBSS, it will be a telescope system, but a free-ranging one that will be able to take pictures of objects of interest, including any damaged U.S. satellite. The third element of this space triad is the rapid attack identification detection and reporting system (RAIDRS). RAIDRS is not a separate spacecraft but rather a program to develop tools which will recognize an attack on a satellite.

5. NEW AND IMPROVED AIRCRAFT

Stealth is a huge advantage to the United States, as has been amply demonstrated by the Lockheed Martin F-117A and the Northrop Grumman B-2A. The next generation of stealth aircraft built upon a strong base. First among these is the Lockheed Martin F-22A Raptor. The Raptor is a fifth-generation stealth aircraft that combines stealth, supercruise, agility, and integrated avionics in a manner that ensures air dominance for the next fifty years. It is a quantum advance over the remarkable Boeing F-15 Eagle, which has dominated the fighter scene for more than thirty years.

The Raptor's stablemate is the Lockheed Martin F-35 Joint Strike Fighter (JSF), the weapon system of the Navy, Air Force, Marines, and our allies for the next generation. Intended to replace the Lockheed Martin F-16 and A-10, the F-35 brings cutting-edge technologies to the battlespace of the future. The JSF draws heavily on—and contributes to the development of—the F-22A's capabilities.

These two truly transformational systems will impart a huge advantage to the United States' capability for asymmetric warfare. Unfortunately, the breadth and depth of these transformational characteristics are not recognized as such by many key decision-makers because their high level of technological prowess is difficult to comprehend. Those same bureaucrats who control the budget also cannot comprehend who the potential enemies might be.

While both the Raptor and the Joint Strike Fighter have enjoyed (or suffered) a great deal of controversy, neither have experienced the white heat of criticism directed toward the Bell/Boeing CV-22 Osprey. This joint-service, medium-lift, multimission, tilt-rotor aircraft is designed to have the maneuverability and lift capability of a helicopter combined with the speed of a conventional aircraft. It is intended for long-range clandestine penetration of enemy territory by Special Operations Forces, and ultimately, for combat search-and-rescue operations as well.

Among the improved aircraft, and after years of calumnious attacks, the Boeing C-17A Globemaster III has so distinguished itself in combat that there are calls being heard in Congress for additional procurement. This need is exacerbated by the retirement of the Lockheed Martin C-141 and the aging of the Lockheed Martin C-5 fleet. Able to deliver troops and large cargo elements to forward bases, the C-17A can also perform tactical airlift and airdrop missions.

The Northrop Grumman B-2A Spirit is a multirole stealth bomber capable of delivering conventional, precision-guided, and nuclear munitions. Like the C-17A, it proved itself in combat to the discomfort of many of its detractors. The incredible range and accuracy of the B-2A has been repeatedly demonstrated in the Middle East, and questions about its vulnerability to weather and wear have been answered positively.

These weapon systems are joined by an array of other assets that include everything from the venerable B-52 to the advanced versions of the E-8 Joint STARS, along with the proven capability of the A-10, F-15, F-16, C-130, and so on. Each of the weapon systems—no matter how old—is benefiting from developments in modern targeting systems and in ever more accurate precision-guided munitions, including the JDAM, the joint stand off weapon (JSOW) and the joint air-to-surface standoff missile (JSSAM).

The targeting systems include the Litening, a precision targeting pod system that significantly increases the combat effectiveness of the aircraft during day, night, and under-the-weather conditions. Similar capability is found in the Sniper XR pod. Both of these pods, and other systems as well, allow maximum use to be made of the capabilities of improved precision-guided munitions.

It is sad to relate that the overall effectiveness of the Air Force has declined in one area, that of its organic tactical electronic capability. With the demise of the Northrop Grumman EF-111A and the Boeing (McDonnell Douglas) F-4G, no replacement has been offered. Instead, reliance has been placed on the increasingly overworked Northrop Grumman EA-6B aircraft, which is both out of production and in short supply. This is both inexplicable and inexcusable. The arrival of Boeing EF-18 Growler aircraft may offer some relief in the future.

Changes in equipment and the addition of new missions inevitably require changes in organization. The USAF has responded rather presciently in recent years to the need for organizational change, despite the pain often involved, and at this writing is engaged in yet another full-scale review that might see as many as five major commands eliminated in an effort to free up no less than 40,000 man power spaces.

ORGANIZATION

Even before the triumphant conclusion of the first Gulf War and the subsequent disappearance of the Soviet Union, the USAF had determined that future budgets would be cut and that a substantial reduction in size would be necessary. As has previously been noted, a general reorganization in 1992 saw the establishment of the current Air Combat Command and Air Mobility Command.

It was also recognized that instead of operating from forward bases as a garrison force, the United States would operate in the future for the most part as an expeditionary force. This gave rise to the creation of the Air Expeditionary Force concept, which was intended to smooth out the force structure and stabilize the tempo of operations. Unfortunately, the pellmell press of real-world events has worked against this. While the AEF is still the best management tool available, the hard fact is that the Air Force is undersized for the many missions laid upon it, and the AEF concept has been somewhat distorted to meet the new requirements.

An absolutely vital part of Air Force organization will be the optimization of the Reserve and the National Guard as part of the Total Force concept. The present limits on the size of the Air Force dictate that full use be made of the incredible amount of unique talent and experience resident in the Guard and the Reserve. Yet these forces must be protected from abuse, including too frequent calls to extended periods of active duty. Some of the solutions will be difficult to agree upon, for they will involve a change of functions from operating first-line fighters to duties which can be done from stateside facilities, such as operating UAS vehicles. It is in the interest of all parties to get the issues resolved as soon as possible.

Sometimes important matters of organization are in the details. The Air Force for years had difficulty sizing the pilot force. It was difficult to predict how long people would stay in the Air Force at any given time, for the service commitment after pilot training was too short. Sophisticated aircraft and tactics meant that a pilot required a year of training and four years of experience to become productive. With a five-year commitment in force, pilot retention was subject to external vagaries, including the attraction of the civil airline industry. This made it difficult to maintain a balance in the experience of the pilot force as a whole.

General Ryan sought to extend the commitment to ten years after graduation from pilot training, but met resistance from those who felt that such a requirement would inhibit people from applying for pilot training. He pushed the idea through, and there was absolutely no decline in applications. The ten years was regarded by potential pilots as a reasonable return on investment by the Air Force. The effect was to stabilize the rate of pilot training, improve retention, and smooth out the experience curve of Air Force pilots.

In general, the Air Force has reorganized often to maximize the use of its diminishing resources. No matter what changes in organization might occur in the future, the priority will always be given to maintaining offensive and defensive strength to meet all emergencies. The pessimistic forecasts of future budgets indicate that meeting the new challenges will depend upon extracting more productivity from an already extended force. The public and the Congress must become aware that this time-honored process of demanding ever more from a shrinking workforce, even though it has worked so well for so long, must ultimately have practical limits.

With strong doctrine, excellent strategy, unparalleled equipment, and suitable organization, the United States Air Force is in a position to conduct ever more sophisticated and effective operations. The organization is at this writing in a state of flux, in anticipation of further cuts in personnel, but the resulting structure will be tailored to meet all requirements.

OPERATIONS

As with doctrine, strategy, and tactics, one must first look to the past to understand current and future operations. In this case, a swift return is warranted to the 1991 Gulf War, where the usefulness of Effect-Based Operations (EBO) was widely perceived. A review of the Gulf War tells us that it began with attacks on more targets in one day than the entire Eighth Air Force hit in all of 1942 and 1943. Moreover, it was not the number of first-day air attacks that was important, but rather the fact that these attacks were planned to achieve specific effects. They were designed to paralyze Saddam Hussein's control of his forces, neutralize both the will and the capacity of those forces to fight, and eliminate its ability to build what proved to be the elusive weapons of mass destruction. This type of operation became known as "parallel warfare" and was based upon achieving specific effects, not absolute destruction of target systems. Unlike previous wars, the focus was not on attrition and annihilation, but rather on achieving specific effects, in this case, the removal of Iraqi forces from Kuwait.

Contemporary technological advances, including the use of nonlethal weapons and improved information operations, enhanced the potential for EBO in the second Gulf War, where the array of space-based systems and the precision operation of the Combined Air Operations Center made it even more efficient. The ability to get the right information to the right decision-makers in the shortest possible time proved invaluable.

The combination of such factors as the Air Expeditionary Force and the confluence of technology represented by USAF C⁴ISR, the CAOC, and other modern means of effecting EBO give the United States a capability that truly fulfills the 1990s promise of *Global Reach, Global Power*.

However, reduced budgets and awareness of increased capability has forced a reliance on a "silver bullet" force, dependent upon our stealth and space-based systems. Unfortunately, these space-based systems and the electronic networks upon which we currently depend are relatively fragile and are insufficiently hardened against an enemy employing an electromagnetic pulse weapon (EMP) against us.

It is necessary to create hardened and redundant space systems, but this will be enormously expensive, and it remains to be seen if the public and the Congress will be sufficiently farsighted to provide funds to prevent an EMP catastrophe.

The advent of some of the technological advances has caused some unexpected side effects on operations, some of them critically important, others merely interesting. Among the critically important, the USAF has experienced a reduction in its electronic combat capability that must be addressed, despite the expense. It has elected to reduce its ICBM forces to a degree that many feel will be inadequate against an emerging power such as China. USAF capability has grown to the extent that it can now conduct operations that require ground forces in support, rather than the reverse. In contrast, fighting irregular wars has placed a great demand on its transport capability at a time when it does not have an adequate number of heavy lifters, nor does it have much prospect of gaining more.

Less crucial but still interesting, the concept of creating "centers of excellence" has spread, and its general effect has been to raise performance standards at every level. USAF airmen are now doing nontraditional roles as a part of the AEF, supporting ground operations at an increasing rate. Simulation and war-gaming are gaining greater currency, and are actually borrowing from the techniques created by commercial simulator game companies. In a similar way, independent think tanks and commercial companies have grown up which specialize in network-centric, effects-based operations and tactical data-link consultations. One of the requirements for these commercial ventures stems from the increase in the utilization of reachback techniques, by which databases in the United States become available to frontline operations.

The techniques established in the first Gulf War were honed in the successive operations of the intervening decade, and were precision-sharp all through operations in the second Gulf War. In the long period after the end of the major combat phase, there has been an inevitable recognition of the effects of irregular warfare and a shift of emphasis to combating it. All of the advances depended upon reaping the rewards of research and development, which will be critically more important in the future as emerging powers accelerate their own technologies.

It is to the credit of the United States that the magnificent military apparatus it has created for defense is applied with such unprecedented generosity to all nations when a natural disaster strikes. One of the most telling examples was in Operation Unified Assistance, the massive relief efforts mounted in December 2004 in response to the South Asian tsunami. The operation demonstrated the efficiency of Air Force and land-based air operations compared to other services, and to sea-based air. Land-based aircraft moved more than 230 percent of the cargo moved by sea-base aircraft, using less than 15 percent of the personnel.

Within seventy-two hours of the disaster, U.S. aircrews were delivering the first load of relief supplies to Thailand as well as ferrying U.S. assessment teams throughout the region to apply triage to the suffering. In this forty-seven-day effort, U.S. aircraft moved an average of about 270 tons of relief-related cargo per day. Over 3,400 sorties were flown, resulting in logging more than a million miles flown across this broad region. This 125,000-ton effort involved over one hundred fixed and rotary-wing aircraft from the Air Force, Marines, Navy, Special Operations, and Coast Guard. The average number of airlift relief tons flown per day during OUA exceeded that of any other humanitarian assistance or disaster relief operation since the Berlin Airlift. The operation was distinguished by the first use of a Joint Forces Air Component Commander in the Pacific Command. Using a hub-and-spoke approach, supplies were smoothly transferred from the heavy airlifters such as the Lockheed Martin C-5 to the intratheater tactical airlifters, the Lockheed Martin C-130s, and thence to rotary-wing aircraft for transport to the areas needing assistance.

These successful disaster relief operations have many implications. One of these is that disaster relief helps counter terrorist ideology. The shared experiences in the tsunami relief efforts helped the United States and Indonesia repair damaged relations both at a formal level, and, informally, in the eyes of Indonesian citizens. In a similar way, the swift aid to Pakistan after the tragic 2005 earthquake had measurable effect upon the opinions of Pakistani citizens. And, not least, the unplanned but swift response of the military to relief efforts of Hurricane Katrina were not only widely applauded, they were seen as an essential part of future planning for disaster relief.

RESEARCH AND DEVELOPMENT

Many of the results of the Air Force's previous expensive, controversial, and demanding research-and-development efforts have already been discussed. But future Air Force research and development is being squeezed relentlessly by four factors.

The first of these is the universal problem of a shrinking budget. The second is the inescapable fact that future developments in technology are more difficult to achieve and hence far more expensive. The third, mentioned once again for emphasis, is that the public and the Congress have become intolerant of failure, and any setback, even a minor one, can precipitate a program cancellation. And the fourth is perhaps most crucial, for it is the declining American industrial base in an era of globalization. We are producing far too few engineers in the aerospace disciplines, in part because of a consolidation (some might say implosion) of the aerospace industry.

With this in mind it is interesting to contemplate what the major challenges in future research and development may be. Many are so highly classified that they remain undisclosed, but there are a number under discussion that represent a breathtaking array of possibilities. They include: the use of lasers for tracking and communications and as a directed-energy weapon; breakthroughs in micro- and nanotechnology; creation of the evolved expendable launch vehicle; increased emphasis on hypersonic flight; the space-based infrared system; the space-based tracking and surveillance system; stealthy, unmanned combat aerial systems; new combat systems, including manned and unmanned bombers; and increased airlift capability, including a stealth airlifter. Each of these would require a book of its own for a full explanation of its capabilities.

Unfortunately, as beckoning as these possibilities are, they represent a breathtaking budget, and the future of the United States as a nation clearly depends upon its willingness to undertake the financial burden, and the risk, of a substantial research and development program. Charged with this task, it is only fair to consider some mistakes that the Air Force may have, in the opinion of some, made in the past.

OPINIONS ON AIR FORCE MISTAKES

Polling Air Force leaders and critics on the mistakes the Air Force may or may not have made has led to some contradictory conclusions. This is not surprising, given the passion with which these points of view are defended, but the items in the following baker's dozen of perceived errors are both insightful and interesting, if not all necessarily correct.

1. The Air Force has not been sufficiently skillful in articulating the significance, efficiency, and effectiveness of air and space power. As a result, it has suffered in the budget battles.

2. The Air Force has tended to "go along to get along" with the other services, and has not been effective in getting the proper priorities from Congress. It has tried to avoid being seen as "too air-centric." An example of this was the self-imposed cut of one-hundred F-22As from the 1997 Quadrennial Defense Review, when none of the other services took a similar cut in force structure.

3. In contrast to the above, others maintain that the Air Force believed its own press reports after the first Gulf War. It pushed its successes in the faces of too many people in too many places, forgetting the old Washington adage related to "conservation of enemies."

4. The Air Force allowed the acquisition oversight process to fall into the hands of too few individuals who abused their positions and caused the tanker leasing fiasco. This reflected adversely upon the Secretary of the Air Force and the Air Staff. Just as the Dreyfus Affair will forever embarrass the French Army, so will the Darleen Druyan Affair be a source of shame to the USAF.

5. The Air Force has forsaken long-term strategy for the masterful execution of day-to-day operations.

6. The Air Force failed to embrace space activities, arguing that the mission was too expensive for the service to pay the bill. The Air Force has never aggressively argued the case for space as it does for air.

7. The Air Force was not as active as it should have been in cultivating the intelligence community, especially on space-related issues.

8. The Air Force failed to focus on some key areas, including human intelligence (HUMINT) and electronic warfare, especially the suppression of enemy air defenses.

9. The Air Force failed to accelerate development of the F-22, letting "the best be the enemy of the good" concept rule.

10. The Air Force has failed to reap the fruits of its investment in hypersonic research.

11. The Air Force has failed to adequately invest in the high-demand, low-density platforms such as JSTARS, AC-130s, and others.

12. The Air Force has contracted out too much important work, depending upon private firms for essential tasks.

13. The emphasis on obtaining the F-22A is a colossal mistake, given potential foreign opposition, and is one that has driven almost all other Air Force funding considerations.

There is no consensus on these errors, but they certainly offer an abundance of problems to think about. Given that some of them might be valid criticisms, and considering all the previous successes of the Air Force, it is fair now to assess what the mission of the Air Force should be.

MISSION STATEMENT

There are a wide variety of opinions on the mission of the United States Air Force. (Technically of course, the Air Force has no mission, but rather a set of functions. These are to organize, train, and equip air and space forces for employment by joint commanders.) The concept of the Air Force mission arouses strong emotions among its leaders, as the following (sometimes redundant) summations indicate:

The mission of the Air Force is

1. Providing air superiority, precision engagement, information superiority, global attack, rapid global mobility, agile combat support, and command and control.
2. Preparing and conducting global air and space operations to assure joint force victory in support of national security objectives.
3. Ensuring dominance of the air and space medium for global awareness, access, and control.
4. Organizing, training, and equipping air forces for prompt, sustained operations, including combat, intelligence, and transport.
5. Maintaining persistent C^4 global mobility and rapid strike capability.
6. To fly, fight, and win.
7. To achieve and maintain air and space superiority.
8. Providing global transparency and attack in real time on any target on, above, or below the surface of the earth.
9. Providing air *dominance*, not air superiority; with air dominance, all other missions of all other services are possible. Without it, nothing is possible.

Two words—"global" and "joint"—illustrate concepts that are explicit in many of the expressions of the Air Force mission, and are implicit in the others.

There is at this writing an official expression of the Air Force mission, issued on December 7, 2005, by Secretary of the Air Force Michael W. Wynne and Air Force Chief of Staff General T. Michael Moseley, in a joint letter to airmen.

After indicating that the mission of the Air Force is its guiding compass, which needs to be clearer than ever before, they defined the Air Force mission as follows:

The mission of the United States Air Force is to deliver sovereign options for the defense of the United States of America and its global interests—to fly and fight in Air, Space, and Cyberspace.

It is not an exaggeration to say that many readers were startled, for two new concepts were introduced, "sovereign options" and "cyberspace."

The term "sovereign options" was defined as being the essence of a superpower. The letter noted, "Our task is to provide the National Command Authority and the combatant commanders with an array of options that are not limited by the tyranny of distance, the urgency of time, or the strength of our enemy's defenses."

The letter went on to say that "With one hand the Air Force can deliver humanitarian assistance to the farthest reaches of the globe, while with the other hand we can destroy a target anywhere in the world."

The term "cyberspace" was defined as including network security, data transmission, and the sharing of information. Secretary Wynne added, "The Air Force is a natural leader in the cyber world, and we thought it would be best to recognize that talent."

Their letter further noted, "Adversaries of the United States will use any method or venue necessary to contest America, and it is an Airman's calling to dominate air, space and cyberspace."

The new mission statement met with mixed (albeit informally expressed) reviews. The idea of delivering "sovereign options" rather than bombs did not sit well with many, although it was obvious that the central mission of the Air Force was maintained in the familiar words "to fly and fight."

But some, on consideration, saw in the words "sovereign options" an implicit acknowledgement that in the event of a world crisis, the first and telling response would be delivered by the Air Force. Further analysis led to the conjecture that the term actually embodied something of a threat—that "sovereign options" hinted at the immense power the Air Force has at its disposal, but has refrained from using for political reasons.

Months pass between finishing a book and its publication, and it may be that the mission of the Air Force will be restated in that interval. But for now, the official mission statement, thoughtfully contemplated, retains all of the confidence in maintaining air and space superiority of the past together with a powerful suggestion of what it might mean in the future to any aggressor.

THE AIR FORCE AND THE FUTURE

Predicting the future is notably difficult. In 1987, no one predicted—or would have any reason to have predicted—any of the following: the fall of the Berlin Wall, the reunification of Germany, the collapse of the Soviet Union, the first Gulf War, the advent of terrorism, two wars in the Balkans, the invasion of Afghanistan, and the two wars in Iraq.

Looking twenty years into the future now is equally difficult. The Air Force must be prepared for all eventualities. It will be able to do so because of its unique characteristics of speed, range, flexibility, precision, and lethality. These make it possible for the Air Force to operate at all levels of warfare simultaneously. Air and space power will continue to determine the nature of America's response to future attacks. The basic missions will not change, but the capabilities of the systems fulfilling those missions will change and perhaps drastically.

Air and space power will continue to be a "combat enabler," permitting the execution of joint operations. It will also continue to be a "combat disenabler," damaging enemy forces until they are ineffective. (It currently possesses an operational excellence that it often accomplishes both objectives simultaneously at the start of operations.)

This capability is especially crucial in the global war on terror (GWOT). In previous wars, finding and identifying adversaries was relatively easy— killing them was the hard part. In the GWOT this is reversed; finding and identifying the enemy is incredibly difficult, but once found, killing them is not difficult. The Air Force cannot focus only on the GWOT, but must be prepared for other events, including the emergence of another, possibly hostile, superpower in the future. China and Russia both have sufficient nuclear weapons to obliterate the United States.

As in previous conflicts, the role of the USAF will depend on the situation. There will be instances where the Air Force is the primary service, supported in a joint operation. In other instances, the Air Force will support the efforts of the Special Forces. But in order to remain relevant and in order to win the vital budgetary battles that enable relevance, the Air Force itself must lead with ideas. It must go well beyond the idea of being a force application provider, and must harness its technology to do so.

There are those who believe that the GWOT has been unnecessarily prolonged by the failure to use the full array of weapons at the disposal of the United States to force Muslim national states to take responsibility for the actions of the radical fundamentalist Muslims. This scenario might result in extensive casualties in the Muslim world, but might be effective in deterring the use of weapons of mass destruction in the United States.

On the other hand, many people believe that the war on terror will not be won by military power. Their view is that the GWOT is not military in nature, but more about politics, sociology, economics, ethics, and deep-rooted religious beliefs. To win it, new political, social, and economic tools will have to be developed. Thus it may be that in addition to maintaining its effectiveness in all combat regimes, the Air Force will have to use its immense communication resources to develop an entirely new weapon, something that might be called a "truth bomb." The Air Force would be a perfect instrument to change the attitude of the hostile Muslim world by transmitting masses of accurate information to each and every individual Muslim. To have any hope of success, it must generate an information explosion of educative material to overcome centuries of cultural stagnation. The Muslim world must be made increasingly aware of the difference between the United States as portrayed in the mosques and the United States as it really is.

CONCLUSION

The United States Air Force has advanced almost as much in the ten years from 1997 to 2007 as it had previously advanced in the fifty years from 1947 to 1997.

Most of the advances derive from changes in technology, but none of them could have been supported without the incredible talent resident in the men and women of the Air Force. This includes those in the Reserve and Air National Guard, operating in concert with the active-duty Air Force.

In the last decade, more and more demands for flexibility have been made on all ranks within the Air Force, which is blessed by having one remarkable generation after another of young people committed to the service of our country. A change to this might have been expected given the increased operational tempo caused by the war in Iraq and the global war on terror. It has not. One of the great incentives to stay in the Air Force has been the increased recognition and responsibility given to the noncommissioned officer corps. And an interesting, and to some an unbelievable, facet of the new flexibility is that civilians are being increasingly integrated into combat operations, even contractor personnel.

The future is unpredictable, but some things are relatively certain. The responsibilities of the United States Air Force will increase at a swift rate, and Air Force personnel will discharge those responsibilities with skill and dedication. No matter what lies ahead, the people of the United States Air Force, in concert with the nation's other military services, will do the planning, improvising, sacrificing, fighting, and, when necessary, dying to keep our country free.

APPENDIX ONE

SECRETARIES OF THE AIR FORCE

CHIEFS OF STAFF OF THE AIR FORCE

CHIEF MASTER SERGEANTS OF THE AIR FORCE

SECRETARIES OF THE AIR FORCE

Name	Dates in Office	
	From	To
Stuart Symington	September 18, 1947	April 24, 1950
Thomas K. Finletter	April 24, 1950	January 20, 1953
Harold E. Talbott	February 4, 1953	August 13, 1955
Donald A. Quarles	August 15, 1955	April 30, 1957
James H. Douglas, Jr.	May 1, 1957	December 10, 1959
Dudley C. Sharp	December 11, 1959	January 20, 1961
Eugene M. Zuckert	January 24, 1961	September 30, 1965
Harold Brown	October 1, 1965	February 15, 1969
Robert C. Seamans, Jr.	February 15, 1969	May 14, 1973
John L. McLucas (Acting)	May 15, 1973	July 18, 1973
John L. McLucas	July 18, 1973	November 23, 1975
James W. Plummer (Acting)	November 24, 1975	January 1, 1976
Thomas C. Reed	January 2, 1976	April 6, 1977
John C. Stetson	April 6, 1977	May 18, 1979
Hans Mark (Acting)	May 18, 1979	July 26, 1979
Hans Mark	July 26, 1979	February 9, 1981
Verne Orr	February 9, 1981	November 30, 1985
Russell A. Rourke	December 9, 1985	April 7, 1986
Edward C. Aldridge, Jr. (Acting)	April 8, 1986	June 8, 1986
Edward C. Aldridge, Jr.	June 9, 1986	December 16, 1988
James F. McGovern (Acting)	December 16, 1988	April 29, 1989
John J. Welch, Jr. (Acting)	April 29, 1989	May 21, 1989
Donald B. Rice	May 22, 1989	January 20, 1993
Michael B. Donley (Acting)	January 20, 1993	July 13, 1993
Gen. Merrill A. McPeak (Acting)	July 14, 1993	August 5, 1993
Sheila E. Widnal	August 6, 1993	October 31, 1997
F. Whitten Peters (Acting)	November 1, 1997	July 30, 1999
F. Whitten Peters	July 30, 1999	January 20, 2001
Lawrence J. Delaney (Acting)	Jan 20, 2001	June 1, 2001
James G. Roche	June 1, 2001	January 20, 2005
Peter B. Teets (Acting)	January 20, 2005	March 25, 2005

Name	Dates in Office From	To
Michael L. Dominquez (Acting)	March 25, 2005	July 29, 2005
Preston M. Geren (Acting)	July 29, 2005	November 3, 2005
Michael W. Wynne	November 3, 2005	

CHIEFS OF STAFF OF THE AIR FORCE

Name	Dates in Office From	To
Gen. Carl A. Spaatz	September 26, 1947	April 29, 1948
Gen. Hoyt S. Vandenberg	April 30, 1948	June 29, 1953
Gen. Nathan F. Twining	June 30, 1953	June 30, 1957
Gen. Thomas D. White	July 1, 1957	June 30, 1961
Gen. Curtis E. Lemay	June 30, 1961	January 31, 1965
Gen. John P. McConnell	February 1, 1965	July 31, 1969
Gen. John D. Ryan	August 1, 1969	July 31, 1973
Gen. George S. Brown	August 1, 1973	June 30, 1974
Gen. David C. Jones	July 1, 1974	June 20, 1978
Gen. Lew Allen, Jr.	July 1, 1978	June 30, 1982
Gen. Charles A. Gabriel	July 1, 1982	June 30, 1986
Gen. Larry D. Welch	July 1, 1986	June 30, 1990
Gen. Michael J. Dugan	July 1, 1990	September 17, 1990
Gen. John M. Loh (Acting)	September 18, 1990	October 29, 1990
Gen. Merrill A. McPeak	October 30, 1990	October 25, 1994
Gen. Ronald R. Fogleman	October 26, 1994	September 1, 1997
Gen. Ralph E. Eberhart (Acting)	September 2, 1997	October 5, 1997
Gen. Michael E. Ryan	October 6, 1997	September 6, 2001
Gen. John P. Jumper	September 6, 2001	September 2, 2006
Gen. T. Michael Moseley	September 2, 2006	

CHIEF MASTER SERCEANTS OF THE AIR FORCE

Name	Dates in Office From	To
CMSAF Paul W. Airey	April 3, 1967	July 31, 1969
CMSAF Donald L. Harlow	August 1, 1969	September 30, 1971
CMSAF Richard D. Kisling	October 1, 1971	September 30, 1973
CMSAF Thomas N. Barnes	October 1, 1973	July 31, 1977
CMSAF Robert D. Gaylor	August 1, 1977	July 31, 1979
CMSAF James M. McCoy	August 1, 1979	July 31, 1981

| | Dates in Office | |
Name	From	To
CMSAF Arthur L. Andrews	August 1, 1981	July 31, 1983
CMSAF Sam E. Parish	August 1, 1983	June 30, 1986
CMSAF James C. Binnicker	July 1, 1986	July 31, 1990
CMSAF Gary R. Pfingston	August 1, 1990	October 25, 1994
CMSAF David J. Campanale	October 26, 1994	November 4, 1996
CMSAF Eric W. Benken	November 5, 1996	July 30, 1999
CMSAF Frederick J. Finch	July 30, 1999	July 1, 2002
CMASF Gerald R. Murray	July 1, 2002	June 30, 2006
CMASF Rodney J. McKinley	July 1, 2006	

APPENDIX TWO

GUARD, RESERVE, AIR FORCE ACADEMY, AND CIVIL AIR PATROL

AIR NATIONAL GUARD (ANG)

The Air National Guard has its headquarters in Washington, D.C. The lineage of the ANG stretches through the Army National Guard all the way to 1636 and the establishment of the Massachusetts National Guard. The first aviation element of the Guard was established on August 2, 1909, when the Missouri National Guard created a fifteen-man aero-detachment. Guard air units were federalized for the Mexican expedition of 1916 and have participated in all major conflicts since that time. However, it was not until 1920 that aviation units of the National Guard were formally established and recognized, with two units in place by 1921. By World War II there were twenty-nine National Guard observation squadrons; all were called to active duty by October 1941. After World War II, National Guard air units were formed with a much broader span of duties, but were primarily equipped with fighters. By 1949, there were 514 individual ANG units equipped with 2,263 aircraft. But, as one observer of the scene said in 1950, the United States had forty-nine air forces—the USAF and forty-eight Air National Guard air forces, one for each state. Sixty-six ANG units were recalled for the Korean War, in which they served with distinction. The standards of equipment and training for the ANG were continuously improved during the Cold War years, so that it was able to make a substantial contribution to the war effort in Vietnam. One direct result of this performance was the incorporation of the Air National Guard into the "Total Force" policy under which ANG, Reserve, and regular Air Force units all trained on similar equipment to the same standards of proficiency. At the present time, the ANG has eighty-eight wings, assigned to Air Combat Command, Air Education and Training Command, Air Force Special Operations Command, Air Mobility Command, and Pacific Air Forces. It has approximately 1,200 aircraft and provides the USAF 100 percent of the fighter interceptor force, 45 percent of tactical airlift, 43 percent of KC-135 air refueling, 33 percent of fighters, 28 percent of rescue, and 8 percent of strategic airlift. The ANG flies the Lockheed Martin C-5A, C-141, and C/HC/EC-130 transports; the Boeing KC-135 tanker; the Fairchild A/OA-10A attack aircraft; the Rockwell B-1B bomber; and the McDonnell Douglas F-15s and Lockheed Martin F-16 fighters.

U.S. Air Force Reserve (USAFR)

The U.S. Air Force Reserve has its headquarters at Robins Air Force Base, Georgia. It traces its origins to the National Defense Act of 1916, which established a Reserve Corps of 2,300 officers and men. It remained a part of the organization of the antecedents of the USAF, and in 1950 it was placed under the Continental Air Command (CONAC). The USAFR became a separate operating agency (SOA) in 1968, and is now a field operating agency It has three numbered air forces, the 4th at McClellan AFB, California; the 10th at Bergstrom Air Reserve Station, Texas; and the 22nd at Dobbins Air Reserve Base, Georgia. There are thirty-seven flying wings operating about 470 aircraft, including the Boeing B-52 H bomber; the Lockheed Martin F-16 fighter: the Fairchild A/OA-10 attack aircraft; the Lockheed Martin C-5A/B, C-141B, and C-130E-H air lifters; the Boeing KC-135 tanker; the Lockheed Martin HC-130H and Sikorsky HH 60G rescue aircraft; and the Lockheed Martin WC-130 H weather and MC-130E special operations aircraft. In its early post–World War II years, the Air Force Reserve was poorly equipped; during the Korean War it had received Curtiss C-46 and Douglas C-47 transports along with Douglas B-26 light bombers. Modernization was slow until after the Vietnam War, when a surplus of equipment made it possible to equip reserve units with Cessna A-37 and Republic F-105 attack aircraft, McDonnell Douglas F-4 and Lockheed Martin F-16 fighters, several versions of the Lockheed Martin C-130 transports, Boeing KC-135 tankers, and Sikorsky H-3 Jolly Green Giant helicopters.

As a part of the Total Force, ten Reserve units participated with distinction in the Persian Gulf War.

U.S. Air Force Academy (USAFA)

The Air Force Academy is located at Colorado Springs, Colorado. Its mission is to develop and inspire air and space leaders for the future, to produce dedicated Air Force officers and leaders, and to instill that leadership through academics, military training, athletic conditioning, and spiritual and ethical development. Because of its specialized nature, it is a direct reporting unit (DRU), reporting to Headquarters USAF. Appointment to the Academy is made by congressional sponsor or by meeting eligibility requirements in other competitive methods. It was established on April 1, 1954, and the first class of 306 cadets entered in July 1955 at a temporary location at Lowry Air Force Base, Colorado. The present complex was completed by August 1958, in time for the first class's graduation in 1959. In 1996, 1,218 cadets entered the program. Over the years, approximately one-third of the cadets leave the program, 75 percent by resignation. About 60 percent of the graduates go on to pilot training. Women entered the Academy as cadets in 1976; almost 3,000 had entered by 1996, with an average graduation rate of about 61 percent. Cadets complete four years of study to obtain a Bachelor of Science degree. The total cadet enrollment is about 4,000, and 1,288 officers, 1,114 enlisted personnel, and 1,861 civilians are required to operate the Academy. An intensive flying familiarization program is given, using ninety-five aircraft of several different types, including gliders.

THE CIVIL AIR PATROL

The Civil Air Patrol has a long and distinguished tradition that reaches back to its founding on December 1, 1941. During World War II, the CAP allowed private pilots and aviation enthusiasts to use their skills in civil defense efforts. It came under the control and direction of the United States Army Air Forces in 1943, and became a permanent peacetime institution on July 1, 1946, when President Harry S. Truman established it as a federally chartered, benevolent, civilian corporation with Public Law 476. In May, 1948, the CAP became an official auxiliary of the United States Air Force.

The mission of the CAP is aerospace education, cadet training, and emergency services. Its members fly 80 percent of the search-and-rescue mission hours directed by the Air Force Rescue and Coordination Center at Langley Air Force Base, Virginia.

Since 1985, it has assisted the U.S. Customs Service in its counter-drug efforts by flying air reconnaissance missions along U.S. boundaries, and now works with the Drug Enforcement Administration and the U.S. Forest Service in a similar capacity.

Membership consists of 19,000 cadets and 34,000 adult volunteers organized into fifty-two wings—one for each state, the District of Columbia, and Puerto Rico. The CAP's members operate more than 5,000 privately owned aircraft and 530 CAP aircraft.

APPENDIX THREE

SIGNIFICANT TERRORIST INCIDENTS, 1961–2003: A BRIEF CHRONOLOGY

This document, based entirely on public sources, was prepared for background information and reference purposes. It is not intended to be a complete or comprehensive account of all terrorist incidents during these years, and it is not an official expression of U.S. policy. Please e-mail questions or comments to History@State.gov.

Courtesy:
Office of the Historian
Bureau of Public Affairs
U.S. Department of State
March 2004
http://www.state.gov/r/pa/ho/pubs/fs/5902.htm

1961–1982
May 1, 1961, First U.S. Aircraft Hijacked. Puerto Rican–born Antuilo Ramierez Ortiz forced at gunpoint a National Airlines plane to fly to Havana, Cuba, where he was given asylum.

August 28, 1968, Ambassador to Guatemala Assassinated. U.S. Ambassador to Guatemala John Gordon Mein was murdered by a rebel faction when gunmen forced his official car off the road in Guatemala City and raked the vehicle with gunfire.

July 30, 1969, Ambassador to Japan Attacked. U.S. Ambassador to Japan A. H. Meyer was attacked by a knife-wielding Japanese citizen.

September 3, 1969, Ambassador to Brazil Kidnapped. U.S. Ambassador to Brazil Charles Burke Elbrick was kidnapped by the Marxist revolutionary group MR-8.

February 10, 1970, Attack on the Munich Airport. Three terrorists attacked El Al passengers in a bus at the Munich Airport with guns and grenades. One passenger was killed and 11 were injured. All three terrorists were captured by airport police. The Action Organization for the Liberation of Palestine and the Popular Democratic Front for the Liberation of Palestine claimed responsibility for the attack.

July 31, 1970, U.S. Agency for International Development Adviser Kidnapped. In Montevideo, Uruguay, the Tupamaros terrorist group kidnapped AID police adviser Dan Mitrione; his body was found on August 10.

July 21, 1972, "Bloody Friday." An Irish Republican Army (IRA) bomb attack killed 11 people and injure 130 in Belfast, Northern Ireland. Ten days later, three IRA car bomb attacks in the village of Claudy left six dead.

September 5, 1972, Munich Olympic Massacre. Eight Palestinian "Black September" terrorists seized eleven Israeli athletes in the Olympic Village in Munich, West Germany. In a bungled rescue attempt by West German authorities, nine of the hostages and five terrorists were killed.

March 2, 1973, Ambassador to Sudan Assassinated. U.S. Ambassador to Sudan Cleo A. Noel and other diplomats were assassinated at the Saudi Arabian embassy in Khartoum by members of the Black September organization.

May 4, 1973, Consul General in Mexico Kidnapped. U.S. Consul General in Guadalajara Terrence Leonhardy was kidnapped by members of the People's Revolutionary Armed Forces.

December 17, 1973, Attack and Hijacking at the Rome Airport. Five terrorists pulled weapons from their luggage in the terminal lounge at the Rome airport, killing two persons. They then attacked a Pan American 707 bound for Beirut and Tehran, destroying it with incendiary grenades and killing 29 persons, including 4 senior Moroccan officials and 14 American employees of ARAMCO. They then herded five Italian hostages into a Lufthansa airliner and killed an Italian customs agent as he tried to escape, after which they forced the pilot to fly to Beirut. After Lebanese authorities refused to let the plane land, it landed in Athens, where the terrorists demanded the release of two Arab terrorists. In order to make Greek authorities comply with their demands, the terrorists killed a hostage and threw his body onto the tarmac. The plane then flew to Damascus, where it stopped for two hours to obtain fuel and food. It then flew to Kuwait, where the terrorists released their hostages in return for passage to an unknown destination. The Palestine Liberation Organization disavowed the attack, and no group claimed responsibility for it.

August 19, 1974, Ambassador to Cyprus Assassinated. U.S. Ambassador to Cyprus Rodger P. Davies and his Greek Cypriot secretary were shot and killed by snipers during a demonstration outside the U.S. Embassy in Nicosia.

January 27–29, 1975, Domestic Terrorism. Puerto Rican nationalists bombed a Wall Street bar, killing 4 and injuring 60; two days later, the Weather Underground claims responsibility for an explosion in a bathroom at the U.S. Department of State in Washington.

June 27, 1976, Entebbe Hostage Crisis. Members of the Baader-Meinhof Group and the Popular Front for the Liberation of Palestine (PFLP) seized an Air France airliner and its 258 passengers. They forced the plane to land in Uganda. On July 3 Israeli commandos successfully rescued the passengers.

September 21, 1976, Assassination of Former Chilean Diplomat. Exiled Chilean Foreign Minister Orlando Letelier was killed by a car bomb in Washington.

March 16, 1978, Kidnapping of Italian Prime Minister. Premier Aldo

Moro was seized by the Red Brigades and assassinated 55 days later.

February 14, 1979, Ambassador to Afghanistan Assassinated. Four Afghans kidnapped U.S. Ambassador Adolph Dubs in Kabul and demanded the release of various "religious figures." Dubs was killed, along with four alleged terrorists, when Afghan police stormed the hotel room where he was being held.

November 4, 1979, Iran Hostage Crisis. After President Carter agreed to admit the Shah of Iran into the U.S., Iranian radicals seized the U.S. Embassy in Tehran and took 66 American diplomats hostage. Thirteen hostages were soon released, but the remaining 53 were held until their release on January 20, 1981.

November 20, 1979, Grand Mosque Seizure. Two hundred Islamic terrorists seized the Grand Mosque in Mecca, Saudi Arabia, taking hundreds of pilgrims hostage. Saudi and French security forces retook the shrine after an intense battle in which some 250 people were killed and 600 wounded.

August 31, 1981, U.S. Installation Bombing. The Red Army Faction exploded a bomb at the U.S. Air Force Base at Ramstein, West Germany.

October 6, 1981, Assassination of Egyptian President. Soldiers who were secretly members of the Takfir Wal-Hajira sect attacked and killed Egyptian President Anwar Sadat during a troop review.

December 4, 1981, Murder of Missionaries. Three American nuns and one lay missionary were found murdered outside San Salvador, El Salvador. They were killed by members of the National Guard, and the killers are currently in prison.

September 14, 1982, Assassination of Lebanese President. President Bashir Gemayel was assassinated by a car bomb parked outside his party's Beirut headquarters.

1983
April 8, 1983, Colombian Hostage-Taking. A U.S. citizen was seized by the Revolutionary Armed Forces of Colombia (FARC) and held for ransom.

April 18, 1983, Bombing of U.S. Embassy in Beirut. Sixty-three people, including the CIA's Middle East director, were killed and 120 were injured in a 400-pound suicide truck-bomb attack on the U.S. Embassy in Beirut, Lebanon. The Islamic Jihad claimed responsibility.

May 25, 1983, Naval Officer Assassinated in El Salvador. A U.S. Navy officer was assassinated by the Farabundo Martí National Liberation Front.

October 9, 1983, North Korean Hit Squad. North Korean agents blew up a delegation from South Korea in Rangoon, Burma, killing 21 persons and injuring 48.

October 23, 1983, Bombing of Marine Barracks, Beirut. Simultaneous suicide truck-bomb attacks were made on American and French compounds in Beirut, Lebanon. A 12,000-pound bomb destroyed the U.S. compound, killing 242 Americans, while 58 French troops were killed when a 400-pound device destroyed a French base. Islamic Jihad claimed responsibility.

November 15, 1983, Naval Officer Assassinated in Greece. A U.S. Navy officer was shot by the November 17 terrorist group in Athens while his car was stopped at a traffic light.

1984

March 16, 1984, Kidnapping of Embassy Official. The Islamic Jihad kidnapped and later murdered Political Officer William Buckley in Beirut, Lebanon. Other U.S. citizens not connected to the U.S. government were seized over a succeeding two-year period.

April 12, 1984, Restaurant Bombing in Spain. Eighteen U.S. servicemen were killed and 83 people were injured in a bomb attack on a restaurant near a U.S. Air Force Base in Torrejón, Spain.

June 5, 1984, Temple Seizure. Sikh terrorists seized the Golden Temple in Amritsar, India. One hundred people died when Indian security forces retook the Sikh holy shrine.

October 31, 1984, Assassination of Indian Prime Minister. Premier Indira Gandhi was shot to death by members of her security force.

1985

February 7, 1985, Kidnapping of U.S. Officials in Mexico. Under the orders of narcotrafficker Rafael Caro Quintero, Drug Enforcement Administration agent Enrique Camarena Salazar and his pilot were kidnapped, tortured, and executed.

June 14, 1985, TWA Hijacking. A Trans-World Airlines flight was hijacked en route to Rome from Athens by two Lebanese Hezbollah terrorists and forced to fly to Beirut. The 8 crew members and 145 passengers were held for seventeen days, during which one American hostage, a U.S. Navy sailor, was murdered. After being flown twice to Algiers, the aircraft was returned to Beirut after Israel released 435 Lebanese and Palestinian prisoners.

June 19, 1985, Attack on a Restaurant in El Salvador. Members of the FMLN (Farabundo Martí National Liberation Front) fired on a restaurant in the Zona Rosa district of San Salvador, killing four Marine security guards assigned to the U.S. Embassy and nine Salvadoran civilians.

June 23, 1985, Air India Bombing. A bomb destroyed an Air India Boeing 747 over the Atlantic, killing all 329 people aboard. Both Sikh and Kashmiri terrorists were blamed for the attack. Two cargo handlers were killed at Tokyo airport, Japan, when another Sikh bomb exploded in an Air Canada aircraft en route to India.

September 30, 1985, Soviet Diplomats Kidnapped. In Beirut, Lebanon, Sunni terrorists kidnapped four Soviet diplomats. One was killed but three were later released.

October 7, 1985, *Achille Lauro* Hijacking. Four Palestinian Liberation Front terrorists seized the Italian cruise liner in the eastern Mediterranean Sea, taking more than 700 hostages. One U.S. passenger was murdered before the Egyptian government offered the terrorists safe haven in return for the hostages' freedom.

November 23, 1985, Egyptian Airliner Hijacking. An EgyptAir airplane bound from Athens to Malta and carrying several U.S. citizens was hijacked by the Abu Nidal Group.

December 27, 1985, Airport Attacks in Rome and Vienna. Four gunmen belonging to the Abu Nidal Organization attacked the El Al and Trans World Airlines ticket counters at Rome's Leonardo da Vinci Airport with grenades and automatic rifles. Thirteen persons were killed and 75 were wounded before Italian police and Israeli security guards killed three of the gunmen and captured the fourth. Three more Abu Nidal gunmen

attacked the El Al ticket counter at Vienna's Schwechat Airport, killing three persons and wounding 30. Austrian police killed one of the gunmen and captured the others.

1986
March 30, 1986, Aircraft Bombing in Greece. A Palestinian splinter group detonated a bomb as TWA Flight 840 approached Athens airport, killing four U.S. citizens.

April 5, 1986, Berlin Discothèque Bombing. Two U.S. soldiers were killed and 79 American servicemen were injured in a Libyan bomb attack on a nightclub in West Berlin, West Germany. In retaliation U.S. military jets bombed targets in and around Tripoli and Benghazi.

September 14, 1986, Kimpo Airport Bombing. North Korean agents detonated an explosive device at Seoul's Kimpo airport, killing 5 persons and injuring 29 others.

1987
April 24, 1987, Bus Attack. Sixteen U.S. servicemen riding in a Greek Air Force bus near Athens were injured in an apparent bombing attack carried out by the revolutionary organization known as November 17.

November 29, 1987, Downing of Airliner. North Korean agents planted a bomb aboard Korean Air Lines Flight 858, which subsequently crashed into the Indian Ocean.

December 26, 1987, Servicemen's Bar Attack. Catalan separatists bombed a Barcelona bar frequented by U.S. servicemen, resulting in the death of one U.S. citizen.

1988
February 17, 1988, Kidnapping of William Higgins. U.S. Marine Corps Lt. Col. W. Higgins was kidnapped and murdered by the Iranian-backed Hezbollah group while serving with the United Nations Truce Supervisory Organization (UNTSO) in southern Lebanon.

April 14, 1988, Naples USO Attack. The Organization of Jihad Brigades exploded a car bomb outside a USO Club in Naples, Italy, killing one U.S. sailor.

June 28, 1988, Attack on U.S. Diplomat in Greece. The defense attaché of the U.S. Embassy in Greece was killed when a car bomb was detonated outside his home in Athens.

December 21, 1988, Pan Am 103 Bombing. Pan American Airlines Flight 103 was blown up over Lockerbie, Scotland, by a bomb believed to have been placed on the aircraft by Libyan terrorists in Frankfurt, West Germany. All 259 people on board were killed.

1989

April 21, 1989, Assassination of U.S. Army Officer. The New People's Army (NPA) assassinated Colonel James Rowe in Manila. The NPA also assassinated two U.S. government defense contractors in September.

September 19, 1989, Bombing of UTA Flight 772. A bomb explosion destroyed UTA Flight 772 over the Sahara Desert in southern Niger during a flight from Brazzaville to Paris. All 170 persons aboard were killed. Six Libyans were later found guilty in absentia and sentenced to life imprisonment.

November 30, 1989, Assassination of German Bank Chairman. The Red Army Faction assassinated Deutsche Bank chairman Alfred Herrhausen in Frankfurt.

1990

January 15, 1990, U.S. Embassy Bombed in Peru. The Tupac Amaru Revolutionary Movement bombed the U.S. Embassy in Lima, Peru.

May 13, 1990, U.S. Soldiers Assassinated in the Philippines. The New People's Army (NPA) killed two U.S. Air Force personnel near Clark Air Force Base in the Philippines.

1991

January 18–19, 1991, Attempted Iraqi Attacks on U.S. Posts. Iraqi agents planted bombs at the home residence of the U.S. Ambassador to Indonesia's and at the USIS library in Manila.

February 13, 1991, Sniper Attack on the U.S. Embassy in Bonn. Three Red Army Faction members fired automatic rifles across the Rhine River at the U.S. Embassy chancery. No one was hurt.

May 21, 1991, Assassination of Former Indian Prime Minister. A female member of the Liberation Tigers of Tamil Eelam killed herself, Prime Minister Rajiv Gandhi, and 16 others by detonating an explosive vest after presenting a garland of flowers to the former prime minister during an election rally in the Indian state of Tamil Nadu.

1992

January 17–21, 1992, Kidnapping of U.S. Businessmen in the Philippines. A senior official of the corporation Philippine Geothermal was kidnapped in Manila by the Red Scorpion Group, and two U.S. businessmen were seized independently by the National Liberation Army and by Revolutionary Armed Forces of Colombia (FARC).

March 17, 1992, Bombing of the Israeli Embassy in Argentina. Hezbollah claimed responsibility for a blast that leveled the Israeli Embassy in Buenos Aires, Argentina, causing the deaths of 29 and wounding 242.

1993

January 31, 1993, Kidnappings of U.S. Citizens in Colombia. FARC terrorists kidnapped three U.S. missionaries.

February 26, 1993, World Trade Center Bombing. The World Trade Center in New York City was badly damaged when a car bomb planted by Islamic terrorists exploded in an underground garage. The bomb left six people dead and a thousand injured. The men carrying out the attack were followers of Umar Abd al-Rahman, an Egyptian cleric who preached in the New York City area.

April 14, 1993, Attempted Assassination of Former President Bush by Iraqi Agents. The Iraqi intelligence service attempted to assassinate former U.S. President George Bush during a visit to Kuwait. In retaliation, the U.S. launched a cruise missile attack two months later on the Iraqi capital Baghdad.

1994
February 25, 1994, Hebron Massacre. Jewish right-wing extremist and U.S. citizen Baruch Goldstein machine-gunned Moslem worshippers at a mosque in West Bank town of Hebron, killing 29 and wounding about 150.

September 23, 1994, FARC Hostage-taking. FARC rebels kidnapped U.S. citizen Thomas Hargrove in Colombia.

December 24, 1994, Air France Hijacking. Members of the Armed Islamic Group seized an Air France flight to Algeria. The four terrorists were killed during a rescue effort.

1995
March 8, 1995, Attack on U.S. Diplomats in Pakistan. Two unidentified gunmen killed two U.S. diplomats and wounded a third in Karachi, Pakistan.

March 20, 1995, Tokyo Subway Station Attack. Twelve persons were killed and 5,700 were injured in a Sarin nerve-gas attack on a crowded subway station in the center of Tokyo. A similar attack occurred nearly simultaneously in the Yokohama subway system. The Aum Shinri-kyo cult was blamed for the attacks.

April 19, 1995, Bombing of the Federal Building in Oklahoma City. Right-wing extremists Timothy McVeigh and Terry Nichols destroyed the Federal Building in Oklahoma City with a massive truck bomb that killed 166 and injured hundreds more in what was up to then the largest terrorist attack on American soil.

July 4, 1995, Kashmiri Hostage-Taking. In India six foreigners, including two U.S. citizens, were taken hostage by Al-Faran, a Kashmiri separatist group. One non-U.S. hostage was later found beheaded.

August 21, 1995, Jerusalem Bus Attack. Hamas claimed responsibility for the detonation of a bomb that killed six and injured over 100 persons, including several U.S. citizens.

September 13, 1995, Attack on U.S. Embassy in Moscow. A rocket-propelled grenade was fired through the window of the U.S. Embassy in Moscow, ostensibly in retaliation for U.S. strikes on Serb positions in Bosnia.

November 13, 1995, Saudi Military Installation Attack. The Islamic Movement of Change planted a bomb in a Riyadh military compound that killed one U.S. citizen, several foreign nationals employed by the U.S. government, and over 40 others.

November 19, 1995, Egyptian Embassy Attack. A suicide bomber drove a vehicle into the Egyptian embassy compound in Islamabad, Pak- istan, killing at least 16 and injuring 60 persons. Three militant Islamic groups claimed responsibility.

1996

January 8, 1996, Papuan Hostage Abduction. In Indonesia, 200 Free Papua Movement (OPM) guerrillas abducted 26 individuals in the Lorenta nature preserve, Irian Jaya Province. Indonesian Special Forces members rescued the remaining nine hostages on May 15.

January 19, 1996, Kidnapping in Colombia. Revolutionary Armed Forces of Colombia (FARC) guerrillas kidnapped a U.S. citizen and demanded a $1 million ransom. The hostage was released on May 22.

January 31, 1996, Tamil Tigers Attack. Members of the Liberation Tigers of Tamil Eelam rammed an explosives-laden truck into the Central Bank in the heart of downtown Colombo, Sri Lanka, killing 90 civilians and injuring more than 1,400 others, including two U.S. citizens.

February 9, 1996, IRA Bombing. An Irish Republican Army (IRA) bomb detonated in London, killing two persons and wounding more than 100 others, including two U.S. citizens.

February 15, 1996, Athens Embassy Attack. Unidentified assailants fired a rocket at the U.S. embassy compound in Athens, causing minor damage to three diplomatic vehicles and some surrounding buildings. Circumstances of the attack suggested it was an operation carried out by the 17 November group.

February 16, 1996, ELN Kidnapping. Six alleged National Liberation Army (ELN) guerrillas kidnapped a U.S. citizen in Colombia. After nine months, the hostage was released.

February 26, 1996, Hamas Bus Attack. In Jerusalem, a suicide bomber blew up a bus, killing 26 persons, including three U.S. citizens, and injuring some 80 persons, including three other U.S. citizens.

March 4, 1996, Dizengoff Center Bombing. Hamas and the Palestine Islamic Jihad (PIJ) both claimed responsibility for a bombing outside of Tel Aviv's largest shopping mall that killed 20 persons and injured 75 others, including two U.S. citizens.

May 13, 1996, West Bank Attack. Arab gunmen opened fire on a bus and a group of Yeshiva students near the Bet El settlement, killing a dual U.S.- Israeli citizen and wounding three Israelis. No one claimed responsibility for the attack, but Hamas was suspected.

May 31, 1996, AID Worker Abduction. A gang of former contra guerrillas kidnapped a U.S. employee of the Agency for International Development who was assisting with election preparations in rural northern Nicaragua. She was released unharmed the next day after members of the international commission overseeing the preparations intervened.

June 9, 1996, Zekharya Attack. Unidentified gunmen opened fire on a car near Zekharya, killing a dual U.S.- Israeli citizen and an Israeli. The Popular Front for the Liberation of Palestine (PFLP) was suspected.

June 15, 1996, Manchester Truck Bombing. An IRA truck bomb detonated at a Manchester shopping center, wounding 206 persons, including two German tourists, and caused extensive property damage.

June 25, 1996, Khobar Towers Bombing. A fuel truck carrying a bomb exploded outside the U.S. military's Khobar Towers housing facility in Dhahran, killing 19 U.S. military personnel and wounding 515 persons, including 240 U.S. personnel. Several groups claimed responsibility for the attack.

July 20, 1996, ETA Bombing. A bomb exploded at Tarragona International Airport in Reus, Spain, wounding 35 persons, including British and Irish tourists. The Basque Fatherland and Liberty (ETA) organization was suspected.

August 1, 1996, Bombing of Archbishop of Oran. A bomb exploded at the home of the French archbishop of Oran, killing him and his chauffeur. The attack occurred after the archbishop's meeting with the French foreign minister. The Algerian Armed Islamic Group (GIA) is suspected.

August 17, 1996, Sudanese Rebel Kidnapping. Sudan People's Liberation Army (SPLA) rebels kidnapped six missionaries in Mapourdit, including a U.S. citizen, an Italian, three Australians, and a Sudanese. The SPLA released the hostages eleven days later.

September 13, 1996, PUK Kidnapping. In Iraq, Patriotic Union of Kurdistan (PUK) militants kidnapped four French workers for Pharmaciens Sans Frontières, a Canadian United Nations High Commissioner for Refugees official, and two Iraqis.

October 1, 1996, Assassination of South Korean Consul. In Vladivostok, Russia, assailants attacked and killed a South Korean consul near his home. No one claimed responsibility, but South Korean authorities believed that the attack was carried out by professionals and that the assailants were North Koreans. North Korean officials denied the country's involvement in the attack.

November 1, 1996, Red Cross Worker Kidnappings. In Sudan a breakaway group from the Sudanese People's Liberation Army (SPLA) kidnapped three International Committee of the Red Cross workers, including a U.S. citizen, an Australian, and a Kenyan. On December 9 the rebels released the hostages in exchange for ICRC supplies and a health survey for their camp.

December 3, 1996, Paris Subway Explosion. A bomb exploded aboard a Paris subway train as it arrived at the Port Royal station, killing two French nationals, a Moroccan, and a Canadian, and injuring 86 persons. Among those injured were one U.S. citizen and a Canadian. No one claimed responsibility for the attack, but Algerian extremists were suspected.

December 11, 1996, Abduction of US. Citizen by FARC. Five armed men claiming to be members of the Revolutionary Armed Forces of Colombia kidnapped and later killed a U.S. geologist at a methane gas exploration site in La Guajira Department.

December 17, 1996, Tupac Amaru Seizure of Diplomats. Twenty-three members of the Tupac Amaru Revolutionary Movement (MRTA) took several hundred people hostage at a party given at the Japanese ambassador's residence in Lima, Peru. Among the hostages were several US officials, foreign ambassadors and other diplomats, Peruvian government officials, and Japanese businessmen. The group demanded the release of all MRTA members in prison and safe passage for them and the hostage takers. The terrorists released most of the hostages in December but held 81 Peruvians and Japanese citizens for several months.

1997
January 2–13, 1997, Egyptian Letter Bombs.
A series of letter bombs with Alexandria, Egypt, postmarks were discovered at *Al-Hayat* newspaper bureaus in Washington, New York City, London, and Riyadh, Saudi Arabia. Three similar devices, also postmarked in Egypt, were found at a prison facility in Leavenworth, Kan. Bomb-disposal experts defused all the devices, but one detonated at the *Al-Hayat* office in London, injuring two security guards and causing minor damage.

February 4–17, 1997, Tajik Hostage Abductions.
Near Komsomolabad, Tajikistan, a paramilitary group led by Bakhrom Sodirov abducted four United Nations military observers. The victims included two Swiss, one Austrian, one Ukrainian, and their Tajik interpreter. The kidnappers demanded safe passage for their supporters from Afghanistan to Tajikistan. In four separate incidents occurring between Dushanbe and Garm, Bakhrom Sodirov and his group kidnapped two International Committee for the Red Cross members, four Russian journalists and their Tajik driver, four UNHCR members, and the Tajik security minister, Saidamir Zukhurov.

February 14, 1997, Venezuelan Abduction.
Six armed Colombian guerrillas kidnapped a U.S. oil engineer and his Venezuelan pilot in Apure, Venezuela. The kidnappers released the pilot on 22 February. According to authorities, FARC was responsible for the kidnapping.

February 23, 1997, Empire State Building Sniper Attack.
A Palestinian gunman opened fire on tourists at an observation deck atop the Empire State Building in New York City, killing a Danish national and wounding visitors from the United States, Argentina, Switzerland, and France before turning the gun on himself. A handwritten note carried by the gunman claimed this was a punishment attack against the "enemies of Palestine."

February 24, 1997, ELN Kidnapping.
National Liberation Army (ELN) guerrillas kidnapped a U.S. citizen employed by a Las Vegas gold corporation who was scouting a gold-mining operation in Colombia. The ELN demanded a ransom of $2.5 million.

March 7, 1997, FARC Kidnapping.
FARC guerrillas kidnapped a U.S. mining employee and his Colombian colleague who were searching for gold in Colombia. On November 16, the rebels released the two hostages after receiving a $50,000 ransom.

July 12, 1997, Hotel Nacional Bombing.
A bomb exploded at the Hotel Nacional in Havana, injuring three persons and causing minor damage. A previously unknown group calling itself the Military Liberation Union claimed responsibility.

September 4, 1997, Israeli Shopping Mall Bombing.
Three suicide bombers of Hamas detonated bombs in the Ben Yehuda shopping mall in Jerusalem, killing eight persons, including the bombers, and wounding nearly 200 others. A dual U.S.–Israeli

citizen was among the dead, and seven U.S. citizens were wounded.

October 23, 1997, OAS Abductions. In Colombia ELN rebels kidnapped two foreign members of the Organization of American States and a Colombian human rights official at a roadblock. The ELN claimed that the kidnapping was intended "to show the international community that the elections in Colombia are a farce."

October 30, 1997, Yemeni Kidnappings. Al-Sha'if tribesmen kidnapped a U.S. businessman near Sanaa. The tribesmen sought the release of two fellow tribesmen who were arrested on smuggling charges and several public works projects they claim the government promised them. They released the hostage on November 27.

November 12, 1997, Murder of U.S. Businessmen in Pakistan. Two unidentified gunmen shot to death four U.S. auditors from Union Texas Petroleum Corporation and their Pakistani driver after they drove away from the Sheraton Hotel in Karachi. The Islami Inqilabi Council, or Islamic Revolutionary Council, claimed responsibility in a call to the U.S. consulate in Karachi. In a letter to Pakistani newspapers, the Aimal Khufia Action Committee also claimed responsibility.

November 17, 1997, Tourist Killings in Egypt. Al-Gama'at al-Islamiyya (IG) gunmen shot and killed 58 tourists and four Egyptians and wounded 26 others at the Hatshepsut Temple in the Valley of the Kings near Luxor. Thirty-four Swiss, eight Japanese, five Germans, four Britons, one French, one Colombian, a dual Bulgarian-British citizen, and four unidentified persons were among the dead. Twelve Swiss, two Japanese, two Germans, one French, and nine Egyptians were among the wounded.

1998
February 19, 1998, U.N. Observer Abductions. Armed supporters of late Georgian president Zviad Gamsakhurdia abducted four UN military observers from Sweden, Uruguay, and the Czech Republic.

March 21–23, 1998, FARC Abduction. FARC rebels kidnapped a U.S. citizen in Sabaneta, Colombia. FARC members also killed three persons, wounded 14, and kidnapped at least 27 others at a roadblock near Bogota. Four U.S. citizens and one Italian were among those kidnapped, as well as the acting president of the National Electoral Council and his wife.

April 15, 1998, Somali Hostage-Takings. Somali militiamen abducted nine Red Cross and Red Crescent workers at an airstrip north of Mogadishu. The hostages included a U.S. citizen, a German, a Belgian, a French, a Norwegian, two Swiss, and one Somali. The gunmen were members of a subclan loyal to Ali Mahdi Mohammed, who controlled the northern section of the capital.

August 1, 1998, IRA Bombing, Banbridge. A 500-pound car bomb planted by the Real IRA exploded outside a shoe store in Banbridge, Northern Ireland, injuring 35 persons and damaging at least 200 homes.

August 7, 1998, U.S. Embassy Bombings in East Africa. A bomb exploded at the rear entrance of the U.S. Embassy in Nairobi, Kenya, killing 12 U.S. citizens, 32 Foreign Service Nationals (FSNs), and 247 Kenyan citizens. Approximately 5,000 Kenyans, 6 U.S. citizens, and 13 FSNs were injured. The embassy building sustained extensive structural damage. Almost simultaneously, a bomb detonated outside the U.S. Embassy in Dar es Salaam, Tanzania, killing seven FSNs and three Tanzanian citizens, and injuring one U.S. citizen and 76 Tanzanians. The explosion caused major structural damage to the embassy facility. The U.S. government held Osama Bin Laden responsible.

August 15, 1998, IRA Bombing, Omagh. A 500-pound car bomb planted by the Real IRA exploded outside a local courthouse in the central shopping district of Omagh, Northern Ireland, killing 29 persons and injuring over 330.

October 18, 1998, Colombian Pipeline Bombing. A bomb planted by the National Liberation Army (ELN) exploded on the Ocensa pipeline in Antioquia Department, killing approximately 71 persons and injuring at least 100 others. The pipeline is jointly owned by the Colombia State Oil Company Ecopetrol and a consortium including U.S., French, British, and Canadian companies.

November 15, 1998, Armed Kidnapping in Colombia. Armed assailants followed a U.S. businessman and his family home in Cundinamarca Department and kidnapped his eleven-year-old son after stealing money, jewelry, one automobile, and two cell phones. The kidnappers demanded $1 million in ransom. On January 21, 1999, the kidnappers released the boy.

1999

January 2, 1999, Angolan Aircraft Downing. A UN plane carrying a U.S. citizen, four Angolans, two Philippine nationals, and one Namibian was shot down, according to a U.N. official. No deaths or injuries were reported. Angolan authorities blamed the attack on National Union for the Total Independence of Angola (UNITA) rebels. UNITA officials denied shooting down the plane.

February 14, 1999, Ugandan Rebel Attack. A pipe bomb exploded inside a bar, killing five persons and injuring 35 others. One Ethiopian and four Ugandan nationals died in the blast, and one U.S. citizen working for US-AID, two Swiss nationals, one Pakistani, one Ethiopian, and 27 Ugandans were injured. Ugandan authorities blamed the attack on the Allied Democratic Forces (ADF).

February 16, 1999, Greek Embassy Seizure. Kurdish protesters stormed and occupied the Greek embassy in Vienna, taking the Greek ambassador and six other persons hostage. Several hours later the protesters released the hostages and left the embassy. The attack followed the Turkish Government's announcement of the successful capture of the Kurdistan Workers' Party (PKK) leader Abdullah Ocalan. Kurds also occupied Kenyan, Israeli, and other Greek diplomatic facilities in France, Holland, Switzerland, Britain, and Germany over the following days.

February 25, 1999, FARC Kidnappings. FARC kidnapped three U.S. citizens working for the Hawaii-based Pacific Cultural Conservancy International. On March 4, the bodies of the three victims were found in Venezuela.

March 1, 1999, Hutu Abductions. One hundred fifty armed Hutu rebels attacked three tourist camps in Uganda, killed four Ugandans, and abducted three U.S. citizens, six Britons, three New Zealanders, two Danish citizens, one Australian, and one Canadian national. Two of the U.S. citizens and six of the other hostages were subsequently killed by their abductors.

March 23, 1999, ELN Hostage-Taking. Armed guerrillas kidnapped a U.S. citizen in Boyaca, Colombia. The National Liberation Army (ELN) claimed responsibility and demanded $400,000 ransom. On 20 July, ELN rebels released the hostage unharmed following a ransom payment of $48,000.

May 30, 1999, ELN Hostage-Taking. In Cali, Colombia, armed ELN militants attacked a church in the neighborhood of Ciudad Jardín, kidnapping 160 persons, including six U.S. citizens and one French national. The rebels released approximately 80 persons, including three U.S. citizens, later that day.

June 27, 1999, Shell Platform Bombing. In Port Harcourt, Nigeria, armed youths stormed a Shell oil platform, kidnapping one U.S. citizen, one Nigerian national, and one Australian citizen and causing undetermined damage. A group calling itself "Enough Is Enough in the Niger River" claimed responsibility. Further seizures of oil facilities followed.

August 4, 1999, AFRC Kidnappings. An Armed Forces Revolutionary Council (AFRC) faction kidnapped 33 UN representatives near Occra Hills, Sierra Leone. The hostages included one U.S. citizen, five British soldiers, one Canadian citizen, one representative from Ghana, one military officer from Russia, one officer from Kyrgyzstan, one officer from Zambia, one officer from Malaysia, a local bishop, two UN officials, two local journalists, and 16 Sierra Leonean nationals.

October 1, 1999, Burmese Embassy Seizure. Burmese dissidents seized the Burmese embassy in Bangkok, Thailand, taking 89 persons hostage, including one U.S. citizen.

December 23, 1999, PLA Kidnapping. Colombian People's Liberation Army (PLA) forces kidnapped a U.S. citizen in an unsuccessful ransoming effort.

December 24, 1999, Indian Airlines Airbus Hijacking. Five militants hijacked a flight bound from Katmandu to New Delhi carrying 189 people. The plane and its passengers were released unharmed on December 31.

2000
January 27, 2000, Car Bombing in Spain. Police officials reported unidentified individuals set fire to a Citroën car dealership in Iturreta, causing extensive damage to the building and destroying 12 vehicles. The attack bore the hallmark of the Basque Fatherland and Liberty (ETA).

May 1, 2000, RUF Attacks on U.N. Mission Personnel. On May 1 in Makeni, Sierra Leone, Revolutionary United Front (RUF) militants kidnapped at least 20 members of the United Nations Assistance Mission in Sierra Leone (UNAMSIL) and surrounded and opened fire on a UNAMSIL facility, according to press reports.

The militants killed five UN soldiers in the attack. RUF militants kidnapped 300 UNAMSIL peacekeepers throughout the country, according to press reports. On 15 May in Foya, Liberia, the kidnappers released 139 hostages. On 28 May, on the Liberia and Sierra Leone border, armed militants released unharmed the last of the UN peacekeepers. In Freetown, according to press reports, armed militants ambushed two military vehicles carrying four journalists. A Spaniard and one U.S. citizen were killed in a May 25 car bombing in Freetown for which the RUF was probably responsible. Suspected RUF rebels also kidnapped 21 Indian UN peacekeepers in Freetown on June 6. Additional attacks by RUF on foreign personnel followed.

June 8, 2000, Diplomatic Assassination in Greece. In Athens, Greece, two unidentified gunmen killed British defense attaché Stephen Saunders in an ambush. The Revolutionary Organization 17 November claimed responsibility.

June 27, 2000, ELN Kidnapping. In Bogotá, Colombia, ELN militants kidnapped a five-year-old U.S. citizen and his Colombian mother, demanding an undisclosed ransom.

August 12, 2000, Kidnappings in Kyrgyzstan. In the Kara-Su Valley, the Islamic Movement of Uzbekistan took four U.S. citizens hostage. The Americans escaped on August 18.

October 1, 2000, Church Bombing in Tajikistan. Unidentified militants detonated two bombs in a Christian church in Dushanbe, killing seven persons and injuring 70 others. The church was founded by a Korean-born U.S. citizen, and most of those killed and wounded were Korean. No one claimed responsibility.

October 12, 2000, Helicopter Hijacking. In Sucumbios Province, Ecuador, a group of armed kidnappers led by former members of defunct Colombian terrorist organization the Popular Liberation Army (EPL) took hostage ten employees of Spanish energy consortium REPSOL. Those kidnapped included five U.S. citizens, one Argentine, one Chilean, one New Zealander, and two French pilots who escaped four days later. On January 30, 2001, the kidnappers murdered American hostage Ronald Sander. The remaining hostages were released on February 23 following the payment of $13 million in ransom by the oil companies.

October 12, 2000, Attack on USS *Cole*. In Aden, Yemen, a small dinghy carrying explosives rammed the destroyer USS *Cole*, killing 17 sailors and injuring 39 others. Supporters of Osama Bin Laden were suspected.

December 30, 2000, Manila Bombing. A bomb exploded in a plaza across the street from the U.S. embassy in Manila, injuring nine persons. The Moro Islamic Liberation Front was likely responsible.

2001
January 17, 2001, Srinagar Airport Attack and Assassination Attempt. In India, six members of the Lashkar-e-Tayyba militant group were killed when they attempted to seize a local airport. Members of Hizbul Mujaheddin fired two rifle grenades at Farooq Abdullah, chief minister for Jammu and Kashmir. Two persons were wounded in the unsuccessful assassination attempt.

March 4, 2001, BBC Studios Bombing. A car bomb exploded at midnight outside of the British Broadcasting Corporation's main production studios in London. One person was injured. British authorities suspected the Real IRA had planted the bomb.

March 4, 2001, Suicide Bombing in Israel. A suicide bomb attack in Netanya killed 3 persons and wounded 65. Hamas later claimed responsibility.

March 9, 2001, ETA Bombing. Two policemen were killed by the explosion of a car bomb in Hernani, Spain.

March 15, 2001, Airliner Hijacking in Istanbul. Three Chechens hijacked a Russian airliner during a flight from Istanbul to Moscow and forced it to fly to Medina, Saudi Arabia. The plane carried 162 passengers and a crew of 12. After a 22-hour siege during which more than 40 passengers were released, Saudi security forces stormed the plane, killing a hijacker, a passenger, and a flight attendant.

April 22, 2001, Bus Stop Bombing. A member of Hamas detonated a bomb he was carrying near a bus stop in Kfar Siva, Israel, killing one person and injuring 60.

May 27, 2001, Philippines Hostage Incident. Muslim Abu Sayyaf guerrillas seized 13 tourists and 3 staff members at a resort on Palawan Island and took their captives to Basilan Island. The captives included three U.S. citizens: Guellermo Sobero and missionaries Martin and Gracia Burnham. Philippine troops fought a series of battles with the guerrillas between June 1 and June 3 during which 9 hostages escaped and two were found dead. The guerrillas took additional hostages when they seized the hospital in the town of Lamitan. On June 12,

Abu Sayyaf spokesman Abu Sabaya claimed that Sobero had been killed and beheaded; his body was found in October. The Burnhams remained in captivity until June 2002, when Martin Burnham was killed by his rescuers.

June 1, 2001, Tel-Aviv Nightclub Bombing. Hamas claimed responsibility for the suicide bombing of a popular Israeli nightclub that caused over 140 casualties.

August 9, 2001, Hamas Restaurant Bombing. A Hamas-planted bomb detonated in a Jerusalem pizza restaurant, killing 15 people and wounding more than 90. The Israeli response included occupation of Orient House, the Palestine Liberation Organization's political headquarters in East Jerusalem.

September 9, 2001, Suicide Bombing in Israel. The first suicide bombing carried out by an Israeli Arab killed three persons in Nahariya. Hamas claimed responsibility.

September 9, 2001, Death of "the Lion of the Panjshir." Two suicide bombers fatally wounded Ahmed Shah Massoud, a leader of Afghanistan's Northern Alliance, which had opposed both the Soviet occupation and the post-Soviet Taliban government. The bombers posed as journalists and were apparently linked to Al Qaeda. The Northern Alliance did not confirm Massoud's death until September 15.

September 11, 2001, Terrorist Attacks on U.S. Homeland. Two hijacked airliners crashed into the twin towers of the World Trade Center. Soon thereafter, the Pentagon was struck by a third hijacked plane. A fourth hijacked plane, suspected of being bound for a high-profile target in Washington, crashed into a field in southern Pennsylvania. The attacks killed 3,025 U.S. citizens and other nationals. President Bush and Cabinet officials indicated that Osama Bin Laden was the prime suspect and that they considered the United States in a state of war with international terrorism. In the aftermath of the attacks, the United States formed the Global Coalition Against Terrorism.

October 1, 2001, Attack on the Jammu and Kashmir Legislature. After a suicide car bomber forced the gate of the state legislature in Srinagar, two gunmen entered the building and held off police for seven hours before being killed. Forty persons died in the incident. Jaish-e-Muhammad claimed responsibility.

October–November 2001, Anthrax Attacks. On October 7 the U.S. Centers for Disease Control and Prevention (CDC) reported that investigators had detected evidence that the deadly anthrax bacterium was present in the building where a Florida man who died of anthrax on October 5 had worked. Discovery of a second anthrax case triggered a major investigation by the Federal Bureau of Investigation (FBI). The two anthrax cases were the first to appear in the United States in twenty-five years. Anthrax subsequently appeared in mail received by television networks in New York and by the offices in Washington of Senate Majority Leader Tom Daschle and other members of Congress. Attorney General John Ashcroft said in a briefing on October 16, "When people send anthrax through the mail to hurt people and invoke terror, it's a terrorist act."

October 17, 2001, Assassination of an Israeli Cabinet Minister. A Palestinian gunman assassinated Israeli Minister of Tourism Rehavam Zeevi in the Jerusalem hotel where he was staying. The Popular Front for the Liberation of Palestine (PFLP) claimed to have avenged the death of PFLP Mustafa Zubari.

October 28, 2001, Attack on a Church in Pakistan. Six masked gunmen shot up a church in Bahawalpur, Pakistan, killing 15 Pakistani Christians. No group claimed responsibility, although various militant Muslim groups were suspected.

December 1, 2001, Suicide Bombings in Jerusalem. Two suicide bombers attacked a Jerusalem shopping mall, killing 10 persons and wounding 170.

December 2, 2001, Suicide Bombing in Haifa. A suicide bomb attack aboard a bus in Haifa, Israel, killed 15 persons and wounded 40. Hamas claimed responsibility for both this attack and those on December 1 to avenge the death of a Hamas member at the hands of Israeli forces a week earlier.

December 13, 2001, Attack on the Indian Parliament. Five gunmen attacked the Indian Parliament in New Delhi shortly after it had adjourned. Before security forces killed them, the attackers killed six security personnel and a gardener. Indian officials blamed Lashkar-e-Tayyiba and demanded that Pakistan crack down on it and on other Muslim separatist groups in Kashmir.

2002
January 15, 2002, Ambush on the West Bank. Palestinian militants fired on a vehicle in Beit Sahur, killing one

passenger and wounding the other. The dead passenger claimed U.S. and Israeli citizenship. The Al-Aqsa Martyrs' Battalion claimed responsibility.

January 17, 2002, Shooting Incident in Israel. A Palestinian gunman killed 6 persons and wounded 25 in Hadera, Israel, before being killed by Israeli police. The Al-Aqsa Martyrs' Brigade claimed responsibility as revenge for Israel's killing of a leading member of the group.

January 22, 2002, Drive-by Shooting at a U.S. Consulate. Armed militants on motorcycles fired on the U.S. consulate in Calcutta, India, killing 5 Indian security personnel and wounding 13 others. The Harakat ul-Jihad-i-Islami and the Asif Raza Commandos claimed responsibility. Indian police later killed two suspects, one of whom confessed to belonging to Lashkar-e-Tayyiba as he died.

January 22, 2002, Bomb Explosion in Kashmir. A bomb exploded in a crowded retail district in Jammu, Kashmir, killing one person and injuring nine. No group claimed responsibility.

January 23, 2002, Kidnapping of Daniel Pearl. Armed militants kidnapped *Wall Street Journal* reporter Daniel Pearl in Karachi, Pakistan. Pakistani authorities received a videotape on February 20 depicting Pearl's murder. His grave was found near Karachi on May 16. Pakistani authorities arrested four suspects. Ringleader Ahmad Omar Saeed Sheikh claimed to have organized Pearl's kidnapping to protest Pakistan's subservience to the United States, and had belonged to Jaish-e-Muhammad, an Islamic separatist group in Kashmir. All four suspects were convicted on July 15. Saeed Sheikh was sentenced to death, the others to life imprisonment.

January 27, 2002, Suicide Bombing in Jerusalem. A suicide bomb attack in Jerusalem killed one other person and wounded 100. The incident was the first suicide bombing made by a Palestinian woman.

February 16, 2002, Suicide Bombing in the West Bank. A suicide bombing in an outdoor food court in Karmei Shomron killed 4 persons and wounded 27. Two of the dead and two of the wounded were U.S. citizens. The Popular Front for the Liberation of Palestine (PFLP) claimed responsibility.

March 7, 2002, Suicide Bombing in the West Bank. A suicide bombing in a supermarket in the settlement of Ariel wounded ten persons, one of whom was a U.S. citizen. The PFLP claimed responsibility.

March 9, 2002, Suicide Bombing in Jerusalem. A suicide bombing in a Jerusalem restaurant killed 11 persons and wounded 52, one of whom was a U.S. citizen. The Al-Aqsa Martyrs' Brigade claimed responsibility.

March 14, 2002, Drive-by Shooting in Colombia. Gunmen on motorcycles shot and killed two U.S. citizens who had come to Cali, Colombia, to negotiate the release of their father, who was a captive of the FARC. No group claimed responsibility.

March 17, 2002, Grenade Attack on a Church in Pakistan. Militants threw grenades into the Protestant International Church in Islamabad, Pakistan, during a service attended by diplomatic and local personnel. Five persons, two of them U.S. citizens, were killed and 46 were wounded. The dead Americans were State Department employee Barbara Green and her daughter Kristen Wormsley. Thirteen U.S. citizens were among the wounded. The Lashkar-e-Tayyiba group was suspected.

March 20, 2002, Car Bomb Explosion in Peru. A car bomb exploded at a shopping center near the U.S. embassy in Lima, Peru. Nine persons were killed and 32 wounded. The dead included two police officers and a teenager. Peruvian authorities suspected either the Shining Path rebels or the Tupac Amaru Revolutionary Movement. The attack occurred three days before President George W. Bush visited Peru.

March 21, 2002, Suicide Bombing in Jerusalem. A suicide bombing in Jerusalem killed 3 persons and wounded 86 more, including two U.S. citizens. The Palestinian Islamic Jihad claimed responsibility.

March 27, 2002, Suicide Bombing in Israel. A suicide bombing in a noted restaurant in Netanya, Israel, killed 22 persons and wounded 140. One of the dead was a U.S. citizen. The Islamic Resistance Movement (Hamas) claimed responsibility.

March 30, 2002, Temple Bombing in Kashmir. A bomb explosion at a Hindu temple in Jammu, Kashmir, killed ten persons. The Islamic Front claimed responsibility.

March 31, 2002, Suicide Bombing in the West Bank. A suicide bombing near an ambulance station in Efrat wounded four persons, including a U.S. citizen. The Al-Aqsa Martyrs' Brigade claimed responsibility.

April 10, 2002, Armed Attack on Kashmir. Armed militants attacked a residence in Gando, Kashmir, killing five persons and wounding four. No group claimed responsibility.

April 11, 2002, Synagogue Bombing in Tunisia. A suicide bomber detonated a truck loaded with propane gas outside a historic synagogue in Djerba, Tunisia. The 16 dead included 11 Germans, one French citizen, and three Tunisians. Twenty-six German tourists were injured. The Islamic Army for the Liberation of the Holy Sites claimed responsibility.

April 12, 2002, Suicide Bombing in Jerusalem. A female suicide bomber killed 6 persons in Jerusalem and wounded 90 others. The Al-Aqsa Martyrs' Brigade claimed responsibility.

May 8, 2002, Car Bombing in Pakistan. A car bomb exploded near a Pakistani navy shuttle bus in Karachi, killing 12 persons and wounding 19. Eleven of the dead and 11 of the wounded were French nationals. Al Qaeda was suspected of the attack.

May 9, 2002, Parade Bombing in Russia. A remotely controlled bomb exploded near a May Day parade in Kaspiisk, Dagestan, killing 42 persons and wounding 150. Fourteen of the dead and 50 of the wounded were soldiers. Islamists linked to Al Qaeda were suspected.

May 14, 2002, Attack on a Bus in India. Militants fired on a passenger bus in Kaluchak, Jammu, killing seven persons. They then entered a military housing complex and killed three soldiers and seven military dependents before they were killed. The al-Mansooran and Jamiat ul-Mujahedin claimed responsibility.

May 17, 2002, Bomb Attacks in Kashmir. A bomb explosion near a civil secretariat area in Srinagar, Kashmir, wounded six persons. In Jammu, a bomb exploded at a fire services headquarters, killing two and wounding 16. No group claimed responsibility for either attack.

June 7, 2002, Hostage Rescue Attempt in the Philippines. Philippine Army troops attacked Abu Sayyaf terrorists on Mindanao Island in an attempt to rescue U.S. citizen Martin Burnham and his wife Gracia, who had been kidnapped more than a year earlier. Burnham was killed, but his wife, though wounded, was freed. A Filipino hostage was killed, as were four of the guerrillas. Seven soldiers were wounded.

June 14, 2002, Car Bombing in Pakistan. A car bomb exploded near the U.S. consulate and the Marriott Hotel in Karachi, Pakistan. Eleven persons were killed and 51 were sounded, including one U.S. and one Japanese citizen. Al Qaeda and al-Qanin were suspected.

June 19, 2002, Suicide Bombing in Jerusalem. A suicide bombing at a bus stop in Jerusalem killed 6 persons and wounded 43, including two U.S. citizens. The Al-Aqsa Martyrs' Brigade claimed responsibility.

July 17, 2002, Suicide Bombing in Tel Aviv. Two suicide bombers attacked the old bus station in Tel Aviv, Israel, killing 5 persons and wounding 38. The dead included one Romanian and two Chinese; another Romanian was wounded. The Islamic Jihad claimed responsibility.

July 31, 2002, Bombing at the Hebrew University. A bomb hidden in a bag in the Frank Sinatra International Student Center of Jerusalem's Hebrew University killed 9 persons and wounded 87. The dead included 5 U.S. citizens and 4 Israelis. The wounded included 4 U.S. citizens, 2 Japanese, and 3 South Koreans. The Islamic Resistance Movement (Hamas) claimed responsibility.

August 4, 2002, Suicide Bombing in Israel. A suicide bomb attack on a bus in Safed, Israel, killed 9 persons and wounded 50. Two of the dead were Philippine citizens; many of the wounded were soldiers returning from leave. Hamas claimed responsibility.

August 5, 2002, Attack on a School in Pakistan. Gunmen attacked a Christian school attended by children of missionaries from around the world. Six persons (two security guards, a cook, a carpenter, a receptionist, and a private citizen) were killed and a Philippine citizen was wounded. A group called al-Intigami al-Pakistani claimed responsibility.

August 6, 2002, Attack on Pilgrims in Kashmir. Armed militants attacked a group of Hindu pilgrims with guns and grenades in Pahalgam, Kashmir. Nine persons were killed and 32 were wounded. The Lashkar-e-Tayyiba claimed responsibility.

September 11, 2002, Assassination in Kashmir. Gunmen killed Kashmir's Law Minister Mushtaq Ahmed Lone and six security guards in Tikipora. Lashkar-e-Tayyiba, Jamiat ul-Mujahedin, and Hizb ul-Mujahedin all claimed responsibility. Other militants attacked the residence of the Minister of Tourism with grenades, injuring four persons. No group claimed responsibility.

September 18, 2002, Ambush on the West Bank. Gunmen ambushed a vehicle on a road near Yahad, killing an Israeli and wounding a Romanian worker. The Al-Aqsa Martyrs' Brigade claimed responsibility.

September 19, 2002, Suicide Bomb Attack in Israel. A suicide bomb attack on a bus in Tel Aviv killed 6 persons and wounded 52. One of the dead was a British subject. Hamas claimed responsibility.

October 6, 2002, Attack on a French Tanker. An explosive-laden boat rammed the French oil tanker *Limburg,* which was anchored about five miles off al-Dhabbah, Yemen. One person was killed and 4 were wounded. Al Qaeda was suspected.

October 12, 2002, Car Bomb Explosion in Bali. A car bomb exploded outside the Sari Club Discotheque in Denpasar, Bali, Indonesia, killing 202 persons and wounding 300 more. Most of the casualties, including 88 of the dead, were Australian tourists. Seven Americans were among the dead. Al Qaeda claimed responsibility. Two suspects were later arrested and convicted. Iman Samudra, who had trained in Afghanistan with Al Qaeda and was suspected of belonging to Jemaah Islamiya, was sentenced to death on September 10, 2003.

October 23–26, 2002, Chechen Rebels Seize a Moscow Theater. Fifty Chechen rebels led by Movsar Barayev seized the Palace of Culture Theater in Moscow, Russia, to demand an end to the war in Chechnya. They seized more than 800 hostages from thirteen countries and threatened to blow up the theater. During a three-day siege, they killed a Russian policeman and five Russian hostages. On October 26, Russian Special Forces pumped an anesthetic gas through the ventilation system and then stormed the theater. All of the rebels were killed, but 94 hostages (including one American) also died, many from the effects of the gas. A group led by Chechen warlord Shamil Basayev claimed responsibility.

October 28, 2002, Assassination of an AID Official. Gunmen in Amman assassinated Laurence Foley, Executive Officer of the U.S. Agency for International Development Mission in Jordan. The Honest People of Jordan claimed responsibility.

November 21, 2002, Suicide Bombing in Jerusalem. A suicide bomb attack on a bus on Mexico Street in Jerusalem killed 11 persons and wounded 50 more. One of the dead was a Romanian. Hamas claimed responsibility.

November 24, 2002, Attack on Temples in Kashmir. Armed militants attacked the Reghunath and Shiv temples in Jammu, Kashmir, killing 13 persons and wounding 50. The Lashkare-e-Tayyiba claimed responsibility.

November 28, 2002, Attacks on Israeli Tourists in Kenya. A three-person suicide car bomb attack on the Paradise Hotel in Mombasa, Kenya, killed 15 persons and wounded 40. Three of the dead and 18 of the wounded were Israeli tourists; the others were Kenyans. Near Mombasa's airport, two SA-7 shoulder-fired missiles were fired at an Arkia Airlines Boeing 757 that was carrying 261 passengers back to Israel. Both missiles missed. Al Qaeda, the Government of Universal Palestine in Exile, and the Army of Palestine claimed responsibility for both attacks. Al-Ittihad al-Islami was also suspected of involvement.

December 26, 2002, Attack on a Bus in the Philippines. Armed militants ambushed a bus carrying Filipino workers employed by the Canadian Toronto Ventures Inc. Pacific mining company in Zamboanga del Norte. Thirteen persons were killed and 10 wounded. Philippine authorities suspected the Moro Islamic Liberation Front (MILF), which had been extorting money from Toronto Ventures. The Catholic charity Caritas-Philippines said that Toronto Ventures had harassed tribesmen who opposed mining on their ancestral lands.

December 27, 2002, Bombing of a Government Building in Chechnya. A suicide bomb attack involving two explosives-laden trucks destroyed the offices of the pro-Russian Chechen government in Grozny. The attack killed over 80 people and wounded 210. According to a Chechen Web site run by the Kavkaz Center, Chechen warlord Shamil Basayev claimed responsibility.

2003

January 5, 2003, Suicide Bombings in Tel Aviv. Two suicide bomb attacks killed 22 and wounded at least 100 persons in Tel Aviv, Israel. Six of the victims were foreign workers. The Al-Aqsa Martyrs' Brigade claimed responsibility.

February 7, 2003, Night Club Bombing in Colombia. A car bomb exploded outside a night club in Bogotá, Colombia, killing 32 persons and wounding 160. No group claimed responsibility, but Colombian officials suspected the Colombian Revolutionary Armed Forces (FARC) of committing the worst terrorist attack in the country in a decade.

February 8, 2003, Assasination of a Kurdish Leader. Members of Ansar al-Islam assassinated Kurdish legislator Shawkat Haji Mushir and captured two other Kurdish officials in Qamash Tapa in northern Iraq.

March 5, 2003, Suicide Bombing in Haifa. A suicide bombing aboard a bus in Haifa, Israel, killed 15 persons and wounded at least 40. One of the dead claimed U.S. as well as Israeli citizenship. The bomber's affiliation was not immediately known.

March 30, 2003, Suicide Bombing in Netanya. A suicide bombing in a cafe in Netanya, Israel, wounded 38 persons. Only the bomber was killed. Islamic Jihad claimed responsibility and called the attack a "gift" to the people of Iraq.

May 5, 2003, Unsuccessful Hostage Rescue Attempt in Colombia. The FARC killed ten hostages when Colombian special forces tried to rescue them from a jungle hideout near Urrao, in Colombia's Antioquia State. The dead included Governor Guillermo Gavira and former Defense Minister Gilberto Echeverri Mejia, who had been kidnapped in April 2002.

May 12, 2003, Truck Bomb Attacks in Saudi Arabia. Suicide bombers attacked three residential compounds for foreign workers in Riyadh, Saudi Arabia. The 34 dead included 9 attackers, 7 other Saudis, 9 U.S. citizens, and one citizen each from the United Kingdom, Ireland, and the Philippines. Another American died on June 1. It was the first major attack on U.S. targets in Saudi Arabia since the end of the war in Iraq. Saudi authorities arrested 11 Al Qaeda suspects on May 28.

May 12, 2003, Truck Bombing in Chechnya. A truck bomb explosion demolished a government compound in Znamenskoye, Chechnya, killing 54 persons. Russian authorities blamed followers of a Saudi-born Islamist named Abu Walid. President Vladimir Putin said that he suspected that there was an Al Qaeda connection.

May 12, 2003, Attempted Assassination in Chechnya. Two female suicide bombers attacked Chechen administrator Mufti Akhmed Kadyrov during a religious festival in Iliskhan

Yurt. Kadyrov escaped injury, but 14 other persons were killed and 43 were wounded. Chechen rebel leader Shamil Basayev claimed responsibility.

May 16, 2003, Suicide Bomb Attacks in Morocco. A team of 12 suicide bombers attacked five targets in Casablanca, Morocco, killing 43 persons and wounding 100. The targets were a Spanish restaurant, a Jewish community, a Jewish cemetery, a hotel, and the Belgian Consulate. The Moroccan Government blamed the Islamist al-Assirat al-Moustaquim (The Righteous Path), but foreign commentators suspected an Al Qaeda connection.

May 18, 2003, Suicide Bomb Attack in Jerusalem. A suicide bomb attack on a bus in Jerusalem's French Hill district killed 7 persons and wounded 20. The bomber was disguised as a religious Jew. Hamas claimed responsibility.

May 19, 2003, Suicide Bombing in Afula. A suicide bomb attack by a female Palestinian student killed 3 persons and wounded 52 at a shopping mall in Afula, Israel. Both Islamic Jihad and the Al-Aqsa Martyrs' Brigade claimed responsibility.

June 11, 2003, Suicide Bombing in Jerusalem. A suicide bombing aboard a bus in Jerusalem killed 16 persons and wounded at least 70, one of whom died later. Hamas claimed responsibility, calling it revenge for an Israeli helicopter attack on Hamas leader Abdelaziz al-Rantisi in Gaza City the day before.

August 1, 2003, Truck Bombing in Northern Ossetia. A suicide truck bomb attack destroyed a Russian military hospital in Mozdok, North Ossetia and killed 50 persons. Russian authorities attributed the attack to followers of Chechen rebel leader Shamil Basayev.

August 5, 2003, Hotel Bombing in Indonesia. A car bomb exploded outside the Marriott Hotel in Jakarta, Indonesia, killing 10 persons and wounding 150. One of the dead was a Dutch citizen. The wounded included an American, a Canadian, an Australian, and two Chinese. Indonesian authorities suspected the Jemaah Islamiah, which had carried out the October 12, 2002 bombing in Bali.

August 7, 2003, Bombing of the Jordanian Embassy in Baghdad. A car bomb exploded outside the Jordanian Embassy in Baghdad, Iraq, killing 19 persons and wounding 65. Most of the victims were apparently Iraqis, including five police officers. No group claimed responsibility.

August 12, 2003, Suicide Bombings in Israel and the West Bank. The first suicide bombings since the June 29 Israeli-Palestinian truce took place. The first, in a supermarket at Rosh Haayin, Israel, killed one person and wounded 14. The second, at a bus stop near the Ariel settlement in the West Bank, killed one person and wounded 3. The Al-Aqsa Martyrs' Brigade claimed responsibility for the first; Hamas claimed responsibility for the second.

August 19, 2003, Bombing of the UN Headquarters in Baghdad. A truck loaded with surplus Iraqi ordnance exploded outside the United Nations Headquarters in Baghdad's Canal Hotel. A hospital across the street was also heavily damaged. The 23 dead included UN Special Representative Sergio Viera de Mello. More than a hundred persons were wounded. It was not clear whether the bomber was a Baath Party loyalist or a foreign Islamic militant. An Al Qaeda branch called the Brigades of the Martyr Abu Hafz al-Masri later claimed responsibility.

August 19, 2003, Suicide Bombing in Jerusalem. A suicide bombing aboard a bus in Jerusalem killed 20 persons and injured at least 100, one of whom died later. Five of the dead were American citizens. Hamas and Islamic Jihad claimed responsibility, although Hamas leader al-Rantisi said that his organization remained committed to the truce while reserving the right to respond to Israeli military actions.

August 29, 2003, Car Bomb Kills Shi'ite Leader in Najaf. A car bomb explosion outside the Shrine of the Imam Ali in Najaf, Iraq killed at least 81 persons and wounded at least 140. The dead included the Ayatollah Mohammed Bakir al-Hakim, one of four leading Shi'ite clerics in Iraq. Al-Hakim had been the leader of the Supreme Council for the Islamic Revolution in Iraq (SCIRI) since its establishment in 1982, and SCIRI had recently agreed to work with the U.S.-sponsored Iraqi Governing Council. It was not known whether the perpetrators were Baath Party loyalists, rival Shi'ites, or foreign Islamists.

September 9, 2003, Suicide Bombings in Israel. Two suicide bombings took place in Israel. The first, at a bus stop near the Tsrifin army base southeast of Tel Aviv, killed 7 soldiers and wounded 14 soldiers and a civilian. The second, at a cafe in Jerusalem's German Colony neighborhood, killed 6 persons and wounded 40. Hamas did not claim responsibility until the next day, although a spokesman called the first attack "a response to Israeli aggression."

September 20, 2003, Assassination of an Iraqi Governing Council Member. Gunmen shot and seriously wounded Akila Hashimi, one of three female members of the Iraqi Governing Council, near her home in Baghdad. She died September 25.

September 22, 2003, A Second Attack on the UN Headquarters in Baghdad. A suicide car bomb attack on the UN Headquarters in Baghdad killed a security guard and wounded 19 other persons.

October 4, 2003, Suicide Bombing in Israel. A Palestinian woman made a suicide bomb attack on a restaurant in Haifa, killing 19 persons and wounding at least 55. Islamic Jihad claimed responsibility for the attack. The next day, Israel bombed a terrorist training camp in Syria.

October 9, 2003, Attacks in Iraq. Gunmen assassinated a Spanish military attaché in Baghdad. A suicide car bomb attack on an Iraqi police station killed 8 persons and wounded 40.

October 12, 2003, Car Bombings in Baghdad. Two suicide car bombs exploded outside the Baghdad Hotel, which housed U.S. officials. Six persons were killed and 32 wounded. Iraqi and U.S. security personnel apparently kept the cars from actually reaching the hotel.

October 15, 2003, Bomb Attack on U.S. Diplomats in the Gaza Strip. A remote-controlled bomb exploded under a car in a U.S. diplomatic convoy passing through the northern Gaza Strip. Three security guards, all employees of DynCorp, were killed. A fourth was wounded. The diplomats were on their way to interview Palestinian candidates for Fulbright scholarships to study in the United States. Palestinian President Arafat and Prime Minister Qurei condemned the attack, while the major Palestinian militant groups denied responsibility. The next day, Palestinian security forces arrested several suspects, some of whom belonged to the Popular Resistance Committees.

October 26, 2003, Rocket Attack on the al-Rashid Hotel in Baghdad. Iraqis using an improvised rocket launcher bombarded the al-Rashid Hotel in Baghdad, killing one U.S. Army officer and wounding 17 persons. The wounded included four U.S. military personnel and seven American civilians. Deputy Secretary of Defense Paul D. Wolfowitz, who was staying at the hotel, was not injured. After visiting the wounded, he said, "They're not going to scare us away; we're not giving up on this job."

October 26, 2003, Assassination of a Deputy Mayor in Baghdad. Two gunmen believed to be Baath Party loyalists assassinated Faris Abdul Razaq al-Assam, one of three deputy mayors of Baghdad. U.S. officials did not announce al-Assam's death until October 28.

October 27, 2003, Wave of Car Bombings in Baghdad. A series of suicide car bombings in Baghdad killed at least 35 persons and wounded at least 230. Four attacks were directed at Iraqi police stations, the fifth and most destructive was directed at the International Committee of the Red Cross headquarters, where at least 12 persons were killed. A sixth attack failed when a car bomb failed to explode and the bomber was wounded and captured by Iraqi police. U.S. and Iraqi officials suspected that foreign terrorists were involved; the unsuccessful bomber said he was a Syrian national and carried a Syrian passport. After a meeting with Administrator L. Paul Bremer, President Bush said, "The more successful we are on the ground, the more these killers will react."

November 8, 2003, Suicide Bombing in Riyadh. In Riyadh, a suicide car bombing took place in the Muhaya residential compound, which was occupied mainly by nationals of other Arab countries. Seventeen persons were killed and 122 were wounded. The latter included four Americans. The next day, Deputy Secretary of State Armitage said al-Qaeda was probably responsible.

November 12, 2003, Truck Bombing in Nasiriyah. A suicide truck bomb destroyed the headquarters of the Italian military police in Nasiriyah, Iraq, killing 18 Italians and 11 Iraqis and wounding at least 100 persons.

November 15, 2003, Synagogue Bombings in Istanbul. Two suicide truck bombs exploded outside the Neve Shalom and Beth Israel synagogues in Istanbul, killing 25 persons and wounding at least 300 more. The initial claim of responsibility came from a Turkish militant group, the Great Eastern Islamic Raiders' Front, but Turkish authorities suspected an Al Qaeda connection. The next day, the London-based newspaper *Al-Quds al-Arabi* received an e-mail in which an Al Qaeda branch called the Brigades of the Martyr Abu Hafz al-Masri claimed responsibility for the Istanbul synagogue bombings.

November 15, 2003, Grenade Attacks in Bogota. Grenade attacks on two bars frequented by Americans in Bogotá killed one person and wounded 72, including four Americans. Colombian authorities suspected FARC. The U.S. Embassy suspected that the attacks had targeted Americans and warned against visiting commercial centers and places of entertainment.

November 20, 2003, More Suicide Truck Bombings in Istanbul. Two more suicide truck bombings devastated the British HSBC Bank and the British Consulate General in Istanbul, killing 27 persons and wounding at least 450. The dead included Consul General Roger Short. U.S., British, and Turkish officials suspected that Al Qaeda had struck again. The U.S. Consulate in Istanbul was closed, and the Embassy in Ankara advised American citizens in Istanbul to stay home.

November 20, 2003, Car Bombing in Kirkuk. A suicide car bombing in Kirkuk killed 5 persons. The target appeared to be the headquarters of the Patriotic Union of Kurdistan. PUK officials suspected the Ansar al-Islam group, which was said to have sheltered fugitive Taliban and al-Qaeda members after the U.S. campaign in Afghanistan.

November 29–30, 2003, Attacks on Other Coalition Personnel in Iraq. Iraqi insurgents stepped up attacks on nationals of other members of the Coalition. On November 29, an ambush in Mahmudiyah killed seven out of a party of eight Spanish intelligence officers. Iraqi insurgents also killed two Japanese diplomats near Tikrit. On November 30, another ambush near Tikrit killed two South Korean electrical workers and wounded two more. A Colombian employee of Kellogg Brown & Root was killed and two were wounded in an ambush near Balad.

December 5, 2003, Train Bombing in Southern Russia. A suicide bomb attack killed 42 persons and wounded 150 aboard a Russian commuter train in the south Russian town of Yessentuki. Russian officials suspected Chechen rebels; President Putin said the attack was meant to disrupt legislative elections. Chechen rebel

leader Aslan Maskhadov denied any involvement.

December 9, 2003, Suicide Bombing in Moscow. A female suicide bomber killed 5 other persons and wounded 14 outside Moscow's National Hotel. She was said to be looking for the State Duma.

December 15, 2003, Suicide Car Bombings in Iraq. Two days after the capture of Saddam Hussein, there were two suicide car bomb attacks on Iraqi police stations. One at Husainiyah killed 8 persons and wounded 20. The other, at Ameriyah, wounded 7 Iraqi police. Guards repelled a second vehicle.

December 19, 2003, Office Bombing in Baghdad. A bomb destroyed the Baghdad office of the Supreme Council of the Islamic Revolution in Iraq, killing a woman and wounding at least seven other persons.

December 24, 2003, Suicide Car Bombing in Irbil. A suicide car bomb attack on the Kurdish Interior Ministry in Irbil, Iraq, killed 5 persons and wounded 101.

December 25, 2003, Attempted Assassination in Rawalpindi. Two suicide truck bombers killed 14 persons as President Musharraf's motorcade passed through Rawalpindi, Pakistan. An earlier attempt on December 14 caused no casualties. Pakistani officials suspected Afghan and Kashmiri militants. On January 6, 2004, Pakistani authorities announced the arrest of six suspects who were said to be members of Jaish-e-Muhammad.

December 25, 2003, Suicide Bombing in Israel. A Palestinian suicide bomber killed four persons at a bus stop near Petah Tikva, Israel. The

Popular Front for the Liberation of Palestine claimed responsibility for the attack in retaliation for Israeli military operations in Nablus that had begun two days earlier.

December 31, 2003, Restaurant Bombing in Baghdad. A car bomb explosion outside Baghdad's Nabil Restaurant killed 8 persons and wounded 35. The wounded included three *Los Angeles Times* reporters and three local employees.

APPENDIX FOUR

UNITED NATIONS RESOLUTION 1441
AND THE BLIX/EL-BARADEI LETTER

**United Kingdom of Great Britain and Northern Ireland and United
States of America: Draft Resolution**
[Adopted as Resolution 1441 at Security Council meeting 4644, 8 November
2002]

The Security Council,
Recalling all its previous relevant resolutions, in particular its resolutions 661
(1990) of 6 August 1990, 678 (1990) of 29 November 1990, 686 (1991) of 2
March 1991, 687 (1991) of 3 April 1991, 688 (1991) of 5 April 1991, 707
(1991) of 15 August 1991, 715 (1991) of 11 October 1991, 986 (1995) of 14
April 1995, and 1284 (1999) of 17 December 1999, and all the relevant
statements of its President,
Recalling also its resolution 1382 (2001) of 29 November 2001 and its intention
to implement it fully,
Recognizing the threat Iraq's non-compliance with Council resolutions and
proliferation of weapons of mass destruction and long-range missiles poses
to international peace and security,
Recalling that its resolution 678 (1990) authorized Member States to use all
necessary means to uphold and implement its resolution 660 (1990) of 2 August 1990 and all relevant resolutions subsequent to resolution 660 (1990)
and to restore international peace and security in the area,
Further recalling that its resolution 687 (1991) imposed obligations on Iraq as a
necessary step for achievement of its stated objective of restoring international peace and security in the area,
Deploring the fact that Iraq has not provided an accurate, full, final, and complete disclosure, as required by resolution 687 (1991), of all aspects of its
programmes to develop weapons of mass destruction and ballistic missiles

with a range greater than one hundred and fifty kilometres, and of all holdings of such weapons, their components and production facilities and locations, as well as all other nuclear programmes, including any which it claims are for purposes not related to nuclear-weapons-usable material,

Deploring further that Iraq repeatedly obstructed immediate, unconditional, and unrestricted access to sites designated by the United Nations Special Commission (UNSCOM) and the International Atomic Energy Agency (IAEA), failed to cooperate fully and unconditionally with UNSCOM and IAEA weapons inspectors, as required by resolution 687 (1991), and ultimately ceased all cooperation with UNSCOM and the IAEA in 1998,

Deploring the absence, since December 1998, in Iraq of international monitoring, inspection, and verification, as required by relevant resolutions, of weapons of mass destruction and ballistic missiles, in spite of the Council's repeated demands that Iraq provide immediate, unconditional, and unrestricted access to the United Nations Monitoring, Verification and Inspection Commission (UNMOVIC), established in resolution 1284 (1999) as the successor organization to UNSCOM, and the IAEA, and regretting the consequent prolonging of the crisis in the region and the suffering of the Iraqi people,

Deploring also that the Government of Iraq has failed to comply with its commitments pursuant to resolution 687 (1991) with regard to terrorism, pursuant to resolution 688 (1991) to end repression of its civilian population and to provide access by international humanitarian organizations to all those in need of assistance in Iraq, and pursuant to resolutions 686 (1991), 687 (1991), and 1284 (1999) to return or cooperate in accounting for Kuwaiti and third country nationals wrongfully detained by Iraq, or to return Kuwaiti property wrongfully seized by Iraq,

Recalling that in its resolution 687 (1991) the Council declared that a ceasefire would be based on acceptance by Iraq of the provisions of that resolution, including the obligations on Iraq contained therein,

Determined to ensure full and immediate compliance by Iraq without conditions or restrictions with its obligations under resolution 687 (1991) and other relevant resolutions and recalling that the resolutions of the Council constitute the governing standard of Iraqi compliance,

Recalling that the effective operation of UNMOVIC, as the successor organization to the Special Commission, and the IAEA is essential for the implementation of resolution 687 (1991) and other relevant resolutions,

Noting the letter dated 16 September 2002 from the Minister for Foreign Affairs of Iraq addressed to the Secretary-General is a necessary first step toward rectifying Iraq's continued failure to comply with relevant Council resolutions,

Noting further the letter dated 8 October 2002 from the Executive Chairman of UNMOVIC and the Director-General of the IAEA to General Al-Saadi of the Government of Iraq laying out the practical arrangements, as a follow-up to their meeting in Vienna, that are prerequisites for the resumption of inspections in Iraq by UNMOVIC and the IAEA, and expressing the gravest concern at the continued failure by the Government of Iraq to provide confirmation of the arrangements as laid out in that letter,

Reaffirming the commitment of all Member States to the sovereignty and territorial integrity of Iraq, Kuwait, and the neighbouring States,

Commending the Secretary-General and members of the League of Arab States and its Secretary-General for their efforts in this regard,

Determined to secure full compliance with its decisions,

Acting under Chapter VII of the Charter of the United Nations,

1. *Decides* that Iraq has been and remains in material breach of its obligations under relevant resolutions, including resolution 687 (1991), in particular through Iraq's failure to cooperate with United Nations inspectors and the IAEA, and to complete the actions required under paragraphs 8 to 13 of resolution 687 (1991);

2. *Decides*, while acknowledging paragraph 1 above, to afford Iraq, by this resolution, a final opportunity to comply with its disarmament obligations under relevant resolutions of the Council; and accordingly decides to set up an enhanced inspection regime with the aim of bringing to full and verified completion the disarmament process established by resolution 687 (1991) and subsequent resolutions of the Council;

3. *Decides* that, in order to begin to comply with its disarmament obligations, in addition to submitting the required biannual declarations, the Government of Iraq shall provide to UNMOVIC, the IAEA, and the Council, not later than 30 days from the date of this resolution, a currently accurate, full, and complete declaration of all aspects of its programmes to develop chemical, biological, and nuclear weapons, ballistic missiles, and other delivery systems such as unmanned aerial vehicles and dispersal systems designed for use on aircraft, including any holdings and precise locations of such weapons, components, subcomponents, stocks of agents, and related material and equipment, the locations and work of its research, development and production facilities, as well as all other chemical, biological, and nuclear programmes, including any which it claims are for purposes not related to weapon production or material;

4. *Decides* that false statements or omissions in the declarations submitted by Iraq pursuant to this resolution and failure by Iraq at any time to comply with, and cooperate fully in the implementation of, this resolution shall constitute a further material breach of Iraq's obligations and will be reported to the Council for assessment in accordance with paragraphs 11 and 12 below;

5. *Decides* that Iraq shall provide UNMOVIC and the IAEA immediate, unimpeded, unconditional, and unrestricted access to any and all, including underground, areas, facilities, buildings, equipment, records, and means of transport which they wish to inspect, as well as immediate, unimpeded, unrestricted, and private access to all officials and other persons whom UNMOVIC or the IAEA wish to interview in the mode or location of UNMOVIC's or the IAEA's choice pursuant to any aspect of their mandates; further decides that UNMOVIC and the IAEA may at their discretion conduct interviews inside or outside of

Iraq, may facilitate the travel of those interviewed and family members outside of Iraq, and that, at the sole discretion of UNMOVIC and the IAEA, such interviews may occur without the presence of observers from the Iraqi Government; and instructs UNMOVIC and requests the IAEA to resume inspections no later than 45 days following adoption of this resolution and to update the Council 60 days thereafter;

6. *Endorses* the 8 October 2002 letter from the Executive Chairman of UNMOVIC and the Director-General of the IAEA to General Al-Saadi of the Government of Iraq, which is annexed hereto, and decides that the contents of the letter shall be binding upon Iraq;

7. *Decides* further that, in view of the prolonged interruption by Iraq of the presence of UNMOVIC and the IAEA and in order for them to accomplish the tasks set forth in this resolution and all previous relevant resolutions and notwithstanding prior understandings, the Council hereby establishes the following revised or additional authorities, which shall be binding upon Iraq, to facilitate their work in Iraq:

- UNMOVIC and the IAEA shall determine the composition of their inspection teams and ensure that these teams are composed of the most qualified and experienced experts available;
- All UNMOVIC and IAEA personnel shall enjoy the privileges and immunities, corresponding to those of experts on mission, provided in the Convention on Privileges and Immunities of the United Nations and the Agreement on the Privileges and Immunities of the IAEA;
- UNMOVIC and the IAEA shall have unrestricted rights of entry into and out of Iraq, the right to free, unrestricted, and immediate movement to and from inspection sites, and the right to inspect any sites and buildings, including immediate, unimpeded, unconditional, and unrestricted access to Presidential Sites equal to that at other sites, notwithstanding the provisions of resolution 1154 (1998);
- UNMOVIC and the IAEA shall have the right to be provided by Iraq the names of all personnel currently and formerly associated with Iraq's chemical, biological, nuclear, and ballistic missile programmes and the associated research, development, and production facilities;
- Security of UNMOVIC and IAEA facilities shall be ensured by sufficient United Nations security guards;
- UNMOVIC and the IAEA shall have the right to declare, for the purposes of freezing a site to be inspected, exclusion zones, including surrounding areas and transit corridors, in which Iraq will suspend ground and aerial movement so that nothing is changed in or taken out of a site being inspected;
- UNMOVIC and the IAEA shall have the free and unrestricted use and landing of fixed- and rotary-winged aircraft, including manned and unmanned reconnaissance vehicles;
- UNMOVIC and the IAEA shall have the right at their sole discretion verifiably to remove, destroy, or render harmless all prohibited

weapons, subsystems, components, records, materials, and other related items, and the right to impound or close any facilities or equipment for the production thereof; and

- UNMOVIC and the IAEA shall have the right to free import and use of equipment or materials for inspections and to seize and export any equipment, materials, or documents taken during inspections, without search of UNMOVIC or IAEA personnel or official or personal baggage;

8. *Decides* further that Iraq shall not take or threaten hostile acts directed against any representative or personnel of the United Nations or the IAEA or of any Member State taking action to uphold any Council resolution;

9. *Requests* the Secretary-General immediately to notify Iraq of this resolution, which is binding on Iraq; demands that Iraq confirm within seven days of that notification its intention to comply fully with this resolution; and demands further that Iraq cooperate immediately, unconditionally, and actively with UNMOVIC and the IAEA;

10. *Requests* all Member States to give full support to UNMOVIC and the IAEA in the discharge of their mandates, including by providing any information related to prohibited programmes or other aspects of their mandates, including on Iraqi attempts since 1998 to acquire prohibited items, and by recommending sites to be inspected, persons to be interviewed, conditions of such interviews, and data to be collected, the results of which shall be reported to the Council by UNMOVIC and the IAEA;

11. *Directs* the Executive Chairman of UNMOVIC and the Director-General of the IAEA to report immediately to the Council any interference by Iraq with inspection activities, as well as any failure by Iraq to comply with its disarmament obligations, including its obligations regarding inspections under this resolution;

12. *Decides* to convene immediately upon receipt of a report in accordance with paragraphs 4 or 11 above, in order to consider the situation and the need for full compliance with all of the relevant Council resolutions in order to secure international peace and security;

13. *Recalls*, in that context, that the Council has repeatedly warned Iraq that it will face serious consequences as a result of its continued violations of its obligations;

14. *Decides* to remain seized of the matter.

Text of Blix/
El-Baradei letter

ANNEX

International Atomic Energy Agency
United Nations Monitoring,
Verification and Inspection
Commission
The Director General
The Executive Chairman

8 October 2002

Dear General Al-Saadi,

During our recent meeting in Vienna, we discussed practical arrangements that are prerequisites for the resumption of inspections in Iraq by UNMOVIC and the IAEA. As you recall, at the end of our meeting in Vienna we agreed on a statement which listed some of the principal results achieved, particularly Iraq's acceptance of all the rights of inspection provided for in all of the relevant Security Council resolutions. This acceptance was stated to be without any conditions attached.

During our 3 October 2002 briefing to the Security Council, members of the Council suggested that we prepare a written document on all of the conclusions we reached in Vienna. This letter lists those conclusions and seeks your confirmation thereof. We shall report accordingly to the Security Council.

In the statement at the end of the meeting, it was clarified that UNMOVIC and the IAEA will be granted immediate, unconditional and unrestricted access to sites, including what was termed "sensitive sites" in the past. As we noted, however, eight presidential sites have been the subject of special procedures under a Memorandum of Understanding of 1998. Should these sites be subject, as all other sites, to immediate, unconditional and unrestricted access, UNMOVIC and the IAEA would conduct inspections there with the same professionalism.

H. E. General Amir H. Al-Saadi
Advisor
Presidential Office
Baghdad
Iraq

We confirm our understanding that UNMOVIC and the IAEA have the right to determine the number of inspectors required for access to any particular site. This determination will be made on the basis of the size and complexity of the site being inspected. We also confirm that Iraq will be informed of the designation of additional sites, i.e. sites not declared by Iraq or previously inspected by either UNSCOM or the IAEA, through a Notification of Inspection (NIS) provided upon arrival of the inspectors at such sites.

Iraq will ensure that no proscribed material, equipment, records or other relevant items will be destroyed except in the presence of UNMOVIC and/or IAEA inspectors, as appropriate, and at their request.

UNMOVIC and the IAEA may conduct interviews with any person in Iraq whom they believe may have information relevant to their mandate. Iraq will facilitate such interviews. It is for UNMOVIC and the IAEA to choose the mode and location for interviews.

The National Monitoring Directorate (NMD) will, as in the past, serve as the Iraqi counterpart for the inspectors. The Baghdad Ongoing Monitoring and Verification Centre (BOMVIC) will be maintained on the same premises and under the same conditions as was the former Baghdad Monitoring and Verification Centre. The NMD will make available services as before, cost free, for the refurbishment of the premises.

The NMD will provide free of cost: (a) escorts to facilitate access to sites to be inspected and communication with personnel to be interviewed; (b) a hotline for BOMVIC which will be staffed by an English speaking person on a 24 hour a day/seven days a week basis; (c) support in terms of personnel and ground transportation within the country, as requested; and (d) assistance in the movement of materials and equipment at inspectors' request (construction, excavation equipment, etc.). NMD will also ensure that escorts are available in the event of inspections outside normal working hours, including at night and on holidays.

Regional UNMOVIC/IAEA offices may be established, for example, in Basra and Mosul, for the use of their inspectors. For this purpose, Iraq will provide, without cost, adequate office buildings, staff accommodation, and appropriate escort personnel.

UNMOVIC and the IAEA may use any type of voice or data transmission, including satellite and/or inland networks, with or without encryption capability. UNMOVIC and the IAEA may also install equipment in the field with the capability for transmission of data directly to the BOMVIC, New York and Vienna (e.g. sensors, surveillance cameras). This will be facilitated by Iraq and there will be no interference by Iraq with UNMOVIC or IAEA communications.

Iraq will provide, without cost, physical protection of all surveillance equipment, and construct antennae for remote transmission of data, at the request of UNMOVIC and the IAEA. Upon request by UNMOVIC through the NMD, Iraq will allocate frequencies for communications equipment.

Iraq will provide security for all UNMOVIC and IAEA personnel. Secure and suitable accommodations will be designated at normal rates by Iraq for these personnel. For their part, UNMOVIC and the IAEA will require that their staff not stay at any accommodation other than those identified in consultation with Iraq.

On the use of fixed-wing aircraft for transport of personnel and equipment and for inspection purposes, it was clarified that aircraft used by UNMOVIC and IAEA staff arriving in Baghdad may land at Saddam International Airport.

The points of departure of incoming aircraft will be decided by UNMOVIC. The Rasheed airbase will continue to be used for UNMOVIC and IAEA helicopter operations. UNMOVIC and Iraq will establish air liaison offices at the airbase. At both Saddam International Airport and Rasheed airbase, Iraq will provide the necessary support premises and facilities. Aircraft fuel will be provided by Iraq, as before, free of charge.

On the wider issue of air operations in Iraq, both fixed-wing and rotary, Iraq will guarantee the safety of air operations in its air space outside the no-fly zones. With regard to air operations in the no-fly zones, Iraq will take all steps within its control to ensure the safety of such operations.

Helicopter flights may be used, as needed, during inspections and for technical activities, such as gamma detection, without limitation in all parts of Iraq and without any area excluded. Helicopters may also be used for medical evacuation.

On the question of aerial imagery, UNMOVIC may wish to resume the use of U-2 or Mirage overflights. The relevant practical arrangements would be similar to those implemented in the past.

As before, visas for all arriving staff will be issued at the point of entry on the basis of the UN Laissez-Passer or UN Certificate; no other entry or exit formalities will be required. The aircraft passenger manifest will be provided one hour in advance of the arrival of the aircraft in Baghdad. There will be no searching of UNMOVIC or IAEA personnel or of official or personal baggage. UNMOVIC and the IAEA will ensure that their personnel respect the laws of Iraq restricting the export of certain items, for example, those related to Iraq's national cultural heritage. UNMOVIC and the IAEA may bring into, and remove from, Iraq all of the items and materials they require, including satellite phones and other equipment. With respect to samples, UNMOVIC and IAEA will, where feasible, split samples so that Iraq may receive a portion while another portion is kept for reference purposes. Where appropriate, the organizations will send the samples to more than one laboratory for analysis.

We would appreciate your confirmation of the above as a correct reflection of our talks in Vienna.

Naturally, we may need other practical arrangements when proceeding with inspections. We would expect in such matters, as with the above, Iraq's co-operation in all respect.

Yours sincerely,
(*Signed*)   (*Signed*)
Hans Blix   Mohamed El-Baradei
Executive Chairman   Director General
United Nations Monitoring,
International Atomic Energy Agency
Verification and Inspection Commission

BIBLIOGRAPHY

27th Fighter Escort Wing. Yearbook. Texas: Taylor Publishing Company.

Air Force Association. *Foundation Forum: Opportunities and Challenges in Acquisition and Logistics.* Washington, D.C.: Aerospace Education Foundation, 1995.

Air Force Materiel Command: A Legacy in Military Aviation Logistics and R & D. Ohio: Air Force Materiel Command, 1993.

Arnold, Henry H. *Global Mission.* New York: Harper & Brothers, 1949.

Ballard, Jack S., Ray L. Bowers, et al. *The United States Air Force in Southeast Asia, 1961–1973: An Illustrated Account.* Washington, D.C.: Office of Air Force History, 1984.

Bergquist, Mayor Ronald E. *The Role of Airpower in the Iran-Iraq War.* Montgomery, AL: Air University Press, 1988.

Blumenson, Martin, Robert W. Coakley, et al. *Command Decisions.* Washington, D.C.: Office of the Chief of Military History, United States Army, 1960.

Bonds, Ray, ed. *The Vietnam War: The Illustrated History of the Conflict in Southeast Asia.* New York: Salamander Books, 1979.

Bowers, Ray L. *The United States Air Force in Southeast Asia: Tactical Airlift.* Washington, D.C.: Office of Air Force History, 1983.

Boyd, Robert J. *SAC Fighter Planes and Their Operations.* Omaha Headquarters, Strategic Air Command, 1988.

Boyne, Walter J. *Silver Wings, A History of the United States Air Force.* New York: Simon & Schuster, 1993.

—— *Clash of Wings, World War II in the Air.* New York: Simon & Schuster, 1994.

—— *Gulf War.* Lincolnwood, Illinois: Publications International, 1991.

—— *Weapons of Desert Storm.* Lincolnwood, Illinois: Publications International 1991.

Bradley, Omar N., and Clay Blair. *A General's Life.* New York: Simon & Schuster, 1983.

Brennan, Mathew. *Headhunters: Stories from the 1st Squadron, 9th Cavalry, in Vietnam, 1965–1971..* Novato, Calif.: Presidio Press, 1987.

Bright, Charles D., ed. *Historical Dictionary of the U.S. Air Force.* New York: Greenwood Press, 1992.

Buckingham, William A., Jr. *Operation Ranch Hand: The Air Force and Herbicides in Southeast Asia, 1961–1971.* Washington, D.C.: Office of Air Force History, 1982.

Burnham, Frank A. *Aerial Search: The CAP Story.* Fallbrook, Calif.: Aero Publishers, 1974.

Bush, Vannevar. *Modern Arms and Free Men: A Discussion of the Role of Science in Preserving Democracy.* New York: Simon & Schuster, 1949.

Chant, Christopher. *A Compendium of Armaments and Military Hardware.* London and New York: Routledge & Kegan Paul, 1987.

Chinnery, Philip D. *Life on the Line.* New York: St. Martin's Press, 1988.

——. *Vietnam: The Helicopter War.* Annapolis, Md.: Naval Institute Press, 1991.

Cooling, Benjamin Franklin. *Case Studies in the Development of Close Air Support.* Washington, D.C.: Office of Air Force History, 1990.

Coyne, James P. *Airpower in the Gulf.* Arlington, VA: Air Force Association, 1992.

Davis, Larry. *Wild Weasel: The SAM Suppression Story.* Carrolton, TX: Squadron/Signal Publications, 1993.

Denton, Senator Jeremiah A. *When Hell Was in Session.* Montgomery, AL.: Traditional Press, 1982.

Department of the Air Force. *Vezzano to Desert Storm (History of the Fifteenth Air Force, 1943–1991).* Washington, D.C.: Department of the Air Force.

The Development of Air Doctrine in the Army Air Arm, 1917–1941. USAF Historical Studies No. 89. Montgomery, Alabama: USAF Historical Division, 1955.

Donald, David, ed., *U.S. Air Force, Air Power Directory.* London: Aerospace Publishing, 1992.

Donnelly, Thomas, Margaret Roth, and Caleb Baker. *Operation Just Cause.* New York: Lexington Books, 1991.

Dorr, Robert F. *Air War Hanoi.* London, New York, and Sydney: Blandford Press, 1988.

——. *Desert Shield, the Buildup: The Complete Story.* Motorbooks International, 1991.

——. *Desert Storm, Air War.* Osceola, WI: Motorbooks International, 1991.

Dorr, Robert F. and Warren Thompson. *The Korean Air War.* Osceola, WI: Motorbooks International, 1994.

Drendel, Lou. *Air War over Southeast Asia: A Pictorial Record.* Vol. 2, 1967–1970. Carrolton, Texas: Squadron/Signal Publications, Inc., 1983.

——. *. . . And Kill MiGs.* Carrolton, Texas: Squadron/Signal Publications, 1984.

——. *TAC (A Pictorial History of the USAF Tactical Air Forces, 1970–1977).* Carrolton, Texas: Squadron/Signal Publications, 1978.

——. *THUD.* Carrolton, Texas: Squadron/Signal Publications, 1986.

Drury, Richard S. *My Secret War.* Fallbrook, Calif.: Aero Publishers, Inc., 1979.

Eschmann, Karl J. *Linebacker: The Untold Story of the Air Raids over North Vietnam.* New York: Ballantine Books, 1989.

Flanagan, John F. *Vietnam Above the Treetops.* New York: Praeger, 1992.

Fogleman, General Ronald R. "A New American Way of War." *Aerospace Education Foundation Forum.* Arlington, VA: Aerospace Education Foundation, 1996.

Forrestal, James V. *The Forrestal Diaries.* Walter Millis and E. S. Duffield, eds. New York: Viking, 1951.

Francillon, Rene J. *McDonnell Douglas Aircraft Since 1920.* Vol. 2. Annapolis, Md.: Naval Institute Press, 1979.

Frisbee, John L., ed. *Makers of the United States Air Force.* Washington, D.C.: Office of Air Force History, 1987.

Futrell, Robert Frank. *Ideas, Concepts, Doctrine: A History of Basic Thinking in the United States Air Force, 1907–1964.* Montgomery, Alabama: Air University Press, 1971.

——. *Ideas, Concepts, Doctrine: Basic Thinking in the United States Air Force. Vol. 1, 1907–1960.* Montgomery, Alabama: Air University Press, 1989.

——. *Ideas, Concepts, Doctrine: Basic Thinking in the United States Air Force. Vol. 2, 1961–1984.* Montgomery, Alabama: Air University Press, 1989.

——. *The United States Air Force in Korea, 1950–1953.* Washington, D.C.: Office of Air Force History.

Gansler, Jacques S. *Affording Defense.* Cambridge, Mass.: MIT Press, 1989.

Gantz, Lieutenant Colonel Kenneth F., ed. *The United States Air Force Report on the Ballistic Missile.* New York: Doubleday, 1958.

Gaston, James C. *Planning the American Air War: Four Men and Nine Days in 1941.* Washington, D.C.: National Defense University Press, 1982.

Gorn, Michael H. *Harnessing the Genie: Science and Technology Forecasting for the Air Force, 1944–1986.* Washington, D.C.: Office of Air Force History, 1988.

Gross, Charles Joseph. *Prelude to the Total Force: The Air National Guard, 1943–1969.* Washington, D.C.: Office of Air Force History, 1985.

Gurney, Colonel Gene. *Vietnam: The War in the Air. A Pictorial History of the U.S. Air Forces in the Vietnam War: Air Force, Army, Navy, and Marines.* New York: Crown, 1985.

Halberstadt, Hans. *The Wild Weasels: History of U.S. Air Force SAM Killers, 1965–Today.* Osceola, Wis.: Motorbooks International, 1992.

Hallion, Richard P. *Storm over Iraq.* Washington, D.C.: Smithsonian Institution Press, 1992.

Harrison, Marshall. *A Lonely Kind of War: Forward Air Controller, Vietnam.* Novato, Calif.: Presidio Press, 1989.

Haulman, Dr. Daniel L., and Colonel William C. Stancik, eds. *Air Force Victory Credits World War I, World War II, Korea, and Vietnam.* Montgomery, AL: United States Air Force Historical Research Center, 1988.

Holley, I. B., Jr. *Ideas and Weapons.* Washington, D.C.: Office of Air Force History, 1983.

Hurley, Colonel Alfred F., and Major Robert C. Ehrhart, eds. *Air Power and Warfare.* Washington, D.C.: Office of Air Force History, 1979.

Kitfield, James. *Prodigal Soldiers.* New York: Simon & Schuster, 1995.

Kohn, Richard H., and Joseph P. Harahan, eds. *USAF Warrior Studies.* Washington, D.C.: Office of Air Force History, 1986.

Kutler, Stanley I. *Encyclopedia of the Vietnam War.* New York: Charles Scribner's Sons, 1996.

Lauer, Colonel Timothy M., and Steven L. Llanso. *Encyclopedia of Modern U.S. Military Weapons.* New York: Berkley Books, 1995.

Lifeline Adrift: The Defense Industrial Base in the 1990s. Arlington, VA: Aerospace Education Foundation, 1991.

Littauer, Raphael, and Norman Uphoff, eds. *The Air War in Indochina.* Boston: Beacon Press, 1972.

Logan, Don. *The 388th Tactical Fighter Wing at Korat Royal Thai Air Force Base, 1972.* Altglen, PA: Schiffer Military/Aviation History, 1995.

Macy, Robert and Melinda. *Destination Baghdad.* Las Vegas: M&M Graphics, 1991.

Manning, Thomas A., Dick J. Bukard, et al. *History of Air Training Command, 1943–1993.* San Antonio. TX: Headquarters, Air Education and Training Command, 1993.

McCarthy, Brigadier General James R., and Lieutenant Colonel George B. Allison. *Linebacker II: A View from the Rock.* Montgomery, AL: Airpower Research Institute, 1979.

McPeak, Merrill A. *Selected Works, 1990–1994.* Montgomery, AL: Air University Press, 1995.

Mesko, Jim. *Airmobile: The Helicopter War in Vietnam.* Carrolton, Texas: Squadron/Signal Publication, Inc., 1984.

Millis, Walter, and E. S. Duffield. *The Forrestal Diaries.* New York: Viking, 1951.

Momyer, General William W. *Air Power in Three Wars: WWII, Korea, Vietnam.* Washington, D.C.: Department of the Air Force.

Moody, Walton S. *Building a Strategic Air Force.* Washington, D.C.: Air Force History and Museums Program, 1996.

Morrocco, John. *Thunder from Above: Air War, 1941–1968.* Boston, Mass.: Boston Publishing, 1984.

Morse, Stan, ed. *Gulf Air War Debrief Described by the Pilots That Fought.* London: Aerospace Publishing, 1991.

Mrozek, Donald J. *The U.S. Air Force After Vietnam: Postwar Challenges and Potential for Responses.* Montgomery, Alabama: Air University Press, 1988.

Murray, Williamson. *Air War in the Persian Gulf.* Baltimore: Nautical & Aviation Publishing Company of America, 1995.

Neufeld, Jacob. *The Development of Ballistic Missiles in the United States Air Force, 1945–1960.* Washington, D.C.: Office of Air Force History, 1990.

——. *Reflections on Research and Development in the United States Air Force.* Washington, D.C.: Center for Air Force History, 1993.

New World Vistas: Air and Space Power for the 21st Century (Summary Volume). Washington, D.C., Department of the Air Force, 1995.

Nordeen, Lon O., Jr. *Air Warfare in the Missile Age.* Washington, D.C.: Smithsonian Institution Press, 1985.

Perry, Mark. *Four Stars.* Boston: Houghton Mifflin, 1989.

Pimlott, John. *Vietnam: The Decisive Battles.* New York: Macmillan, 1990.

Pogue, Forrest C. *George C. Marshall: Ordeal and Hope.* New York: Viking, 1966.

——. *George C. Marshall: Statesman, 1945–1949.* New York: Viking, 1987.

Polmar, Norman, and Timothy Laurer. *Strategic Air Command.* Baltimore: Nautical & Aviation Publishing Company of America, 1970.

Puryear, Edgar F., Jr. *George S. Brown, General, U.S. Air Force: Destined for Stars.* Novato, Calif.: Presidio Press, 1983.

Ravenstein, Charles A. *The Organization and Lineage of the United States Air Force*. Washington, D.C.: Office of Air Force History, 1986.

Reinberg, Linda. *In the Field: The Language of the Vietnam War*. New York and Oxford: Facts on File, 1991.

Ralston, Major General Joseph, W. "Fighter Modernization in the 1990's." *Foundation Forum*, January 31–February 1, 1991.

Rich, Ben R., and Leo Janos. *Skunk Works: A Personal Memoir of My Years at Lockheed*. Boston: Little, Brown, 1994.

Robbins, Christopher. *The Ravens*. New York: Crown, 1987.

Roberts, Michael. *The Illustrated Directory of the United States Air Force*. New York: Crescent Books, 1989.

Schlight, John. *The War in South Vietnam: The Years of the Offensive, 1965–1968*. Washington, D.C.: Office of Air Force History, 1988.

Shapley, Deborah. *Promise and Power: The Life and Times of Robert McNamara*. Boston: Little, Brown, 1993.

Skinner, Michael. *U.S.A.F.E.: A Primer of Modern Air Combat in Europe*. Novato, Calif.: Presidio Press, 1983.

Smith, Barry D. *Air Rescue: Saving Lives Stateside*. London: Osprey Publishing, 1989.

Smith, Harvey H., Donald W. Bernier, et al. *Area Handbook for South Vietnam*. Washington, D.C.: Foreign Areas Studies Division, American University, 1967.

Spector, Ronald H. *Researching the Vietnam Experience*. Washington, D.C.: Analysis Branch, U.S. Army Center of Military History, 1984.

Strategic Air Command History Office. *From Snark to Peacekeeper: A Pictorial History of Strategic Air Command Missiles*. Omaha, NE: Office of the Historian, Headquarters Strategic Air Command, 1990.

Sturm, Thomas A. *The USAF Scientific Advisory Board: Its First Twenty Years, 1944–1964*. Washington, D.C.: Office of Air Force History, 1986.

Summers, Harry G., Jr. *On Strategy: A Critical Analysis of the Vietnam War*. New York: Dell, 1982.

Swanborough, Gordon, and Peter M. Bowers. *United States Military Aircraft Since 1909*. Washington, D.C.: Smithsonian Institution Press, 1989.

Termena, Bernard J., Layne B. Peiffer, and H. P. Carlin. *Logistics: An Illustrated History of AFLC and its Antecedents, 1921–1981*. Dayton, Ohio: Headquarters, Air Force Logistics Command.

Tilford, Earl H., Jr. *Search and Rescue in Southeast Asia, 1961–1975*. Washington, D.C.: Office of Air Force History, 1980.

———. *Setup: What the Air Force Did in Vietnam and Why*. Montgomery, Alabama: Air University Press, 1991.

Tolson, Lieutenant General John J. *Vietnam Studies: Airmobility, 1961–1971*. Washington, D.C.: Department of the Army, 1973.

Trotti, John. *Phantom over Vietnam*. New York: Berkley Books, 1985.

Van Staaveren, Jacob. *Interdiction in Southern Laos, 1960–1968*. Washington, D.C.: Center for Air Force History, 1993.

Venkus, Colonel Robert E. *Raid on Qaddafi*. New York: St. Martin's Press, 1992.

Waddell, Colonel Dewey, and Major Norm Wood, eds. *Air War—Vietnam.* New York: Arno Press, 1978.

Warden, John A., III. *The Air Campaign: Planning for Combat.* Washington, D.C.: National Defense University Press, 1988.

Watson, George M., Jr. *The Office of the Secretary of the Air Force, 1947–1965.* Washington, D.C.: Center for Air Force History, 1993.

Werrell, Kenneth P. *The Evolution of the Cruise Missile.* Montgomery, Alabama: Air University Press, 1985.

Wolf, Richard I. *United States Air Force Basic Documents on Roles and Missions.* Washington, D.C.: Office of Air Force History, 1987.

Wolk, Herman S. *Planning and Organizing the Postwar Air Force, 1943–1947.* Washington, D.C.: Office of Air Force History, 1984.

Woodward, Bob. *The Commanders.* New York: Simon & Schuster, 1991.

Wright, Lieutenant Colonel Monte D., and Lawrence J. Paszek. *Science, Technology, and Warfare.* Proceedings of the Third Military History Symposium, United States Air Force Academy. Washington, D.C.: Office of Air Force History, 1969.

Yarborough, Colonel Tom. *Da Nang Diary: A Forward Air Controller's Year of Combat over Vietnam.* New York: St. Martin's Press, 1990.

Yenne, Bill. *The History of the U.S. Air Force.* New York: Exeter Books, 1984.

A CHRONOLOGY OF AEROSPACE POWER SINCE 1903

1903–1913
March 23, 1903. First Wright brothers airplane patent, based on their 1902 glider, is filed in America.

August 8, 1903. The Langley gasoline-engine model plane is successfully launched from a catapult on a houseboat.

December 8, 1903. Second and last trial of Langley airplane, piloted by Charles M. Manly, is wrecked in launching from a houseboat on Potomac River in Washington, D.C.

December 17, 1903. At Kitty Hawk, N. C., Orville Wright achieves the world's first manned, powered, sustained, and controlled flight by a heavier-than-air vehicle. His fourth and longest flight of the day is 852 feet in 59 seconds. Three days earlier, Wilbur Wright achieved the world's first powered airplane flight—105 feet in 3.5 seconds—but crashed soon after takeoff, and his flight is not regarded as being either sustained or controlled.

January 18, 1905. The Wright brothers open negotiations with the U.S. government to build an airplane for the Army, but nothing comes of this first meeting.

February 5, 1905. T. S. Baldwin takes part in a 10-mile race between his dirigible and an automobile. The dirigible and its pilot win by a three-minute margin.

June 23, 1905. The first flight of the Wright Flyer III is made at Huffman Prairie, outside Dayton, Ohio. The Wright brothers' first fully controllable aircraft is able to turn and bank and remain aloft for up to thirty minutes.

May 22, 1906. After turning down two previous submissions, the U.S. government issues the Wright brothers the first patent on their flying machine.

November 12, 1906. Brazilian Alberto Santos-Dumont sets the first recognized absolute speed record of 25.66 mph in the Santos-Dumont Type 14-bis at Bagatelle, France. However, this speed is slower than speeds posted by the Wright brothers in the United States.

August 1, 1907. The Aeronautical Division of the U.S. Army Signal Corps, forerunner of U.S. Air Force, is established.

October 26, 1907. Henri Farman sets the recognized absolute speed record of 32.74 mph in a Voisin-Farman biplane at Issy-les-Moulineaux, France.

December 23, 1907. The Army's Chief Signal Officer, Brig. Gen. James Allen, issues the first specification for a military airplane.

January 13, 1908. Henri Farman wins the 50,000-franc Deutsch-Archdeacon Prize for the first officially observed 1-kilometer circular flight in Europe.

May 14, 1908. The first passenger flight takes place in the Wright plane at Kitty Hawk in preparation for delivery of a government airplane. Wilbur Wright pilots the machine, with Charles Furnas, an employee, as the first passenger.

May 19, 1908. Signal Corps Lt. Thomas E. Selfridge becomes the first soldier to fly a heavier-than-air machine.

July 4, 1908. Glenn H. Curtiss wins the *Scientific American* trophy with his *June Bug* biplane by flying for more than a mile over Hammondsport, N.Y. Speed for the trip is 39 mph.

August 8, 1908. At Camp d'Auvours, France, Wilbur Wright surpasses French flight records for duration, distance, and altitude.

September 3, 1908. First test flight of an Army flying machine is made at Fort Myer, Va., by Orville Wright.

September 17, 1908. Lt. Thomas E. Selfridge becomes the first person killed in a powered aircraft accident when a Wright Flyer crashes at Fort Myer, Va. Orville Wright, at the controls, suffers serious injuries.

November 13, 1908. Wilbur Wright, in a Wright biplane at Camp d'Auvours, France, and Henri Farman, in a Voisin at Issy, France, concurrently set a world altitude record of 82 feet.

April 24, 1909. Wilbur Wright pilots a Wright biplane at Centocelle, Italy, from which the first aerial motion picture is taken.

July 27, 1909. Orville Wright, with Lt. Frank P. Lahm as passenger, makes the first official test flight of the Army's first airplane at Fort Myer, Va.

August 2, 1909. The Army accepts its first airplane, bought from the Wright brothers for $25,000, plus a $5,000 bonus because the machine exceeds the speed requirement of 40 mph.

August 23, 1909. At the world's first major air meet in Reims, France, Glenn Curtiss becomes the first American to claim the recognized absolute speed record as he flies at 43.385 mph in his Reims Racer biplane.

August 25, 1909. Land for the first Signal Corps airfield is leased at College Park, Md.

October 23, 1909. Lt. Benjamin D. Foulois takes his first flying lesson from Wilbur Wright at College Park, Md.

October 26, 1909. Lt. Frederick E. Humphreys becomes the first Army pilot to solo in the Wright Military Flyer at College Park, Md.

November 3, 1909. Lt. George C. Sweet becomes the first Navy officer to fly, as a passenger in the Wright Military Flyer.

January 19, 1910. Signal Corps Lt. Paul Beck, flying as a passenger with Louis Paulhan in a Farman biplane, drops three 2-pound sandbags in an effort to hit a target at the Los Angeles Flying Meet. This is the first bombing experiment by an Army officer.

March 2, 1910. Benjamin Foulois becomes the first Army officer to fly an Army airplane.

March 19, 1910. Orville Wright opens the first Wright Flying School at Montgomery, Ala., on a site that will later become Maxwell AFB.

May 25, 1910. In Dayton, Ohio, Wilbur and Orville Wright fly together for the first time.

July 10, 1910. Walter Brookins becomes the first airplane pilot to fly at an altitude greater than a mile. He reaches 6,234 feet in a Wright biplane over Atlantic City, N.J.

July 10, 1910. Leon Morane pushes the recognized absolute speed record to 66.181 mph in a Bleriot monoplane at Reims, France.

August 20, 1910. Army Lt. Jacob Fickel fires a .30 caliber Springfield rifle at the ground while flying as a passenger in a Curtiss biplane over Sheepshead Bay Track near New York City. This is the first time a military firearm has been discharged from an airplane.

September 2, 1910. Blanche Scott becomes the first American woman to solo, flying a Curtiss pusher at the Curtiss company field in Hammondsport, N.Y. She is not granted a pilot's license, however.

October 11, 1910. Former President Theodore Roosevelt becomes the first Chief Executive to fly. He goes aloft as a passenger in a Wright biplane over St. Louis, Mo.

November 7, 1910. Phillip O. Parmalee performs the world's first air cargo mission, flying a bolt of silk from Dayton to Columbus, Ohio.

November 14, 1910. Navy Lt. Eugene Ely, in a Curtiss biplane, takes off from the deck of a modified cruiser, USS *Birmingham*.

January 18, 1911. Navy Lt. Eugene Ely, flying a Curtiss pusher, makes the first landing on a ship. He touches down on a 119-foot-long wooden platform on the stern of the cruiser USS *Pennsylvania*, riding at anchor in San Francisco Bay.

February 1, 1911. The first licensed aircraft manufacturer in the U.S., the Burgess and Curtis Co. (no relation to the company founded by Glenn Curtiss), of Marblehead, Mass., receives authorization from the Wright Co.

March 3, 1911. The first appropriation for Army air operations—$25,000—is authorized for Fiscal Year 1912.

April 11, 1911. The Army's first permanent flying school is established at College Park, Md.

May 8, 1911. The first Navy airplane, A-1, an amphibian, is ordered from Glenn Curtiss. This date has been officially proclaimed the birthday of naval aviation.

May 12, 1911. Edward Nieuport sets the recognized absolute speed record of 74.415 mph in a Nieuport monoplane at Chalons, France. On June 16, he will push the speed record to 80.814 mph.

September 17-December 10, 1911. Calbraith Perry Rodgers, in the Wright EX biplane *Vin Fiz*, makes the first transcontinental flight, from Sheepshead Bay, N.Y., to Long Beach,

Calif. He makes 76 stops and crashes 20 times.

February 22, 1912. Jules Vedrines pushes the recognized absolute speed record past the 100 mph barrier, as he hits 100.22 mph in a Deperdussin racer at Pau, France.

February 23, 1912. First official recognition of the rating "Military Aviator" appears in War Department Bulletin No. 2.

June 5, 1912. Lt. Col. C. B. Winder of the Ohio National Guard becomes the first National Guard pilot. He was taught at the Army Aviation School.

June 14, 1912. Cpl. Vernon Burge becomes the Army's first enlisted pilot.

July 5, 1912. Capt. Charles DeF. Chandler and Lts. T. D. Milling and H. H. Arnold become the first flyers to qualify as "Military Aviators."

November 5, 1912. First artillery adjustments directed from a plane begin at Fort Riley, Kan., by Lts. H. H. Arnold, pilot, and Follett Bradley, observer.

November 27, 1912. The Army Signal Corps purchases the first of three Curtiss-F two-seat biplane flying boats.

December 11, 1912. A French pilot, Roland Garros, sets an altitude record of 18,406 feet in a Morane airplane at Tunis.

February 11, 1913. The first bill for a separate aviation corps, HR 28728, is introduced in Congress by Rep. James Hay of W. Va. It fails to pass.

March 2, 1913. First flight pay is authorized: 35 percent over base pay for officers detailed on aviation duty.

April 27, 1913. Pilot Robert G. Fowler and cameraman R. A. Duhem make the first flight across the Isthmus of Panama. They are arrested by Panamanian authorities upon publication in a newspaper of the story and pictures of the flight.

May 13, 1913. The first flight of the world's first four-engine airplane, the *Russian Knight*, affectionately called "Le Grand," takes place in Russia. The aircraft is designed by Igor I. Sikorsky.

May 30, 1913. The Massachusetts Institute of Technology begins teaching aerodynamics.

June 21, 1913. Eighteen-year-old Georgia "Tiny" Broadwick becomes the first woman to make a parachute jump in the U.S. Her 1,000-foot leap takes place over Los Angeles, Calif.

June 30, 1913. The first Navy aviator is killed: Ens. W. D. Billingsley is thrown from a seaplane.

July 19, 1913. In the skies over Seattle, Wash., Milton J. Bryant begins a new form of advertising—skywriting.

August 27, 1913. Lt. Petr Nikolaevich Nesterov of the Imperial Russian Army performs history's first inside loop while flying a Nieuport Type IV over Kiev.

November 30, 1913. In late November or early December, the first known aerial combat takes place over Naco, Mexico, between Phil Rader, flying for Gen. Victoriano Huerta, and Dean Ivan Lamb, with Venustiano Carranza. Details are unknown, except that a dozen pistol shots are exchanged.

1914–1923

January 1, 1914. America's first regularly scheduled airline starts operation across Tampa Bay between St. Petersburg and Tampa, Fla., with one Benoist flying boat. It lasts three months.

January 20, 1914. The Navy's aviation unit from Annapolis, Md., arrives at Pensacola, Fla., to set up the first naval air station.

February 24, 1914. In the wake of a rash of accidents, an Army investigative board condemns all pusher-type airplanes.

April 25, 1914. Navy Lt. (j.g). P.N.L. Bellinger, flying a Curtiss AB-3 flying boat from the battleship USS *Mississippi* (BB-23), makes the first U.S. operational air sortie against another country when he searches for sea mines during the Veracruz incident.

May 5, 1914. A patent is issued for hinged inset trailing-edge ailerons.

July 18, 1914. The Aviation Section of the Signal Corps is created by Congress. Sixty officers and students and 260 enlisted men are authorized.

August 25, 1914. Stephan Banic, a coal miner in Greenville, Pa., is issued a patent for a workable parachute design.

August 26, 1914. The first air battle of World War I on the eastern front takes place. Staff Capt. Petr Nikolaevich Nesterov records the first aerial ramming in combat.

December 1–16, 1914. Two-way air-to-ground radio communication is demonstrated in a Burgess-Wright biplane by Army Signal Corps Lts. H. A. Dargue and J. O. Mauborgne over Manila, the Philippines.

January 19–20, 1915. Germany launches the first zeppelin bombing

raids on England. One airship, the L.6, turns back, but two others, the L.3 and L.4, drop their bombs on Great Yarmouth and King's Lynn.

March 3, 1915. Congress approves the act establishing the National Advisory Committee for Aeronautics. NACA is to "supervise and direct the scientific study of flight with a view to [its] practical solution." The committee, initially given a budget of $5,000, will evolve into the National Aeronautics and Space Administration.

April 1, 1915. French Lt. Roland Garros shoots down a German Albatros two-seater with a Hotchkiss machine gun fixed on the nose of his Morane-Saulnier Type L monoplane. The airplane's propeller is fitted with wedge-shaped steel deflector plates that protect the blades from damage as the rounds pass through the propeller arc.

November 6, 1915. Navy Lt. Cmdr. Henry C. Mustin makes the first airplane catapult launching from a moving vessel, USS *North Carolina*, in Pensacola Bay, Fla.

December 11, 1915. The first foreign students to enter a U.S. flying training program—four Portuguese Army officers—report to the Signal Corps Aviation School at San Diego, Calif.

March 15, 1916. The 1st Aero Squadron begins operations with Gen. John J. Pershing in a punitive expedition against Mexico and Pancho Villa.

March 21, 1916. The French government authorizes the formation of the Escadrille Américaine. The unit, made up of American volunteer pilots, is later renamed the Lafayette Escadrille.

June 18, 1916. H. Clyde Balsey of the Lafayette Escadrille is shot down near

Verdun, France, the first American-born aviator shot down in World War I.

April 30, 1917. During the month, Maj. William "Billy" Mitchell becomes the first American Army officer to fly over the German lines.

November 27, 1917. Brig. Gen. Benjamin D. Foulois takes over as Chief of the Air Service for the American Expeditionary Force (AEF). He replaces Brig. Gen. William L. Kenly.

January 19, 1918. The U.S. School of Aviation Medicine begins operations at Hazelhurst Field, Mineola, N.Y.

January 23, 1918. The first ascent by an AEF balloon is made at the balloon school in Cuperly, France.

February 5, 1918. While flying as a substitute gunner with a French squadron, Lt. Stephen W. Thompson becomes the first American to record an aerial victory while in a U.S. uniform. He shoots down a German Albatros D.III but is credited with only half the victory, sharing the kill with the French pilot.

February 18, 1918. The first American fighter unit proper, the 95th Aero Squadron, arrives in France.

February 28, 1918. Using a radiotelephone, human voice is transmitted from an aircraft to the ground for the first time. The flight took place in San Diego, Calif.

February 28, 1918. Regulation of the airways begins with an order by President Woodrow Wilson requiring licenses for civilian pilots or owners. More than 800 licenses are issued.

March 11, 1918. Lt. Paul Baer becomes the first AEF Air Service member awarded the Distinguished Service Cross.

March 19, 1918. The 94th Aero Squadron makes the first U.S. operational flights across the front lines in France.

April 14, 1918. Lts. Alan Winslow and Douglas Campbell, flying Nieuport 28s of the 94th Aero Squadron, down two German fighters in a 10-minute battle. Lieutenant Winslow is the first pilot in the American sector of the front to down an airplane; Lieutenant Campbell is the first U.S.-trained pilot to score a victory.

April 21, 1918. Rittmeister Manfred von Richthofen, the Red Baron, is shot down in action over France by Capt. A. Roy Brown, a Canadian. The German ace, killed in the battle, had 80 aerial victories.

May 7, 1918. Flying a Nieuport 28, 1st Lt. Edward V. Rickenbacker, who would go on to be the leading American ace of World War I, records his first solo victory, downing a German Pfalz. Flying with the 94th Aero Squadron, he had recorded a half victory, his first, on April 29.

May 15, 1918. The Aviation Section of the Signal Corps begins regular airmail service from Washington, D.C., to New York City.

May 20, 1918. The Division of Military Aeronautics is established, with Maj. Gen. William L. Kenly as director.

May 24, 1918. U.S. Army Air Service organized.

June 12, 1918. The 96th Aero Squadron bombs the Dommary-Baroncourt railway yards in France in the first daylight bombing raid carried out by the AEF.

August 2, 1918. The 135th Corps Observation Squadron makes its first wartime patrol in U.S.-assembled DH-4s powered by American-made Liberty engines.

September 12, 1918. Lt. Frank Luke shoots down his first enemy observation balloon. By the time he is killed seventeen days later, he has shot down nearly 16 balloons and airplanes. In his last mission, near Murvaux, France, he shoots down three observation balloons but comes under attack by eight German pilots and from ground batteries. Severely wounded, he makes a strafing pass on some enemy ground troops before making a forced landing. Surrounded, he defends himself with his automatic pistol until he is killed by enemy troops. He is posthumously awarded the Medal of Honor for his actions.

September 25, 1918. Capt. Edward V. Rickenbacker of the 94th Aero Squadron attacks seven enemy aircraft, shooting down two of them near Billy, France, and wins the first Medal of Honor given for air activity.

October 6, 1918. 2nd Lts. Harold E. Goettler (pilot) and Erwin R. Bleckley (observer) are killed by ground fire while attempting to drop supplies to a battalion of the Army's 77th Division, which had been cut off in the Argonne Forest near Binarville, France. Having been subjected to heavy ground fire on their first attempt, they flew at a lower altitude on the second trip in order to get the packages more precisely on the designated spot. The duo are posthumously awarded the Medal of Honor for their actions.

October 30, 1918. Flying a Spad VII, Capt. Edward V. Rickenbacker, America's Ace of Aces, records his last two aerial victories, an observation balloon and a Fokker DVII, over France. Captain Rickenbacker, who finished the war with 24.33 victories, recorded 12.83 confirmed victories in the month of October alone.

November 7, 1918. Dr. Robert H. Goddard demonstrates tube-launched solid-propellant rockets at Aberdeen Proving Ground, Md.

November 10, 1918. The Air Service records its last two aerial victories of World War I, as Maj. Maxwell Kirby of the 94th Aero Squadron tallies the last solo (and his only) kill, and two crews from the 104th Observation Squadron team up for the other victory.

December 4–22, 1918. Under the command of Maj. Albert D. Smith, four JN-4s fly from San Diego, Calif., to Jacksonville, Fla., to complete the Army's first transcontinental flight. Only Major Smith's plane manages to make the entire trip.

January 24, 1919. Army Air Service pilot 1st Lt. Temple M. Joyce makes 300 consecutive loops in a Morane fighter at Issoudun, France.

May 16–27, 1919. Navy Lt. Cmdr. Albert C. "Putty" Read and a crew of five fly from Trepassey Bay, Newfoundland, to Lisbon, Portugal, via the Azores, in the Curtiss NC-4 flying boat, spending 53 hours, 58 minutes aloft. This is the first crossing of the Atlantic Ocean by air. Two other NCs start the trip but do not complete it.

June 14–15, 1919. Capt. John Alcock and Lt. Arthur Whitten Brown of the United Kingdom make the first nonstop flight across the Atlantic in 16 hours, 12 minutes.

September 1, 1919. Dive bombing is demonstrated at Aberdeen Proving Ground, Md.

October 30, 1919. The reversible-pitch propeller is tested for the first time at McCook Field near Dayton, Ohio.

February 27, 1920. Maj. R. W. "Shorty" Schroeder sets a world altitude record of 33,114 feet in the Packard-LePere LUSAC-11 biplane over McCook Field.

June 4, 1920. The Army reorganization bill is approved, creating an Air Service with 1,514 officers and 16,000 enlisted men.

June 5, 1920. A provision in the Fiscal Year 1921 appropriations bill restricts the Army Air Service to operating from land bases.

February 22, 1921. American transcontinental airmail service begins. The route between San Francisco and Mineola, N.Y., is flown in fourteen segments by pilots flying U.S.-built de Havilland DH-4s. The first flight, made mostly in bad weather, takes 33 hours, 20 minutes.

July 13–21, 1921. In a series of tests off the mouth of the Chesapeake Bay, Army airplanes from Langley Field, Va., sink three ships, including the captured German battleship *Ostfriesland*, demonstrating the vulnerability of naval craft to aerial attack.

September 26, 1921. Sadi Lecointe pushes the recognized absolute speed record past 200 mph, as he hits 205.223 mph in the Nieuport-Delage Sesquiplane at Ville-Sauvage, France.

November 12, 1921. Wesley May, with a five-gallon can of gasoline strapped to his back, climbs from the wing of one aircraft to the wing of another in the first "air-to-air" refueling.

March 20, 1922. USS *Langley* (CV-1), the Navy's first aircraft carrier, is commissioned in Norfolk, Va. The ship is the converted collier *Jupiter*.

September 4, 1922. Lt. James H. Doolittle makes the first transcontinental crossing in an aircraft in a single day—2,163 miles in 21 hours, 20 minutes.

October 17, 1922. The first carrier takeoff in U.S. Navy history is made by Lt. V. C. Griffin in a Vought VE-7SF from USS *Langley* (CV-1), at anchor in the York River in Virginia.

October 18, 1922. Gen. William H. "Billy" Mitchell becomes the first U.S. military pilot to hold the recognized absolute speed record, as he sets a mark of 222.97 mph in the Curtiss R-6 at Selfridge Field, Mich. This is · also the first time the world speed record has been certified outside of France.

May 2–3, 1923. Lt. Oakley G. Kelly and Lt. John A. Macready complete the first nonstop transcontinental flight. The trip from New York to San Diego takes 26 hours, 50 minutes, 3 seconds in a Fokker T-2.

September 4, 1923. First flight of the airship USS *Shenandoah* (ZR-1) is made at NAF Lakehurst, N.J. The airship will make 57 flights in two years before it is destroyed by a storm near Marietta, Ohio.

1924–1933
February 5, 1924. 2nd Lt. Joseph C. Morrow, Jr., qualifies as the 24th and last Military Aviator under the rules set up for that rating.

March 4, 1924. The Army Air Service takes on a new mission: aerial icebreaking. Two Martin bombers and two DH-4s bomb the frozen Platte River at North Bend,

Neb., for six hours before the ice clears.

April 6–September 28, 1924. The Army Air Service completes the first circumnavigation of the globe. Four crews in Douglas World Cruisers begin the voyage in Seattle, Wash., but only two aircraft (*Chicago* and *New Orleans*) and their crews complete the trip.

September 28, 1923. At Cowes, on the Isle of Wight, off England's southern coast, Navy Lt. David Rittenhouse claims the Schneider Cup for the United States for the first time. Flying a Curtiss CR-3, Lieutenant Rittenhouse wins the prestigious seaplane race with an average speed of 177.37 mph.

October 12–15, 1924. As part of World War I reparations, the German zeppelin LZ-126 is flown from Friedrichshafen, Germany, to NAF Lakehurst, N.J. The Navy will later christen the airship USS *Los Angeles* (ZR-3).

October 28, 1924. Army Air Service airplanes break up cloud formations at 13,000 feet over Bolling Field, D.C., by "blasting" them with electrified sand.

January 24, 1925. The Navy airship USS *Los Angeles* (ZR-3), with 25 scientists and astronomers on board, is used to make observations of a solar eclipse.

February 2, 1925. President Calvin Coolidge signs the Kelly Act, authorizing the air transport of mail under contract. This is the first major legislative step toward the creation of a U.S. airline industry.

July 15, 1925. The Dr. A. Hamilton Rice Expedition, the first group of explorers to use an airplane, returns to the U.S. The expedition, which used a Curtiss Seagull floatplane, discovered the headwaters of the Amazon River.

October 26, 1925. Lt. James H. Doolittle, flying the Curtiss R3C-2 floatplane racer, wins the Schneider Cup race in Baltimore, Md., with an average speed of 232.57 mph. This marked back-to-back wins for the United States and the only time the Army had competed in a seaplane race. The next day, he sets a world seaplane record of 245.713 mph over a 3-kilometer course.

December 17, 1925. Air power pioneer Billy Mitchell is found guilty of violating the 96th Article of War ("conduct of a nature to bring discredit on the military service") and is sentenced to a five-year suspension of rank, pay, and command. Already demoted from brigadier general, Colonel Mitchell decides instead to resign from the Army.

January 16, 1926. The Daniel Guggenheim Fund for the Promotion of Aeronautics is founded.

March 16, 1926. Dr. Robert H. Goddard launches the world's first liquid-fueled rocket at Auburn, Mass.

July 2, 1926. U.S. Army Air Service becomes U.S. Army Air Corps.

July 2, 1926. Congress establishes the Distinguished Flying Cross (made retroactive to April 6, 1917).

May 20–21, 1927. The first solo non-stop transatlantic flight is completed by Charles A. Lindbergh in the Ryan NYP *Spirit of St. Louis*: New York to Paris in 33 hours, 32 minutes.

September 16, 1927. In a staged publicity event, MGM Studios attempts to make the first nonstop flight across the United States with an animal on board an aircraft. Noted pilot

Martin Jensen is chosen to fly Leo, MGM's trademark lion, from San Diego, Calif., to New York City for a promotional tour. Man and beast never arrive, however. After a nationwide search and three days of front-page headlines, Jensen and Leo are found unhurt in the Arizona desert. A storm had forced Jensen down and the Ryan BI monoplane (which had been fitted with a steel cage for Leo) was heavily damaged on landing.

November 16, 1927. The U.S. Navy's second true aircraft carrier—USS *Saratoga* (CV-3)—is commissioned. The ship will later be deliberately destroyed during a 1946 atomic bomb test.

January 27, 1928. The Navy airship USS *Los Angeles* (ZR-3) lands on the aircraft carrier USS *Saratoga* (CV-3) during a fleet exercise near Newport, R.I., and resumes its patrol after replenishment.

February 15, 1928. President Coolidge signs a bill authorizing acceptance of a new site near San Antonio, Tex., to become the Army Air Corps training center. This center is now Randolph AFB.

March 1–9, 1928. USAAC Lt. Burnie R. Dallas and Beckwith Havens make the first transcontinental flight in an amphibious airplane. Total flight time in the Loening Amphibian is 32 hours, 45 minutes.

March 30, 1928. Italian Maj. Mario de Bernardi pushes the recognized absolute speed record past 300 mph, as he hits 318.624 mph in the Macchi M.52R at Venice, Italy.

April 15–21, 1928. Sir George Hubert Wilkins and Lt. Carl B. Eielson fly from Point Barrow, Alaska, across the Arctic Ocean to Spitsbergen, Norway,

in a Lockheed Vega. This first west-to-east trip over the top of the world takes only 21 hours of flying, but the duo are delayed by weather.

May 12, 1928. Lt. Julian S. Dexter of the Air Corps Reserve completes a 3,000-square-mile aerial mapping assignment over the Florida Everglades. The project takes 65 hours of flying, spread over two months.

June 9, 1928. For the third consecutive year, Army Air Corps Lt. Earle E. Partridge wins the distinguished gunnery badge at the Air Corps Machine Gunning Matches at Langley Field, Va.

June 15, 1928. Lts. Karl S. Axtater and Edward H. White, flying in an Air Corps blimp directly over an Illinois Central train, dip down and hand a mailbag to the postal clerk on the train, thus completing the first airplane-to-train transfer.

August 1, 1928. Airmail rates rise to 5 cents for the first ounce and 10 cents for each additional ounce.

September 22, 1928. The number of people whose lives have been saved by parachutes exceeds 100 when Lt. Roger Q. Williams bails out over San Diego, Calif.

October 11–15, 1928. The German *Graf Zeppelin* (LZ-127) makes the first transoceanic voyage by an airship carrying paying passengers. *Graf Zeppelin* travels from Friedrichshafen, Germany, to NAF Lakehurst, N.J., in nearly 112 hours, with 20 passengers and a crew of 37.

November 11, 1928. In a Lockheed Vega, Sir George Hubert Wilkins and Lt. Carl B. Eielson make the first flight over Antarctica.

January 1–7, 1929. *Question Mark*, a Fokker C-2 commanded by Maj. Carl

A. "Tooey" Spaatz and including Capt. Ira C. Eaker and Lt. Elwood R. Quesada among its crew, sets an endurance record for a refueled aircraft of 150 hours, 40 minutes, 14 seconds.

January 23–27, 1929. The aircraft carriers USS *Lexington* (CV-2) and USS *Saratoga* (CV-3) participate in fleet exercises for the first time.

February 10–11, 1929. Evelyn Trout sets a women's solo flight endurance record of 17 hours, 21 minutes, 37 seconds in the monoplane *Golden Eagle*.

April 24, 1929. Elinor Smith, seventeen years old, sets a women's solo endurance record of 26 hours, 21 minutes, 32 seconds in a Bellanca CH monoplane at Roosevelt Field, Long Island, N.Y.

May 16, 1929. At the first Academy Award ceremonies in Los Angeles, Calif., the Paramount movie *Wings* wins the Oscar for Best Picture for 1927–28. The World War I flying epic stars Richard Arlen, Buddy Rogers, and Clara Bow. A young Gary Cooper has a minor role.

September 24, 1929. Lt. James H. Doolittle makes the first blind, all-instrument flight at Mitchel Field, N.Y., in a completely covered cockpit (accompanied by check pilot). He took off, flew a short distance, and landed.

September 30, 1929. At Frankfurt, Germany, Fritz von Opel travels just over a mile in the world's first flight of a rocket-powered airplane. The Rak-1 tops 85 mph but crashes.

November 23, 1929. After visiting Dr. Robert H. Goddard, Charles A. Lindbergh arranges a grant of $50,000 from the Daniel Guggenheim Fund for the Promotion of Aeronautics to support Dr. Goddard's work with rockets.

November 29, 1929. Navy Cmdr. Richard E. Byrd, Bernt Balchen, Army Capt. Ashley McKinley, and Harold June make the first flight over the South Pole. Mr. Balchen is the pilot of the Ford Trimotor *Floyd Bennett*.

December 31, 1929. The Daniel Guggenheim Fund for the Promotion of Aeronautics ends its activities.

April 12, 1930. Led by Capt. Hugh Elmendorf, nineteen pilots of the 95th Pursuit Squadron set an unofficial world record for altitude formation flying over Mather Field, Calif. The P-12 pilots reach 30,000 feet, shattering the old record of 17,000 feet.

May 3, 1930. Laura Ingalls performs 344 consecutive loops. Shortly afterward, she tries again and does 980. In another flight during 1930, she does 714 barrel rolls, setting a pair of records that few people have cared to challenge.

May 15, 1930. Ellen Church, a registered nurse, becomes the world's first airline stewardess, as she serves sandwiches on a Boeing Air Transport flight between San Francisco, Calif., and Cheyenne, Wyo. She sits in the jumpseat of the Boeing Model 80A.

October 25, 1930. Transcontinental commercial air service between New York and Los Angeles begins.

March 10, 1931. Air Corps Capt. Ira Eaker attempts to set the transcontinental speed record in the Lockheed Y1C-17, a special version of the civilian Vega. Taking off from Long Beach, Calif., Captain Eaker gets as far as Tolu, Ky., before he has to make a forced landing in a field because of air in the fuel lines. Captain Eaker had traveled 1,740 miles at an average speed of 237 mph, which, if he had been able to complete the flight, would have shattered the existing coast-to-coast speed mark.

September 4, 1931. Jimmy Doolittle wins the first Bendix Trophy transcontinental race, flying the Laird *Super Solution* from Los Angeles to Cleveland with an average speed of 223.058. Total flying time is 9 hours and 10 minutes. He then flies on to New York to complete a full flight across the continent.

September 29, 1931. Flying in the same aircraft that won the last Schneider Cup seaplane race, Royal Air Force Flt. Lt. George Stainforth pushes the recognized absolute speed record past 400 mph as he hits 407.001 mph in the Supermarine S.6b at Lee-on-Solent, England.

October 3–5, 1931. Americans Clyde "Upside Down" Pangborn and Hugh Herndon, Jr., make the first nonstop transpacific flight from Japan to America, in a Bellanca monoplane. The trip takes 41 hours, 13 minutes.

December 22, 1931. Maj. Gen. Benjamin D. Foulois takes oath as Chief of Air Corps.

August 25, 1932. Amelia Earhart becomes the first woman to complete a nonstop transcontinental flight.

November 19, 1932. National monument to Wilbur and Orville Wright is dedicated at Kitty Hawk, N.C.

April 4, 1933. The Navy dirigible USS *Akron* (ZRS-4) hits the sea during a training flight off the East Coast and breaks up. Of a crew of nearly 80, only three survive. Among the casualties is Rear Adm. William A. Moffett, head of the Navy's Bureau of Aeronautics.

July 15–22, 1933. Famed aviator Wiley Post, flying the Lockheed Vega *Winnie Mae*, becomes the first person to fly around the world solo. The 15,596-mile flight takes 7 days, 18 hours, 49 minutes at an average speed of 134.5 mph.

September 4, 1933. Jimmy Wedell sets a world landplane speed record of 304.98 mph in the Wedell-Williams racer over Glenview, Ill.

December 31, 1933. The prototype Soviet Polikarpov I-16 Mosca is flown for the first time. When the type enters service in 1934, it is the first monoplane fighter to have an enclosed cockpit and fully retractable landing gear.

1934–1939

February 19, 1934. President Franklin D. Roosevelt issues an Executive Order canceling existing airmail contracts because of fraud and collusion. The Army Air Corps is designated to take over airmail operations.

May 1, 1934. Navy Lt. Frank Akers makes a blind landing in a Berliner-Joyce OJ-2 at College Park, Md., in a demonstration of a system intended for aircraft carrier use. In subsequent flights, he makes takeoffs and landings between NAS Anacostia, D.C., and College Park under a hood without assistance.

May 19, 1934. The first flight of the Ant-20 *Maxim Gorki*, at this time the world's largest aircraft, is made in the Soviet Union. The aircraft was designed by Andrei Tupolev.

June 1, 1934. Army Air Corps airmail operations are terminated.

June 18, 1934. Boeing begins company-funded design work on the Model 299, which will become the B-17.

July 18, 1934. Lt. Col. Henry H. "Hap" Arnold leads a flight of 10 Martin B-10 bombers on a six-day photographic mapping mission to Alaska.

December 31, 1934. Helen Richey, flying a Ford Trimotor from Washington, D.C., to Detroit, Mich., becomes the first woman in the U.S. to pilot an airmail transport aircraft on a regular schedule.

February 12, 1935. The Navy airship USS *Macon* (ZRS-5) crashes off the California coast with two fatalities out of a crew of 83. This loss effectively ends the Navy's rigid airship program.

March 1, 1935. General Headquarters (GHQ) Air Force is created at Langley Field, Va. It is a compromise for those seeking a completely independent Air Force and the War Department's General Staff, which wants to retain control of what is thought of as an auxiliary to the ground forces.

March 9, 1935. Future Reichsmarshal Hermann Göring announces the existence of the Luftwaffe in an interview with London *Daily Mail* correspondent Ward Price. This statement implies a gross violation of the Versailles Treaty, which prohibits Germany from having an air force.

March 21, 1935. Company pilot Bill Wheatley, with chief engineer I. M. Mac Laddon as a passenger, makes the first flight of the Consolidated XP3Y-1, the forerunner to the Catalina patrol bomber/rescue aircraft, at NAS Anacostia, D.C. The P-Boat would be produced for more than 10 years and would become the most numerous, (3,200+, including more than 300 for the USAAF) and quite possibly, the most famous flying boat ever.

August 15, 1935. Famed pilot Wiley Post and humorist Will Rogers are killed in a crash of the hybrid Lockheed Orion-Explorer shortly after takeoff near Point Barrow, Alaska.

September 15, 1935. Alexander P. de Seversky sets a recognized class for record speed over a 3-kilometer course (piston-engined amphibians) of 230.41 mph in a Seversky N3PB at Detroit, Mich. This is the oldest certified aviation record still standing.

November 22, 1935. First transpacific airmail flight, in *China Clipper*, by Capt. Edwin C. Musick, takes place from San Francisco to Honolulu, Midway Island, Wake Island, Guam, and Manila.

December 17, 1935. First flight of the Douglas Sleeper Transport, the first of 10,654 DC-3s and derivatives Douglas will build between 1935 and 1947. The U.S. military uses C-47s in three wars, and some "Gooney Birds" are still in use today. The DC-3 is one of the most famous airplanes of all time.

February 19, 1936. Airpower advocate Billy Mitchell dies in New York City at the age of 57. He is buried in Milwaukee, Wis.

March 5, 1936. Vicker's chief test pilot "Mutt" Summers makes the first flight of the Supermarine Type 300 from Eastleigh Airport in Hampshire, England. The brainchild of designer R. J. Mitchell, this prototype is the first of 18,298 Merlin-powered Spitfires of all marks to be built by 1945.

September 4, 1936. Louise Thaden and Blanche Noyes become the first women to win the Bendix Trophy transcontinental race from New York to Los Angeles in a Beech Model 17 Staggerwing with an average speed of 165.346 mph. Total flying time is 14 hours and 55 minutes.

April 12, 1937. Frank Whittle bench-tests the first practical jet engine in laboratories at Cambridge University, England.

May 6, 1937. The German dirigible *Hindenburg* (LZ-129) burns while mooring at Lakehurst, N.J., killing 36 people.

May 21, 1937. Amelia Earhart and Fred Noonan leave from San Francisco in a Lockheed Electra on a round-the-world flight that ends on July 2, 1937, when they disappear in the Pacific.

July 20, 1937. First shoulder-sleeve insignia authorized for an independent American air unit—for General Headquarters Air Force.

September 1, 1937. Air Corps 1st Lt. Ben Kelsey makes the first flight of the Bell XFM-1 Airacuda multiplace fighter at Buffalo, N.Y. Both the plane and the concept prove to be dismal failures. The Airacuda turns out to be a maintenance nightmare, and the multiplace fighter concept is just not practical.

October 15, 1937. The Boeing XB-15 makes its first flight at Boeing Field in Seattle, Wash., under the control of test pilot Eddie Allen.

February 17, 1938. Six Boeing B-17 Flying Fortresses, under the command of Lt. Col. Robert Olds, leave Miami, Fla., on a goodwill flight to Buenos Aires, Argentina. The return trip to Langley Field, Va., is the longest non-stop flight in Air Corps history.

April 6, 1938. Company pilot James Taylor makes the first flight of the Bell XP-39 Airacobra at Wright Field, near Dayton, Ohio. Nearly 4,800 Lend-Lease P-39s will be used to particularly good effect by Soviet pilots to destroy German tanks.

April 22, 1938. World War I ace Edward V. Rickenbacker buys a majority stake in Eastern Air Lines from North American Aviation for $3.5 million. That sum would roughly cover the cost of a single engine for a Boeing 757 today.

May 15, 1938. U.S. Secretary of the Interior Harold L. Ickes refuses to allow inert helium to be exported to Germany for use in Zeppelins. Secretary Ickes feels that the gas might be diverted to military purposes.

July 10–14, 1938. Howard Hughes, Harry H. P. Conner, Army Lt. Thomas Thurlow, Richard Stoddard, and Ed Lund set a round-the-world flight record of 3 days, 19 hours, 8 minutes, 10 seconds in a Lockheed Model 14 Super Electra passenger aircraft. The crew travels 14,791 miles.

July 17–18, 1938. Ostensibly aiming for California, Douglas "Wrong-Way" Corrigan, flying a Curtiss Robin, lands in Dublin, Ireland, after a nonstop 28-hour flight from Floyd Bennett Field in Brooklyn, N.Y.

August 22, 1938. The Civil Aeronautics Act goes into effect. The Civil Aeronautics Authority will now coordinate all nonmilitary aviation. (The Federal Aviation Act, which created the Federal Aviation Administration, will be passed August 15, 1958.)

September 29, 1938. Brig. Gen. H. H. "Hap" Arnold is named Chief of the Army Air Corps, succeeding Maj. Gen. Oscar Westover, who was killed in a plane crash September 21.

October 14, 1938. Company test pilot Edward Elliott makes the first flight of the Curtiss XP-40 at Buffalo, N.Y. Almost 14,000 P-40s will be built before production ends in 1944.

December 31, 1938. The Boeing Model 307 Stratoliner, the first passenger plane to have a pressurized cabin, makes its first flight.

January 27, 1939. Lt. Benjamin Kelsey makes the first flight of the Lockheed XP-38 at March Field, Calif. The two leading American aces of all time, Maj. Richard Bong (40 victories) and Maj. Thomas McGuire (38 victories), would fly P-38s.

March 5, 1939. Using a hook trailing from their Stinson Reliant, Norman Rintoul and Victor Yesulantes demonstrate a nonstop airmail system by picking a mail sack off a pole in Coatesville, Pa.

March 30, 1939. Flugkapitan Hans Dieterle sets a world speed record of 463.82 mph in the Heinkel He-100V-8. The flight is made at Oranienburg, Germany.

April 3, 1939. President Roosevelt signs the National Defense Act of 1940, which authorizes a $300 million budget and 6,000 airplanes for the Army Air Corps and increases AAC personnel to 3,203 officers and 45,000 enlisted troops.

April 26, 1939. Flugkapitan Fritz Wendel sets the last recognized absolute speed record before World War II as he pilots the Messerschmitt Bf-209V-1 to a speed of 469.224 mph at Augsburg, Germany.

May 20, 1939. Regularly scheduled transatlantic passenger and airmail service begins.

June 20, 1939. The German Heinkel He-176, the first aircraft to have a throttle-controlled liquid-fuel rocket engine, makes its first flight at Peenemünde with Flugkapitan Erich Warsitz at the controls.

August 27, 1939. The first jet-powered aircraft, the Heinkel He-178, makes its first flight. Flugkapitan Erich Warsitz is the pilot.

September 1, 1939. At 4:34 A.M., Lt. Bruno Dilley leads three Junkers Ju-87 Stuka dive bombers in an attack against the Dirschau Bridge. The German invasion of Poland, the first act of World War II, begins six minutes later.

October 8, 1939. A Lockheed Hudson crew from the Royal Air Force's No. 224 Squadron shoots down a German Do-18 flying boat. This is the first victory recorded by an American-built aircraft in World War II.

October 13, 1939. Evelyn Pinchert Kilgore becomes the first woman to be issued a Civil Aeronautics Authority instructor's certificate.

December 29, 1939. The prototype Consolidated XB-24 Liberator makes a 17-minute first flight from Lindbergh Field in San Diego, Calif., with company pilot Bill Wheatley at the controls. More than 18,100 B-24s will be built in the next five and a half years, making for the largest military production run in US history.

1940–1945

February 21, 1940. Henry A. H. Boot and John T. Randall, working at the University of Birmingham, England, create the first practical magnetron. The magnetron, a resonant-cavity microwave generator, is vital in the development of airborne radar.

May 16, 1940. President Roosevelt calls for 50,000 airplanes a year.

July 10, 1940. The Luftwaffe attacks British shipping in the English Channel docks in South Wales. These actions are the first in what will become the Battle of Britain.

August 13–October 5, 1940. Against overwhelming odds, Royal Air Force pilots fend off the Luftwaffe during the Battle of Britain and ward off German invasion of the British Isles. The Luftwaffe loses 1,733 aircraft and crews.

September 17, 1940. Adolf Hitler announces that Operation Sea Lion, the German invasion of Great Britain, "has been postponed indefinitely." This effectively marks the end of the Battle of Britain, although fighting continues.

October 8, 1940. The Royal Air Force announces formation of the first Eagle Squadron, a Fighter Command unit to consist of volunteer pilots from the U.S.

March 21, 1941. The first black flying unit, the 99th Pursuit Squadron, is activated. As part of the 332nd Pursuit Squadron, it will become known as the Tuskegee Airmen.

April 11, 1941. With the possibility that the U.S. would be drawn into World War II and that all of Europe could be in Axis hands, the Army Air Corps invites Consolidated and Boeing to submit design studies for a bomber capable of achieving 450 mph at 25,000 feet, a range of 12,000 miles at 275 mph, and a payload of 4,000 pounds of bombs at maximum range. This study results in the Convair B-36.

May 6, 1941. Company test pilot Lowery Brabham makes the first flight of the Republic P-47 Thunderbolt at Farmingdale, Long Island, N.Y. The P-47, the heaviest single-engine fighter ever built in the U.S., will see action in every theater in World War II as both a high-altitude escort fighter and as a low-level fighter-bomber.

June 20, 1941. The Army Air Forces are established, comprising the Office of the Chief of Air Corps and the Air Force Combat Command, with Maj. Gen. H. H. Arnold as Chief.

July 8, 1941. The RAF makes a daylight attack on Wilhelmshaven, Germany, using Boeing Fortress Is. This is the first operational use of the B-17 Flying Fortress.

August 12, 1941. First successful rocket-assisted takeoff of an airplane takes place.

December 1, 1941. Civil Air Patrol is established.

December 7, 1941. Imperial Japanese forces attack Pearl Harbor.

December 8, 1941. The day after the Japanese attack on Pearl Harbor, company test pilot Robert Stanley makes the first flight of the Bell XP-63 Kingcobra, a bigger and more powerful version of the P-39, at Buffalo, N.Y.

December 10, 1941. Five B-17s of the 93rd Bomb Squadron, 19th Bomb Group, carry out the first heavy bomb mission of World War II, attacking a Japanese convoy near the Philippines and also sinking the first enemy vessel by U.S. aerial combat bombing.

December 16, 1941. Lt. Boyd "Buzz" Wagner becomes the first American USAAF ace of World War II by shooting down his fifth Japanese plane over the Philippines.

December 20, 1941. The American Volunteer Group (Claire L. Chennault's Flying Tigers), in action over Kunming, China, enters combat for the first time.

February 23, 1942. B-17s attack Rabaul, the first Allied raid on the newly established Japanese base.

February 22, 1942. First American air headquarters in Europe in World War II, U.S. Army Bomber Command, is established in England, with Brig. Gen. Ira C. Eaker commanding.

March 7, 1942. The first five African-American pilots graduate from training at Tuskegee Army Air Field in Alabama. By the end of the war, the "Tuskegee Airmen" would number 950 pilots and open the door to the armed forces for other African-Americans.

March 9, 1942. The War Department is reorganized into three autonomous forces: Army Air Forces, Ground Forces, and Services of Supply.

April 8, 1942. The first flight of supplies takes place over "the Hump"—a 500-mile air route from Assam, India, over the Himalayas, to Kunming, China, where the Chinese continue to resist Japanese forces. By August, Tenth Air Force will be ferrying over 700 tons a month to these troops, who were cut off by the Japanese control of the Burma Road.

April 18, 1942. Sixteen North American B-25s, commanded by Lt. Col. James H. Doolittle, take off from USS *Hornet* (CV-8) and bomb Tokyo. For planning and successfully carrying out this daring raid, Doolittle is promoted to brigadier general and is awarded the Medal of Honor.

May 4–8, 1942. The Battle of the Coral Sea becomes the first naval engagement fought solely by aircraft.

May 26, 1942. Contract test pilot Vance Breese makes the first flight of the Northrop XP-61 Black Widow from Northrop Field in Hawthorne, Calif. The Black Widow is the Army Air Forces' first purpose-designed night fighter.

June 3–4, 1942. In the Battle of Midway, three U.S. carriers destroy four Japanese carriers while losing one of their own, inflicting a major defeat on the Japanese fleet.

June 12, 1942. In the first mission against a European target, 13 B-24s of HALPRO Detachment fly from Egypt against the Ploesti, Romania, oil fields.

July 4, 1942. The first Army Air Forces bomber mission over Western Europe (flown in Douglas A-20s) in World War II is flown against four airdromes in the Netherlands.

July 4, 1942. The Flying Tigers are incorporated into the AAF as the 23rd Pursuit Group.

July 7, 1942. A B-18 of 396th Bombardment Squadron sinks a German submarine off Cherry Point, N.C., in first sure kill off the Atlantic coast by aircraft.

August 17, 1942. The first American heavy bomber mission in Western Europe in World War II is flown by B-17s of the 97th Bombardment Group against the Rouen-Sotteville railyards in France.

October 2, 1942. The Bell XP-59A lifts off from Muroc Dry Lake Bed, Calif., with Bell test pilot Robert Stanley at the controls. It is the first flight of a jet airplane in the United States. The next day, Col. Lawrence C. Craigie makes the first flight by a USAAF pilot.

November 2, 1942. NAS Patuxent River, Md., is established as the Navy's test center for aircraft and equipment.

November 8–11, 1942. Army pilots take off from carriers to support the invasion of North Africa. The P-40 pilots then touch down at land bases.

December 1942. The first issue of *Air Force* magazine is published. It succeeds the *Army Air Forces Newsletter*.

December 4, 1942. Ninth Air Force B-24 Liberator crews, based in Egypt, bomb Naples—the first American attacks in Italy.

December 27, 1942. 2nd Lt. Richard I. Bong, who would later go on to be America's leading ace of all time and win the Medal of Honor, records his first aerial victory. Bong, who recorded all of his victories while flying the Lockheed P-38, would score more than half of his kills while flying with the 9th Fighter Squadron.

January 5, 1943. Army Air Forces Maj. Gen. Carl A. Spaatz is appointed Commander in Chief of the Allied Air Forces in North Africa.

January 9, 1943. Famed Boeing test pilot Eddie Allen and Lockheed test pilot Milo Burcham make the first flight of the Lockheed C-69 transport (the military version of the Model 49 Constellation) at Burbank, Calif. Allen was on loan to Lockheed for the occasion.

January 27, 1943. The first American air raid on Germany is made by Eighth Air Force B-17 crews against Wilhelmshaven and other targets in the northwestern part of the country.

February 15, 1943. It is announced that Maj. Gen. Ira C. Eaker will succeed Maj. Gen. Carl A. Spaatz as commander of USAAF's Eighth Air Force.

February 18, 1943. First class of 39 flight nurses graduate from AAF School of Air Evacuation, Bowman Field, Ky.

February 27, 1943. RAF Bomber Command announces that the Allied air forces have made 2,000 sorties in the past 48 hours.

March 2–4, 1943. A Japanese attempt to reinforce Lae, New Guinea, is foiled by aircraft of the Southwest Pacific Air Forces during the Battle of the Bismarck Sea. Modified B-25s are used for the first time in low-level skip-bombing techniques. More than 60 enemy aircraft are destroyed and some 40,000 tons of Japanese shipping are sunk.

March 10, 1943. Fourteenth Air Force is formed under the command of Maj. Gen. Claire L. Chennault.

March 19, 1943. Lt. Gen. H. H. Arnold is promoted to four-star rank, a first for the Army Air Forces.

April 4, 1943. The B-24 *Lady Be Good*, returning from a bombing mission, overshoots its base at Soluch, Libya, and is not heard from again. In 1959, the wreckage will be found by an oil exploration party 440 miles into the Libyan desert.

April 18, 1943. P-38 pilots from Henderson Field, Guadalcanal, intercept and shoot down two Mitsubishi "Betty" bombers over Bougainville. The aerial ambush kills Japanese Adm. Isoroku Yamamoto, who planned the Pearl Harbor attack.

May 30, 1943. All organized Japanese resistance ceases on Attu in the Aleutian Islands off Alaska. Attu was recaptured by American forces at a fearful cost in lives; all but 28 members of the Japanese garrison sacrificed themselves.

June 15, 1943. The 58th Bombardment Wing, the Army Air Forces' first B-29 unit, is established at Marietta, Ga.

June 15, 1943. The world's first operational jet bomber, the German Arado Ar-234V-1 Blitz, makes its first flight.

July 2, 1943. Lt. Charles Hall shoots down a German FW-190 over Sicily, becoming the first black U.S. flyer to down an Axis plane.

July 19, 1943. Rome is bombed for the first time. Flying from Benghazi, Libya, 158 B-17 crews and 112 B-24 crews carry out a morning raid. A second attack is staged in the afternoon.

August 1, 1943. Staging from Benghazi, 177 Ninth Air Force B-24s drop 311 tons of bombs from low level on the oil refineries at Ploesti during Operation Tidal Wave. Forty-nine aircraft are lost, and seven others land in Turkey. This is the first large-scale, minimum-altitude attack by AAF heavy bombers on a strongly defended target. It is also the longest major bombing mission to date in terms of distance from base to target. Four officers, Col. Leon W. Johnson, Col. John R. Kane, Maj. John L. Jerstad, and 2nd Lt. Lloyd H. Hughes, are awarded the Medal of Honor for their actions. More Air Force Medals of Honor are awarded for this mission than for any other in the service's history.

August 17, 1943. Eighth Air Force bombers attack the Messerschmitt works at Regensburg, Germany, and ball-bearing plants at Schweinfurt in a massive daylight raid. German fighters down 60 of the 376 American aircraft.

August 31, 1943. The Grumman F6F Hellcat goes into operational use with VF-5 off USS *Yorktown* (CV-10) in an attack on Marcus Island, 700 miles south of Japan. Hellcat pilots will account for nearly three-fourths of all Navy air-to-air victories in World War II.

September 12, 1943. German commandos, led by Capt. Otto Skorzeny, help Italian dictator Benito Mussolini break out of a hotel in Gran Sasso where he is being held prisoner. Captain Skorzeny and Il Duce escape in a Fieseler Fi-156 Storch observation plane.

September 27, 1943. P-47s with belly tanks go the whole distance with Eighth Air Force bombers for a raid on Emden, Germany.

October 14, 1943. Eighth Air Force conducts the second raid on the ball-bearing factories at Schweinfurt, Germany. As a result, the Germans will disperse their ball-bearing manufacturing, but the cost of the raid is high; 60 of the 291 B-17s launched do not return, and 138 more are damaged.

October 31, 1943. Over New Georgia in the Solomon Islands, a Chance Vought F4U-2 Corsair aviator accomplishes the Navy's first successful radar-guided interception.

November 22–26; 1943. At the Cairo Conference, Roosevelt and Churchill, along with Chiang Kai-Shek, agree that B-29s will be based in the China-Burma-India theater for strikes on the Japanese home islands.

December 5, 1943. P-51 pilots begin escorting US bombers to European targets. Ninth Air Force begins Operation Crossbow raids against German bases where secret weapons are being developed.

December 24, 1943. First major Eighth Air Force assault on German V-weapon sites is made when 670 B-17s and B-24s bomb the Pas de Calais area of France.

January 8, 1944. Developed in only 143 days, the prototype Lockheed XP-80 Shooting Star, *Lulu Belle*, makes its first flight at Muroc Dry Lake (later Edwards AFB), Calif., with Milo Burcham at the controls. It is the first American fighter to exceed 500 mph in level flight.

January 11, 1944. The first U.S. use of forward-firing rockets is made by Navy TBF-1C Avenger crews against a German submarine.

January 22, 1944. Mediterranean Allied Air Forces fly 1,200 sorties in support of Operation Shingle, the amphibious landings at Anzio, Italy.

February 15, 1944. The Nazi-occupied Abbey of Monte Cassino, Italy, is destroyed by 254 American B-17 crews, B-25 crews, and B-26 crews attacking in two waves. The ruins of the abbey will not be captured by Fifth Army until May 18, 1944.

February 20, 1944. The first mission of "Big Week"—six days of strikes by Eighth Air Force (based in England) and Fifteenth Air Force (based in Italy) against German aircraft plants—is flown.

March 4, 1944. B-17s of the Eighth Air Force conduct the first daylight bombing raid on Berlin.

March 5, 1944. British Brig. Gen. Orde Wingate's Raiders, popularly known as Chindits, land at "Broadway," a site near Indaw, Burma, in a daring night operation. General Wingate will be killed 19 days later in an airplane crash.

March 6, 1944. In the first major USAAF attack on Berlin, 660 heavy bombers unload 1,600 tons of bombs.

March 16, 1944. NACA proposes that a jet-propelled transonic research airplane be developed. This ultimately leads to the Bell X-1.

March 25, 1944. Fifteenth Air Force crews temporarily close the Brenner Pass between Italy and Austria. This mission, against the Aviso viaduct, is the first operational use of the VB-1 Azon (azimuth only) radio-controlled bomb.

April 11, 1944. Led by Royal Air Force Wing Commander R. N. Bateson, six de Havilland Mosquitos of No. 613 Squadron bomb an art gallery at The Hague where population records are kept. These records, many of which were destroyed, were used by the Gestapo to suppress the Dutch resistance.

April 12, 1944. Maj. Richard I. Bong records three aerial kills in a single mission to bring his personal tally to 28, for which he is recognized amid much hoopla as surpassing the total of America's World War I Ace of Aces, Captain Edward Rickenbacker. (Captain Eddie even sent Major Bong a case of scotch.) However, when the Air Force revises its aerial victory credits in the late 1960s, Captain Rickenbacker's long-accepted total of 26 kills is reduced to 24.33, which means that Major Bong actually passed Captain Rickenbacker on April 3, 1944.

May 11, 1944. Operation Strangle (March 19 to May 11) ends. Mediterranean Allied Air Forces' operations against enemy lines of communication in Italy total 50,000 sorties, with 26,000 tons of bombs dropped.

May 21, 1944. Operation Chattanooga Choo-Choo—systematic Allied air attacks on trains in Germany and France—begins.

June 2, 1944. The first shuttle bombing mission, using Russia as the eastern terminus, is flown. Lt. Gen. Ira C. Eaker, head of Mediterranean Allied Air Forces, flies in one of the B-17s.

June 6, 1944. Allied pilots fly approximately 15,000 sorties on D day. It is an effort unprecedented in concentration and size.

June 9, 1944. Allied units begin operations from bases in France.

June 13, 1944. The first German V-1s fired in combat are launched against England. Four of 11 strike London.

June 15, 1944. Forty-seven B-29 crews based in India and staging through Chengdu, China, attack steel mills at Yawata in the first B-29 strike against Japan.

June 19–20, 1944. "The Marianas Turkey Shoot": in two days of fighting, the Japanese lose 476 aircraft. American losses are 130 planes.

June 22, 1944. The GI Bill is signed into law.

July 5, 1944. The Northrop MX-324, the first US rocket-powered airplane, is flown for the first time by company pilot Harry Crosby at Harper Dry Lake, Calif.

July 9, 1944. Part of wrecked and captured Fiesler Fi-103 buzz bombs are delivered to Wright Field, Ohio, for evaluation. Seventeen days later Ford Motor Company finishes building a copy of the Argus pulse jet motor, and by October, Republic is chosen to build copies of the bomb's airframe. The U.S.-built duplicates are called JB-1 Loons.

July 17, 1944. Napalm incendiary bombs are dropped for the first time by American P-38 pilots on a fuel depot at Coutances, near Saint-Lô, France.

July 22, 1944. In the first all-fighter shuttle, Italy-based U.S. P-38 Lightnings and P-51 Mustangs of Fifteenth Air Force attack Nazi airfields at Bacau and Zilistea, northeast of Ploesti. The planes land at Russian bases.

July 27, 1944. The executive committee of the NACA discusses robots and their possibilities for military and other uses.

August 4, 1944. The first Aphrodite mission (a radio-controlled B-17 carrying 20,000 pounds of TNT) is flown against V-2 rocket sites in the Pas de Calais section of France.

August 14, 1944. Capt. Robin Olds records his first victory while flying with the 434th Fighter Squadron in the ETO. He would go on to tally 11 more kills by July 4, 1945. His next aerial victory would come on January 2, 1967, making him the only American ace to record victories in nonconsecutive wars.

August 28, 1944. Eighth Force's 78th Fighter Group claims the destruction of an Me-262, the first jet to be shot down in combat.

September 1, 1944. Company pilot Robert Stanley makes the first flight of the Bell RP-63A Kingcobra, a highly unusual modification to the P-63 that allowed the aircraft to be used as a piloted target. These Pinball aircraft were heavily armored (even the cockpit glazing was extra-thick) and gunnery students would fire frangible bullets made of lead and plastic at these aircraft in flight.

September 8, 1944. The German V-2, the world's first ballistic missile, is first used in combat. Two strike Paris; two more are launched against London.

September 14, 1944. Col. Floyd B. Wood, Maj. Harry Wexler, and Lt. Frank Reckord fly into a hurricane in a Douglas A-20 to gather scientific data.

September 17, 1944. Operation Market Garden begins: 1,546 Allied aircraft and 478 gliders carry parachute and glider troops in an airborne assault between Eindhoven and Arnhem in the Netherlands in an effort to secure a Rhine crossing at Arnhem.

October 24, 1944. Navy Capt. David McCampbell, who will go on to be the Navy's leading ace of all time, sets the U.S. record for aerial victories in a single engagement when he shoots down nine Japanese fighters.

November 1, 1944. A Boeing F-13 (photoreconnaissance B-29) crew makes the first flight over Tokyo since the 1942 Doolittle raid. The first XXI Bomber Command raid will be made on November 24, when 88 B-29s bomb the city.

November 3, 1944. The Japanese start their "Fu-Go Weapon" offensive against the United States. These balloon weapons are carried across the Pacific on the jet stream and release their bomblets over the U.S.

November 10, 1944. Thirty-six B-25s of Fifth Air Force attack a Japanese convoy near Ormoc Bay, Leyte, sinking three ships.

December 15, 1944. Bound for France, famed bandleader Maj. Glenn Miller and two others take off from England in a Noorduyn C-64 Norseman and are never heard from again. Several possible causes for the disappearance have been formulated, but none is ever proved.

December 15, 1944. President Roosevelt signs legislation creating the five-star ranks of General of the Army and Admiral of the Fleet.

December 17, 1944. The 509th Composite Group, assembled to carry out atomic bomb operations, is established at Wendover, Utah.

December 17, 1944. On the forty-first anniversary of the Wright brothers' historic first flight, Maj. Richard I. Bong, America's leading ace of all time, records his fortieth and final aerial victory.

December 21, 1944. Gen. H. H. Arnold becomes General of the Army—the first airman to hold five-star rank.

December 26, 1944. Maj. Thomas B. McGuire, Jr., records four aerial victories in a single mission in the Southwest Pacific. These kills bring Major McGuire's victory total to 38, making him the second leading American ace of all time. Major McGuire, a Medal of Honor recipient, is killed in combat 12 days later.

January 20, 1945. Army Air Forces Maj. Gen. Curtis E. LeMay succeeds Brig. Gen. Haywood "Possum" Hansell as commander of XXI Bomber Command in the Mariana Islands.

February 3, 1945. A total of 959 B-17 crews carry out the largest raid to date against Berlin by American bombers.

February 19, 1945. The Marine V Amphibious Corps, with air and sea support, lands on Iwo Jima. The capture of this small spit of volcanic rock has important considerations for the Army Air Forces, as the island's three airfields will be used as emergency landing fields for Marianas-based B-29s and as a base for fighter operations. By March 26, the island will be secured, at a cost of more than 19,000 Japanese and 6,520 American lives.

February 20, 1945. Secretary of War Henry Stimson approves plans to establish a rocket proving ground near White Sands, N. M.

February 25, 1945. B-29 crews begin night incendiary raids on Japan; 334 aircraft drop 1,667 tons of firebombs and destroy 15 square miles of Tokyo.

March 9, 1945, In a change of tactics in order to double bomb loads, Twentieth Air Force sends more than 300

B-29s from the Marianas against Tokyo in a low-altitude incendiary night raid, destroying about one-fourth of the city.

March 11, 1945. The greatest weight of bombs dropped in a USAAF strategic raid on a single target in Europe falls on Essen, Germany, as 1,079 bomber crews release 4,738 tons of bombs.

March 14, 1945. The first Grand Slam (22,000-pound) bomb is dropped from an Avro Lancaster flown by Royal Air Force Squadron Leader C. C. Calder. Two spans of the Bielefeld railway viaduct in Germany are destroyed.

March 18, 1945. Some 1,250 U.S. bombers, escorted by 670 fighters, deal Berlin its heaviest daylight blow— 3,000 tons of bombs on transportation and industrial areas.

March 27, 1945. B-29 crews begin night mining missions around Japan, eventually establishing a complete blockade.

April 9, 1945. The last B-17 rolls off the line at Boeing's Seattle, Wash., plant.

April 10, 1945. The last Luftwaffe wartime sortie over Britain is made by an Arado Ar-234B pilot on a reconnaissance mission out of Norway.

April 10, 1945. Thirty of 50 German Me-262 jet fighters are shot down by U.S. bombers and their P-51 escorts. The German fighters shoot down ten bombers—the largest loss of the war in a single mission to jets.

April 17, 1945. *Flak Bait*, a Martin B-26B Marauder, completes a record 200th bombing mission. The aircraft, which has now flown more missions over Europe than any other Allied aircraft in World War II, will go on to complete two more missions.

April 23, 1945. Flying Consolidated PB4Y-2 Privateers, Navy crews from VPB-109 launch two Bat missiles against Japanese ships in Balikpapan Harbor, Borneo. This is the first known use of automatic homing missiles during World War II.

May 8, 1945. V-E Day. The war ends in Europe.

June 22, 1945. Okinawa is declared captured by U.S. forces. The price paid to capture this island—16,000 men, 36 ships, and 800 aircraft—is a key consideration in the decision to use the atomic bombs on Japan.

June 26, 1945. B-29 crews begin nighttime raids on Japanese oil refineries.

July 16, 1945. The world's first atomic bomb is successfully detonated at Trinity Site, a desert location near Alamagordo, N.M. The weapon (referred to as "the gadget") is the prototype of the "Fat Man" plutonium bomb and has an explosive yield of 19 kilotons.

August 6, 1945. The "Little Boy" (uranium) atomic bomb is dropped on Hiroshima from the B-29 *Enola Gay*, commanded by Col. Paul W. Tibbets, Jr.

August 6, 1945. Maj. Richard I. Bong, America's all-time leading ace, is killed in a P-80 accident. He had 40 confirmed victories.

August 9, 1945. The "Fat Man" atomic bomb is dropped on Nagasaki from the B-29 *Bockscar*, commanded by Maj. Charles W. Sweeney.

August 14, 1945. Lt. Robert W. Clyde (pilot) and Lt. Bruce K. Leford (radar operator) record the last aerial victory of World War II. Flying a Northrop P-61 nicknamed *Lady in the Dark*, the crew gets behind a Nikajima Oscar, and, in an attempt to escape from its pursuer, the Japanese fighter crashes into the Pacific without a shot being fired.

September 2, 1945. V-J Day. On board USS *Missouri* in Tokyo Bay, Japanese Foreign Minister Mamoru Shigemitsu and Chief of Staff Gen. Yoshijiro Umezu sign instruments of surrender. (*NOTE*: Alternatively, V-J Day is regarded by some to be August 15, the date upon which Emperor Hirohito broadcast his radio message, the Imperial Rescript of Surrender, touching off the celebrations normally associated with V-J Day in allied Nations.)

November 6, 1945. The first landing of a jet-powered aircraft on a carrier is made by Ens. Jake C. West in the Ryan FR-1 Fireball, a fighter propelled by both a turbojet and a reciprocating engine. The landing on USS *Wake Island* (CVE-65) is inadvertent; the plane's piston engine fails, and Ensign West comes in powered only by the turbojet.

November 7, 1945. Royal Air Force Group Capt. Hugh Wilson sets the first postwar recognized absolute speed record and breaks the 600-mph barrier at the same time, as he flies a Gloster Meteor F.4 to a speed of 606.26 mph at Herne Bay, England.

1946–1950

February 4, 1946. The Air Force Association is incorporated.

February 9, 1946. Gen. Carl A. Spaatz is designated Commanding General, Army Air Forces, succeeding Gen. H. H. Arnold.

February 15, 1946. Thirty-five movie stars, studio executives, and reporters board a Lockheed Constellation piloted by Howard Hughes for the inauguration of TWA daily nonstop service between Los Angles, Calif., and New York City. Among the stars are Paulette Goddard, Veronica Lake, and Edward G. Robinson.

February 28, 1946. Maj. William Lien makes the first flight of the Republic XP-84 at Muroc Dry Lake, Calif. The Thunderjet is the Air Force's first postwar fighter and will be used extensively for ground attack missions in the Korean War. Later designated F-84, the Thunderjet is the first fighter to carry a tactical nuclear weapon.

March 21, 1946. Strategic Air Command, Tactical Air Command, and Air Defense Command are activated.

April 24, 1946. The first flights of the Soviet-designed-and-built Yak-15 and MiG-9 prototypes are made.

May 4–16, 1946. Five separate recognized class records for altitude with payload in piston-engine aircraft are set by five different USAAF crews flying Boeing B-29A Superfortresses at Harmon Field, Guam. Col. J. B. Warren also sets a separate record for greatest load carried to 2,000 meters. These records still stand.

May 17/19, 1946. Eight separate recognized class records for speed over a closed course (1,000 and 2,000 km) with payload in piston-engine aircraft are set by two different USAAF crews flying Boeing B-29A Superfortresses at Dayton, Ohio. These records still stand.

June 21/28. 1946. Six separate recognized class records for speed over a closed course (5,000 km) with payload in piston-engine aircraft are set by two different USAAF crews flying Boeing B-29A Superfortresses at Dayton, Ohio. These records still stand.

June 26, 1946. "Knot" and "nautical mile" are adopted by the Army Air

Forces and the Navy as standard aeronautical units of speed and distance.

July 21, 1946. Navy Lt. Cmdr. James Davidson makes the first successful takeoff and landing of a jet-powered aircraft from an aircraft carrier. He is flying a McDonnell FH-1 Phantom from the USS *Franklin D. Roosevelt* (CVB-42).

July 1946. *Air Force* magazine becomes the official journal of the Air Force Association.

August 8, 1946. Almost five years after the prototype was ordered, company test pilots Beryl A. Erickson and G. S. Gus Green and a crew of seven make the first flight of the mammoth Convair XB-36 prototype at Fort Worth, Tex. .

August 15, 1947. U.S. Air Forces in Europe is established as a major command.

August 31, 1946. Famed Hollywood stunt pilot Paul Mantz wins the first postwar Bendix Trophy transcontinental race from Los Angeles to Cleveland in a North American P-51 Mustang with an average speed of 435.501 mph. Total flying time is 4 hours and 42 minutes. Col. Leon Gray wins the first Bendix Trophy Jet Division race, flying a Lockheed P-80 Shooting Star over the same course with an average speed of 494.779 mph. Total flying time is 4 hours and 8 minutes.

September 18, 1946. Company pilot Sam Shannon makes the first official flight of the Convair XF-92 at Muroc Dry Lake, Calif. (A short hop had been made on June 9.) The first true delta-winged aircraft, the XF-92 will prove invaluable as a testbed for delta-wing research.

December 9, 1946. Company pilot Chalmers Slick Goodlin makes the first powered flight of the Bell X-1 supersonic research aircraft. He reaches Mach .75 and an altitude of 35,000 feet after being released from a Boeing B-29 mother ship.

March 16, 1947. Company pilots Sam Shannon and Russell R. Rogers make the first flight of the Convair 240 airliner prototype at San Diego, Calif. Versions of the 240 will be used by the Air Force as the T-29 navigator trainer and as the C-131 Samaritan medical evacuation/transport aircraft. One aircraft, the NC-131 variable stability testbed, keeps flying into the 1990s.

June 19, 1947. Col. Albert Boyd sets the recognized absolute speed record, as he flies the Lockheed P-80R to a speed of 623.608 mph at Muroc Dry Lake, Calif.

July 29–30, 1947. Lt. Col. O. F. Lassiter sets a recognized class record for speed over a 10,000-km closed circuit without payload (piston-engine aircraft) of 273.194 mph in a Boeing B-29A Superfortress at Dayton, Ohio. The record still stands.

August 25, 1947. Marine Maj. Marion Carl breaks the recognized absolute speed record set two months previously as he pilots the Douglas D-558-I Skystreak to a speed of 650.8 mph at Muroc Dry Lake, Calif.

September 18, 1947. The U.S. Air Force is established as a separate service, with W. Stuart Symington as its first Secretary. Gen. Carl A. Spaatz, Commanding General of the AAF, becomes the first Chief of Staff on September 26.

October 1, 1947. Company test pilot George S. "Wheaties" Welch, who was one of the few AAF fighter pilots who was able to get airborne during the Pearl Harbor attack, makes the first flight of the North American XP-86 Sabre at Muroc Dry Lake, Calif. The Sabre is the Air Force's first swept-wing fighter.

October 14, 1947. The first super-sonic flight is made by Capt. Charles E. Yeager in the rocket-powered Bell XS-1 (later redesignated X-1) over Muroc Dry Lake.

October 21, 1947. The first flight of the Northrop YB-49 flying-wing jet bomber is made. The Air Force's Northrop B-2 stealth bomber bears a family resemblance to this plane when it debuts in 1989.

November 2, 1947. Howard Hughes's wooden H-4 *Hercules* (the "Spruce Goose") makes its first (and only) flight over Los Angeles harbor. Distance traveled is about a mile.

November 23, 1947. The world's largest landplane, the Convair XC-99, the cargo version of the B-36 bomber, makes its first flight at Lindbergh Field in San Diego, Calif., with company test pilots Russell R. Rogers and Beryl A. Erickson at the controls. This aircraft will lift a record 100,000 pound pay-load on April 15, 1949.

December 17, 1947. The prototype Boeing XB-47 Stratojet bomber makes its first flight from Boeing Field in Seat-tle, Wash., with company pilots Bob Robbins and Scott Osler at the controls.

December 30, 1947. The Soviet MiG-15 is flown for the first time.

January 30, 1948. Orville Wright dies in his hometown of Dayton, Ohio, at age 76.

February 20, 1948. The first Boeing B-50 Superfortress is delivered to Strategic Air Command (SAC).

April 21, 1948. Secretary of Defense James V. Forrestal assigns the primary responsibility for air defense of the United States to the Air Force.

April 26, 1948. The Air Force an-nounces a policy of racial integration— the first service to do so—well before President Truman's Executive Order on equal opportunity in July 1948.

April 30, 1948. Gen. Hoyt S. Vanden-berg is designated to succeed Gen. Carl A. Spaatz as Air Force Chief of Staff.

June 26, 1948. Operation Vittles, the Berlin Airlift, begins with Douglas C-47 crews bringing 80 tons of sup-plies into the city on the first day. By the time it ends, on September 30, 1949, the Anglo-American airlift will have delivered a total of 2.3 million tons of food, fuel, and supplies to the beleaguered city.

August 16, 1948. Company pilot Fred C. Bretcher makes the first flight of the Northrop XF-89 Scorpion all-weather interceptor at Muroc AFB, Calif.

August 23, 1948. The prototype McDonnell XF-85 Goblin parasite fighter makes its first free flight. It is intended to be carried in the bomb bay of a B-36 for fighter support over a tar-get, but the project will be abandoned a year later when air refueling of fight-ers proves eminently more practical.

September 15, 1948. Air Force Maj. Richard L. Johnson, flying a North American F-86, recaptures the world speed record for the U.S., streaking over a 3-kilometer course at Muroc AFB, Calif., at 670.981 mph.

October 15, 1948. Maj. Gen. William H. Tunner assumes command of the newly created Combined Airlift Task Force during the Berlin Airlift.

December 7–8, 1948. On the sev-enth anniversary of the Japanese attack on Pearl Harbor, Hawaii, a 7th Bomb Wing crew flies a Convair B-36B "Peacemaker" on a 35.5-hour mission from Carswell AFB, Tex., to Hawaii and back. The B-36 is undetected by local air defenses at Pearl Harbor.

December 8, 1948. A six-engine B-36 completes 9,400-mile nonstop flight from Fort Worth, Tex., to Hawaii and back to Fort Worth without refueling.

December 16, 1948. Company pilot Charles Tucker makes the first flight of the Northrop X-4 Bantam at Muroc AFB, Calif. The X-4 is designed to study flight characteristics of small, swept-wing semitailless aircraft at transonic speeds.

December 17, 1948. The 45th anniversary of the first powered flight is commemorated by the donation of the original Wright Flyer to the Smithsonian Institution. The Flyer was displayed in Britain for many years because of a dispute between the Wrights and the Smithsonian.

December 29, 1948. Defense Secretary Forrestal says the U.S. is working on an "earth satellite vehicle program," a project to study the operation of guided rockets beyond Earth's pull of gravity.

December 31, 1948. The 100,000th flight of the Berlin Airlift is made.

January 25, 1949. The U.S. Air Force adopts blue uniforms.

February 4, 1949. The Civil Aeronautics Administration sanctions the use of ground-controlled approach as a "primary aid" for commercial airline crews.

February 26–March 2, 1949. *Lucky Lady II*, a SAC B-50A, is flown on the first nonstop flight around the world. The 23,452-mile flight takes 94 hours,

1 minute and requires four midair refuelings.

March 4, 1949. The U.S. Navy's Martin JRM-2 flying boat *Caroline Mars* carries a record 269 passengers from San Diego to San Francisco, Calif.

March 4, 1949. Crews flying in the Berlin Airlift exceed 1 million tons of cargo hauled.

March 15, 1949. Military Air Transport Service establishes Global Weather Central at Offutt AFB, Neb., for support of SAC.

April 4, 1949. Meeting in Washington, D.C., the foreign ministers of Belgium, Britain, Canada, Denmark, France, Iceland, Italy, Luxembourg, the Netherlands, Norway, and Portugal, along with the U.S. Secretary of State, sign the North Atlantic Treaty.

April 16, 1949. Company test pilot Tony LeVier and flight test engineer Tony Faulkerson makes the first flight of the YF-94 Starfire prototype from Van Nuys, Calif. The Starfire, actually a modified TP-80, is designed to serve as an interim all-weather interceptor.

May 9, 1949. Republic chief test pilot Carl Bellinger makes the first flight of the XF-91 Thunderceptor jet/rocket hybrid at Muroc AFB, Calif. This unusual aircraft has variable-incidence wings of inverse-taper design (wider at the tips than at the roots).

May 11, 1949. President Harry S. Truman signs a bill providing for a 3,000-mile-long guided-missile test range for the Air Force. The range is subsequently established at Cape Canaveral, Fla.

June 2, 1949. Gen. H. H. Arnold is given the permanent rank of General of the Air Force by a special act of Congress.

August 9, 1949. Navy Lt. J. L. Fruin makes another emergency escape with an ejection seat in the U.S. near Walterboro, S.C. His McDonnell F2H-1 Banshee is traveling at more than 500 knots at the time.

August 10, 1949. President Truman signs the National Security Act Amendments of 1949, renaming the National Military Establishment the Department of Defense.

September 23, 1949. President Truman announces that the Soviet Union has successfully exploded an atomic bomb.

September 30, 1949. The Berlin Airlift, gradually reduced since May 12, 1949, officially ends. Results show 2,343,301.5 tons of supplies carried on 277,264 flights. U.S. planes carried 1,783,826 tons.

October 4, 1949. A Fairchild C-82 Packet crew airdrops an entire field artillery battery by parachute at Fort Bragg, N.C.

November 18, 1949. A crew flying a Douglas C-74 Globemaster I, *The Champ*, lands at RAF Marham, England, after a 23-hour flight from Mobile, Ala. On board are a transatlantic-record 103 passengers and crew.

1950–1953

January 23, 1950. USAF establishes Air Research and Development Command, which in 1961 will be redesignated Air Force Systems Command.

January 31, 1950. President Truman announces that he has directed the Atomic Energy Commission "to continue its work on all forms of atomic-energy weapons, including the so-called hydrogen or super bomb." This is the first confirmation of U.S. H-bomb work.

March 15, 1950. The Joint Chiefs of Staff, in a statement of basic roles and missions, give the Air Force formal and exclusive responsibility for strategic guided missiles.

April 21, 1950. Piloted by Navy Lt. Cmdr. R. C. Starkey, a Lockheed P2V-3C Neptune weighing 74,668 pounds becomes the heaviest aircraft ever launched from an aircraft carrier. The Neptune is flown off USS *Coral Sea* (CV-43).

April 24, 1950. Thomas K. Finletter becomes Secretary of the Air Force.

June 25, 1950. North Korea attacks South Korea to begin Korean War.

June 27, 1950. President Truman announces he has ordered the USAF to aid South Korea, which has been invaded by North Korean Communist forces.

June 27, 1950. Flying a North American F-82, 1st Lt. William G. Hudson destroys a Yak-11 near Seoul, the first enemy plane shot down in the Korean War.

June 30, 1950. President Truman authorizes General Douglas MacArthur to dispatch air forces against targets in North Korea.

July 1, 1950. Carrier aircraft go into action in Korea with strikes in and around Pyongyang. Also Lt. (j.g.) L. H. Plog and Ensign E. W. Brown each down a Yak-9, the first U.S. Navy kills in air combat in Korea.

September 22, 1950. Air Force Col. David Schilling makes the first nonstop transatlantic flight in a jet aircraft, flying a Republic F-84E from Manston, England, to Limestone (later Loring) AFB, Me., in 10 hours, 1 minute. The trip requires three in-flight refuelings.

November 8, 1950. 1st Lt. Russell J. Brown, Jr., flying a Lockheed F-80 Shooting Star, downs a North Korean MiG-15 in history's first all-jet aerial combat.

April 6, 1951. The Labor Department announces that employment in aircraft and parts plants increased by 100,000 people in the first six months of the Korean War.

May 20, 1951. Capt. James Jabara becomes the Air Force's first Korean War ace. He eventually downs 15 enemy planes in Korea.

June 20, 1951. Company pilot Jean Skip Ziegler makes the first flight of the Bell X-5 at Edwards AFB, Calif. The world's first aircraft to have variable-sweep wings. On the plane's ninth flight, the wings are moved to the full 60-degree sweepback.

August 18, 1951. Col. Keith Compton wins the first USAF jets-only Bendix Trophy transcontinental race, flying from Muroc AFB, Calif., to Detroit, Mich., in a North American F-86A Sabre with an average speed of 553.761 mph. Total flying time is 3 hours, 27 minutes.

August 21, 1951. The Medal of Honor is awarded posthumously to Maj. Louis J. Sebille, USAF, who was killed August 5 near Hamch'ang, Korea. Sebille attacked Red troops in his damaged plane until it crashed. This is the first Air Force Medal of Honor awarded in the Korean War.

September 14, 1951. Flying a night intruder mission, Capt. John A. Walmsley attacks a North Korean supply train near Yangdok, North Korea. His bombs hit an ammunition car, and the train breaks in two. He then makes a strafing attack on the remaining cars, but his guns jam after the first pass. Using the newly installed searchlight in the Douglas B-26 Intruder's nose, he lights the way for another pilot to finish off the train. Captain Walmsley's aircraft is hit by groundfire and crashes. Captain Walmsley will be posthumously awarded the Medal of Honor for his actions.

September 20, 1951. The Air Force makes the first successful recovery of animals from rocket flight when a monkey and 11 mice survive an Aerobee flight to 236,000 feet.

October 2, 1951. Col. Francis S. Gabreski of the 51st Fighter Wing downs a MiG-15, which gives him 6.5 victories in Korea. Since he had 28 victories in World War II, he is the highest-scoring Air Force ace with victories in two wars.

November 30, 1951. Maj. George A. Davis, Jr., becomes another USAF ace of two wars—World War II (7) and Korea (14).

February 1, 1952. The Air Force acquires its first general-purpose computer (a Univac I).

February 10, 1952. Despite being outnumbered 12 to 2, Maj. George A. Davis, Jr., and his wingman attack a formation of MiG-15s over the Sinuiju-Yalu River area of Korea in order to protect a force of U.S. fighter bombers. Major Davis, who had recorded 7 air-to-air victories in World War II and had added 14 more in Korea, shoots down two of the MiGs (although these are not confirmed kills) before being shot down himself. His wingman manages to escape. For his unselfish action, Major Davis is posthumously awarded the Medal of Honor.

April 15, 1952. The Boeing YB-52 Stratofortress bomber prototype makes its maiden flight from its facility in Seattle, Wash. Company pilot A.M. "Tex" Johnston is at the controls.

June 23–24, 1952. Combined air elements of the Air Force, Navy, and Marines virtually destroy the electrical power potential of North Korea. The two-day attack involves 1,200 sorties and is the largest single air effort since World War II.

July 14, 1952. The Ground Observer Corps begins its round-the-clock sky-watch program as part of a nationwide air defense effort.

November 1, 1952. The United States tests its first thermonuclear device at Eniwetok in the Marshall Islands. The device, codenamed Mike, has a yield of 10.4 million tons of TNT, 1,000 times more powerful than the bomb dropped on Hiroshima in World War II.

November 22, 1952. While leading a flight of four Lockheed F-80s on a mission to dive-bomb enemy gun positions that are harassing friendly ground troops near Sniper Ridge, North Korea, Maj. Charles J. Loring's aircraft is hit repeatedly as he verifies the position of the enemy guns. His aircraft badly damaged, he turns and deliberately crashes into the gun positions, destroying them completely. For this selfless action, Major Loring is posthumously awarded the Medal of Honor.

January 2, 1953. Cessna Aircraft is declared the winner of the Air Force's primary jet trainer competition. This Cessna, later designated T-37, beats out 14 entries.

January 14, 1953. 1st Lt. Joseph M. McConnell, Jr., who will go on to become the leading American ace in Korea, records his first aerial victory, a MiG-15. Assigned to the 39th Fighter Squadron, he is flying a North American F-86 at the time.

January 26, 1953. Chance Vought Aircraft completes the last F4U Corsair. The Corsair was production for 13 years (and built by two other manufacturers during World War II), and almost 12,700 were built in a number of versions, making for one of the longest and largest production runs in history.

February 4, 1953. Harold E. Talbott becomes Secretary of the Air Force.

March 16, 1953. Republic delivers the 4,000th F-84 Thunderjet to the Air Force. The F-84 has been in production since 1946.

April 7, 1953. The Atomic Energy Commission reveals that it is using QF-80 drone aircraft at the Nevada Proving Ground. The drones are flown directly through atomic bomb blast clouds to collect samples for later examination.

May 12, 1953. Secretary of Defense Charles E. Wilson reveals that projected Air Force strength has been revised downward to 120 wings, instead of the 143 previously planned.

May 18, 1953. Capt. Joseph M. McConnell, Jr., downs three MiG-15 fighters in two separate engagements. These victories give Captain McConnell a total of 16 kills in just five months of action and make him the leading American ace of the Korean War.

May 23, 1953. Company pilot Geoge S. "Wheaties" Welch makes the first flight of the North American YF-100 Super Sabre prototype at the Air Force Flight Test Center at Edwards AFB, Calif. He exceeds Mach 1 on this first flight.

June 8, 1953. Officially activated just a week before, USAF's 3600th Air Demonstration Flight, the Thunderbirds, perform their first aerial demonstration. Flying Republic F-84G Thunderjets, the team flies the show at their home, Luke AFB, Ariz.

June 16, 1953. North American delivers the 1,000th T-28 Trojan tandem-seat trainer to the Air Force.

June 30, 1953. Gen. Nathan F. Twining becomes Air Force Chief of Staff.

July 16, 1953. Lt. Col. William Barnes pushes the recognized absolute speed record past 700 mph, hitting 715.751 mph in a North American F-86D over the Salton Sea in California.

July 27, 1953. Capt. Ralph S. Parr, a member of the 335th Fighter Interceptor Squadron, flying a North American F-86, records the last aerial victory in the Korean War when he shoots down an Il-2 near Hohadong shortly after midnight. It was his 10th aerial victory.

July 27, 1953. The Korean armistice goes into effect. (It was actually signed the day before.)

July 29, 1953. Two days after the armistice ending the Korean War, the Air Force announces that the Far East Air Force shot down 839 MiG-15 jet fighters, probably destroyed 154 more, and damaged 919 others during the 37 months of war. United Nations air forces lost 110 aircraft in air-to-air combat, 677 to enemy ground fire, and 213 airplanes to "other causes."

August 21, 1953. Flying the Douglas D-558-II Skyrocket, Marine Corps Lt. Col. Marion Carl sets an altitude record of 83,235 feet after being dropped from a Boeing P2B (B-29) flying at 34,000 feet over Edwards AFB, Calif.

September 1, 1953. The first jet-to-jet air refueling takes place between a Boeing KB-47 and a "standard" B-47.

September 11, 1953. A Grumman F6F-5K Hellcat drone is destroyed in the first successful interception test of the N-7 (AIM-9) Sidewinder air-to-air missile at China Lake, Calif. The Naval Ordnance Test Station, which had fashioned the missile basically out of spare parts, conducts the test. More than 150,000 Sidewinders have been produced since.

September 21, 1953. North Korean pilot Lt. Noh Kum Suk defects and flies his MiG-15 to Kimpo AB, South Korea. He is granted asylum and given $100,000.

October 3, 1953. Navy Lt. Cmdr. James B. Verdin establishes a world speed record of 752.94 mph in the Douglas XF4D-1 Skyray over the Salton Sea in California. This is the first time a jet-powered carrier plane has set the speed record.

October 19, 1953. Assistant Secretary of the Air Force Roger Lewis reveals that Boeing B-52 bombers will cost approximately $3.6 million each in production, but the first four aircraft will cost about $20 million each to amortize the design, development, and tooling costs.

October 24, 1953. Company pilot Richard L. Johnson makes the first flight of the Convair XF-102 prototype at Edwards AFB, Calif. Performance of this aircraft is found to be lacking, and the greatly redesigned YF-102A will fly in early 1954. The supersonic Delta Dagger is the USAF's first production delta-winged aircraft, and it will be the first interceptor to become operational armed only with missiles and unguided rockets.

October 29, 1953. Lt. Col. Frank K. Pete Everest, Jr., sets a new world speed record of 755.149 mph in the North American YF-100 prototype over the Salton Sea in California. He breaks the record set just a few weeks earlier by Navy Lt. Cmdr. James B. Verdin.

November 1, 1953. The Air Reserve Personnel Center is established at Lowry AFB, Colo.

November 6, 1953. A Boeing B-47 Stratojet is flown from Limestone (later Loring) AFB, Me., to RAF Brize Norton, England, in 4 hours, 53 minutes to establish a new transatlantic speed record from the continental U.S.

November 20, 1953. NACA test pilot Scott Crossfield becomes the first pilot to exceed Mach 2. His Douglas D-588-II Skyrocket research plane is dropped from a Navy P2B-1S (B-29) at an altitude of 32,000 feet over Edwards AFB.

December 12, 1953. Maj. Charles E. Yeager pilots the rocket-powered Bell X-1A to a speed of Mach 2.435 (approximately 1,650 mph) over Edwards AFB.

1954–1963

February 15, 1954. President Dwight D. Eisenhower nominates Charles A. Lindbergh to be a brigadier general in the Air Force Reserve.

February 24, 1954. President Eisenhower approves the National Security Council's recommendation for construction of the Distant Early Warning (DEW) Line. Operational control of the DEW Line will be transferred from the Air Force to the Royal Canadian Air Force on February 1, 1959.

March 1, 1954. In the Marshall Islands, the U.S. successfully explodes its first deliverable hydrogen bomb.

March 7, 1954. Company test pilot Tony LeVier makes the first flight of the Lockheed XF-104 Starfighter at Edwards AFB, Calif. A first attempt on February 28 was cut short after the aircraft experienced gear retraction problems. Designed as a supersonic air superiority fighter, the F-104 will set a number of records for the U.S., but it will find greater utility for a number of other countries than it will for the USAF.

March 18, 1954. Boeing rolls out the first production B-52A Stratofortress at its plant in Seattle, Wash. Production will continue until 1962.

April 1, 1954. President Eisenhower signs into law a bill creating the U.S. Air Force Academy.

May 25, 1954. A Navy ZPG-2 airship lands at NAS Key West, Fla., after staying aloft for 200.1 hours. Cmdr. M. H. Eppes, the airship captain, is later awarded the Distinguished Flying Cross.

June 22, 1954. The Douglas A4D (A-4) Skyhawk makes its first flight from Edwards AFB with company pilot Robert Rahn at the controls. Some 2,960 aircraft later, "Scooters" will still be flying with the Navy as trainers and with several foreign countries as front-line equipment into the mid-1990s.

July 15, 1954. The Boeing Model 367–80 makes its first flight, with company pilot A. M. "Tex" Johnston in command. The aircraft is the prototype for the Air Force's C/KC-135 series and the progenitor of the 707, which will become the first civilian jetliner to see wide use.

August 23, 1954. Lockheed pilots Stanley Beltz and Roy Wimmer crew the first flight of the YC-130 Hercules at Burbank, Calif. More than 2,100 aircraft later, the C-130 will still be in production at Marietta, Ga., and it is expected to be produced beyond the turn of the century.

August 26, 1954. Maj. Arthur "Kit" Murray reaches a record height of 90,443 feet in the Bell X-1A, which was released from a B-29 over Edwards AFB.

September 24, 1954. Company test pilot Robert C. Little makes the first flight of the McDonnell F-101A Voodoo at Edwards AFB, Calif. The One-oh-Wonder hits Mach 1.2 on its first flight and will go on to fill several roles for a number of Air Force commands.

October 12, 1954. The Cessna XT-37 Tweet trainer is flown for the first time at Wichita, Kan. The T-37 will still be soldiering on nearly 40 years later as the Air Force's primary trainer.

October 27, 1954. Benjamin O. Davis, Jr., son of the first black general officer in the U.S. Army, becomes the first black general officer in the U.S. Air Force. He will retire April 30, 1965, as a lieutenant general.

November 2, 1954. Company test pilot J. F. Coleman, flying in the radical tail-sitting Convair XFY-1, makes a vertical takeoff, changes to horizontal flight, and then returns to vertical for a landing in San Diego, Calif.

November 7, 1954. The Air Force announces plans to build a $15.5 million research laboratory for atomic aircraft engines. To be built in Connecticut, the plant is to be run by Pratt & Whitney and will be finished in 1957.

December 10, 1954. To determine if a pilot can eject from an airplane at supersonic speed and live, Lt. Col. John Paul Stapp, a flight surgeon, rides a rocket sled to 632 mph, decelerates to zero in 125 seconds, and survives more than 35 times the force of gravity.

February 7, 1955. After 131 shows, the Thunderbirds, the Air Force's aerial demonstration team, perform their last show in the Republic F-84G Thunderjet at Webb AFB, Tex. In April, the team will convert to swept-wing F-84F Thunderstreaks.

February 23, 1955. The Army picks Bell Helicopter from a list of 20 competing companies to build its first turbine-powered helicopter. The winning design, designated XH-40, will become the HU-1 (and later still, UH-1) Iroquois, the renowned "Huey."

February 26, 1955. North American Aviation test pilot George Smith becomes the first person to survive ejection from an aircraft flying at supersonic speed. His F-100 Super Sabre is traveling at Mach 1.05 when the controls jam and he is forced to punch out.

July 11, 1955. The first class (306 cadets) is sworn in at the Air Force Academy's temporary location at Lowry AFB, Colo.

August 4, 1955. Company pilot Tony LeVier makes the first official flight of the Lockheed U-2 spyplane at Groom Lake, Nev. An inadvertent hop had been made on July 29.

August 15, 1955. Donald A. Quarles becomes Secretary of the Air Force.

October 22, 1955. Company test pilot Rusty Roth makes the first flight of the Republic YF-105 Thunderchief at Edwards AFB, Calif. The aircraft, commonly known as the Thud (among other things), is the largest single-engine, single-seat fighter ever built.

November 26, 1955. Secretary of Defense Charles Wilson assigns responsibility for development and operations of land-based intercontinental ballistic missiles to the Air Force.

January 17, 1956. Department of Defense reveals the existence of SAGE, an electronic air defense system.

February 17, 1956. Company test pilot Tony LeVier inadvertently makes the first flight of the Lockheed F-104A Starfighter as the plane skips off the runway during high-speed taxi tests at Edwards AFB, Calif. The first official flight takes place March 4.

March 10, 1956. The recognized absolute speed record passes the 1,000-mph barrier as company pilot Peter Twiss hits 1,132.13 mph in the Fairey Delta 2 research aircraft at Sussex, England.

May 20, 1956. After 91 shows in a little more than a year, the Thunderbirds perform their last demonstration in the Republic F-84F Thunderstreak at Bolling AFB, D.C.

May 21, 1956. An Air Force crew flying Boeing B-52B Stratofortress at 40,000 feet airdrops a live hydrogen bomb over Bikini Atoll in the Pacific. The bomb has a measured blast of 3.75 megatons.

May 28, 1956. Company pilot Pete Girard makes the first flight of the Ryan X-13 Vertijet VTOL research aircraft in hover mode at Edwards AFB, Calif. He had also made the type's first conventional flight on December 10, 1955.

June 30, 1956. The Thunderbirds, the Air Force's aerial demonstration squadron, fly their first show in the supersonic North American F-100 Super Sabre, the type the team would fly for most of the next 13 years.

September 27, 1956. Capt. Milburn Apt, USAF, reaches Mach 3.196 in the Bell X-2, becoming the first pilot to fly three times the speed of sound. Captain Apt is killed, however, when the aircraft tumbles out of control.

October 1, 1956. NASA awards its Distinguished Service Medal to Dr. Richard T. Whitcomb, inventor of the area rule concept, which results in aircraft (such as the Convair F-102) having Coke-bottle-shaped fuselages in order to reduce supersonic drag.

October 26, 1956. Less than 16 months after design work began, and, ironically, the same day that legendary plane-maker Larry Bell dies, company pilot Floyd Carlson makes the first flight of the Bell XH-40 at Fort Worth, Tex. Later redesignated UH-1, the Iroquois, or Huey as it is more popularly known, will go on to be one of the significant helicopters of all time.

November 11, 1956. With company pilot Beryl A. Erickson at the controls, the USAF's first supersonic bomber, the delta-winged Convair B-58 Hustler, capable of flying at speeds of more than 1,000 mph, makes its first flight at Fort Worth, Tex.

December 26, 1956. Company pilot Richard L. Johnson makes the first flight of the first Convair F-106 Delta Dart at Edwards AFB, Calif. The F-106, a substantially redesigned and much improved version of the F-102 interceptor, remains in service until 1988 and is later modified into target drones.

January 18, 1957. Commanded by Maj. Gen. Archie J. Old, Jr., USAF, three B-52 Stratofortresses complete a 24,325-mile round-the-world nonstop flight in 45 hours, 19 minutes, with an average speed of 534 mph. It is the first globe-circling nonstop flight by a jet aircraft.

April 11, 1957. With company pilot Pete Girard at the controls, the Ryan X-13 Vertijet makes its first full-cycle flight. Girard takes off vertically from the aircraft's mobile trailer, transitions to horizontal flight, performs several maneuvers, and then lands vertically.

May 1, 1957. James H. Douglas, Jr., becomes Secretary of the Air Force.

July 1, 1957. Gen. Thomas D. White becomes Air Force Chief of Staff.

July 1, 1957. Pacific Air Forces established.

July 13, 1957. President Dwight D. Eisenhower becomes the first chief executive to fly in a helicopter as he takes off from the White House lawn in a Bell UH-13J Sioux. Maj. Joseph E. Barrett flies the president a short distance to a military command post at a remote location as part of a military exercise.

July 19, 1957. A Douglas MB-1 Genie aerial rocket is fired from a Northrop F-89J Scorpion, marking the first time in history that an air-to-air rocket with a nuclear warhead is launched and detonated. The test takes place at 20,000 feet over the Nevada Test Site.

July 31, 1957. The DEW Line, a distant early warning radar defense instal-lation extending across the Canadian Arctic, is reported to be fully operational.

August 1, 1957. NORAD, the joint U.S.-Canadian North American Air Defense Command, is informally established.

August 15, 1957. Gen. Nathan F. Twining becomes Chairman of the Joint Chiefs of Staff, the first USAF officer to serve in this position.

October 4, 1957. The space age begins when the Soviet Union launches Sputnik I, the world's first artificial satellite, into Earth orbit.

November 3, 1957. The first animal in space, a dog named Laika, is carried aboard Sputnik II. The satellite is carried aloft by a modified ICBM.

November 11–13, 1957. Gen. Curtis E. LeMay and crew fly a Boeing KC-135 from Westover AFB, Mass., to Buenos Aires, Argentina, to set a world jet-class record distance in a straight line of 6,322 miles. The crew will set a class speed record on the trip back.

December 6, 1957. The first U.S. attempt to orbit a satellite fails when a Vanguard rocket loses thrust and explodes.

December 12, 1957. Flying a McDonnell F-101A Voodoo, USAF Maj. Adrian Drew sets a world record of 1,207.34 mph at Edwards AFB, Calif.

December 17, 1957. The Convair HGM-16 Atlas intercontinental ballistic missile (ICBM) makes its first successful launch and flight.

January 31, 1958. Explorer I, the first U.S. satellite, is launched by the Army at Cape Canaveral. The satellite, launched on a Jupiter-C rocket, will later play a key role in the discovery of the Van Allen radiation belt.

February 4, 1958. The keel of the world's first nuclear-powered aircraft carrier, USS *Enterprise* (CVN-65), is laid at the Newport News Shipbuilding and Drydock Co. yards in Virginia.

February 27, 1958. Approval is given to USAF to start research and development on an ICBM program that will later be called Minuteman.

March 6, 1958. The first production Northrop SM-62 Snark intercontinental missile is accepted by the Air Force after four previous successful launchings.

April 8, 1958. An Air Force KC-135 Stratotanker crew flies 10,229.3 miles nonstop and unrefueled from Tokyo to Lajes Field, Azores, in 18 hours, 50 minutes.

May 7, 1958. USAF Maj. Howard C. Johnson sets a world altitude record of 91,243 feet in a Lockheed F-104A Starfighter. Nine days later, USAF Capt. Walter W. Irwin sets a world speed record of 1,404.09 mph, also in an F-104.

May 14, 1958. Trans World Airlines becomes the first air carrier to hire a black stewardess.

May 27, 1958. The first flight of the McDonnell F4H-1 (F-4) Phantom II is made by company pilot Robert Little (who was wearing street shoes at the time) at the company's facility in St. Louis, Mo. On May 20, 1978, McDonnell Douglas will deliver the 5,000th F-4.

June 17, 1958. Boeing and Martin are named prime contractors to develop competitive designs for the Air Force's X-20 Dyna-Soar boost-glide space vehicle. This project, although later canceled, is the first step toward the space shuttle.

July 23, 1958. The Boeing Vertol VZ-2A tilt-wing research aircraft makes the first successful transition from vertical to horizontal flight and vice versa.

July 26, 1958. Capt. Iven C. Kincheloe, Jr., USAF, holder of the world altitude record (126,200 feet, set in the Bell X-2, September 7, 1956), is killed in an F-104 crash.

August 6, 1958. A Department of Defense Reorganization Act removes operational control of combat forces from the individual services and reassigns the missions to unified and specified commands on a geographic or functional basis. The main role of the services becomes to organize, train, and equip forces.

September 1, 1958. A new enlisted supergrade, senior master sergeant (E-8), is created.

September 26, 1958. A Boeing B-52D crew sets a world distance record of 6,233.98 miles and a speed record of 560.75 mph (over a 10,000-meter course) during a two-lap fright from Ellsworth AFB, S.D., to Douglas, Ariz., to Newburg, Ore., and back.

October 1, 1958. The National Aeronautics and Space Administration (NASA) is officially established, replacing NACA.

December 16, 1958. The Pacific Missile Range begins launching operations with the successful flight of the Chrysler PGM-19 Thor missile, the first ballistic missile launched over the Pacific Ocean. It is also the first free world firing of a ballistic missile under simulated combat conditions.

December 18, 1958. Project Score, an Atlas booster with a communications repeater satellite, is launched into Earth orbit. The satellite carries a

Christmas message from President Eisenhower that is broadcast to Earth, the first time a human voice has been heard from space.

January 8, 1959. NASA requests eight Redstone-type launch vehicles from the Army for Project Mercury development flights. Four days later, McDonnell Aircraft Co. is selected to build the Mercury capsules.

January 22, 1959. Air Force Capt. William B. White sets a record for the longest nonstop flight between points in the U.S., as he flies a Republic F-105 Thunderchief 3,850 miles from Eielson AFB, Alaska, to Eglin AFB, Fla., in 5 hours, 27 minutes.

February 6, 1959. USAF successfully launches the first Martin HGM-25A Titan ICBM.

February 28, 1959. USAF successfully launches the Discoverer I satellite into polar orbit from Vandenberg AFB, Calif.

April 2, 1959. Chosen from a field of 110 candidates, seven test pilots—Air Force Capts. L. Gordon Cooper, Jr., Virgil I. "Gus" Grissom, and Donald K. "Deke" Slayton; Navy Lt. Cmdrs. Walter M. Schirra, Jr., and Alan B. Shepard, Jr., and Lt. M. Scott Carpenter; and Marine Lt. Col. John H. Glenn, Jr.—are announced as the Project Mercury astronauts.

April 12, 1959. The Air Force Association's World Congress of Flight is held in Las Vegas, Nev.—the first international air show in U.S. history. Fifty-one foreign nations participate. NBC-TV telecasts an hour-long special, and *Life* magazine gives it five pages of coverage.

April 15, 1959. USAF Capt. George A. Edwards sets a speed record of

816.279 mph in a McDonnell RF-101C Voodoo on a 500-km closed course at Edwards AFB.

April 20, 1959. The prototype Lockheed UGM-27A Polaris sea-launched ballistic missile successfully flies a 500-mile trajectory in a Navy test. Three days later, the Air Force carries out the first flight test of the North American GAM-77 Hound Dog air-launched strategic missile at Eglin AFB.

May 28, 1959. Astrochimps Able and Baker are recovered alive in the Atlantic after their flight to an altitude of 300 miles in the nose cone of a PGM-19 Jupiter missile launched from Cape Canaveral Missile Test Annex, Fla.

June 3, 1959. The first class is graduated from the Air Force Academy.

June 8, 1959. The Post Office enters the missile age, as 3,000 stamped envelopes are carried aboard a Vought RGM-6 Regulus I missile launched from the submarine USS *Barbero* (SSG-317) in the Atlantic. The unarmed missile lands 21 minutes later at the Naval Auxiliary Air Station at Mayport, Fla.

June 8, 1959. After several attempts, North American Aviation pilot Scott Crossfield makes the first nonpowered flight in the X-15.

July 1, 1959. The first experimental reactor (Kiwi-A) in the nuclear space rocket program is operated successfully in a test at Jackass Flats, Nev.

August 7, 1959. First intercontinental relay of voice message by satellite takes place. The voice is that of Maj. Robert G. Mathis, later USAF Vice Chief of Staff.

August 7, 1959. Two USAF F-100Fs make the first flight by jet fighter aircraft over North Pole.

September 9, 1959. The Atlas missile is fired for the first time by a SAC crew from Vandenberg AFB, Calif., and the missile type is declared operational by the commander in chief SAC. The shot travels about 4,300 miles at 16,000 mph.

September 12, 1959. The Soviet Union launches Luna 2, the first man-made object to reach the moon.

November 16, 1959. Air Force Capt. Joseph W. Kittinger, Jr., after ascending to an altitude of 76,400 feet in *Excelsior I,* an open-gondola balloon (setting three unofficial altitude records on the way), makes the longest free-fall parachute jump in history (64,000 feet) in 2 minutes, 58 seconds at White Sands, N. M.

December 1, 1959. A new enlisted grade E-9, chief master sergeant, is created.

December 11, 1959. Dudley C. Sharp becomes Secretary of the Air Force.

December 15, 1959. Maj. Richard W. Rogers regains the world speed record for the U.S., piloting his Convair F-106 Delta Dart to a speed of 1,525.6 mph at an altitude of 40,550 feet at Edwards AFB, Calif.

December 15, 1959. Maj. Joseph Rogers sets the recognized absolute speed record of 1,525.965 mph in a Convair F-106A at Edwards AFB, Calif.

December 30, 1959. The first U.S. ballistic-missile-carrying submarine, USS *George Washington* (SSBN-598), is commissioned at Groton, Conn.

January 25, 1960. In what is billed as the "first known kill of a ballistic missile," an Army MIM-23 Hawk antiaircraft missile downs an unarmed MGR-1 Honest John surface-to-surface unguided rocket.

March 22, 1960. The Civil Aeronautics Board reports that slightly more than 10 percent of revenue passenger miles flown in scheduled domestic operations during 1959 were flown by pure jet aircraft.

March 29, 1960. The Naval Weapons Station Annex at Charleston, S. C., opens. It will provide a final assembly capability for UGM-27 Polaris sea-launched ballistic missiles and also a capability for loading them on submarines.

April 1, 1960. The RCA-built TIROS 1 (Television Infrared Observation Satellite), the world's first meteorological satellite, is successfully launched from Cape Canaveral Missile Test Annex atop a Thor launch vehicle.

April 4, 1960. Project Ozma is initiated at the National Radio Astronomy Observatory at Green Bank, W.Va., to listen for possible signal patterns from outer space other than "natural" noise.

April 22, 1960. A federal court of appeals upholds a Federal Aviation Administration order that automatically grounds pilots over 60 years old.

May 1, 1960. Central Intelligence Agency pilot Francis Gary Powers, flying a Lockheed U-2 reconnaissance aircraft, is shot down over the Soviet Union near Sverdlovsk. He is captured and later put on trial for espionage. The incident creates an international furor, and a superpower summit scheduled for later in the month is canceled. In 1962, Mr. Powers will be exchanged for Soviet KGB agent Rudolf Abel.

May 20, 1960. The Air Force launches from Cape Canaveral Missile Test Annex a Convair HGM-16 Atlas ICBM that carries a 1.5-ton payload

9,040 miles to the Indian Ocean. This is the greatest distance ever flown by a U.S. ICBM.

May 21, 1960. The last World War II–era North American B-25 Mitchell is retired from active Air Force service at Eglin AFB.

July 20, 1960. The first underwater launch of a Lockheed UGM-27 Polaris ballistic missile is successfully carried out from USS *George Washington* (SSBN-598) off Cape Canaveral Missile Test Annex.

August 16, 1960. At an altitude of 102,800 feet over Tularosa, N. M., Air Force Capt. Joseph W. Kittinger, Jr., makes the ultimate leap of faith. In the four and a half minutes between stepping out of the balloon's open gondola and opening his parachute, he free-falls 84,700 feet, reaching a speed of 614 mph. Captain Kittinger lands unharmed 13 minutes, 45 seconds after jumping. This the highest jump and longest free fall ever recorded.

September 21, 1960. Tactical Air Command formally accepts the first Republic F-105D Thunderchief all-weather fighter in ceremonies at Nellis AFB, Nev. The aircraft will not officially enter service until the following year, when deliveries to Seymour Johnson AFB, N.C., begin.

October 1, 1960. Ballistic Missile Early Warning System radar post at Thule, Greenland, begins regular operations, part of chain of three planned installations to warn of air or missile attacks on North America over an Arctic route.

January 12, 1961. A B-58 Hustler piloted by Maj. Henry J. Deutschendorf, Jr., sets six international speed and payload records on a single flight, thus breaking five previous records held by

the Soviet Union. On January 14, another B-58 from the same wing breaks three of the records set on January 12.

January 24, 1961. Eugene M. Zuckert becomes Secretary of the Air Force.

January 31, 1961. A chimpanzee named Ham is launched atop a Redstone booster from Cape Canaveral Missile Test Annex in a test of the Mercury manned capsule.

February 1, 1961. The first Boeing LGM-30A Minuteman ICBM is launched from Cape Canaveral Missile Test Annex. It travels 4,600 miles and hits the target area. This is the first time a first-test missile is launched with all systems and stages functioning.

February 3, 1961. SAC's Boeing EC-135 Airborne Command Post begins operations. Dubbed "Looking Glass," the planes and their equipment provide a backup means of controlling manned bombers and launching land-based ICBMs in case a nuclear attack wipes out conventional command-and-control systems.

April 12, 1961. The Soviet Union stuns the world with the first successful manned spaceflight. Cosmonaut Yuri Gagarin is not only history's first spaceman, he is also the first person to orbit the Earth.

May 5, 1961. Cmdr. Alan B. Shepard, Jr., USN, becomes the first Project Mercury astronaut to cross the space frontier. His flight in *Freedom 7* lasts 15 minutes, 28 seconds, reaches an altitude of 116.5 miles, and ends 303.8 miles downrange.

May 25, 1961. President John F. Kennedy, at a joint session of Congress, declares a national space objective: "I believe that this nation should commit itself to achieving the goal, before this decade is out, of landing a man on the moon and returning him safely to Earth."

June 30, 1961. Gen. Curtis E. LeMay becomes Air Force Chief of Staff.

July 21, 1961. Capt. Virgil I. Grissom becomes the first Air Force astronaut in space. He attains an altitude of 118.3 miles on the second Mercury mission.

August 6–7, 1961. Flying in the Vostok 2 spacecraft, Soviet Air Force Capt. Gherman Titov becomes the first person to orbit the Earth for more than a day. He also becomes the first person to get spacesick.

January 10–11, 1962. Maj. Clyde P. Evely sets a recognized class record for great circle distance without landing (jet aircraft) of 12, 532.28 miles from Kadena AB, Okinawa, to Madrid, Spain, in a Boeing B-52H Stratofortress. The record still stands.

January 12, 1962. Maj. H. J. Deutschendorf, Jr., sets two recognized class records for 2,000-km speed over a closed circuit with payload (jet aircraft) of 1,061.81 mph in a Convair B-58A Hustler at Edwards AFB, Calif. The records still stand.

February 2, 1962. A C-123 Ranch Hand aircraft crashes while spraying defoliant on a Viet Cong ambush site. It is the first U.S. Air Force plane lost in South Vietnam.

February 20, 1962. Marine Lt. Col. John H. Glenn, Jr., becomes the first U.S. astronaut to orbit the Earth. His *Friendship* 7 flight lasts nearly five hours.

March 5, 1962. Capts. Robert G. Sowers, Robert MacDonald, and John T. Walton, flying in a Convair B-58A Hustler bomber, are the only contest-

ant in the 21st and last Bendix Trophy transcontinental race. The crew completes the Los Angeles to New York course with an average speed of 1214.71 mph and total elapsed time is 2 hours and 56 seconds. This is still the certified speed record over a recognized course between the two cities.

April 30, 1962. Company pilot Lou Schalk makes the first official flight of the Lockheed A-12, the forerunner of the SR-71 high-speed reconnaissance aircraft, at Groom Lake, Nev. Two earlier hops had been made on April 25 and 26.

May 24, 1962. Navy Lt. Cmdr. Scott Carpenter makes the fourth flight of the Mercury space program. The flight is less than perfect, as a number of in-flight problems lead to the astronaut's overshooting the recovery ship, the USS *Intrepid* (CVS-11), by more than 250 miles.

July 17, 1962. Maj. Robert White pilots the North American X-15 to an altitude of 314,750 feet, thus making the first spaceflight in a manned aircraft. After the 11-minute fight, Major White lands at Edwards AFB, Calif.

September 12, 1962. Navy Lt. Cmdrs. Don Moore and Fred Fanke separately set two recognized class records for altitude with 1,000- and 2,000-kg payloads (piston-engine amphibians) of 29,475 feet and 27,404.93 feet respectively in a Grumman UF-2G Albatross at Floyd Bennett Field, N.Y. Both records still stand.

September 14, 1962. Maj. F. L. Fulton sets a recognized class record for altitude with 5,000-kg payload (jet aircraft) of 85,360.8 feet in a Convair B-58A Hustler at Edwards AFB, Calif. The record still stands.

October 3, 1962. Navy Cmdr. Walter M. "Wally" Schirra, Jr., makes what is described as a textbook orbital flight during the fifth flight in the Mercury program. He flies in a 100×176 mile orbit, the highest to date, and completes nearly six orbits. He is also the first astronaut to splash down in the Pacific Ocean.

October 14, 1962. An Air Force reconnaissance flight photographs nuclear-armed Soviet missiles in Cuba. Moscow subsequently agrees to remove the missiles under threat of U.S. invasion of Cuba.

October 25, 1962. Coast Guard Cmdr. W. Fenlon sets a recognized class record for great circle distance without landing (piston-engine amphibians) of 3,571.65 miles from Kodiak, Alaska, to Pensacola, Fla., in a Grumman UF-2G Albatross. The record still stands.

November 30, 1962. The first tethered hovering flight is made by the Lockheed XV-4A Hummingbird vertical takeoff and landing airplane at Marietta, Ga.

December 14, 1962. NASA's Mariner II satellite scans the surface of Venus for 35 minutes as it flies past the planet at a distance of 21,642 miles.

January 17, 1963. NASA pilot Joe Walker qualifies for astronaut wings by flying the North American X-15 to an altitude of 271,700 feet or 51.46 miles. He is the 11th man to pass the 50-mile mark.

February 28, 1963. The first Minuteman squadron, the 10th Strategic Missile Squadron (SMS) at Malmstrom AFB, Mont., is declared operational.

March 20, 1963. Capt. Henry E. Erwin, Jr., sets two recognized class records for altitude with 5,000-kg payload

(19,747 feet) and greatest payload carried to an altitude of 2,000 meters (12,162.90 pounds) in a Grumman HU-16B Albatross at Eglin AFB, Fla. Both records still stand.

April 11, 1963. The first successful launch of a Boeing LGM-30 Minuteman I ICBM is conducted at Vandenberg AFB, Calif.

May 15, 1963. Maj. L. Gordon Cooper becomes the second Air Force astronaut in space as he makes nearly twenty-two orbits in his spacecraft, *Faith* 7. He is the last American to be launched into space alone, he is the first to spend a complete day in orbit, and because of a failure of the automatic system, he is the first to perform an entirely manual reentry. This is the last Project Mercury space mission.

June 16–19, 1963. Cosmonaut Jr. Lt. Valentina Tereshkova, a former cotton mill worker, becomes the first woman in space. Her Vostok 6 flight lasts nearly three days.

August 22, 1963. NASA pilot Joe Walker achieves an unofficial world altitude record of 354,200 feet in the X-15.

October 17, 1963. The first LGM-30A Minuteman I operational test launch is carried out at Vandenberg AFB, Calif., by a crew from Malmstrom AFB. The shot is a partial success. The reentry vehicle overshoots the target.

October 30, 1963. Navy Lt. James H. Flatley lands a Lockheed KC-130F Hercules on the aircraft carrier USS *Forrestal* (CVA-59) in the Atlantic off Boston, Mass., in a test to see if the Hercules could be used as a Super COD (carrier on-board delivery) aircraft. Lieutenant Flatley and crew will eventually make 21 unarrested full-stop landings and a like number of unassisted takeoffs from the carrier.

November 7, 1963. The Northrop-developed three-parachute landing system for the Apollo command module is successfully tested at White Sands, N.M.

December 17, 1963. With company pilots Leo Sullivan and Hank Dees at the controls, the Lockheed C-141A StarLifter, USAF's first jet-powered transport, makes its first flight at Marietta, Ga., on the sixtieth anniversary of the Wright brothers' first flight.

December 17, 1963. The Thunderbirds, the Air Force's aerial demonstration squadron, fly their 690th and last show in the North American F-100C Super Sabre.

1964–1973

January 8, 1964. The newest Air Force decoration, the Air Force Cross, is posthumously awarded to reconnaissance pilot Maj. Rudolf Anderson, Jr., the only combat casualty of the 1962 Cuban missile crisis.

February 1, 1964. The Boeing 727 passenger liner enters revenue service with Eastern Air Lines.

February 3, 1964. Four airmen locked in a spaceship simulator exhibit no ill effects after exposure to a pure oxygen atmosphere 30 days.

February 29, 1964. President Lyndon B. Johnson announces the existence of the Lockheed A-11 (YF-12A), with a cruising speed of more than Mach 3 at altitudes above 70,000 feet. The plane was ordered as a single-seat reconnaissance aircraft for the CIA in 1960. Only three YF-12A interceptors are built, and the SR-71 program for the Air Force takes precedence.

April 26, 1964. At Norfolk, Va., the Thunderbirds, the Air Force's aerial demonstration team, fly their first show in the Republic F-105B Thunderchief. The team will perform only six shows in the Thud, as it will soon be determined that it is not a suitable show aircraft.

May 11, 1964. The North American XB-70 Valkyrie is rolled out at Palmdale, Calif. Designed to fly at three times the speed of sound and at altitudes above 70,000 feet, the XB-70 is originally planned as a manned bomber, but funding limitations allow for only two aircraft, to be used strictly for testing and research.

August 1964. USAF moves into Southeast Asia in force. B-57s from Clark AB in the Philippines deploy to Bien Hoa in South Vietnam and additional F-100s move to Da Nang on August 5. Eighteen F-105s deploy from Japan to Korat Royal Thai Air Base beginning August 6.

August 2, 1964. The destroyer USS *Maddox* (DD-731) is attacked by North Vietnamese patrol boats in the Gulf of Tonkin. A second incident, involving the *Turner Joy* (DD-951), reportedly occurs two days later. Congress passes the Gulf of Tonkin Resolution on August 7.

August 19, 1964. The Hughes Syncom III satellite is launched by a Thor-Delta launch vehicle. After several weeks of maneuvers, it becomes the world's first geosynchronous satellite.

September 21, 1964. The North American XB-70A Valkyrie makes its first flight, with company pilot Alvin White and USAF pilot Col. Joseph Cotton at the controls.

September 28, 1964. USS *Daniel Webster* (SSBN-626), the first submarine equipped with the Lockheed UGM-27C (A3) Polaris sea-launched ballistic missile, departs Charleston, S. C., on its first patrol.

November 17–26, 1964. C-130s flown by U.S. Air Force in Europe crews deliver Belgian paratroopers to the Congo for a rescue operation credited with saving the lives of nearly 2,000 hostages at Stanleyville threatened by rebels.

December 14, 1964. U.S. Air Force flies the first "Barrel Roll" armed reconnaissance mission in Laos.

December 21, 1964. Company pilots Richard Johnson and Val Prahl make the first flight of the variable-geometry General Dynamics F-111A from Air Force Plant 4 in Fort Worth, Tex. The flight lasts 22 minutes.

December 22, 1964. Lockheed gets approval to start development for the Air Force of the CX-HLS transport, which will become the C-5A. Also on this date, company pilot Bob Gilliland makes the first flight of the Lockheed SR-71A Blackbird strategic reconnaissance aircraft from Palmdale, Calif. He takes the aircraft to an altitude exceeding 45,000 feet and a speed of more than 1,000 mph on the flight.

February 1, 1965. The first Boeing LGM-30F Minuteman II ICBM unit, the 447th Strategic Missile Squadron at Grand Forks AFB, N.D., is activated.

February 1, 1965. Gen. John P. McConnell becomes Air Force Chief of Staff.

February 8, 1965. The Air Force performs its first retaliatory air strike in North Vietnam. A North American F-100 Super Sabre flies cover for attacking South Vietnamese fighter aircraft, suppressing ground fire in the target area.

February 18, 1965. First Air Force jet raids are flown against an enemy concentration in South Vietnam. American Pilots fly Martin B-57 Canberra bombers and North American F-100 fighters against the Viet Cong in South Vietnam, near An Khe.

March 1, 1965. An unarmed Boeing LGM-30B Minuteman I ICBM is successfully launched from an underground silo 10 miles north of Newell, S.D. It is the first time a site other than Vandenberg AFB or Cape Kennedy AFS, Fla., is used for an ICBM launch.

March 2, 1965. Capt. Hayden J. Lockhart, flying an F-100 in a raid against an ammunition dump north of the Vietnamese demilitarized zone, is shot down and becomes the first Air Force pilot to be taken prisoner by the North Vietnamese. He will not be released until February 12, 1973.

March 23, 1965. Air Force Maj. Virgil I. Grissom becomes the first astronaut in the manned spaceflight program to go aloft a second time, as he and Navy Lt. Cmdr. John W. Young are launched on the first Gemini mission, Gemini 3. This three-orbit, 4-hour-and-53-minute shakedown flight is also the first time a spacecraft's orbit is changed in space.

May 1, 1965. Using two Lockheed YF-12As, three Air Force crews set six class and absolute records at Edwards AFB, Calif. Col. Robert Stevens and RSO Lt. Col. Daniel Andre set the recognized absolute speed record with a mark of 2,070.115 mph over the 10.1-mile straight course.

June 3–7, 1965. Air Force Maj. Edward H. White makes the first U.S.

spacewalk. The Gemini 4 mission is the first U.S. spaceflight to be controlled from the Manned Spaceflight Center in Houston, Tex., and the crew, which also includes Air Force Maj. James A. McDivitt, stays aloft for a record 62 orbits.

June 18, 1965. SAC B-52s are used for the first time in Vietnam, when 28 aircraft strike Viet Cong targets near Saigon.

July 10, 1965. Capt. Thomas S. Roberts, with his backseater, Capt. Ronald C. Anderson, and Capt. Kenneth E. Holcombe and his backseater, Capt. Arthur C. Clark, both flying McDonnell Douglas F-4C Phantom Iis, shoot down two MiG-17s, the first Air Force air-to-air victories of the Vietnam War.

August 11, 1965. Flying in North American F-100D Super Sabres, the Thunderbirds, the Air Force's aerial demonstration squadron, fly their 1,000th show at Waukeegan, Ill.

August 21–29, 1965. The Gemini 5 crew of Air Force Lt. Col. L. Gordon Cooper and Navy Lt. Cmdr. Charles Conrad carry out the U.S.'s first long-duration spaceflight, ending one orbit short of eight full days.

October 1, 1965. Dr. Harold Brown is sworn in as Secretary of the Air Force.

October 18, 1965. New York's Air National Guard 107th Tactical Fighter Group becomes the first tactical guard unit to be deployed in peacetime to the Pacific for a joint-service exercise.

December 15, 1965. In a first for the U.S. space program, the crews of Gemini 6 and Gemini 7 rendezvous in space. Unlike the Soviets who had not managed earlier to get two spacecraft in close proximity to one another in or-

bit, the Gemini 6 crew of Navy Capt. Walter Schirra and USAF Maj. Tom Stafford maneuver to within 4 inches of Gemini 7.

January 1, 1966. Military airlift units of the Air National Guard (ANG) begin flying about 75 cargo flights a month to Southeast Asia. These flights are in addition to the more than 100 overseas missions a month flown by the ANG in augmenting the Military Airlift Command's global airlift mission.

January 17, 1966. A B-52 loaded with four hydrogen bombs collides with a KC-135 while refueling near Palomares, Spain. Seven of the 11 crew members involved are killed. Three of the four weapons are quickly recovered. The fourth, which falls into the Mediterranean Sea, is not recovered until early spring.

January 23, 1966. The newly renamed (as of January 1) Military Airlift Command completes Operation Blue Light, the airlift of the Army's 3d Brigade, 25th Infantry Division, from Hawaii to Pleiku, South Vietnam, to offset the buildup of Communist forces there. The airlift begins on December 23, 1965, and its 231 C-141 sorties move approximately 3,000 troops and 4,700 tons of equipment.

February 28, 1966. The U.S. space program suffers its first fatalities: the Gemini 9 prime crew of Elliot See and Charles Basset are killed as their Northrop T-38 crashes in St. Louis in bad weather. They were on a trip to inspect their spacecraft at the McDonnell Douglas plant at Lambert Field.

March 4, 1966. A flight of Air Force F-4C Phantoms is attacked by three MiG-17s in the first air-to-air combat of the war over North Vietnam. The MiGs make unsuccesful passes before fleeing to the sanctuary of the Communist capital area.

March 10, 1966. Maj. Bernard F. Fisher, a 1st Air Commando Squadron A-1E pilot, lands on the A Shau airstrip, after it has been overrun by North Vietnamese regulars, to rescue downed A-1E pilot Maj. D. Wayne "Jump" Myers. Major Fisher is later awarded the Medal of Honor for his heroic act.

March 16, 1966. The Gemini 8 crew, Neil Armstrong and USAF Maj. David R. Scott, successfully carry out the first docking with another vehicle in space.

April 1, 1966. Seventh Air Force, with headquarters at Saigon, is activated as a subcommand of Pacific Air Forces.

April 12, 1966. Strategic Air Command B-52 bombers strike targets in North Vietnam for the first time. They hit a supply route in the Mu Gia Pass, about 85 miles north of the border.

April 26, 1966. Maj. Paul J. Gilmore and 1st Lt. William T. Smith became the first Air Force pilots to destroy a MiG-21. Flying escort for F-105 Thunderchiefs near Hanoi when the flight is attacked, the F-4C pilots down the MiG with an AIM-9 Sidewinder missile.

June 17, 1966. Army Lt. Col. E. L. Nielsen sets a recognized class record for 100-km speed over a closed course (turboprop aircraft) of 293.41 mph in a Grumman OV-1A Mohawk at Peconic River, L.I., N.Y. The record still stands.

October 7, 1966. The Air Force selects the University of Colorado to conduct independent investigations into unidentified flying object (UFO) reports.

November 11, 1966. The Gemini program comes to an end as Navy Cmdr. James Lovell and Air Force Maj. Edwin Buzz Aldrin complete a successful mission on Gemini 12. Astronaut Aldrin makes three spacewalks on the 59-orbit mission.

January 2, 1967. USAF fighter pilots, in the famous MiG Sweep mission, down seven North Vietnamese MiG-21s over the Red River Valley in North Vietnam.

January 2, 1967. By shooting down a MiG-21, Col. Robin Olds becomes the first and only USAF ace with victories in World War II and Vietnam. Flying with Colonel Olds in the backseat of the McDonnell Douglas F-4C was 1st Lt. Charles Clifton.

January 27, 1967. Astronauts USAF Lt. Col. Virgil I. Grissom, Navy Lt. Cmdr. Roger B. Chaffee, and USAF Lt. Col. Edward H. White are killed in a flash fire aboard their Apollo 1 command module during a ground test. The disaster sets the moon landing effort back two years.

February 24, 1967. USAF Capt. Hilliard A. Wilbanks, a forward-air controller, resorts to firing an M-16 rifle out the side window of his Cessna O-1 Bird Dog in order to try to cover the retreat of a South Vietnamese Ranger battalion caught in an ambush near Dalat. Severely wounded by ground fire, Captain Hilliard crashes in the battle area, but is rescued by the Rangers. He dies while being evacuated to a hospital. Captain Hilliard is later posthumously awarded the Medal of Honor for his actions.

March 10, 1967. Air Force F-105 Thunderchief and F-4C Phantom II crews bomb the Thai Nguyen steel plant in North Vietnam for the first time. Capt. Merlyn H. Dethlefsen, an F-105 pilot, is later awarded the Medal of Honor for his actions this day in suppressing enemy air defenses.

March 10, 1967. Capt. Mac C. Brestel, an F-105 pilot with the 355th Tactical Fighter Squadron, Takhli RTAFB, Thailand, becomes the first Air Force combat crewman to down two MiGs during a single mission.

April 3, 1967. CMSgt Paul W. Airey becomes the first Chief Master Sergeant of the Air Force.

April 19, 1967. Over North Vietnam, Maj. Leo K. Thorsness (along with his electronic warfare officer, Capt. Harold E. Johnson) destroys two enemy SAM sites, then shoots down a MiG-17 before escorting search and rescue helicopters to a downed aircrew. Although the Republic F-105 was very low on fuel, Major Thorsness attacks four MiG-17s in an effort to draw the enemy aircraft away from area. He then lands at a forward air base. Awarded the Medal of Honor for his actions this day, Major Thorsness will not receive his medal until 1973, as on April 30, 1967, he is shot down and spends the next six years as a POW.

May 13, 1967. For the second time, pilots of the 8th Tactical Fighter Wing, Ubon RTAFB, Thailand, shoot down seven MiGs in a single day's action over North Vietnam.

May 20, 1967. Col. Robin Olds (pilot) and backseater 1st Lt. Steven Croaker down two MiG-17s over the Bak Le railyards, giving Olds four aerial victories in Vietnam. He also recorded 12 victories in World War II, making him the only ace to down enemy aircraft in two nonconsecutive wars.

June 1, 1967. Using air refueling, two Sikorsky HH-3E crews complete the first nonstop transatlantic helicopter flight.

August 26, 1967. Badly injured after his North American F-100F is shot down over North Vietnam, Maj. George E. Day is captured and severely tortured. He manages to escape and eventually makes it to the Demilitarized Zone. After several attempts to signal U.S. aircraft, he is ambushed and recaptured, and is later moved to prison in Hanoi, where he continues to offer maximum resistance to his captors Finally released in 1973, Major Day is awarded the Medal of Honor for his conspicuous gallantry while a POW.

September 9, 1967. Sgt. Duane D. Hackney is presented with the Air Force Cross for bravery in rescuing an Air Force pilot in Vietnam. He is the first living enlisted man to receive the award.

October 3, 1967. Maj. William Knight flies the North American X-15A-2 to the unofficial absolute world speed record of Mach 6.72 (4,534 mph) over Edwards AFB.

October 24, 1967. U.S. planes attack North Vietnam's largest airbase, Phuc Yen, for the first time in a combined Air Force, Navy, and Marine strike. During the attack, the Air Force downs its 69th MiG.

November 8, 1967. While attempting to rescue an Army Reconnaissance team, Capt. Gerald O. Young's Sikorsky HH-3E is shot down in Laos. Badly burned, he gives aid to a crew member who also escaped from the wreckage. After 17 hours of leading enemy forces away from his injured crewman and himself evading capture, the two are rescued. Captain Young is later awarded the Medal of Honor for his actions.

November 9, 1967. While on a flight over Laos, Capt. Lance P. Sijan ejects from his disabled McDonnell Douglas F-4C and successfully evades capture for more than six weeks. He is caught, but manages to escape. Recaptured and tortured, he later contracts pneumonia and dies. For his conspicuous gallantry as a POW, Captain Sijan is posthumously awarded the Medal of Honor.

December 11, 1967. The Aerospatiale-built Concorde supersonic jetliner prototype rolls out at the company's plant in Toulouse, France.

January 12, 1968. The Air Force announces a system for tactical units to carry with them everything they need to operate at "bare" bases equipped only with runways, taxiways, parking areas, and a water supply.

February 29, 1968. Jeanne M. Holm, WAF Director, and Helen O'Day, assigned to Office of the Air Force Chief of Staff, become the first women promoted to colonel.

March 2, 1968. The first of 80 C-5A Galaxy transports rolls out at Lockheed's Marietta, Ga., facility.

March 25, 1968. F-111s fly their first combat mission against military targets in North Vietnam.

March 31, 1968. President Lyndon Johnson announces a partial halt of bombing missions over North Vietnam and proposes peace talks.

May 12, 1968. Lt. Col. Joe M. Jackson, flying an unarmed Fairchild C-123 transport, lands at a forward outpost at Kham Duc, South Vietnam, in a rescue attempt of a combat control team. After a rocket-propelled grenade fired directly at his aircraft proves to be a dud, Colonel Jackson takes off with the CCT on board and lands at Da Nang. He is later awarded the Medal of Honor for his actions.

May 18, 1968. In response to a massive flood, the Air Force airlifts 88.5 tons of food and other supplies to Ethiopia.

June 30, 1968. The world's largest aircraft, the Lockheed C-5A Galaxy, makes its first flight, as company pilots Leo Sullivan and Walt Hensleigh use only 4,500 feet of Dobbins AFB's 10,000-foot runway to get airborne

July 1, 1968. The first WAF in the Air National Guard is sworn in as a result of passage of Public Law 90–130, which allows ANG to enlist women.

August 16, 1968. The first test launch of a Boeing LGM-30G Minuteman III ICBM is carried out from Cape Kennedy AFS.

August 21, 1968. NASA pilot William H. Dana becomes the last pilot to fly into space in the North American X-15 research aircraft. One of seven pilots to earn their astronaut wings in the X-15, Mr. Dana attains an altitude of 264,000 feet and a speed of Mach 4.71 in the flight over Edwards AFB, Calif.

September 1, 1968. Lt. Col. William A. Jones III leads a rescue mission near Dong Hoi, North Vietnam. Finding the downed pilot, Colonel Jones attacks a nearby gun emplacement. On his second pass, Colonel Jones's aircraft is hit and the cockpit of his Douglas A-1H is set ablaze. He tries to eject, but the extraction system fails. He then returns to base and reports the exact position of the downed pilot (who is rescued the next day) before receiving medical treatment for his burns. Colonel Jones will die in an aircraft accident in the U.S. before he can be presented the Medal of Honor for his actions the day of the rescue.

October 11–22, 1968. Apollo 7, the first test mission following the disastrous

Apollo 1 fire, is successfully carried out. Navy Capt. Walter M. Schirra, Jr., USAF Maj. Donn F. Eisele, and R. Walter Cunningham stay in Earth orbit for 10 days, 20 hours, 9 minutes.

October 24, 1968. With NASA test pilot William H. Dana at the controls, the North American X-15 makes the type's 199th and final flight, completing 10 years of flight testing. The plane reaches a speed of Mach 5.04 and an altitude of 250,000 feet.

November 26, 1968. While returing to base, 1st Lt. James P. Fleming and four other Bell UH-1F helicopter pilots get an urgent message from an Army Special Forces team pinned down near a riverbank. One helicopter is downed and two other leave the area because of low fuel, but Lieutenant Fleming and another pilot flying in an armed Huey press on with the rescue effort. The first try fails, but not willing to give up, Lieutenant Fleming lands again and is successful in picking the team up. He then lands at his base near Duc Co, South Vietnam, nearly out of fuel. Lieutenant Fleming is later awarded the Medal of Honor for his actions.

November 30, 1968. The Air Force's aerial demonstration squadron, the Thunderbirds, fly their 471st and last show in the North American F-100D Super Sabre. Except for six shows in 1964 when they flew F-105s, the team had been performing in Huns for 13 years.

December 21–27, 1968. Apollo 8 becomes the first manned mission to use the Saturn V booster. Astronauts USAF Col. Frank Borman, Navy Cmdr. James A. Lovell, and USAF Maj. William Anders become the first humans to orbit the moon.

December 31, 1968. The Soviet Union conducts the first flight of the Tu-144, the world's first supersonic transport.

February 9, 1969. Boeing conducts the first flight of the 747. The jumbo jet, with standard seating for 347 passengers, introduces high passenger volume to the world's airways.

February 15, 1969. Robert C. Seamans, Jr., becomes Secretary of the Air Force.

February 24, 1969. After a North Vietnamese mortar shell rocks their Douglas AC-47 gunship, A1C John L. Levitow, stunned and wounded by shrapnel, flings himself on an activated, smoking magnesium flare, drags himself and the flare to the open cargo door, and tosses it out of the aircraft just before the flare ignites. For saving his fellow crew members and the gunship, Airman Levitow is later awarded the Medal of Honor. He is the only enlisted man to win the CMH in Vietnam and is one of only four enlisted airmen ever to win the award.

February 27, 1969. The aerobics physical fitness program developed by Lt. Col. Kenneth H. Cooper, of Air Force Systems Command's Aerospace Medical Laboratory, is adopted by the Air Force to replace the 5BX program.

March 3–13, 1969. Air Force astronauts Col. James A. McDivitt and Col. David R. Scott, along with civilian Russell L. Schweickart, carry out the first in-space test of the lunar module while in Earth orbit during the Apollo 9 mission. The flight also marks the first time a crew transfer is made between space vehicles useing an internal connection.

May 18–26, 1969. In a dress rehearsal for the moon landing, Apollo 10 astronauts Col. Thomas P. Stafford, USAF, and Cmdr. Eugene A. Cernan,

USN, fly the lunar module *Snoopy* to within nine miles of the lunar surface. Astronaut Cmdr. John W. Young, USN, remains in orbit aboard *Charlie Brown*, the command module.

June 1, 1969. The Thunderbirds, the Air Force's aerial demonstration squadron, demonstrate McDonnell Douglas F-4E Phantom II for the graduating seniors at the Air Force Academy. The F-4 is the team's sixth show aircraft.

June 4, 1969. The Air Force Air Demonstration Squadron, the Thunderbirds, fly their first show in their new McDonnell Douglas F-4E Phantom IIs.

July 1, 1969. Air Force service numbers are replaced by Social Security account numbers for military personnel.

July 20, 1969. Man sets foot on the moon for the first time. At 10:56 P.M. EDT, Apollo 11 astronaut Neil Armstrong puts his left foot on the lunar surface. He and lunar module pilot Col. Edwin "Buzz" Aldrin, Jr., USAF, spend just under three hours walking on the moon. Command module pilot Lt. Col. Michael Collins, USAF, remains in orbit.

August 1, 1969. Gen. John D. Ryan is appointed Air Force Chief of Staff.

August 1, 1969. CMSgt. Donald L. Harlow becomes Chief Master Sergeant of the Air Force.

October 1969. *Air Force* magazine cover story "The Forgotten Americans of the Vietnam War" ignites national concern for the prisoners of war and the missing in action. It is reprinted in condensed form as the lead article in the November 1969 issue of *Reader's Digest*, is read in its entirety on the floor of Congress, and is inserted into the *Congressional Record* on six differ-

ent occasions. This article stirs the conscience of the nation and rallies millions to the cause of the POWs and MIAs. *Air Force* magazine publishes an MIA/POW Action Report from June 1970 until September 1974.

November 3, 1969. The Air Force issues a request for proposal for a new bomber to meet its advanced manned strategic aircraft requirement. Its designation will be B-1.

November 14–24, 1969. Apollo 12 is hit by lightning on liftoff, but Cmdrs. Charles Conrad and Alan Bean make the second manned lunar landing with pinpoint accuracy. The lunar module *Intrepid* touches down 1,000 yards from the Surveyor 3 probe, on the moon since 1967. The all-Navy crew, which also includes Cmdr. Richard F. Gordon, is recovered in the Pacific Ocean by USS *Hornet* (CVS-12).

December 17, 1969. Air Force Secretary Robert Seamans announces the termination of Project Blue Book, the service's program to investigate reports of unidentified flying objects (UFOs).

March 15, 1970. The overseas portion of the Automatic Voice Network (AUTOVON) is completed, making it possible to call any U.S. military installation in the world without leaving one's desk.

March 19, 1970. Air Force Maj. Jerauld Gentry makes the first successful powered flight of the Martin Marietta X-24A lifting-body research aircraft over Edwards AFB.

April 11–17, 1970. Thirteen proves an unlucky number for the Apollo program. An explosion in the service module cripples the spaceship and forces the crew to use the lunar module as a lifeboat to get back to Earth. After a tense four days, the Apollo 13 crew safely splashes down in the Pacific.

May 5, 1970. The Air Force Reserve Officers Training Corps admits women after test programs at Ohio State, Auburn University, Drake University and East Carolina University prove successful.

May 15, 1970. Sgt. John L. Levitow is awarded the Medal of Honor for heroic action on February 24, 1969, over Long Binh Army Post, South Vietnam. He is the first Air Force enlisted recipient of the award since World War II.

June 6, 1970. The first operational Lockheed C-5A Galaxy transport is delivered to the 437th Military Airlift Wing at Charleston, S. C. The debut, made before Rep. L. Mendel Rivers (D–S.C.) and most of the House Armed Services Committee, is less than auspicious: the giant aircraft loses a wheel, and several other tires are punctured on landing.

August 21, 1970. Defense Secretary Melvin Laird announces the "Total Force" policy, leading to much greater reliance by the services on Guard and Reserve units.

August 24, 1970. Two Air Force crews complete the first nonstop transpacific helicopter flight as they land their Sikorsky HH-53Cs at Da Nang AB, South Vietnam, after a 9,000-mile flight from Eglin AFB. The helicopters were refueled in flight during the trip.

November 21, 1970. A special task force of Air Force and Army volunteers makes a daring attempt to rescue American servicemen from the Son Tay prisoner of war camp 20 miles west of Hanoi.

January 27, 1971. Navy Cmdr. D. H. Lilienthal sets a recognized class

record for speed over a 15/25-km course (turboprop aircraft) of 501.44 mph in a Lockheed P-3C Orion at NAS Patuxent River, Md. The record still stands.

March 2, 1971. A policy is announced which allows Air Force women who become pregnant to request a waiver to remain on active duty or to be discharged and return to duty within 12 months of discharge.

March 8, 1971. Capt. Marcelite C. Jordan becomes the first woman aircraft maintenance officer after completion of the Aircraft Maintenance Officer's School. She was previously an administrative officer.

July 16, 1971. Jeanne M. Holm becomes the first female general officer in the Air Force.

July 26, 1971. Apollo 15 blasts off with an all–Air Force crew: Col. David R. Scott, Lt. Col. James B. Irwin, and Maj. Alfred M. Worden. The mission is described as the most scientifically important and, potentially, the most perilous lunar trip since the first landing. Millions of viewers throughout the world watch as color TV cameras cover Scott and Irwin as they explore the lunar service using a moon rover vehicle for the first time.

September 3, 1971. President Richard Nixon dedicates the new Air Force Museum building at Wright-Patterson AFB, Dayton, Ohio. A drive to raise private funds for the new museum building had begun in 1960.

October 1, 1971. CMSgt. Richard D. Kisling becomes the third Chief Master Sergant of the Air Force.

October 26–November 4, 1971. Army CWO James K. Church sets one recognized turbine engine helicopter class record for altitude in horizontal flight (36, 122 feet), Capt. B. P. Blackwell sets a record for altitude with 1,000-kg payload (31.165 feet), CWO Eugene E. Price sets two records for altitude with 2,000-kg and 5,000-kg payloads (31,480 feet and 25,518 feet), and CWO Delbert V. Hunt sets a record for time-to-climb to 9,000 meters (5: 58 minutes), all in the same Sikorsky CH-54B Tarhe at Stratford, Conn. These records still stand.

February 20, 1972. Lt. Col. Edgar Allison sets a recognized class record for great circle distance without landing (turboprop aircraft) of 8,732.09 miles, flying from Ching Chuan Kang AB, Taiwan, to Scott AFB, Ill., in a Lockheed HC-130. The record still stands.

April 1, 1972. The Community College of the Air Force is established.

April 12, 1972. Army Maj. John C. Henderson sets recognized turbine engine helicopter class time-to-climb records to 3,000 meters and 6,000 meters (1: 22 minutes and 2: 59 minutes) in a Sikorsky CH-54B Tarhe at Stratford, Conn. The records still stand.

April 27, 1972. Four Air Force fighter crews, releasing Paveway I "smart bombs," knock down the Thanh Hoa Bridge in North Vietnam. Previously, 871 conventional sorties resulted in only superficial damage to the bridge.

May 10, 1972. Capt. Charles B. DeBellevue (WSO), flying with Capt. Richard S. Ritchie (pilot), in a McDonnell Douglas F-4D, records his first aerial kill. Captain DeBellevue, who would go on to be the leading American ace of the Vietnam War, recorded four of his victories with Captain Ritchie. Both airmen flew with the 555th Tactical Fighter Squadron.

May 10–11, 1972. F-4 Phantoms from the 8th Tactical Fighter Wing

drop "smart bombs" on the Paul Doumer Bridge, causing enough damage to keep this mile-long highway and rail crossing at Hanoi out of use. It will not be rebuilt until air attacks on North Vietnam cease in 1973.

June 29, 1972. Capt. Steven L. Bennett attempts to assist a friendly ground unit being overrun near Quang Tri, South Vietnam. Captain Bennett strafes the North Vietnamese regulars with his Rockwell OV-10 Bronco, but is hit by a SAM. Unable to eject because the parachute of his backseater, a Marine artillery spotter, has been shredded by shrapnel, Captain Bennett ditches the aircraft in the Gulf of Tonkin. The observer escapes, but Captain Bennett is trapped and sinks with the wreckage. Captain Bennett is posthumously awarded the Medal of Honor.

July 27, 1972. One month ahead of schedule, company pilot Irv Burrows makes the first flight of the McDonnell Douglas F-15A Eagle air superiority fighter at Edwards AFB, Calif. The F-15 is the first USAF fighter to have a thrust-to-weight ratio greater than 1:1, which means it can accelerate going straight up.

August 28, 1972. Capt. Richard S. Ritchie, with his backseater, Capt. Charles B. DeBellevue, shoots down his fifth MiG-21 near Hanoi, becoming the Air Force's first ace since the Korean War. Two weeks later, Capt. DeBellevue also shoots down his fifth MiG.

September 9, 1972. Capt. Charles B. DeBellevue (WSO), flying with Capt. John A. Madden, Jr. (pilot) in a McDonnell Douglas F-4D, shoots down two MiG-19s near Hanoi. These were the fifth and sixth victories for Captain DeBellevue, which made him the leading

American ace of the war. All of his victories came in a four-month period. Captain Madden would record a third kill two months later.

November 4, 1972. Navy Cmdr. Philip R. Hite sets a recognized class record for distance in a closed circuit (turboprop aircraft) of 6,278.05 miles at NAS Patuxent River, Md., in a Lockheed RP-3D Orion. The record still stands.

December 7–19, 1972. The Apollo 17 mission is the last of the moon landings. It is also the first U.S. manned launch to be conducted at night. Mission commander Navy Cmdr. Eugene A. Cernan and lunar module pilot/geologist Harrison Schmitt spend a record 75 hours on the lunar surface.

December 18, 1972. The U.S. begins Operation Linebacker II, the 11-day bombing of Hanoi and Haiphong. Massive air strikes help persuade North Vietnam to conclude Paris peace negotiations, which will be finalized January 27, 1973.

December 18, 1972. In a throwback to past aerial combat, SSgt. Samuel O. Turner, the tail gunner on a Boeing B-52D bomber, downs a trailing MiG-21 with a blast of .50 cal. machine guns near Hanoi. Six days later, AlC Albert E. Moore, also a B-52 gunner, shoots down a second MiG-21 after a strike on the Thai Nguyen railyard. These were the only aerial gunner kills of the war.

January 8, 1973. Capt. Paul D. Howman (pilot) and 1st Lt. Lawrence W. Kullman (WSO), flying in a McDonnell Douglas F-4D, record the last USAF victory in the Vietnam War as they shoot down a MiG-21 near Hanoi. It was the duo's only aerial victory.

January 27, 1973. Cease-fire agreements ending the war in Vietnam are signed in Paris.

February 12, 1973. Operation Homecoming, the return of 591 American POWs from North Vietnam, begins.

April 10, 1973. First flight of the Boeing T-43A navigation trainer occurs. The T-43 is developed from the 737-200 civil transport.

May 25–June 22, 1973. An all-Navy crew of Capt. Pete Conrad and Cmdrs. Joseph Kerwin and Paul Weitz salvage the Skylab program, as they repair the space station (which had been damaged on launch) in orbit. Their 28-day, 404-orbit mission is the longest in history to this point.

July 1, 1973. Authorization for the military draft ends.

July 18, 1973. John L. McLucas becomes Secretary of the Air Force.

July 28–September 25, 1973. The Skylab 3 crew of Navy Capt. Alan Bean, Marine Maj. Jack Lousma, and scientist Dr. Owen Garriott perform valuable science experiments and Earth observations during their 59-day, 892-orbit stay on the space station.

August 1, 1973. Gen. George S. Brown becomes Air Force Chief of Staff.

October 1, 1973. CMSgt. Thomas N. Barnes becomes Chief Master Sergeant of the Air Force.

November 10, 1973. The Thunderbirds, the Air Force's aerial demonstration squadron, fly their 518th and last show in the McDonnell Douglas F-4E Phantom II at New Orleans, La.. The team will convert to the Northrop T-38A Talon for the 1974 show season.

November 14, 1973. The U.S. ends its major airlift to Israel. In a 32-day operation during the Yom Kippur War,

Military Airlift Command (MAC) airlifts 22,318 tons of supplies.

November 14, 1973. The first production McDonnell Douglas F-15A Eagle is delivered to the Air Force at Luke AFB, Ariz.

November 16, 1973–February 8, 1974. A crew of space rookies,

Marine Lt. Col. Gerald Carr, Air Force Lt. Col. William Pogue, a former Thunderbird pilot, and Dr. Edward Gibson from the third and final Skylab crew. At 84 days, this crew, which observes the Comet Kohoutek during the mission, will hold the American space mission duration record until 1995.

1974–1983
January 21, 1974. The General Dynamics YF-16 prototype makes a first, unplanned, flight at Edwards AFB, Calif. Company test pilot Phil Oestricher was conducting high-speed taxi tests and the aircraft lifted off the runway, and rather than risk damage to the aircraft, the pilot elected to lift off and go around to come in for a normal landing. The first official flight is made on February 2, also by Mr. Oestricher.

June 9, 1974. Company pilot Henry E. Hank Chouteau makes the first flight of the Northrop YF-17 at Edwards AFB, Calif. Although the YF-17 would not be selected as the winner of the Air Force's Lightweight Fighter Technology evaluation program, the YF-17 would become the progenitor of the Navy's F/A-18 Hornet.

July 1, 1974. Gen. David C. Jones becomes Air Force Chief of Staff.

September 1, 1974. Maj. James V. Sullivan and Maj. Noel Widdifield set a New York–London speed record of 1,806.964 mph in a Lockheed SR-71A. The trip takes 1 hour, 54 minutes, 55 seconds.

October 24, 1974. The Air Force's Space and Missile Systems Organization carries out a midair launch of a Boeing LGM-30A Minuteman I from the hold of a Lockheed C-5A.

December 23, 1974. Company pilot Charles Bock, Jr., USAF Col. Emil Sturmthal, and flight test engineer Richard Abrams make the first flight of the Rockwell B-1A variable-geometry bomber from Palmdale, Calif.

January 13, 1975. The General Dynamics YF-16 is announced as the winner of the Air Force's Lightweight Fighter Technology evaluation program. The F-16 is also the leading candidate to become the Air Force's new air combat fighter. The YF-17 becomes the predecessor of the Navy's F/A-18 Hornet.

January 16–February 1, 1975. Three USAF pilots set eight recognized class records for time-to-climb (jet aircraft) in the McDonnell Douglas F-15A *Streak Eagle* at Grand Forks AFB, N.D. One of the records, time-to-climb to 20,000 meters (2:02.94 minutes) set by Maj. Roger J. Smith, still stands.

January 26, 1975. The Force Modernization program, a nine-year effort to replace all Boeing LGM-30B Minuteman Is with either Minuteman IIs (LGM-30F) or Minuteman IIIs (LGM-30G), is completed, as the last 10 LGM-30Gs are turned over to SAC at F. E. Warren AFB. Wyo.

February 1, 1975. Maj. Roger Smith sets a world time-to-climb record to 30,000 meters (98,425 feet)

in 3 minutes, 27.8 seconds in the Mc-Donnell Douglas F-15A Streak Eagle.

May 15, 1975. Carrying 175 Marines, Air Force special operations helicopters land on Kho Tang Island, off the Cambodian coast, to begin rescue of the crew of the U.S. merchant ship *Mayaguez*, which had been seized in international waters by the Cambodian navy three days earlier.

June 30, 1975. The last Douglas C-47A Skytrain in routine Air Force use is retired to the U.S. Air Force Museum at Wright-Patterson AFB, Ohio.

July 15–24, 1975. U.S. astronauts Brig. Gen. Thomas P. Stafford, USAF, Vance D. Brand, and Donald K. Slayton rendezvous, dock, and shake hands with Soviet cosmonauts Alexei Leonov and Valeri Kubasov in orbit during the Apollo-Soyuz Test Project.

August 20, 1975. The Viking 1 mission to Mars is launched from Cape Canaveral AFS, Fla., on a Titan III booster. The spacecraft enters Mars orbit on June 19, 1976, and the lander, which takes soil samples and performs rudimentary analysis on them, soft-lands on July 20, 1976.

September 1, 1975. Gen. Daniel "Chappie" James, Jr., USAF, becomes the first black officer to achieve four-star rank in the U.S. military.

October 21, 1975. Fairchild Republic Co.'s A-10A Thunderbolt II makes its first flight. The first combat-ready A-10A wing will be the 354th Tactical Fighter Wing at Myrtle Beach, S.C., which will begin taking delivery of the fighters in March 1977.

November 29, 1975. The first Red Flag exercise at Nellis AFB, Nev., begins a new era of highly realistic training for combat aircrews.

January 2, 1976. Thomas C. Reed becomes Secretary of the Air Force.

May 8, 1976. The Thunderbirds, the Air Force's aerial demonstration squadron, fly the 2,000th show in their 23-year history at Mountain Home AFB, Idaho. The team's Northrop T-38A Talons are sporting a special paint scheme for America's Bicentennial celebration.

July 3, 1976. In an Israeli commando assault on Entebbe airport in Ugnada, the Israelis destroy four MiG-17s and seven MiG-21s on the ground and rescue 105 mostly Israeli and Jewish hostages held by pro-Palestinian terrorists.

July 27–28, 1976. Three different SR-71 pilots (Maj. Adolphus H. Bledsoe, Capt. Robert C. Helt, and Capt. Eldon W. Joersz) set three absolute world flight records over Beale AFB, Calif: altitude in horizontal flight (85,068.997 feet), speed over a straight course (2,193.16 mph), and speed over a closed course (2,092.294 mph). The records are still standing in 1996.

March 24, 1977. Boeing delivers the first basic production version of the E-3A Sentry (AWACS) to Tinker AFB, Okla.

April 6, 1977. John C. Stetson becomes Secretary of the Air Force.

June 30, 1977. President Jimmy Carter, citing the continued ability of the B-52 fleet and the development of cruise missiles, announces he is canceling the B-1A variable-geometry bomber program. Testing of the four B-1A prototypes will continue, however.

August 1, 1977. CMSgt. Robert D. Gaylor becomes Chief Master Sergeant of the Air Force.

August 23, 1977. Cyclist/pilot Bryan Allen wins the $95,000 Kremer Prize for successfully demonstrating sustained, maneuverable, man-powered flight in the MacReady *Gossamer Condor*. Allen pedals the aircraft, which is made of thin aluminum tubes covered with Mylar plastic and braced with stainless steel, over a 1.15-mile course at Shafter Airport, Shafter, Calif.

August 31, 1977. Alexander Fedotov, flying in the MiG E-266M, a modified MiG-25 Foxbat, sets the recognized absolute record for altitude, reaching 123,523.58 feet at Podmosconvnoe, USSR. This record still stands as of 1996.

October 1, 1977. Volant Oak, the quarterly rotation of six Air Force Reserve and Air National Guard transports to Howard AFB, Panama, for in-place tactical airlift in Central and South America, begins.

December 1, 1977. In total secrecy, company test pilot Bill Park makes the first flight of the Lockheed XST Have Blue demonstrator at Groom Lake, Nev.. Developed in only twenty months, Have Blue is designed as a testbed for stealth technology.

February 22, 1978. The first test satellite in the Air Force's Navstar Global Positioning System is successfully launched into orbit.

March 23, 1978. Capt. Sandra M. Scott becomes the first female aircrew member to pull alert duty in SAC.

July 1, 1978. Gen. Lew Allen, Jr., becomes Air Force Chief of Staff.

August 11–17, 1978. Ben Abruzzo, Maxie Anderson, and Larry Newman complete the first crossing of the Atlantic Ocean by balloon. Flying in the helium-filled *Double Eagle II*, the trio makes the 3,100-mile flight from Presque Isle, Me., to Miserey, France, in 137 hours, 6 minutes.

November 30, 1978. The last Boeing LGM-30G Minuteman III ICBM is delivered to the Air Force at Hill AFB, Utah.

January 6, 1979. The 388th Tactical Fighter Wing at Hill AFB, Utah, receives the first operational General Dynamics F-16A fighters. The first Air Force Reserve F-16s are delivered to the 419th TFW at Hill on January 28, 1984.

June 12, 1979. Pilot/cyclist Bryan Allen makes the first human-powered flight across the English Channel in the *Gossamer Albatross*.

July 9, 1979. The Voyager 2 space probe, launched in 1977, flies within 399,560 miles of Jupiter's cloud tops. Voyager 2 will pass Neptune in 1989.

July 26, 1979. Hans Mark becomes Secretary of the Air Force.

August 1, 1979. CMSgt. James M. McCoy becomes Chief Master Sergeant of the Air Force.

October 1, 1979. All atmospheric defense assets and missions of Aerospace Defense Command are transferred to Tactical Air Command (TAC). Also on this date, the Aerospace Audiovisual Service becomes the single manager for Air Force combat audiovisual documentation.

March 12–14, 1980. Two B-52 crews fly nonstop around the world in 43½ hours, covering 21,256 statute miles, averaging 488 mph, and carrying out sea surveillance/reconnaissance missions.

April 24, 1980. In the middle of an attempt to rescue U.S. citizens held hostage in Iran, mechanical difficulties force several Navy RH-53 helicopter crews to turn back. Later, one of the RH-53s collides with an Air Force HC-130 in a sandstorm at the Desert One refueling site. Eight U.S. servicemen are killed.

May 28, 1980. The Air Force Academy graduates its first female cadets. Ninety-seven women are commissioned as second lieutenants. Lt. Kathleen Conly graduates eighth in her class.

February 9, 1981. Verne Orr becomes Secretary of the Air Force.

April 12, 1981. The space shuttle orbiter *Columbia*, the world's first reusable manned space vehicle, makes its first flight with astronauts John Young and Navy Capt. Robert Crippen aboard.

June 7, 1981. Eight Israeli Air Force F-16s, escorted by F-15s, attack the Osirak nuclear reactor near Baghdad, Iraq, disabling its core. As a result, the U.S. imposes a temporary embargo on the supply of new F-16s to Israel.

June 18, 1981. In total secrecy, company pilot Hal Farley makes the first flight of the Lockheed F-117A stealth fighter at Tonopah Test Range, Nev. The existence of this aircraft would not be publicly revealed until 1988.

June 26, 1981. The first production Grumman/General Dynamics EF-111A, a specially developed ECM tactical jamming aircraft, makes its first flight.

August 1, 1981. CMSgt. Arthur L. Andrews becomes Chief Master Sergeant of the Air Force.

October 2, 1981. President Ronald Reagan reinstitutes the B-1 bomber program canceled by the Carter administration in 1977.

November 10, 1981. For the first time, U.S. Air Forces in Europe and the German Air Force test a section of the autobahn for emergency landings.

July 1, 1982. U.S. Air Force activates its first ground-launched cruise missile (GLCM) wing, the 501st Tactical Missile Wing, at Greenham Common in England.

July 1, 1982. Gen. Charles A. Gabriel becomes Air Force Chief of Staff.

September 1, 1982. Air Force Space Command established.

September 1–30, 1982. H. Ross Perot, Jr. and Jay Coburn complete the first circumnavigation of the globe by helicopter. Flying a modified Bell 206L Longranger, the duo average 117 mph during their 246.5 hours of flight time. The trip starts and ends at Fort Worth, Tex.

November 11, 1982. Vance D. Brand, Robert F. Overmyer, Joseph P. Allen IV, and William B. Lenoir lift off in the space shuttle *Columbia*. STS-5 is the first mission to send four astronauts aloft at one time.

February 9, 1983. The first re-winged C-5A makes its first flight at Marietta, Ga. It will be delivered to the Air Force at the end of the month.

February 10, 1983. The Cruise Pact is signed by the U.S. and Canada, allowing testing of U.S. cruise missiles in northern Canada.

March 23, 1983. Flight testing of the Rockwell B-1A resumes at Edwards AFB. This aircraft is modified for the B-1B development effort.

May 9, 1983. A C-141 crew from the 18th Military Airlift Squadron, McGuire AFB, N.J., becomes USAF's first all-female crew to fly a round-trip mission across the Atlantic.

June 17, 1983. The first LGM-118A Peacekeeper (originally MX) ICBM is test-launched from Vandenberg AFB.

June 18, 1983. The first American woman to go into space, Sally K. Ride, is aboard *Challenger* on the seventh space shuttle mission (STS-7).

July 4, 1983. Flying in their new General Dynamics F-16A Fighting Falcons, the Thunderbirds, the Air Force's aerial demonstration squadron, perform before an estimated crowd of 2 million people at Coney Island, N.Y.

July 22, 1983. Australian Dick Smith, flying a Bell JetRanger, completes the first solo flight around the world in a helicopter. The 35,258-mile trip began August 5, 1982.

August 1, 1983. CMSgt. Sam E. Parish becomes Chief Master Sergeant of the Air Force.

August 30, 1983. Two milestones are recorded on the STS-8 space shuttle mission: The oldest astronaut, William E. Thornton, 54, and the first black astronaut, Lt. Col. Guion S. Bluford, USAF, are sent aloft on the Space Shuttle *Challenger* with three others.

October 25, 1983. Operation Urgent Fury, the rescue of American medical students on the Caribbean island of Grenada, begins. The operation will last until November 2.

November 28, 1983. The ninth space shuttle mission (STS-9) is launched. Mission Commander John W. Young becomes the first person to make six spaceflights, and *Columbia* is the first spacecraft to be launched with a crew of six. The 10-day flight is also the first to use the European Spacelab module.

1984–1989
February 3–11, 1984. Navy Capt. Bruce McCandless becomes the first human satellite as he takes the self-contained Manned Maneuvering Unit (MMU) out for a spin while in Earth orbit on space shuttle mission 41-B.

April 6–13, 1984. The 11th U.S. space shuttle mission (41-C) is a spectacular success as the defective Solar Maximum Mission satellite (Solar Max) is repaired in orbit. After mission specialist George Nelson fails to capture the satellite on his spacewalk, Terry J. Hart uses Challenger's remote manipulator arm to catch Solar Max on the fly. George Nelson and James D. A. van Hoften repair the satellite in the shuttle's payload bay before it is released.

May 22, 1984. The Chiefs of Staff of the Army and the Air Force sign a memorandum of agreement titled

"Joint Force Development Process," also known as "The 31 Initiatives."

August 14, 1984. Boeing rolls out the 1, 832d and last 727, a 727-252F freighter for Federal Express. The 727 is the only commercial transport to exceed the 1,500 mark in aircraft built.

September 4, 1984. The first production Rockwell B-1B bomber is rolled out at Air Force Plant 42 in Palmdale, Calif.

October 4, 1984. After her pilot husband dies of a heart attack, Elaine Yadwin takes the controls of their Piper Cherokee and manages to land at Dade-Collier Airport in South Florida. She is talked down by ground controllers.

October 5–13, 1984. On the 13th space shuttle mission, *Challenger* lifts off for the first time with a crew of seven. Mission 41-G is the first to have two female astronauts (Sally K. Ride and Kathryn D. Sullivan, who will become the first American woman to make a spacewalk) and the first to have a Canadian astronaut aboard (Marc Garneau). Commander Robert L. Crippen becomes the first to fly on the shuttle four times. Aloft, the crew refuels a satellite in orbit for the first time.

October 18, 1984. Company pilot M. L. Evenson and USAF Lt. Col. L. B. Schroeder make the first flight of the Rockwell B-1B variable-geometry bomber at Palmdale, Calif., and land at Edwards AFB. This is the first of 100 aircraft to be built in the revitalized B-1 bomber program.

December 14, 1984. Grumman pilot Chuck Sewell makes the first flight of the X-29A forward-swept-wing demonstrator at Edwards AFB. The X-29s, two of the most unusual aircraft ever built, are designed to prove the aerodynamic benefits of wings that appear to have been put on backwards.

January 24–27, 1985. The 15th space shuttle mission (51-C) is the first dedicated Department of Defense flight. The *Discovery* crew of Navy Capt. Thomas K. Mattingly (mission commander), Air Force Lt. Col. Loren J. Shriver (pilot), and Air Force mission specialists Lt. Col. Ellison S. Onizuka and Maj. Gary E. Payton, along with Marine Corps Lt. Col. James F. Buchli, deploys a classified payload, believed to be a signals intelligence satellite.

September 13, 1985. The first test of the LTV-Boeing ASM-135A air-launched antisatellite weapon against a target is successfully carried out over the Western Missile Test Range. Launched from an F-15, the missile destroys a satellite orbiting at a speed of 17,500 mph approximately 290 miles above Earth.

October 25–November 2, 1985. USAF units take part in joint operations against Cubans and Marxists in Grenada.

December 9, 1985. Russell A. Rourke becomes Secretary of the Air Force.

December 16, 1985. After 20 years of operation, the Pioneer 6 satellite becomes the longest-running spacecraft in history. When launched in 1965, the solar-orbiting satellite had a life expectancy of six months.

January 28, 1986. The space shuttle *Challenger* explodes 73 seconds after liftoff, killing all seven astronauts, including schoolteacher Christa McAuliffe. Others on Mission 51-L include Francis R. Scobee, Navy Cmdr. Mike Smith, Judith Resnik, Ronald E. McNair, Air Force Lt. Col. Ellison S. Onizuka, and Gregory Jarvis. The manned space program will be halted for two years while vehicular and management flaws are corrected.

April 15, 1986. In Operation Eldorado Canyon, 18 USAF F-111s flying from RAF Lakenheath in England are joined by carrier-based Navy aircraft in air strikes against Libya in response to state-sponsored terrorism.

April 24–May 7, 1986. Veterans of three wars attend the Air Force Association's "Gathering of Eagles" in Las Vegas, Nev.

June 9, 1986. Edward C. Aldridge, Jr., becomes Secretary of the Air Force.

July 1, 1986. Gen. Larry D. Welch becomes Air Force Chief of Staff.

July 1, 1986. CMSgt. James C. Binnicker becomes Chief Master Sergeant of the Air Force.

October 1, 1986. The Goldwater-Nichols Act gives theater commanders increased control of forces from all services.

December 23, 1986. Richard Rutan and Jeana Yeager complete the first nonstop unrefueled around-the-world trip in their experimental *Voyager*, starting and stopping at Mojave, Calif. The trip sets recognized absolute records for speed around the world, nonstop, nonrefueled (115.65 mph); great circle distance without landing; and distance in a closed circuit without landing (both 24,986.727 miles).

July 4, 1987. Lt. Col. Robert Chamberlain (and crew) sets a dozen recognized class records for speed with payload (jet aircraft) in a Rockwell B-1B at Palmdale, Calif. The brand-new aircraft was on an acceptance flight and flew a 500-mile closed course near Vandenberg AFB, Calif. The records still stand.

July 17–31, 1987. Mike Hance becomes the first pilot to consecutively take off and land in all 50 states and the District of Columbia as he flies his Mooney 252 private plane from Honolulu, Hawaii, to Oshkosh, Wis., via the United States—all the rest of them, obviously.

September 17, 1987. Maj. Brent A. Hedgpeth (and crew) sets nine recognized class records for 5,000-km speed with and without payload (jet aircraft) of 655.09 mph in a Rockwell B-1B at Palmdale, Calif. The records still stand.

September 24, 1987. The Air Force's Thunderbirds fly for a crowd of 5,000 in Beijing. It has been nearly 40 years since a U.S. combat aircraft flew over and landed on Chinese soil.

January 1, 1988. SAC changes missile crew assignment policy to permit mixed male/female crews in Minuteman and Peacekeeper launch facilities.

January 20, 1988. The 100th and final B-1B bomber rolls off the line at Rockwell's plant in Palmdale.

February 10, 1988. The 2,000th F-16 fighter built is accepted by Singapore.

March 3, 1988. The Pioneer 8 solar orbiter, which was launched November 8, 1968, with a six-month life expectancy, is finally declared defunct.

May 23, 1988. The Bell-Boeing V-22 Osprey, the world's first production tiltrotor aircraft, is rolled out at Bell Helicopter Textron's plant in Arlington, Tex.

August 2, 1988. As evidence of thawing superpower relations. U.S. Secretary of Defense Frank C. Carlucci is given the opportunity to inspect the Soviet Tu-160 Blackjack strategic bomber during a visit to Kubinka AB, near Moscow.

September 29, 1988. Launch of the space shuttle *Discovery* ends the long stand-down of the U.S. manned space program in the wake of the *Challenger* disaster.

October 25, 1988. A U.S. Navy S-3 Viking antisubmarine warfare aircraft from the carrier USS *Theodore Roosevelt* (CVN-71) is given a $21 parking ticket after the crew overshoots a runway at a base in southern England and lands on a public road.

November 6, 1988. The Air Force launches its last Martin Marietta Titan 34D booster from Vandenberg AFB, Calif. It carries a classified payload.

November 7, 1988. The U.S. Postal Service issues a 65-cent commemorative stamp bearing the likeness of Gen. H. H. "Hap" Arnold in ceremonies at the Arnold Engineering and Development Center at Arnold AFB, Tenn.

November 10, 1988. The Air Force reveals the existence of the Lockheed F-117A stealth fighter, operational since 1983.

November 12, 1988. Soviet cosmonauts Vladimir Titov and Musa Manarov break the world space endurance record as they remain on board the space station Mir ("peace") for their 326th day in orbit.

November 19, 1988. Boeing KC-135R tanker crews from the 19th Air Refueling Wing (Robins AFB, Ga.), 340th ARW (Altus AFB, Okla.), 319th Bomb Wing (Grand Forks AFB, N.D.), and 384th BW (McConnell AFB, Kan.) set 16 class time-to-climb records in flights from Robins AFB. Nine of the records still stand.

November 22, 1988. Northrop and the Air Force roll out the B-2 Stealth bomber at Air Force Plant 42 in Palmdale.

November 30, 1988. The Soviets roll out the An-225 transport, the world's largest airplane.

December 9, 1988. The first Sierra Research/de Havilland Canada E-9A airborne telemetry data relay aircraft is delivered to the Air Force's 475th Weapons Evaluation Group at Tyndall AFB, Fla.

December 29, 1988. The first operational dual-role (air superiority and deep interdiction) McDonnell Douglas F-15E fighter is delivered to the Air Force.

January 4, 1989. Two Libyan MiG-23 Flogger fighters, displaying hostile intentions, are shot down over international waters by an element of U.S. Navy F-14 Tomcats operating from the carrier USS *John F. Kennedy* (CVN-67).

February 14, 1989. The first McDonnell Douglas Delta II space booster is launched from Cape Canaveral AFS. The 128-foot-tall rocket boosts the first operational NS-7 Navstar Block II Global Positioning System satellite into orbit.

February 16, 1989. Northrop completes the 3,806th and final aircraft in the F-5/T-38 series. The milestone aircraft, an F-5E, will later be delivered to Singapore.

March 1, 1989. The first General Dynamics F-16A modified under the Air Force's air defense fighter program is delivered to the Air National Guard's 114th Tactical Fighter Training Squadron at Kingsley Field, Ore.

March 19, 1989. Bell pilot Dorman Canon and Boeing pilot Dick Balzer make the first flight of the Bell-Boeing V-22 Osprey at Bell Helicopter Textron's Flight Research Center in Arlington, Tex.

March 21, 1989. NASA completes the flight test of the Mission Adaptive Wing, a modification to the advanced fighter technology integration (AFTI) F-111 that allows the curvature of the aircraft's leading and trailing edges to be varied in flight. The MAW completes 144.9 hours on 59 flights.

March 30, 1989. Fairchild delivers the first of 10 C-26A operational support aircraft to the Air National Guard's 147th Fighter Interceptor Group at Ellington ANGB, Tex. The C-26 is the military version of the Metro III commuter aircraft.

April 17, 1989. Lockheed delivers the 50th and last C-5B Galaxy transport to the Air Force in ceremonies at Marietta, Ga.

April 17–18, 1989. Lockheed pilots Jerry Hoyt and Ron Williams set 16 class time-to-climb and altitude records in separate flights in a NASA U-2C at the Dryden Flight Research Facility at Edwards AFB. The 32-year-old aircraft, which was loaned to NASA in 1971, is retired to a museum after the flights. The records still stand.

May 4, 1989. Air Force Maj. Mark C. Lee releases the Magellan probe from the payload bay of the space shuttle orbiter *Atlantis* during the first day of the four-day STS-30 space mission. The 21-foot-tall, 7,604-pound Magellan probe is designed to map Venus with its synthetic aperture radar.

May 22, 1989. Donald B. Rice becomes Secretary of the Air Force.

June 10, 1989. Capt. Jacquelyn S. Parker becomes the first female pilot to graduate from the Air Force Test Pilot School at Edwards AFB.

June 14, 1989. The first Martin Marietta Titan IV heavy-lift space booster is successfully launched from Launch Complex 40 at Cape Canaveral AFS. The booster, nearly 20 stories tall, carries a classified military payload.

July 6, 1989. The nation's highest civilian award, the Presidential Medal of Freedom, is presented to retired Air Force Gen. James H. Doolittle in White House ceremonies.

July 6, 1989. The 169th and last MGM-31 Pershing 1A intermediate-range ballistic missile is destroyed at the Longhorn Army Ammunition Plant near Karnack, Tex., under the terms of the Intermediate Nuclear Forces (INF) Treaty.

July 17, 1989. Northrop chief test pilot Bruce Hinds and Air Force Col. Richard Couch, director of the B-2 Combined Test Force, make the first flight of the Northrop B-2A advanced technology bomber, flying from Air Force Plant 42 in Palmdale, to the Air Force Flight Test Center at Edwards AFB.

August 2, 1989. The Navy successfully carries out the first undersea launch of the Lockheed UGM-133A Trident II (D5) sea-launched ballistic missile. The missile is launched from USS *Tennessee* (SSBN-734) while cruising off Florida.

August 6, 1989. As further evidence of the thaw in U.S.-Soviet relations, two MiG-29 fighters and the giant An-225 transport land and refuel at Elmendorf AFB, Alaska, on their way to an air show in Canada.

August 8–13, 1989. The 30th mission in the U.S. space shuttle program is carried out, as the crew of five service astronauts launches a classified payload from the orbiter *Columbia*. It is the longest military shuttle flight to date.

August 24, 1989. The Voyager 2 space probe completes its grand tour of the solar system as the 1,787-pound vehicle passes within 3,000 miles of Neptune. Voyager 2 was launched in August 1977.

September 14, 1989. The Bell-Boeing V-22 Osprey tiltrotor aircraft achieves its first conversion from helicopter mode to airplane mode while in flight.

September 15, 1989. McDonnell Douglas delivers the 500th AH-64 Apache helicopter to the U.S. Army at the company's plant in Mesa, Ariz.

October 1, 1989. Air Force Gen. Hansford T. Johnson, pinning on his fourth star and assuming command of U.S. Transportation Command and MAC, becomes the first Air Force Academy graduate to attain the rank of full general. He is a member of the Academy's first graduating class of 1959.

October 3, 1989. The last of 37 Lockheed U-2R/TR-1A/B high-altitude reconnaissance aircraft is delivered to the Air Force.

October 4, 1989. A crew from the 60th Military Airlift Wing, Travis AFB, Calif., lands a Lockheed C-5B transport at McMurdo Station in Antarctica. This is the first time an aircraft so large has landed on the ice continent. The C-5B, carrying 72 passengers and 168,000 pounds of cargo (including two fully assembled Bell UH-1N helicopters), lands without skis.

October 7, 1989. Wayne Handley sets the recognized U.S. record for longest inverted flat spin with the most rotations (67) in a Pitts Special acrobatic aircraft at Salinas, Calif.

December 3, 1989. Solar Max, the first satellite to be repaired in orbit, is destroyed as it reenters the atmosphere over Sri Lanka.

December 14, 1989. MAC approves a policy change that will allow female aircrew members to serve on C-130 and C-141 airdrop missions.

December 20, 1989. Operation Just Cause begins in Panama. The Air Force plays a major role, ranging from airlift, airdrops, and aerial refueling to bringing Panamanian dictator Manuel Noriega to the U.S. In Just Cause, the Lockheed F-117A stealth fighter is used operationally for the first time.

1990–1995

January 25, 1990. The Lockheed SR-71 Blackbird high-altitude, high-speed reconnaissance aircraft is retired from SAC service in ceremonies at Beale AFB, Calif. SR-71 crews flew more than 65 million miles, half at speeds above Mach 3.

January 31, 1990. Coronet Cove, the Air National Guard's rotational deployments to defend the Panama Canal, ends after more than 11 years. More than 13,000 sorties, totaling 16,959 hours, have been flown since the operation began.

February 21, 1990. The Air Force returns to dual-track pilot training. The team of McDonnell Douglas. Beech, and Quintron is selected over two other teams to provide the Tanker/Transport Training System. This turnkey operation will train pilots going on to fly "heavies" using the T-1A Jayhawk.

March 1, 1990. The Rockwell/MBB X-31A enhanced fighter maneuverability (EFM) demonstrator rolls out at Rockwell's facility at Air Force Plant 42 in Palmdale, Calif. A joint venture between the U.S. and West Germany, the X-31 is designed to prove technologies that will allow close-in aerial combat beyond normal flying parameters.

March 6, 1990. Lt. Col. Ed Yielding (pilot) and Lt. Col. J. T. Vida (reconnaissance systems officer) set four speed records, including a transcontinental mark of 2,112.52 mph (1 hour, 8 minutes, 17 seconds elapsed time) over the 2,404.05-statue-mile course from Oxnard, Calif., to Salisbury, Md., on what was at the time the last Air Force flight of the Lockheed SR-71.

March 26, 1990. Grumman rolls out the first production-standard version of the improved F-14D Tomcat for the U.S. Navy at its plant in Calverton, Long Island, N.Y.

April 2, 1990. Air Force pilot Maj. Erwin "Bud" Jenschke demonstrates in-flight thrust reversing for the first time while flying the McDonnell

Douglas NF-15B S/MTD (STOL/Maneuvering Technology Demonstrator) aircraft over Edwards AFB.

April 4, 1990. McDonnell Douglas turns over the last of 60 KC-10A Extender tanker/cargo aircraft to the Air Force at its plant in Long Beach, Calif.

April 5, 1990. The first launch of the Orbital Sciences Corp./Hercules Aerospace Pegasus air-launched space booster, the first all-new booster in two decades, is successfully carried out off the California coast.

April 24, 1990. The space shuttle *Discovery*, with a crew of five, lifts off on the 35th mission in the shuttle program. The next day, astronaut Steven A. Hawley releases the Hubble Space Telescope, an on-orbit observatory with great scientific promise. Although the telescope gathers unprecedented images, it proves to be somewhat myopic (a 2-micron-wide spherical aberration—less than the width of a human hair—is found) and will have to be repaired on a 1993 shuttle flight.

April 25, 1990. Boeing delivers the 200th reengined and upgraded KC-135R tanker to the Air Force. It is delivered to the 340th Air Refueling Group at Altus AFB, Okla.

April 30, 1990. USAF announces that Air Force Special Operations Command, the first new command since 1982, will be established by early summer. This component of the U.S. Special Operations Command will be composed primarily of Twenty-third Air Force assets.

May 4, 1990. The Hughes/Raytheon AIM-120A advanced medium-range air-to-air missile (AMRAAM) passes its "final exam"—demonstration of its ability to achieve multiple kills against multiple targets. There are three direct hits and a lethal near miss in the four-missile vs. four-target test near Eglin AFB.

May 17, 1990. An Air Force crew from McGuire AFB, N.J., lands a Lockheed C-141B transport at Moscow's Sheremetyevo Airport to deliver an inoperative MGM-31 Pershing II missile that will go into a museum in Moscow. The crew then picks up an inoperative Soviet SS-20 for display at the National Air and Space Museum in Washington, D.C.

May 22, 1990. Air Force Special Operations Command is established.

May 22, 1990. Company pilot Larry Walker and Air Force pilot Maj. Erwin Jenschke land the McDonnell Douglas NF-15B S/MTD test bed in a mere 1,650 feet at the Air Force Flight Test Center at Edwards AFB. Pratt & Whitney two-dimensional thrust-reversing engine nozzles are the main method of stopping the aircraft.

June 1, 1990. SAC turns over the first pair of General Dynamics FB-111As to TAC. With one internal modification, the aircraft will be redesignated F-111Gs.

June 22, 1990. The Northrop/McDonnell Douglas YF-23A advanced tactical fighter prototype is rolled out in ceremonies at the ATF Combined Test Force Facility at Edwards AFB. It is powered by two Pratt & Whitney YF119-PW-100 engines. Northrop pilot Paul Metz will make the first flight August 27, 1990.

July 1, 1990. Gen. Michael J. Dugan becomes Air Force Chief of Staff.

July 11, 1990. Four Air National Guard F-16 pilots from the 177th Fighter Interceptor Group at Atlantic City IAP, N.J., escort two Soviet MiG-29 fighters and an Il-76 transport in

U.S. airspace, flying from Kalamazoo, Mich., to Rockford, Ill., as part of the Soviet Union's first U.S. air show tour.

July 12, 1990. The last of 59 Lockheed F-117A stealth fighters is delivered to the Air Force in ceremonies at the company's Palmdale facility.

July 13, 1990. Alaskan Air Command ceases to exist. The former command now becomes a numbered (Eleventh) Air Force and is made part of Pacific Air Forces.

July 24, 1990. SAC ends "Looking Glass," more than 29 years of continuous airborne alert, as a Boeing EC-135C Airborne Command Post aircraft lands at Offutt AFB, Neb.

August 1, 1990. CMSgt. Gary R. Pfingston becomes Chief Master Sergeant of the Air Force.

August 7, 1990. The U.S. begins Operation Desert Shield, the large-scale movement of U.S. forces to the Middle East in response to Iraq's August 2 invasion of Kuwait and threat to Saudi Arabia.

August 8, 1990. A C-141 carrying Airlift Control Element lands in Dhahran, the first USAF aircraft into the crisis zone. F-15s from 1st Tactical Fighter Wing, Langley AFB, Va., and elements of the 82d Airborne Division, Fort Bragg, N.C., arrive in Saudi Arabia. U.S. AWACS aircraft augment Saudi AWACS aircraft orbiting over Saudi Arabia.

August 17, 1990. The first stage of the Civil Reserve Air Fleet is activated for the first time to increase the availability of airlift to the Middle East.

August 21, 1990. By this date, 1 billion pounds of materiel have arrived in or are en route to Saudi Arabia. Six fighter wings are deployed, and SAC steps up refueling efforts and RC-135 reconnaissance flights in the area. By late August, more than 40,000 reserve components of all services have been called up.

August 23, 1990. The first of two Boeing VC-25A Presidential transport aircraft is delivered to the 89th Military Airlift Wing at Andrews AFB, Md. The new aircraft, a modified 747-200B commercial transport, will replace the VC-137C aircraft currently used as Air Force One.

August 29, 1990. The Lockheed/Boeing/General Dynamics YF-22A ATF prototype is unveiled in ceremonies at Lockheed Plant 10 in Palmdale. This aircraft is powered by two General Electric YF120-GE-100 turbofan engines. Lockheed pilot Dave Ferguson makes the first flight of the YF-22 September 29, 1990.

September 6, 1990. The U.S. Postal Service issues a 40-cent postage stamp honoring Lt. Gen. Claire L. Chennault.

September 18, 1990. Gen. John M. Loh becomes Acting Air Force Chief of Staff.

October 11, 1990. Rockwell pilot Ken Dyson makes the first flight of the Rockwell/MBB X-31A enhanced fighter maneuverability (EFM) demonstrator at Air Force Plant 42. The flight lasts 38 minutes.

October 30, 1990. Gen. Merrill A. McPeak becomes Air Force Chief of Staff.

November 9, 1990. Col. (Dr.) Thomas C. Cook, believed to be the Air Force's last World War II combat veteran still serving, retires. He saw action as a B-24 navigator in Europe and transferred to Reserve status in 1948. He returned to active duty in 1976.

December 17, 1990. The Lockheed/Boeing/General Dynamics YF-22 prototype is flown to an unprecedented 60° angle of attack (AOA) attitude and remains in full control in a test flight over Edwards AFB.

January 7, 1991. Saying that nobody could tell him how much it would cost to keep the program going, Secretary of Defense Dick Cheney announces that he is cancelling the McDonnell Douglas-General Dynamics A-12 Avenger attack aircraft program for default. The A-12 would have been the Navy's first stealth aircraft.

January 16, 1991. At 6:35 A.M. local time, B-52G crews from the 2nd Bomb Wing, Barksdale AFB, La., take off to begin what will become the longest bombing mission in history. Carrying 39 AGM-86C air-launched cruise missiles (a conventional version of the nuclear-armed General Dynamics AGM-86B ALCM), the bomber crews fly to the Middle East and launch their missiles against high-priority targets in Iraq.

January 17, 1991. War begins in the Persian Gulf. Operation Desert Shield becomes Operation Desert Storm. More than 1,200 combat sorties are flown, and 106 cruise missiles are launched against targets in Iraq and Kuwait during the first 14 hours of the operation.

January 18, 1991. Eastern Air Lines, one of the oldest U.S. commercial carriers, goes out of business. The air line operated for 64 years.

January 25, 1991. In one of the fastest development and fielding of weapons in modern history, Air Force Systems Command's Armament Division asks the Army to machine 8-inch cannon barrels to the shape of a bomb. On February 24, the first of these Lockheed/Texas Instruments GBU-28/B bombs is tested at Tonopah Test Range, Nev., and penetrates so deeply the weapon is never found. Within five hours of delivery to Saudi Arabia, two of the 4,700-pound weapons are dropped from an F-111 on February 27.

February 6, 1991. Capt. Robert Swain of the 706th Tactical Fighter Group (AFRES), NAS New Orleans, La., shoots down an Iraqi helicopter in the first air-to-air victory for the Fairchild A-10 Thunderbolt II attack aircraft. He uses the plane's 30mm cannon for the kill.

February 15, 1991. In one of the most unusual air-to-air victories ever, Capt. Tim Bennett and Capt. Dan Bakke of the 4th Tactical Fighter Wing at Seymour Johnson AFB, N.C., shoot down an Iraqi helicopter (probably an Mi-24 Hind) with a GBU-10 2,000-pound laser-guided bomb dropped from their F-15E.

February 22, 1991. Soviet cosmonaut Musa Manarov sets a record for accumulated time in space, amassing his 447th day in orbit. Cosmonaut Manarov is on the 83rd day of his Soyuz TM-11 mission, working aboard the space station Mir, when he breaks the record.

February 28, 1991. Iraq surrenders to the U.S.-led coalition. In the 43-day, round-the-clock war, the Air Force flew 59 percent of all sorties with less than 50 percent of the assets, flew more than 50,000 combat sorties, offloaded more than 800 million pounds of fuel, and transported 96,465 passengers and 333 million pounds of cargo.

March 8, 1991. The first Martin Marietta Titan IV heavy-lift space booster to be launched from Vandenberg AFB lifts off. The booster carries a classified payload.

April 6, 1991. Operation Provide Comfort begins, humanitarian air operations to protect and supply Kurds in northern Iraq threatened by Saddam Hussein after the Gulf War, begins.

April 18, 1991. The Air Force carries out the first successful flight test of the Martin Marietta/Boeing MGM-134A small ICBM. The missile flies 4,000 miles from Vandenberg AFB to its assigned target area in the Army's Kwajalein Missile Range in the Pacific Ocean.

April 23, 1991. Air Force Secretary Donald B. Rice announces that the Lockheed/Boeing/General Dynamics F-22 and the Pratt & Whitney F119 engine are the winners in the Advanced Technical Fighter competition.

May 6, 1991. The U.S. destroys the last of 846 MGM-31 Pershing II missiles prohibited by the INF Treaty. On May 12, the Soviet Union destroys the last of 1,846 SS-20 missiles.

June 6, 1991. The Air Force reveals the existence of the Northrop AGM-137A Triservice Standoff Attack Missile (TSSAM), a stealthy ground attack weapon with a range of less than 600 kilometers.

July 1, 1991. The Warsaw Pact, the military coalition of Soviet Bloc countries, formally disbands.

July 2, 1991. McDonnell Douglas Helicopter Co. announces the first flight of the first production helicopter built without a tail rotor. The MD520N uses a blown-air system for antitorque and directional control.

September 15, 1991. The McDonnell Douglas C-17A transport makes its first flight. The crew of four takes off from the company's plant in Long Beach, Calif., and lands 2 hours, 23 minutes later at the Air Force Flight Test Center at Edwards AFB.

September 27, 1991. Strategic bomber crews stand down from their decades-long, round-the-clock readiness for nuclear war.

November 26, 1991. Clark AB, the Philippines, is officially turned over to the Philippine government, ending nearly 90 years of US occupancy. It was the largest overseas USAF base.

December 17–19, 1991. Four naval aviators set 16 recognized class records for altitude, speed, and time-to-climb with and without payload (turboprop aircraft) in a Grumman E-2C Hawkeye at NAS Patuxent River, Md. The records still stand.

December 19, 1991. Navy Lt. Cmdrs. Eric Hinger and Matt Klunder set a recognized class record for altitude with a 1,000-kg payload (turboprop aircraft) of 41,253.6 feet in a Grumman E-2C Hawkeye at NAS Patuxent River, Md. The records still stand.

December 21, 1991. The first Rockwell new-generation AC-130U gunship is flown for the first time.

December 25, 1991. The Soviet Union ceases to exist.

January 31, 1992. The Navy takes delivery of the last production Grumman A-6 Intruder attack aircraft, closing out 31 years of Intruder production.

February 10, 1992. Operation Provide Hope, the delivery of food and medical supplies to the former Soviet Union, begins.

February 28–29, 1992. Four Air Force crews set recognized class time-to-climb records (jet aircraft) in a Rockwell B-1B at Grand Forks AFB, N.D. Twelve of the records still stand.

April 9, 1992. The Air Force's new variable-stability in-flight simulator test aircraft (VISTA), a modified General Dynamics F-16, designated NF-16, that will replace the 40-plus-year-old NT-33, is flown for the first time at the General Dynamics facility in Fort Worth, Tex. The flight lasts 53 minutes.

May 12, 1992. Lockheed Aeronautical Systems Co. Delivers the 2,000th C-130 Hercules transport in ceremonies at Marietta, Ga. The milestone aircraft, a C-130H, is later delivered to the Air National Guard's 123rd Airlift Wing at Standiford Field, Ky.

June 1, 1992. SAC, TAC, and MAC are deactivated. Bomber, fighter, attack, reconnaissance, and electronic combat/electronic warfare aircraft and all ICBMs regroup under Air Combat Command (ACC). Lifter and tanker aircraft regroup under Air Mobility Command (AMC).

June 1, 1992. U.S. Strategic command is established, with responsibility for planning, targeting, and command of U.S. strategic forces.

July 1, 1992. Air Force Systems Command and Air Force Logistics Command are merged to create Air Force Materiel Command, which is to provide "cradle-to-grave" management of weapon systems.

July 3, 1992. Air Force begins Operation Provide Promise, flying humanitarian relief missions into Croatia and Bosnia-Herzegovina. It is the longest-running air supply effort in history, officially ending January 4, 1996.

August 26, 1992. Air Force begins Operation Southern Watch to enforce a ban on Iraqi aircraft operations south of the 32d parallel.

December 9, 1992. Operation Restore Hope, an international humanitarian operation in Somalia, begins. More than 28,000 troops are sent to safeguard food, supplies, and aid workers, from armed factions trying to seize power. Thirty-three Air Force active-duty and Reserve units take part in the initial deployment.

December 16/18, 1992. Capts. Pamela A. Melroy and John B. Norton along with company pilots William R. Casey and Charles N. Walls set a number of recognized altitude records with payload (for two different subclasses of jet aircraft) in a McDonnel Douglas C-17A Globemaster III at Edwards AFB, Calif. A class record for greatest load carried to 2,000 meters of 133,422 pounds is also set on the flight. Thirteen of the records still stand.

December 19, 1992. Capt. Jeff Kennedy and crew set a recognized class record for great circle distance without landing (jet aircraft) of 10,083.11 miles in a Boeing KC-135R. The record still stands.

December 19, 1992. An AMC KC-135R crew from the 97th Air Mobility Wing, Altus AFB, Okla., flies more than 8,700 miles from Kadena AB, Japan, to McGuire AFB, N.J., to set an aircraft class record for nonstop unrefueled flight.

December 27, 1992. While flying combat air patrol in Operation Southern Watch, two F-16 pilots from the 363rd Fighter Wing, Shaw AFB, S.C., intercept a pair of Iraqi MiG-25s flying in the United Nations-imposed no-fly zone over southern Iraq. One of the pilots, flying an F-16D, fires an AIM-120A AMRAAM and downs one of the MiGs, marking the first use of the Aim-120A in combat and the first USAF F-16 air-to-air victory.

January 13, 1993. USAF Maj. Susan Helms, flying aboard *Endeavour*, becomes the first U.S. military woman in space.

March 1, 1993. Lockheed Corp. completes acquisition of General Dynamics' Fort Worth Division. The $1.5 billion purchase gives Lockheed control of the F-16 fighter line and increases the corporation's share of the F-22 program to 67.5 percent.

March 9, 1993. A Lockheed SR-71A Blackbird reconnaissance aircraft comes out of retirement to fly its first scientific flight for NASA at the Dryden Flight Research Center at Edwards AFB. The aircraft, fitted with an ultraviolet video camera in the nose bay, is flown to an altitude of approximately 83,000 feet and collects more than 140,000 images of stars and comets.

April 12, 1993. NATO Operation Deny Fight begins, enforcing a ban ordered by the UN Security Council on aircraft operations in the no-fly zone of Bosnia-Herzegovina. The operation ends December 20, 1995.

April 28, 1993. Secretary of Defense Les Aspin lifts the long-standing ban on female pilots flying U.S. combat aircraft, including Army and Marine Corps attack helicopters.

April 29, 1993. German test pilot Karl Lang makes the first demonstration of a high-angle-of-attack, post-stall, 180° turn known as a Herbst maneuver while flying the Rockwell/MBB X-31A EFM demonstrator. The turn is completed in a 475-foot radius.

May 22, 1993. Lt. Cmdr. Kathryn P. Hire, the first woman in the Navy to be assigned to a combat unit, flies her first mission as a tactical crew member on a Lockheed P-3C Update III maritime patrol aircraft during a bombing exercise. Commander Hire flies with VP-62, a Reserve unit based at NAS Jacksonville, Fla. The first Air Force female combat pilot will be 1st Lt. Jeannie Flynn, who will take her place in an F-15E cockpit later in 1993.

May 25-August 3, 1993. The first successful demonstration of aerobraking (using atmospheric drag to slow a spacecraft) puts the Magellan Venus probe in a lower orbit. The probe suffers no ill effects.

June 14, 1993. The first operational McDonnell Douglas C-17A Globemaster III transport is delivered to the 437th Airlift Wing at Charleston AFB, S.C.

June 17, 1993. Lt. Col. Patricia Fornes becomes the first woman to lead an Air Force ICBM unit. She assumes command of the 740th Missile Squadron at Minot AFB, S.D., a squadron once commanded by her father.

June 29, 1993. The Air Force rolls out the first Boeing OC-135B Open Skies Treaty observation aircraft at Wright-Patterson AFB, Ohio. It is the first of three that will be used by the U.S. to verify foreign compliance with arms treaties.

July 1, 1993. Air Education and Training Command established.

July 1, 1993. Day-to-day control of ICBMs passes to Air Force Space Command.

July 8, 1993. Slingsby Aviation Ltd. rolls out the first T-3A Enhanced Flight Screener for the Air Force at its plant in York, England.

July 30, 1993. The multiaxis thrust-vectoring system installed on the VISTA NF-16 is employed for the first time in a test at the Air Force Flight

Test Center. By September 1993, the aircraft will achieve a transient angle of attack of 110° and a sustained AOA of 80°.

August 5, 1993. The AFTI F-16 completes its 600th mission at the Air Force Flight Test Center. The flight collects data for the AFTI/F-16 Ground Collision Avoidance System test effort.

August 6, 1993. Sheila E. Widnall, associate provost and professor of aeronautics and astronautics at the Massachusetts Institute of Technology, to be Secretary of the Air Force. Dr. Widnall becomes the first female Secretary for any of the armed services. After Senate confirmation, she is sworn in on August 6.

August 11–14, 1993. Global Enterprise, an ACC exercise to train aircrews for long-distance power-projection missions, is carried out from Ellsworth AFB, S.D. Two Rockwell B-1B Lancers are flown to Europe, across the Mediterranean and Red Seas around the Arabian Peninsula, and land at a staging base in southwest Asia. After exchanging crews, the B-1s are flown from southwest Asia, via Japan, over the Aleutians, and then back to South Dakota. Total flight time is 37.3 hours, and the 24-hour first leg is the longest flight ever made by a B-1B crew.

August 17, 1993. The first of 350 early-model Boeing B-52 bombers is cut into five pieces with a 13,000-pound steel guillotine at Davis-Monthan AFB, Ariz. The bombers were destroyed under the terms of the Strategic Arms Reduction Talks II Treaty.

August 18, 1993. McDonnell Douglas's Delta Clipper Experimental (DC-X) subscale single-stage-to-orbit prototype makes a 60-second first flight at the White Sands Missile Range, N.M. The 42-foot-tall vehicle takes off vertically, hovers at about 150 feet, moves laterally approximately 350 feet, and lands tail-down.

August 24/26, 1993. Two mixed Air Force and contractor crews set recognized class time-to-climb and altitude records (jet aircraft) in a McDonnell Douglas C-17A Globemaster III at Long Beach, Calif. All four of the records still stand.

September 10, 1993. Boeing rolls out the 1,000th 747 commercial jetliner in ceremonies at its Seattle, Wash., plant. The milestone aircraft, a 747-400, will be delivered to Singapore Airlines. The first jumbo jet was rolled out in September 1968.

September 15, 1993. Boeing announces that work on the first B-52H bomber to be adapted for conventional warfare missions has been completed at its facility in Wichita, Kan.

October 8, 1993. Capt. Pamela A. Melroy and company pilot Richard M. Cooper set two recognized jet aircraft class records for altitude with a 70,000-kg payload (32,169 feet) and greatest mass carried to a height of 2,000 meters (161,023 pounds) in a McDonnell Douglas C-17A Globemaster III at Edwards AFB, Calif. The records still stand.

December 17, 1993. On the 90th anniversary of the Wright brothers' first sustained flight, the first operational Northrop B-2 stealth bomber, *The Spirit of Missouri*, is delivered to the 509th Bomb Wing at Whiteman AFB, Mo.

February 10, 1994. Lt. Jeannie Flynn, the first female selected for USAF combat pilot training, completes her F-15E training.

February 28, 1994. Air Force F-16s, operating under NATO command, shoot down four Bosnian Serb Super Galeb attack aircraft after twice warning the Serb jets to leave Bosnian airspace. It is NATO's first combat in its 45-year history.

April 7, 1994. Capt. Michael S. Menser (and crew) sets a recognized class record for 10,000-km speed without payload (jet aircraft) of 599.59 mph flying from Grand Forks AFB, N.D., to Monroeville, Ala., to Mullan, Idaho, in a Rockwell B-1B Lancer. At the same time, Capt. R. F. Lewandowski (and crew) sets the recognized record for a different class for 10,000-km speed without payload (jet aircraft) of 594.61 mph over the same course, also in a B-1. Both records still stand.

April 10, 1994. In NATO's first air attacks on ground positions since the Alliance was founded 45 years previously, two Air Force F-16C fighters destroy a Bosnian Serb Army command post with Mk. 82 500-pound bombs.

May 3, 1994. Col. Silas Johnson, Jr., 93rd Wing Commander, flies the last B-52G to the "Boneyard" at Davis Monthan AFB, Ariz., thus removing this series from the active inventory.

June 3, 1994. Maj. Andre A. Gerner and company pilot John D. Burns set a recognized record for STOL aircraft for greatest mass carried to a height of 2,000 meters (44,088 pounds) in a McDonnell Douglas C-17A Globemaster III at Edwards AFB, Calif. The record still stands.

June 24, 1994. The F-117 stealth aircraft is officially named Nighthawk.

June 29, 1994. First visit of a U.S. space shuttle to a space station, the Russian *Mir*.

July 1994. The 184th Bomb Group, Kansas Air National Guard, becomes the first Guard unit to be equipped with the B-1B.

August 2, 1994. During a Global Power mission to Kuwait, two B-52s from the 2nd Bomb Wing, Barksdale AFB, La., sets a world record while circumnavigating Earth. Flying 47.2 hours, the bombers set a world record not only for the longest B-52 flight but also for the longest jet aircraft flight in history. Dropping 54 bombs over a range located 25 miles from the Iraqi border, the aircraft demonstrate their global reach and power on the fourth anniversary of the Iraqi invasion of Kuwait.

August 4, 1994. Two B-1Bs (one from the 384th Bomb Group and one from the 184th Bomb Group of the Kansas Air National Guard) complete a 19-hour nonstop Global Power mission to Hawaii. This is the first time the 184th, the first ANG unit to receive the B-1B, flies a Global Power mission.

October 10, 1994. USAF responds to hostile movements in the Persian Gulf area by Iraq's Saddam Hussein by deploying 122 combat aircraft to augment the 67 already in place. Four bombers fly nonstop from bases in the United States to deliver 55,000 pounds of bombs on target, on time, within audible range of Saddam's forces. The Iraqis withdraw northward. Secretary of Defense William Perry later says, "The Air Force really has deterred a war."

October 14, 1994. The first-ever operational C-17 mission lands in the Persian Gulf area, delivering a 5-ton "rolling command post," five vehicles, and assorted supplies for the Army. The 17.2-hour flight was the longest mission to date for a C-17.

October 26, 1994. Gen. Ronald R. Fogleman becomes Air Force Chief of Staff.

October 26, 1994. CMSgt. David J. Campanale becomes Chief Master Sergeant of the Air Force.

November 21–23, 1994. In Project Sapphire, Air Mobility Command C-5s transport more than 1,300 pounds of highly enriched uranium from the former Soviet republic of Kazakhstan to Dover AFB, Del., to protect this large supply of nuclear materials from terrorists, smugglers, and unfriendly governments. From Dover, the uranium is taken to Oak Ridge, Tenn., to await conversion to commercial nuclear fuel.

February 7, 1995. A crew from Whiteman AFB, Mo., makes the first drop of live bombs from the Northrop Grumman B-2A Spirit stealth bomber. The two Mk. 84 bombs were dropped as part of the B-2's first Red Flag exercise at Nellis AFB, Nev.

March 15, 1995. Lockheed Corporation and Martin Marietta complete their merger that was announced the previous August 29. The newly created Lockheed Martin Corporation, with $23 billion in annual sales, becomes the world's largest aerospace and defense contractor.

April 7, 1995. 2nd Lt. Kelly Flinn, the first woman to join a bomber crew, begins student pilot training with the 11th Bomb Squadron, 2nd Bomb Wing, Barksdale AFB, Louisiana.

April 27, 1995. The Global Positioning System (GPS) satellite constellation is declared to have achieved full operational capability (FOC) by Air Force Space Command.

June 1, 1995. Lockheed Martin and Boeing roll out the stealthy DarkStar

Tier III Minus high-altitude unmanned aerial vehicle in ceremonies at Palmdale, Calif.

June 2, 1995. Air Force F-16 pilot Capt. Scott F. O'Grady is shot down over northwest Bosnia on an Operation Deny Flight mission. Rescued after an eight-day ordeal during which he subsisted on bugs and rainwater, Captain O'Grady returns home to a hero's welcome.

June 3, 1995. Two 7th Wing (Dyess AFB, Tex.) B-1Bs land after completing a historic 36-hour, 13-minute, 20,100-mile, nonstop around-the-world flight. This Global Power mission, called Coronet Bat, requires six air refuelings using assets from ACC, AMC, USAFE, PACAF, U.S. Central Command, ANG, and AFRES. To mirror a realistic training scenario for wartime taskings, Coronet Bat incorporates bombing runs over the Pachino Range, Italy; the Torishima Range, near Kadena AB, Japan; and the Utah Test and Training Range.

June 6, 1995. Astronaut Norman Thagard, flying on the Russian Mir space station, set the U.S. record for spaceflight endurance, passing 84 days, 1 hour, and 17 minutes in space. The previous U.S. record-holders were the three astronauts on the third Skylab mission in 1974.

June 22, 1995. Secretary of the Air Force Sheila Widnall announces that Beech Aircraft has been selected to develop and deliver the Joint Primary Aircraft Training System (JPATS) for the Air Force and Navy. The new trainer, a modified version of the Swiss Pilatus PC-9 turboprop trainer, will replace the USAF's Cessna T-37Bs and the Navy's Beech T-34Cs.

June 28, 1995. The National Air and Space Museum of the Smithsonian

Institution finally puts the *Enola Gay*, the B-29 that dropped the first atomic bomb on Japan, on display. The exhibition program is straightforward and factual. Earlier, amid major controversy, the museum canceled plans to show the *Enola Gay* as a prop in a politicized horror show after Congress and the public—alerted by reports from the Air Force Association—took strong objection.

July 7–August 5, 1995. The C-17 airlifter, earlier beset by troubles so severe that program cancellation was a possibility, produces outstanding results in a month-long wartime surge test. In November, the Department of Defense says the C-17 has bounced back from its problems and authorizes the Air Force to buy up to 120 of these aircraft.

July 8, 1995. The Minuteman III ICBM achieves 100 million hours of operational duty.

July 29, 1995. Air Combat Command activates the 11th Reconnaissance Squadron, an unmanned aerial vehicle (UAV) unit, and assigns it to the 57th Operations Group at Nellis AFB, Nev. Equipped with the Tier II Predator—and later Tier II+ and Tier III—types, the 11th RS is tasked to explore the use of remotely piloted aircraft.

August 25, 1995. A 2nd Bomb Wing B-52H and its five-member crew set an aviation world record from Edwards AFB, Calif.—flying 5,400 nautical miles, unrefueled, with a payload of 11,000 pounds in 11 hours, 23 minutes with an average speed of 556 mph.

August 30, 1995. U.S. Air Force, Navy, and Marine aircraft lead Operation Deliberate Force, a NATO bombing campaign responding to Bosnian Serb mortar attacks that killed 38 civilians at an outdoor market in Sarajevo.

September 1, 1995. Officially emerging from mothballed status, the SR-71 is declared operationally capable by Air Combat Command.

September 10, 1995. The *First Lady*, the first production Lockheed C-130 Hercules, is retired in ceremonies at Duke Field, Fla. This aircraft, which was first flown on April 7, 1955, had a distinguished career, including more than 4,500 combat hours in Southeast Asia after it was converted into a gunship. The *First Lady* was later retired to the USAF Armament Museum at Eglin AFB, Fla.

October 1, 1995. Air Combat Command activates the 609th Information Warfare Squadron at Shaw AFB, S. C.

November 2, 1995. Lt. Col. Greg Feest becomes the first pilot to log 1,000 hours of flight time in the Lockheed F-117A Nighthawk stealth fighter.

December 6, 1995. A crew from the 37th Airlift Squadron at Ramstein AB, Germany, marks the beginning of Operation Joint Endeavor by flying their Lockheed C-130E into Tuzla, Bosnia. Operation Joint Endeavor is the ongoing NATO effort to enforce the Bosnian peace treaty signed at Dayton, Ohio.

December 7, 1995. Literally going out in a blaze of glory, the Galileo spacecraft's atmospheric probe separates from the orbiter and plunges into Jupiter's atmosphere. The probe sends a stream of data back to the orbiter and manages to survive 58.5 minutes before it is crushed by the intense pressure. Galileo was launched from the space shuttle in 1989.

December 19, 1995. A federal judge rules in favor of General Dynamics and McDonnell Douglas, establishing that the Navy's A-12 Avenger stealth attack aircraft was canceled for the convenience of the government, rather than default. The A-12 was canceled in 1991.

December 20, 1995. NATO air operation Decisive Endeavor begins to monitor and enforce peace implementation in Bosnia.

January 4, 1996. Sikorsky test pilot Rus Stiles and Boeing test pilot Bob Gradle make the first flight of the Army's YRAH-66 Comanche helicopter prototype at West Palm Beach, Fla. The RAH-66, designed for armed reconnaissance/light attack missions, is the first helicopter to employ stealth technologies.

February 14, 1996. A crew flying the Northrop Grumman E-8A Joint STARS surveillance platform over Bosnia makes the type's 50th mission in support of Operation Joint Endeavor, breaking a sortie record set in Operation Desert Storm in 1991. Despite the E-8's successes in these two operations, Joint STARS is still officially in development.

March 22, 1996. STS-76 *Atlantis* lifts off from Cape Canaveral with Dr. Shannon Lucid, enroute to a record six-month stay aboard Russian MIR spacecraft.

March 30, 1996. Air Force Special Operations Command receives Cheney

Award for military flight safety achievements.

April 28, 1996. Boeing T-43 crash in Croatia kills Commerce Secretary Ron Brown.

June 1, 1996. USAF entertains unsolicited proposal to reengine B-52 fleet of 94 aircraft.

July 1, 1996. Boeing delivers first AGM-86 Block 1 Conventional Air Launched Cruise Missile (CALCM).

September 3, 1996. Operation Desert Strike begins against Iraqi forces that are threatening Kurdish population.

October, 1996. 93rd Air Control Wing takes two E-8C aircraft to Europe to support peacekeeping forces in Bosnia-Herzegovina.

October 1, 1996. General Joseph W. Ashy, Comander in Chief, United States Space Command, retires.

November 12, 1996. Boeing, Northrop Grumman, and Lockheed Martin are awarded a $1.1 billion contract for a prototype Airborne Laser weapon system.

November 17, 1996. USAF completes flight testing of conformal tanks on the Lockheed Martin F-16.

November 19, 1996. First flight of Spacehab Double Module on STS 80, *Columbia*

1997–2005
January 1, 1997. Operation Northern Watch replaces Operation Provide Comfort, enforcing a no-fly zone in northern Iraq above the 36th parallel.

January 16, 1997. An AFRES C-141 aircrew from the 446th Airlift Wing at McChord AFB, Wash., leaves Beijing with the remains of five Americans,

who died on August 13, 1944, in a B-24J Liberator crash after bombing Japanese ships near Taiwan. Villagers searching the mountainous area for herbs discovered the crash site.

January 19, 1997. A 349th Air Mobility Wing (AFRES) C-141 aircrew from Travis AFB, Calif., airlifts 40,000 pounds of winter clothing to Ellsworth

AFB, S.D., where the cargo was trucked to the Cheyenne River Sioux Reservation at Eagle Butte, S.D.

January 22, 1997. An improved F-117A Nighthawk stealth fighter touches down at Holloman AFB, N.M., as Lockheed Martin delivers the first "RNIP-Plus" aircraft to the 49th Fighter Wing. The F-117's new Ring Laser Gyro/Global Positioning System Navigation Improvement Program reduces navigational drift during flight. Plans called for the entire F-117 fleet to be modified by October 1999.

January 31, 1997. This date marks the activation of the 31st Air Expeditionary Wing, the first expeditionary wing in the Air Force. Such wings would be available for rotational deployments worldwide.

February 1, 1997. Two 509th Bomb Wing B-2 pilots emerge from perhaps the longest simulator flight in Air Force history at Whiteman AFB, Mo. The 44.4-hour simulator flight, called Vigilant Spirit II, end on 1 February for Maj. Steve Moulton and Capt. Jeff Long. The test tried to determine how fatigue impacted pilot abilities to perform a long mission safely. It used sleep, dietary strategies, and seat exercises to reduce fatigue.

February 17, 1997. The Air Force Reserve becomes a major command— the Air Force Reserve Command— reflecting the growing integration of Reserve and regular forces and the increasing role of the Reserve in regular USAF operations.

February 18–March 3, 1997. In Operation Assured Lift, the Air Force airlifts African peacekeeping forces to Liberia to restore order after a civil war there. Five C-130s of the 37th Airlift Squadron, 3rd Air Expeditionary Group, airlifted 1,160 peacekeepers

and 452 tons of their cargo from various African countries to Liberia.

March 2, 1997. Operation Deep Freeze. The 60th Air Mobility Wing's last scheduled C-141 mission in support of leaves Travis AFB, Calif. For more than 40 years, Travis aircrews flew Deep Freeze missions from Christchurch, New Zealand, to McMurdo Station to resupply scientists near Antarctica's southern polar ice cap. The 62nd Airlift Wing at McChord AFB, Wash., took over the mission.

March 17, 1997. Operation Guardian Retrieval evacuates U.S. citizens and other foreigners from Zaire, where civil unrest threatens them. Special Operations Command provides personnel and aircraft for the joint task force, including MC-130s and MH-53s. Air Mobility Command C-5s, C-17s, C-141s, and KC-135s also take part. By the end of March, the operation has flown 57 missions and transported 532 passengers.

March 19, 1997. Exercise Coronet 97-2. Through 20 March, the 509th Bomb Wing conducts the longest B-2 Spirit Global Power mission to date when *The Spirit of Florida* flies from Whiteman AFB, Mo., to Puerto Rico. Maj. Robert O'Neal and Capt. Scott Hughes of the 393rd Bomb Squadron fly the nearly 30-hour round-trip flight with its four aerial refuelings and a live conventional weapon release at the Vieques Range, Puerto Rico.

March 21, 1997. At Pensacola, Fla., Lt. Col. Marcelyn A. Atwood becomes the first woman to command a training squadron and the first USAF officer to command a Navy squadron. The squadron trains both Air Force and Navy pilots.

March 27, 1997. An Air Mobility Command C-141 crew from Scott

AFB, Ill., flies the first U.S. aircraft through the new Reduced Vertical Separation Minimum (RVSM) airspace over the North Atlantic from Gander, Newfoundland, to Stuttgart, Germany. The RVSM cuts vertical separations between aircraft from 2,000 to 1,000 feet at altitudes between 33,000 and 37,000 feet in an effort to deal with the tremendous growth in air traffic.

April 1, 1997. At Langley AFB, Gen. Richard E. Hawley, the Air Combat Command Commander, announces the initial operational capability of the 509th Bomb Wing's B-2A Spirit at Whiteman AFB, Mo., after earlier, successful tests of the Joint Direct Attack Munition (JDAM) and the GPS-Aided Munition (GAM).

April 1, 1997. The USAF transfers all operational C-130s from Air Combat Command to Air Mobility Command, except for the 314th Airlift Wing and its C-130 schoolhouse. Those units go to Air Education and Training Command (AETC).

April 1, 1997. The 375th Airlift Wing at Scott AFB, Ill., regains control of C-21 flying operations in the U.S., when the USAF consolidates eight C-21 units at various bases into two airlift squadrons and one wing. The reorganization standardizes C-21 operations, with the 458th Airlift Squadron operating from Scott and the 457th Airlift Squadron from Andrews AFB. Each squadron has three airlift flights based at other USAF installations in the U.S.

April 2, 1997. A C-141 assigned to the 459th Airlift Wing (Air Force Reserve Command) at Andrews AFB, Md., leaves Scott AFB, Ill., for Sucre, Bolivia, with enough donated medical equipment to set up two surgical rooms. The Diocese of Joliet Peace and Justice at Romeoville, Ill., donated the cargo.

April 18, 1997. Through August 8, Grand Forks AFB, N.D., opens its doors to 3,500 homeless flood victims after a heavy winter snow melt breaks through dikes in the Red River Valley. By the end of the emergency in early August, the Air Mobility Command has flown 13 missions to Grand Forks to airlift 146 tons of cargo and 143 passengers to support the flood relief operation.

May 23, 1997. Through May 25, the 509th Bomb Wing completes the longest B-2 global power mission to date. The 29.9-hour, record-setting flight starts from and returns to Whiteman AFB, Mo. It featured a stop at RAF Mildenhall, United Kingdom, and simulated bomb runs over mid-America.

May 29, 1997. A B-2 from the 509th Bomb Wing at Whiteman AFB, Mo., drops a GAM-113 bomb near Edwards AFB, Calif. This drop marks the first time the 4,700-pound conventional penetrating weapon is paired with the B-2.

June 4, 1997. The Air Force reach a milestone in advanced missile propulsion with the successful ground testing of the Variable Flow Ducted Rocket (VFDR) at Atlantic Research Corp., Gainesville, Va. A solid-fuel, ducted rocket engine, the VFDR increases the capabilities of conventional rockets by burning fuel-rich rocket-exhaust gases with air, thereby doubling or tripling the total energy produced by the engine.

June 9, 1997. During the June 9–20 Central Enterprise exercise, for the first time in the 10-year history of the B-1B bomber, two units become one force at a forward operating location at

RAF Fairford, United Kingdom. The 7th Bomb Wing from Dyess AFB, Tex., and the 28th Bomb Wing from Ellsworth AFB, S.D., along with 350 people, fuse together to form the 7th Expeditionary Bomb Group.

June 10, 1997. An MC-130 crew from the 352nd Special Operations Group evacuates 56 people, including 30 U.S. citizens, from Brazzaville, Congo, to save them from civil strife. Under the command of Lt. Col. Frank J. Kisner, USAF, the crew earn the Mackay Trophy for the outstanding flight of the year.

June 17, 1997. A C-5A Galaxy from the 436th Airlift Wing at Dover AFB, Del., flies from Andrews AFB, Md., to Tashkent, Uzbekistan, on the 500th humanitarian airlift mission supporting Operation Provide Hope.

June 24, 1997. EF-111A Ravens from the 429th Electronic Combat Squadron (ECS) at Cannon AFB, N.M., leave Turkey for home after nearly 2,100 days of deployment in support of Operations Provide Comfort and Northern Watch.

July 30, 1997. Capt. Dewey Gay flies a F-16C Fighting Falcon (No. 83-1164) from the 62nd Fighter Squadron at Luke AFB, Ariz., into history when he touches down after a 1.3-hour sortie that pushed the F-16 over 4,000 hours. It was the first C-model to reach the 4,000-hour mark. The feat took 14 years.

August 1, 1997. Test pilot Paul Metz flies the F-22 Raptor on its first flight at Dobbins Air Reserve Base, Ga.

August 5, 1997. After a 747 Korean Airlines jetliner crashes on Guam, a C-141 from the 305th Air Mobility Wing at McGuire AFB, N.J., flies a 31-member team of the National Trans-portation Safety Board from Andrews AFB, Md., to Fairchild AFB, Wash., where they board a 92nd Air Refueling Wing KC-135 for the flight to Andersen AFB, Guam. A second C-141 from the 305th Air Mobility Wing takes medical equipment and seven physicians from Marine Corps Air Station Kaneohe, Hawaii, to Guam to help treat the crash survivors, while a C-141 from the 62nd Airlift Wing at McChord AFB, Wash., airlifts Red Cross, Federal Bureau of Investigation, and FAA representatives from Hawaii to Guam.

August 9, 1997. C-141 aircrews from the 446th Airlift Wing (AFRC) at McChord AFB, Wash., arrive at Kelly AFB, Tex., with four victims from the August 5 Korean Airlines 747 crash in Guam. Flight nurses and medical technicians from the 349th Aeromedical Evacuation Squadron, 349th Air Mobility Wing (AFRC) at Travis AFB, Calif., support the flight.

August 22, 1997. A C-5 aircrew from Dover AFB, Del., delivers a NASA satellite from Andrews AFB, Md., to Kagoshima, Japan, after a 16-hour flight. The $250 million satellite for the Tropical Rainfall Measuring Mission is the first joint space project between the U.S. and Japan.

September 6, 1997. From Charleston AFB, S.C., a 315th Aeromedical Evacuation Squadron aircrew flies the first C-17 aeromedical evacuation training flight aboard the *The Spirit of Charleston* on a round trip to Bermuda.

September 7, 1997. At Dobbins ARB, Ga., test pilot Paul Metz pilots the extremely maneuverable F-22 Raptor in its first flight.

September 14, 1997. In Exercise Centrazbat 97, eight C-17 Globemaster IIIs airdrop 500 members of the 82nd Airborne Division at Fort Bragg, N.C., into Kazakhstan. This operation covers 7,897 miles, the longest distance ever in an airdrop, to support the two-phase U.S. Atlantic Command exercise for U.S. and Central Asian troops in the former Soviet Union. The C-17s are airborne for over 19 hours as they cross the Atlantic and southern Europe to the drop zone. KC-10 Extenders from the 60th Air Mobility Wing at Travis AFB, Calif., and the 305th Air Mobility Wing at McGuire AFB, N.J., along with KC-135 Stratotankers from the 22nd Air Refueling Wing at McConnell AFB, Kans., and the 6th Air Refueling Wing at MacDill AFB, Fla., provide refueling support.

October 3, 1997. A C-141 and aircrew from the 305th Air Mobility Wing at McGuire AFB, N.J., airlift 35,000 pounds of excess DOD medical supplies to Sofia, Bulgaria.

October 12, 1997. The Air Force begins a relief airlift to Indonesia to help fight forest fires there. Three Wyoming Air National Guard C-130s from the 153d Airlift Wing, equipped with the Modular Airborne Fire Fighting System, drop 685,000 gallons of water and fire retardant on fires in Java and Sumatra, each airplane dropping up to 3,000 gallons at a time. The operation concludes in early December after 194 sorties.

October 14, 1997. Retired Brig. Gen. Charles "Chuck" Yeager breaks the sound barrier again in front of a crowd of more than 5,000 people at Edwards AFB, Calif., to commemorate the 50th anniversary of his historic supersonic flight. Yeager flies an F-15 Eagle while his first supersonic flight chase pilot and longtime friend Bob Hoover and Air Force Flight Test Center Vice Commander James Doolittle III flies an F-16 Fighting Falcon for chase.

November 19, 1997. In Operation Phoenix Scorpion I, to augment Operation Southern Watch, the U.S. deploys additional forces to Southwest Asia and the Indian Ocean to answer Iraq's continued intransigence over U.N. inspections of weapons of mass destruction sites. Over the Pacific, the Air Mobility Command establishes an air bridge and airlift support to deploy B-52s and KC-10s to Diego Garcia. Over the Atlantic, Air Mobility Command provides airlift and refueling support to deliver an Air Expeditionary Force with F-15s, F-16s, B-1s, and KC-135s to Shaikh Isa, Bahrain, a partial Patriot battery to Shaikh Isa, and F-117s to Al Jabar, Kuwait. Iraq's apparent willingness to resume the inspection program ends the deployment on November 24 and within days most forces return home. To close the deployment, the Air Mobility Command flies 51 airlift and 208 tanker missions, delivering over 3,000 short tons of cargo and 7.4 million pounds of fuel.

December 18, 1997. Through January 14, 1998, C-5s, C-141s, KC-10s, and KC-135s flown by active duty, guard and reserve air crews deliver relief supplies to Guam after Super Typhoon Paka with winds up to 236 mph hits the island on December 15. Despite being restricted to daylight flying and limited navigational aids, C-5s from the 436th Airlift Wing at Dover AFB, Del., KC-10s and C-141s from the 60th Air Mobility Wing and the 349th Air Mobility Wing at Travis AFB, Calif., and C-141s from the 62nd Airlift Wing at McChord AFB, Wash., transport more than 200 emergency workers and critical provisions to Andersen AFB, Guam.

December 22, 1997. With the simultaneous turning of seven keys, a Cold War icon becomes obsolete as the last Minuteman II missile silo is destroyed. The implosion of the silo is in accordance with the July 31, 1991, Strategic Arms Reduction Treaty (START). The site, commonly known as Hotel-11, was located north of Dederick, Mo., more than 90 miles southwest of Whiteman AFB.

December 27, 1997. Crews from five Air National Guard C-130s airdrop 50,000 pounds of hay to cattle stranded by blizzards in southeastern New Mexico. The 137th Airlift Wing from Will Rogers World Airport, Oklahoma City, Okla., the 136th Airlift Wing from Naval Air Station Dallas, Tex., and the 153rd Airlift Wing from Cheyenne Municipal Airport/Air Guard Station, Wyo., provide the aircraft.

January 2, 1998. Five Air National Guard C-130 Hercules aircraft respond to a presidential call to support disaster response efforts in New Mexico. Aircrews provide more than 50,000 pounds of feed to cattle stranded by blizzards in southeast New Mexico. More than 25,000, cattle, or 1,000 daily, die in the storms because ranchers cannot reach them.

January 8, 1998. After originating Combat Echo in 1973, the Tactical Air Command combined its program in July 1984 with the Air Defense Command's Combat Pike to form the Combat Archer Weapon System Evaluation Program. This program reaches its 25th anniversary milestone in a continuing effort to develop and validate USAF weapons systems.

January 10, 1998. In Operation Recuperation, through 14 January, C-17 aircrews from the 437th Airlift Wing at Charleston AFB, S.C., fly four missions to deliver snow relief supplies to eastern Canada. The Globemaster IIIs transport emergency workers and 181 tons of cargo (vehicles, generators, water purification equipment, field kitchens, and other supplies) from Edmonton, Alberta, to Montreal, Quebec.

January 15, 1998. Through January 18, 5 C-17s and 13 C-5s deploy civilian utility workers, trucks, chain saws, and other gear to the northeastern U.S. after a major ice storm brings down power lines and trees and leaves some 500,000 homes without electricity. The aircraft pick up most of their passengers and cargo at Pope AFB, N.C., for distribution to Naval Air Station Brunswick and Bangor International Airport, Me., and Stewart International Airport, N.Y.

January 16, 1998. After a devastating earthquake hit China's Northern Hebei Province, a C-17 leaves Kadena AB, Japan, for Beijing with 40 tons of relief supplies, consisting of blankets, sleeping bags, medical supplies, rations, and cold-weather clothing.

January 28, 1998. Through January 20, Air Mobility Command takes part in Exercise Purple Dragon/Operation Big Drop. Joint Task Force exercise by supporting the Big Drop airborne and air assault operation at Fort Bragg, N.C. Some 60 C-17s, C-130s, and C-141s fly troops and heavy equipment to the Fort Bragg drop zones. More than 30,000 service members from the Air Force, Army, Navy, Marines, and Coast Guard participate.

February 8, 1998. In Operation Phoenix Scorpion II, through 3 March, the Air Mobility Command conducts a second deployment to augment forces sent to Southwest Asia in November 1997. The command supports the westward flow of B-52s and KC-10s from the U.S. to Diego Garcia, while airlifters and tankers move F-16s, F-117s, and a B-1B, along with rescue and special operations assets, eastward from the U.S. to the Persian Gulf area. In all, Air Mobility Command forces (Active, Reserve, and Guard) fly 296 airlift missions to carry 11,368 passengers and 11,321 short tons of cargo, while 191 air refueling missions offload 4.7 million pounds of fuel.

February 11, 1998. A B-1B bomber first drops a Joint Direct Attack Munition, a conventional bomb fitted with satellite-guidance equipment, over a test range at China Lake, Calif.

February 11, 1998. The first C-32A, a modified Boeing 757-200, flies its maiden flight from the Renton Municipal Airport in Renton, Wash. The Air Force purchased four C-32As to replace the 89th Airlift Wing's C-137 fleet, which had flown the Vice President, cabinet members, and congressional delegations around the world since the 1950s.

February 23, 1998. B-2s deploy overseas for the first time, flying from Whiteman AFB, Mo., to Andersen AFB, Guam.

February 28, 1998. The RQ-4 Global Hawk first flies. This new unmanned aerial vehicle, designed for high-altitude, long-range, long-endurance reconnaissance missions, takes off from Edwards AFB, Calif., on a 56-minute flight.

March 6, 1998. The Secretary of Defense approve the permanent retirement of the SR-71 Blackbird. With that decision the USAF will retire two SR-71A models (returned to active duty in 1995), a SR-71A and the SR-71B trainer model on loan to NASA, and two from returnable storage.

While in the Air Force inventory, the SR-71 set numerous world high-altitude and speed records.

March 9, 1998. Through April 9, Air Mobility Command performs 104 airlift and 110 air refueling missions for a six-nation tour of Africa. On 22 March, President Bill Clinton—with First Lady Hillary Rodham Clinton, Secretary of State Madeleine K. Albright, and a large official party on Air Force One—leave Andrews AFB, Md., on a 12-day journey through sub-Saharan Africa. It was the first time a sitting American president had visited Ghana, Uganda, Rwanda, South Africa, Botswana, and Senegal.

March 12, 1998. On the silver anniversary of Operation Homecoming a C-141 from the 445th Airlift Wing (AFRC) at Wright-Patterson AFB, Ohio, lands at Randolph AFB, Tex., with more than 50 former U.S. prisoners of war. On February 12, 1973, the same C-141 (no. 66-177) airlifted Americans from Gia Lam Airport, Hanoi, North Vietnam, on the first Operation Homecoming mission to repatriate American servicemen from Southeast Asia.

March 17, 1998. The USAF accepts the first of two C-38A Courier aircraft. Two Air National Guard pilots from the 201st Airlift Squadron fly the aircraft from St. Louis, Mo., to Andrews AFB, Md., to replace the older C-21.

March 17, 1998. The 20th Special Operations Squadron at Hurlburt Field, Fla., receives the Air Force's first production-modified MH-53J Pave Low III helicopter from Lockheed Martin.

June 2, 1998. The 22nd Air Refueling Wing at McConnell AFB, Kans., receives the first production-modified KC-135 Pacer CRAG (Compass,

Radar, and Global Positioning System) aircraft (no. 57-1435). The Pacer CRAG equipment serves as the centerpiece of several KC-135 modernization efforts. Using the CRAG system, a pilot is able to detect cloud formations, wind shear, and other weather hazards. A state-of-the-art flight management system, FMS 800, forms the heart of the CRAG system. It gives pilots the ability to plot courses in a training facility, download the information to a data card, and upload it into the aircraft. That new capability eliminates the need for a navigator.

June 10, 1998. Through July 8, in support of President William J. Clinton's nine-day trip to China, the Air Mobility Command flies 33 C-5, 2 C-141, and 7 KC-135 missions. In this first trip to China by a U.S. President since 1989, President Clinton meets with Chinese President Jiang Zemin in Beijing on June 27 before visiting Shanghai and Xian, China's ancient capital.

June 19, 1998. The first of four C-32 As leave Boeing's plant in Seattle, Wash., and fly to the 89th Airlift Wing at Andrews AFB, Md. The military version of Boeing's 757-200 replaces the VC-137 aircraft in the presidential airlift fleet.

June 24, 1998. Air Mobility Command announces a modernization program to reengine its entire C-5 fleet over 10- to 12-year period. Besides new engines, the modernization will include a comprehensive avionics upgrade with the new All-Weather Flight Control System and Global Air Traffic Management equipment.

July 2, 1998. In Operation Phoenix Flame, through 7 July, the Air Mobility Command flies 10 C-5 and 2 C-141 missions to assist Federal Emergency Management Agency (FEMA) firefighting efforts in Florida. The airlifters deliver more than 300 people and 740 tons of equipment to NAS Jacksonville and Patrick AFB, Fla. Active duty, Reserve, and Guard units support efforts to fight wildfires in some 200,000 acres of Florida.

July 8, 1998. An upgraded T-38C Talon advanced trainer aircraft flies for the first time over Mesa, Ariz. The T-38 received improved avionics under the Avionics Upgrade Program (AUP) and the Pacer Classic initiative. The T-38C takes off from the Williams Gateway Airport Runway for a 1 hour 20 minute flight with Boeing test pilot Ed Wilson sitting in the front seat and USAF test pilot, Capt. John Deems, in the rear seat.

July 8, 1998. In Exercise Baltic Challenge '98, through July 9, a 445th Airlift Wing (AFRC) C-141 aircrew from Wright-Patterson AFB, Ohio, airlifts five dolphins, their handlers, and veterinarians from NAS North Island to Palanga International Airport, Lithuania, to participate in this joint military land and sea exercise. The dolphins find and mark mines on the Baltic seafloor and search for live ordnance from World War II. On July 20, another 445th Airlift Wing aircrew returns the dolphins to California.

July 15, 1998. Raytheon Aircraft Company's first T-6A Texan II aircraft, or the Joint Primary Aircraft Training System (JPATS), successfully completes its first flight at Wichita, Kans. To meet Air Force and Navy primary aircraft training needs, the DOD decided to buy 740 T-6A aircraft, along with the accompanying JPATS Ground Based Training System.

July 30, 1998. At Edwards AFB, Calif., testing on the C-141A Electric

Starlifter comes to an end. The joint Air Force Research Laboratory Propulsion Directorate and Lockheed Martin/Lucas Aerospace testing program had outfitted a large military cargo aircraft with electrically operated fly-by-wire, power-by-wire flight controls for the first time in aviation history. The C-141A aircraft, assigned to the 418th Flight Test Squadron, flew over 1,000 hours in the program. It was the last C-141A in operational service. With the completion of the test program, the Air Force retired the aircraft to Davis-Monthan AFB, Ariz.

August 9, 1998. A 305th Air Mobility Wing C-141 from McGuire AFB, N.J., flies 15 seriously injured State Department employees from Nairobi, Kenya, to Ramstein AB, Germany, on their way to Landstuhl Region Medical Center. The 11 Americans and 4 Kenyan patients received injuries in the August 7 bombing of the U.S. embassy in Nairobi, which claimed at least 200 lives and wounded more than 1,000.

August 13, 1998. At Altus AFB, Okla., a 54th Air Refueling Squadron KC-135 receives the first Pacer CRAG and Traffic Collision Avoidance System (TCAS) modifications. The TCAS modification gave pilots the ability to see other aircraft and get advance warning of possible mid-air collisions.

August 13, 1998. After 12 Americans die in a terrorist bombing of the U.S. embassy in Nairobi on August 7, a 437th Airlift Wing C-17 from Charleston AFB, S.C., returns ten victims to Andrews AFB, Md. In a hangar at Andrews, President William J. Clinton presides over a national ceremony of remembrance. A day earlier, on August 12, a 436th Airlift Wing C-5 from Dover AFB, Del., returned the eleventh victim, SMSgt Sherry Lynn

Olds, from Ramstein AB, Germany, to Dover, where a 375th Airlift Wing C-9 from Scott AFB, Ill., flew her remains to Panama City, Fla., her hometown. The twelfth American victim, married to a Kenyan, was buried in Kenya.

August 20, 1998. The United States launches more than 75 Tomahawk land-attack missiles from ships and submarines against terrorist targets in Afghanistan and Sudan in retaliation for the bombings of U.S. embassies in Kenya and Tanzania on August 7 that killed 257 people.

August 21, 1998. Maj. Michael J. Brill, a full-time Air Reserve technician with the 419th Fighter Wing, 466th Fighter Squadron, at Hill AFB, Utah, becomes the first Air Force pilot to amass more than 4,000 total flying hours in the F-16 Fighting Falcon. According to Lockheed Martin officials, more than 3,000 pilots had topped the 1,000-hour mark in an F-16; more than 400 pilots had surpassed 2,000 hours; and 21 had more than 3,000 hours.

September 9, 1998. In Operation Keiko Lift, through September 10, a 437th Airlift Wing C-17 aircrew from Charleston AFB, S.C., flies Keiko, the Orca killer whale who starred in the movie *Free Willy*, on a 8,630-mile, 10-hour nonstop flight from the Oregon Coast Aquarium at Newport, Oregon, to Iceland's Vestmannaeyjar Island. A KC-10 from Travis AFB, Calif., and KC-135s from McGuire AFB, N. J., refuel the C-17 during its flight to Iceland.

September 21, 1998. Hurricane Georges. After devastates Puerto Rico, the Dominican Republic, and the Virgin Islands, and makes landfall on the Gulf Coast, Air Mobility Command, Air Force Reserve Command, and Air National Guard units provided extensive relief support. Through October 12, the Air Mobility Command flies more than 190 missions to move medical teams, generators, refrigerator units, construction equipment, and similar cargo to the disaster areas, along with 450 passengers and 8,500 short tons of cargo from numerous on-load places in the U.S.

October 11, 1998. The violent repression of Albanians in Kosovo by the Federal Republic of Yugoslavia threatens to spark a war between Yugoslavia and the NATO states. In Operation Phoenix Duke. Air Mobility Command, Air Force Reserve Command, and Air National Guard airlifters and tankers move into position. Yugoslavia averts conflict by agreeing to cease operations against the Kosovar Albanians. Most of the command's assets returned to home station in early November.

October 12, 1998. The 53rd Weather Reconnaissance Squadron, 403rd Wing (Air Force Reserve Command), at Keesler AFB, Miss., receives the USAF's first WC-130J Hercules aircraft. The WC-130J, outfitted as a special weather reconnaissance version of Lockheed Martin's C-130J cargo plane, has a mission to fly into the eye of hurricanes to retrieve critical information about active storms.

October 14, 1998. The 89th Airlift Wing at Andrews AFB, Md., receives the first C-37A, the military version of the Gulfstream V business jet.

October 23, 1998. Two Marine Corps V-22 Ospreys land at Eglin AFB, Fla., for a week of preoperational testing. The V-22 will eventually be transformed into the CV-22 and delivered to the Air Force Special Operations Command in 2004. The Air Force plans to purchase 50 CV-22s to replace

the command's existing rotary and fixed-wing fleet.

November 6, 1998. USAF transport crews and aircraft begin airlifting 7.4 million pounds of relief cargo to Honduras, Nicaragua, Guatemala, and El Salvador after those countries are devastated by Hurricane Mitch, which claims more than 10,000 lives. By the time the airlift ends on March 19, 1999, the crews will fly more than 200 missions.

November 12, 1998. In Operation Phoenix Scorpion III, through November 15, after Saddam Hussein expels UN weapons inspectors from Iraq, the DOD orders more U.S. forces to Southwest Asia using Air Mobility Command airlifters. The Clinton administration accepted Iraq's peace overtures on November 14 to end the deployment. During the four-day effort, the Air Mobility Command completes 257 airlift and tanker missions to move more than 3,000 passengers and 2,700 short tons of cargo. Tankers refuel 90 aircraft, offloading 9.3 million pounds of fuel.

December 4–15, 1998. The crew of the space shuttle *Endeavor* completes the first International Space Station assembly mission, which involves delivery and attachment of *Unity*, the first U.S. module, to Russia's previously launched *Zarya* module. The crewmembers also launches the USAF *MightySat I* to evaluate composite materials, advanced solar cells, and other technology.

December 9, 1998. President William J. Clinton awards an honorary fourth star to Gen. Benjamin O. Davis, Jr., the first black USAF general and the leader of the Tuskegee Airmen in World War II.

December 9, 1998. The threat Iraq poses to the Middle East leads to another deployment, Phoenix Scorpion IV, and Emergency Deployment Readiness Exercise Noble Shirley to deploy Patriot missiles to Israel. Through December 21, C-17s airlift the Patriots from Ramstein AB, Germany, to Nevatim AB, Israel. USAFE assumes tactical control of both the C-17s and the Air Mobility Command's Tanker Airlift Control Element at Nevatim Air Base. This is the first time that tactical control of strategic airlift assets had passed from U.S. Transportation Command and Air Mobility Command to another command.

December 12, 1998. At Beale AFB, a 1st Reconnaissance Squadron pilot, Maj. Alan Zwick, shatters a world record by flying his U-2 and payload to an altitude more than 12.5 miles above the earth. Zwick breaks the previous record of Rudolf Fiam, a Czechoslovakian pilot, who flew a Yakovlev 40 on February 24, 1979, to an altitude of 28,513 feet with a payload of 4,400 pounds. Zwick surpassed that record just 12 minutes into his flight, and he continued to climb to over 66,800 feet, or 38,287 feet higher than the Czechoslovakian record.

December 12, 1998. A KC-135R, modified with Pacer CRAG, flies its first overseas mission from Grand Forks AFB, N.D. The 905th Air Refueling Squadron aircraft transfers 95,000 pounds of fuel to a B-52 in Alaskan airspace during the mission.

December 16, 1998. The U.S. and Great Britain start Operation Desert Fox/Phoenix Scorpion IV after Iraq prevents UN weapons inspector from continuing their work. In the 4-day operation, cruise missiles and aircraft strikes hit about 100 Iraqi weapons-production facilities. It is the largest air campaign against the Iraqis since Desert Storm in 1991, and it features

the first use of the B-1B in combat. Under Phoenix Scorpion IV, the Air Mobility Command deploys forces that were placed on alert a month earlier as a CONUS Crisis Response Force. In the 4-day deployment, Air Mobility Command aircraft flies 159 missions, airlifting 2,462 passengers and 1,940 short tons of cargo.

January 12, 1999. The last three of ten C-27 Spartans with the 310th Airlift Squadron flies from Howard AFB, Panama, to Davis-Monthan AFB, Ariz., for retirement. The C-27s fly humanitarian assistance, peacekeeping, and counterdrug missions in Panama. The Spartan is a smaller, twin-engine version of the C-130.

January 22, 1999. Lockheed Martin receives a $450 million contract to install advanced flight control systems, improved communications, and navigation gear aboard the USAF's C-5s under an Avionics Modernization Program (AMP).

February 7, 1999. From Cape Canaveral, Florida, a Delta II rocket launches *Stardust*, the first spacecraft designed to intercept a comet and return samples to Earth. The National Aeronautics and Space Administration schedules *Stardust* to rendezvous with Comet Wild 2 on January 2, 2004.

February 17, 1999. The first C-130J, a new transport with six-bladed propellers, advanced avionics, and shorter takeoff and landing capability, arrives at Keesler AFB, Miss., to serve with the 403rd Wing.

March 24, 1999. To prevent the forced eviction and genocide of ethnic Albanians living in Kosovo, NATO launchs Operation Allied Force/Noble Anvil/Phoenix Duke II against Yugoslavia. It is exclusively an air campaign and NATO's first combat operation against a sovereign nation. The U.S. names its part of the Operation Noble Anvil. Additionally, the USAF uses the B-2 for the first time in combat, while F-15 pilots shoot down two MiG-29s on the first day of combat. Under Operation Phoenix Duke II, Air Mobility Command's role in the operation precedes the bombing and continues afterward. From 18 February, when the first airlift mission is flown, through 3 July, when the tanker redeployment come to an end, Air Mobility Command flies 2,130 airlift missions to transport 32,111 passengers and 52,645 short tons of cargo. In the air campaign, which ends on June 9, KC-10s and KC-135s fly 9,001 missions to deliver 348.5 million pounds of fuel to receiving aircraft. Air Mobility Command's last air refueling aircraft returns to the U.S. on July 20, while the final airlift missions finishes on July 26.

March 26, 1999. On the third day of Operation Allied Force, one F-15C pilot shorts down two MiG-29s in aerial combat over Yugoslavia.

March 27, 1999. For the first time, an F-117 stealth fighter is shot down in combat. Enemy ground fire brings down the F-117 over Yugoslavia. Capt. John A. Cherrey, an A-10 pilot, earns a Silver Star for locating the downed pilot, who is rescued by helicopter the same day.

March 27, 1999. During Operation Allied Force/Distinguished Flying Cross, four KC-135 aircrew members—Capt. Clifton G. Janney, Maj. Kevin Torres, Capt. Russell O'Dell, and SSgt. Matthew DellaLucca—from the 6th Air Refueling Wing at MacDill AFB, Fla., risk their lives to air-refuel two A-10s, which had nearly run out of fuel. On August 16, 2000, Gen. Charles T. Robertson, Jr., Commander of the Air

Mobility Command awards the Distinguished Flying Cross (DFC) to the crewmembers.

April 4, 1999. A C-17 Globemaster III starts Operation Sustain Hope or Shining Hope by airlifting relief supplies from Dover AFB, Del., to Tirana, Albania. By the end of the humanitarian airlift for Operation Allied Force on July 8, the Air Mobility Command will have used C-5s, C-17s, and C-130s to airlift 913 passengers and 5,939 short tons of food and supplies to refugee camps in Albania, Macedonia, and Montenegro.

April 17, 1999. For the first time, the Air Force sends the RQ-1 Predator on flights in a combat zone. The unmanned aerial vehicle performs reconnaissance over Serbia during Operation Allied Force.

April 27, 1999. President Clinton approves a Selected Reserve Call-Up to support NATO operations over Kosovo Operation Allied Force. The initial call-up focuses on air-refueling resources, which include 47 tankers and 2,116 personnel.

May 1, 1999. The first Air Force Reserve Command units mobilized in support of Operation Allied Force enter on active duty. Eventually, six tanker wings and one rescue wing are called to active duty.

May 2, 1999. Serbian ground forces shoot down an F-16 over Yugoslavia, the second and last USAF aircraft lost to enemy fire in Operation Allied Force; an MH-60 helicopter crew rescues the pilot.

May 4, 1999. An F-16CJ pilot shoots down a Yugoslavian MiG-29 fighter over Kosovo, the fifth and final USAF aerial victory in Operation Allied Force.

May 5, 1999. In Operation Provide Refuge, a DOD-chartered Tower Air Boeing 747 lands at McGuire AFB, N.J., with the first group of refugees from Kosovo, southernmost province in the Federal Republic of Yugoslavia. In this operation, up to 3,000 Kosovar Albanians are flown to McGuire and temporary quarters at neighboring Fort Dix until arrangements can be made for permanent resettlement with relatives or sponsors in the U.S.

June 10, 1999. NATO suspends airstrikes against Yugoslavia after its president agrees to withdraw Serbian forces from Kosovo. That capitulation made Allied Force the first war won by airpower alone. Meanwhile, through July 8 Air Mobility Command airlifters deploy Task Force Falcon and elements of Task Force Hawk to Macedonia to help constitute the U.S. military portion of the Kosovo Force, the multinational peacekeeping force in Kosovo.

June 13, 1999. A KC-135R and aircrew from the 319th Air Refueling Wing at Grand Forks AFB, N.D., flying with the 92nd Air Expeditionary Wing in southern Europe Operation Allied Force, perform the first-ever combat sortie in a Pacer CRAG–modified KC-135. Although Pacer CRAG tankers had been deployed to Southwest Asia to support Operation Southern Watch, the June 13 mission marks the first time a Pacer CRAG aircraft has logged actual combat time.

June 25, 1999. An all-Air Force crew flies the V-22 Osprey for the first time in a 20-minute sortie from Marine Corps Air Facility Quantico, Va., to NAS Patuxent River, Md.

July 11, 1999. A C-141 from the 62nd Airlift Wing at McChord AFB, Wash., and a KC-135 (Hawaii ANG) leave Christchurch, New Zealand, on a 6,375-mile round-trip journey to the South Pole. The C-141 airdrops emergency medical supplies near the Amundsen-Scott South Pole Station to support the diagnosis and treatment of a physician who discovered a lump in her breast. The continual darkness and extreme cold of the Antarctic winter prevented her evacuation until October 16, when a ski-equipped LC-130 from 109th Airlift Wing (ANG) from Schenectady Airport, N.Y., made the trip from Amundsen-Scott South Pole Research Station to McMurdo NAS.

July 23, 1999. Col. Eiieen M. Collins, USAF, the first woman pilot of a space shuttle, in 1995, become the first woman commander of a space-shuttle mission on a flight of the *Columbia*.

July 31, 1999. Two improved T-38C fighter trainers are transferred from Edwards AFB, Calif., to Columbus AFB, Miss., for testing. At Edwards, the T-38s complete a development test and evaluation of the aircraft's Avionics Upgrade Program, while the move to Columbus takes the planes into initial operational testing and evaluation for a specialized undergraduate pilot training course and introduction to fighter fundamentals.

August 18, 1999. After a devastating earthquake rocks western Turkey, the Air Mobility Command begins its support of Operation Avid Response, an international relief effort. A C-5 from the 436th Airlift Wing at Dover AFB, Del., leaves the U.S. for Istanbul with a 70-person search-and-rescue team. The team, with five search-and-rescue dogs, 56,000 pounds of equipment, and three vehicles, sets up operations in nearby Izmit. Two KC-10s from the 305th Air Mobility Wing at McGuire AFB, N.J., refuel the C-5 during its nonstop flight to Turkey. By September

10, when the Avid Response task force at Topel NAS, Turkey, disbands, Air Mobility Command aircraft has flown 20 missions to support the earthquake relief effort.

September 20, 1999. In UN Operation Stabilise C-130 airlift U.S. troops from Australia to Dili, East Timor, after that nation declares its independence from Indonesia and a bloody war follows. Eventually, one C-141, two C-130s, three KC-135s, and two C-17s take part in the operation to move peacekeeping troops into the area.

October 1, 1999. Aerospace Expeditionary Force 1 deploys to Southwest Asia—the first of 10 such forces to rotate availability for deployment duty. The new system allows the Air Force to respond more effectively with ready forces for operations overseas and increases morale by making deployments more predictable.

October 6, 1999. At Grand Forks AFB, N.D., contractors implode the first of 450 Minuteman III ICBM silos to be destroyed under the Strategic Arms Reduction Treaty (START I).

October 13, 1999. The USAF terminates its Enhanced Flying Screening Program for new pilots and ends its use of the T-3A Firefly. At this time, there are 110 T-3As in the USAF inventory.

October 13, 1999. The X-43A, a hypersonic flight research vehicle and the world's first hypersonic air-breathing free-flight vehicle, arrives at Edwards AFB, Calif., for testing by NASA's Dryden Flight Research Center. The X-43 involves supersonic combustible ramjet, or "scramjet," technology. A successful flight in 2001 would make it the first vehicle not powered by a rocket engine to fly at hypersonic speeds.

NASA plans three test flights, two at Mach 7 and one at Mach 10, over the Western Test Range off the coast of southern California. Micro Craft, Inc., of Tullahoma, Tenn., built the 12-foot-long, unpiloted vehicle for NASA's Hyper-X program.

October 15, 1999. The 62nd Airlift Wing from McChord AFB, Wash., lands a C-17 on an airstrip of frozen seawater at McMurdo NAS in Antarctica. This first flight of a C-17 to McMurdo delivers a cargo of mail, fresh fruit, vegetables, and a telescope for the National Science Foundation scientists to use at the South Pole to study black holes in the Milky Way galaxy.

October 21, 1999. The USAF awards the Global Air Traffic Management equipment contract for the entire KC-135 fleet to Rockwell Collins. The $600 million contract covers 544 KC-135s, 20 flight simulators, and 43 other C-135-derived aircraft.

November 2, 1999. The Air Force announces the official transfer of Howard Air Base to Panama in accordance with the terms of the Panama Canal Treaty of 1977. The United States had operated the base in the Canal Zone for 82 years.

November 17, 1999. A KC-10 from the 60th Air Mobility Wing at Travis AFB, Calif., refuels a F-22 Raptor for the first time. Operating from Edwards AFB, Calif., over a four-day period, the KC-10 flies four sorties, each lasting nearly three hours. During the test, the KC-10 makes more than 250 boom connections with the F-22 and off-loads 35,000 pounds of fuel.

December 20–28, 1999. At least six C-130s, one C-5, and two MH-60 USAF aircraft take part in a humanitarian operation to Venezuela to relieve the victims of a flood that left an estimated 200,000 people homeless.

January 4, 2000. Final tallies of the flood devastation near Caracas, Venezuela, reveals 30,000 people are dead and another 400,000 left homeless. Through 10 March, 11 C-17 missions and 5 C-5 missions of Joint Task Force Fundamental Response airlifted 189 passengers and 527 short tons of cargo to Simon Bolivar International Airport near Caracas to support relief efforts.

January 17, 2000. Successful testing of a C-5 equipped with the new Traffic and Collision Avoidance System (TCAS) allows Lockheed Martin to begin retrofitting the remaining 126 C-5s in Air Mobility Command, Air Force Reserve Command, and Air National Guard units under the C-5 Avionics Modernization Program. The C-5 fleet will be equipped with TCAS by late fiscal year 2002.

January 22, 2000. The Air Mobility Command performs its first regularly scheduled C-130 aeromedical evacuation mission when the 911th Airlift Wing from Pittsburgh International Airport Air Reserve Station, Pa., began flying a weekly mission to the Caribbean to support U.S. Southern Command. Normally, the Air Mobility Command uses C-9s and C-141 for its peacetime aeromedical mission, but the phased C-141 retirement creates the C-130 requirement. Two Air Mobility Command-gained C-130 units volunteer to fly regular aeromedical missions, the 911th Airlift Wing and the Air National Guard's 187th Airlift Wing from F. E. Warren AFB, Wyo., to Peterson AFB, Col., on 1 February 2000.

January 25, 2000. Air Combat Command's Commander, General Ralph E.

Eberhart, dedicates the first Block D upgraded B-1s at Dyess AFB, Tex.

March 1, 2000. In Operation Atlas Response, after flooding in Mozambique left about one million people homeless, a C-17 from the 315th Airlift Wing (Air Force Reserve Command) at Charleston AFB, S.C., augmented by crewmembers from the 446th Airlift Wing (Air Force Reserve Command) at McChord AFB, Wash., flew the first planeload of emergency relief supplies to that nation. By the end of the operation on April 16, the Air Mobility Command system had flown 29 strategic airlift missions with C-5s and C-17s to transport 720 passengers and 910 short tons of cargo.

April 17–20, 2000. A 437th Airlift Wing C-17 from Charleston AFB, S.C., airlifts Polish soldiers and equipment from Strachowice AB in southwestern Poland to Mitrovica, Kosovo, to augment NATO peacekeeping forces in the Yugoslav province. In four days, the Air Mobility Command moves 130 Polish troops and 205,000 pounds of equipment. A single C-17, flown by several aircrews, performs the shuttle missions between Ramstein AB, Germany, Strachowice, and Mitrovica.

May 3, 2000. Gen. Joseph W. Ralston becomes the first USAF officer in 37 years to serve as Supreme Allied Commander, Europe, head of all North Atlantic Treaty Organization forces.

May 23, 2000. The first production-model T-6A Texan II aircraft arrives at Randolph AFB, Tex., to replace the Air Force's T-37 and the Navy's T-34 as primary training aircraft.

July 15, 2000. The final B-2 arrives at Whiteman AFB, Mo.

July 25, 2000. Through September 23, in the worst fire season in decades, fire consumes over 6.5 million acres in the American West, more than twice the annual average. For the fire suppression effort, from August 1 to September 23, the Air Mobility Command, the Air National Guard, the Air Force Reserve Command, and commercial carriers fly 48 missions to move 339 short tons of cargo and 5,967 Army and Marine troops to augment civilian firefighters in Montana and Idaho. Guard and Reserve C-130s equipped with modular airborne firefighting systems flew 774 sorties from July 25 to September 6 to drop 970,500 gallons of fire retardant on wildfires in California, Idaho, and Montana.

August 5, 2000. Air Force Reserve Command C-141s from the 452d Air Mobility Wing at March Air Reserve Base, Calif., and 445th Airlift Wing at Wright-Patterson AFB, Ohio, join two active-duty C-141s from the 62nd Airlift Wing at McChord AFB, Wash., to move firefighters and equipment to Idaho Falls, Ida., where they are bussed to Clear Creek, Ida., to fight raging wildfires. Two Air Force Reserve Command C-130s from the 302nd Airlift Wing at Peterson AFB, Colo., and two Air National Guard C-130s from the 146th Airlift Wing at Channel Islands Air National Guard Station, Calif., drop fire retardant on wildfires near Los Angeles and Fresno. Two other Air National Guard units, the 145th Airlift Wing at Charlotte, N.C., and the 153rd Airlift Wing at Cheyenne, Wyo., fly sorties from Hill AFB to drop fire retardant on the wildfires in California.

September 9, 2000. In Operation Centrazbat 2000, a C-17 from Charleston AFB, S.C., and one from Altus AFB, Okla., pick up 160 82nd Airborne Division paratroopers at Pope AFB, N.C., and then fly 20 hours nonstop to a drop zone near Almaty, Kazakhstan. The paratroopers joined troops from Kazakhstan, Uzbekistan, Kyrgyzstan, and other NATO partners in this humanitarian assistance and peacekeeping field training operation. Each C-17 receives two aerial refuelings on the 6,700-mile flight to Kazakhstan.

September 18, 2000. The first USAF CV-22 Osprey arrives at Edwards AFB, Calif.

September 18, 2000. At Edwards AFB, Calif., the Boeing X-32A Joint Strike Fighter concept demonstrator makes its first flight.

October 15, 2000. Crews from the 75th Airlift Squadron and the 86th Aeromedical Evacuation Squadron earn the 2000 Mackay Trophy for evacuating survivors of a terrorist attack in Europe across the Atlantic Ocean to Norfolk, Va. The 28 victims were injured in a terrorist attack on the Navy warship USS *Cole* on October 12, when it was docked at Aden in Yemen. The attack left six dead and a 20×40-foot hole in the ship's hull.

December 4, 2000. At Edwards AFB, Calif., the Boeing X-32A Joint Strike Fighter concept demonstrator successfully completes low-speed approach aircraft carrier variant tests, one of three main goals for the JSF flight test program.

December 16, 2000. Lockheed Martin's X-35C Joint Strike Fighter concept demonstrator for the Navy arrives at Edwards AFB to begin flight testing on aircraft carrier-suitable flying qualities and aircraft performance. The X-35C is similar to the Air Force's X-35A, but has larger wing and control surfaces, ailerons, and a special structure for high-impact landings necessary for carrier operations.

December 19, 2000. A KC-10 Extender from the 6th Air Refueling Squadron at Travis AFB, Calif., refuels Boeing's X-32A Joint Strike Fighter concept demonstrator for the first time.

December 21, 2000. At Edwards AFB, Boeing's X-32A Joint Strike Fighter concept plane makes its first supersonic flight, when Lt. Col. Edward Cabrera, the Air Force's lead X-32 test pilot, takes the jet to 30,000 feet and exceeds Mach 1 (660 mph).

January 4, 2001. A C-17 from the 315th Airlift Wing at Charleston AFB, S.C., flies the 2001 Mars Odyssey spacecraft from Buckley AFB, Colo., to the Kennedy Space Center at Cape Canaveral.

January 5, 2000. An F-22 Raptor, equipped with combat-capable avionics, flies for the first time at Lockheed Martin's facility in Marietta, Ga.

January 31, 2001. Lockheed Martin's X-35C Joint Strike Fighter concept demonstrator completes flight testing at Edwards AFB, Calif.

January 31, 2001. Through February 3, two C-5s and four C-17s flies 115 short tons of humanitarian cargo to Ahmedabad, India, after a 7.7 earthquake devastated western India on 26 January. Each C-5 Galaxy flies nonstop from Travis AFB, Calif., to Andersen AFB, Guam. Hawaii Air National Guard KC-135s refuel each C-5 en route to Guam. At Andersen, workers transfer the C-5 cargo onto four 62nd Airlift Wing C-17s from McChord AFB, Wash., because Ahmedabad's airfield is too small for the C-5s. KC-135Rs from the 18th Wing at Kadena AB refuel the C-17s on their 13-hour flight between Guam and India.

February 21, 2001. An RQ-1 Predator fires a Hellfire missile to hit a tank at Nellis AFB, Nev., becoming the first unmanned aerial vehicle to destroy a ground target. Until then, the remotely controlled airplane had been used only for reconnaissance.

February 24, 2001. Lt. Col. Stayce D. Harris becomes the first black woman to command a USAF flying squadron, the 729th Airlift Squadron at March Air Reserve Base, Calif.

February 27, 2001. The USAF successfully launches a Titan IV-B rocket from Cape Canaveral, Fla., with a MILSTAR satellite, to its intended orbit 22,300 miles above the equator.

March 5, 2001. The Space Shuttle *Columbia* lands at Cape Canaveral AFS, Fla., for the first time. NASA's modified 747 carried the craft there from Palmdale, Calif.

March 29, 2001. At Edwards AFB, Calif., Boeing's X-32B Joint Strike Fighter demonstrator makes its first flight. The 50-minute flight features a series of initial airworthiness tests.

April 1, 2001. A Chinese fighter pilot rams a Navy EP-3 reconnaissance aircraft over international waters of the South China Sea, forcing the Navy plane to land with severe damage on Hainan Island in China. After 11 days, diplomatic efforts persuade China to release the 24-man crew, which includes one USAF member.

April 23, 2001. The RQ-4A Global Hawk completes the first nonstop crossing of the Pacific Ocean by an unmanned aerial vehicle. It flies from Edwards AFB, Calif., to Edinburgh, Australia—a distance of 7,500 miles—in about 23 hours. The flight demonstrates the potential of the world's most advanced high-altitude, long-range, remotely operated aircraft.

May 8, 2001. Secretary of Defense Donald H. Rumsfeld designates the USAF executive agent for the Pentagon's space activities.

May 16, 2001. Through May 24, five C-5 missions airlifts 201 U.S. troops and 217 short tons of cargo from Fort Bragg, N.C., and Ramstein AB, Germany, to Ghana and Senegal, where they train indigenous military personnel to support UN peacekeeping efforts in Sierra Leone. From August to October 2000, in Operation Focus Relief I, AMC flies American forces and equipment to Nigeria to help train Nigerian forces for a peacekeeping role in Sierra Leone. By Late 2001, the Air Mobility Command flies another four C-5, four C-17, and two C-141 missions for peacekeeper training in Ghana, Senegal, and Nigeria. Seven airlift sorties from 11 September through 2 October are flown as Focus Relief III missions.

May 23, 2001. The first of four C-17s for the British Royal Air Force arrive at RAF Brize Norton, UK.

May 30, 2001. A C-17 from Charleston AFB flies a Thermosphere, Ionosphere, Mesosphere, Energetics, and Dynamics (TIMED) spacecraft from Andrews AFB, Md., to Vandenberg AFB, Calif. The TIMED would be launched later in 2001 on a two-year mission. Johns Hopkins University's Applied Physics Laboratory in Laurel, Md., which designed and built the TIMED, would operate the spacecraft to study the sun's influences on the Mesophere and Lower Thermosphere/Ionosphere, some 40 to 110 miles above the earth.

June 10, 2001. To support flood relief efforts in Houston, a C-17 from Charleston AFB, S.C., flies 25 federal relief workers and 15 tons of relief supplies from Albuquerque, N.M., to Ellington Field.

June 18, 2001. At McGuire AFB, N.J., SM Jere Garvin, a 2nd Air Refueling Squadron flight engineer, reaches 10,000 flying hours. His 24-year career includes flight time in C-130s, C-141s, the E-3 Sentry, and KC-10 Extender in over 2,400 sorties. He is the only active-duty flight engineer to reach that milestone by this date.

July 13, 2001. Kelly AFB, Tex., and McClellan AFB, Calif., are closed.

July 14, 2001. A prototype Minuteman ICBM interceptor successfully targets and destroys an unarmed Minuteman II ICBM over the central Pacific. Ten minutes after launch, the interceptor destroys the warhead, traveling some 15,000 mph at more than 140 miles in altitude above the earth. The 30th Space Wing at Vandenberg AFB, Calif., and the Ballistic Missile Defense Organization conduct the Ground-based Midcourse Defense Segment, formerly called the National Missile Defense Program test.

July 25, 2001. United States Transportation Command (US-TRANSCOM) Commander in Chief and Air Mobility Command Commander, Gen. Charles T. Robertson, Jr., flies the first of three C-37A special airlift support aircraft, a military version of the Gulfstream V business jet, from the Gulfstream plant in Savannah, Ga., to MacDill AFB, Fla., for duty with the 6th Air Mobility Wing. The USAF purchased the C-37A to replace older EC-135 and CT-43 aircraft. Two other C-37As are to be based at Andrews AFB, Md., one at Chièvres, Belgium, and a seventh at Hickam AFB, Hawaii.

July 28, 2001. Aircrews from the 509th Bomb Wing at Whiteman AFB fly back-to-back 21-hour sorties in just 45.5 hours using the B-2 *Spirit of Kansas*. It flies over the Royal International Air Tattoo at RAF Cottesmore, England, on July 28 and 29. The crews perform three aerial refuelings, one with a KC-10 and the others with KC-135s, and they practice a Global Power bombing mission. The maintenance crews at Whiteman perform a quick turn on the aircraft in 3.5 hours.

July 28, 2001. The DOD awards Boeing a $485 million contract to engineer and manufacture an Avionics Modernization Program (AMP) and a separate $1 billion contract for aircraft modifications, support, and other services to complement the AMP enhancements for 519 C-130s. The program will equip the C-130 cockpits with flat panels, digital displays, multifunctional radar, and a state-of-the-art communications system.

July 28, 2001. Air Education and Training Command's C-141 aircrew training school at Altus AFB, Okla., is officially closed, ending more than 25 years of C-141 training there. The closure comes with the phased retirement of more than 265 C-141Bs. The Air Force, however, will modify 56 C-141Bs with state-of-the-art glass cockpits and redesignate them as C-141Cs.

August 6, 2001. The 93rd Air Control Wing at Robins AFB, Ga., receives the first updated Block 20 E-8C Joint Surveillance Target Attack Radar System production aircraft. The Computer Replacement Program enhancement in the Block 20 aircraft reduces the number of on-board computers from five to two, while increasing processing power and speed. The system also had several engineering changes to improve maintainability.

August 13, 2001. The Helios, an experimental, solar-powered, remotely piloted National Aeronautics and Space Administration aircraft, reaches a record altitude of 96,500 feet on a flight that lasts almost 17 hours.

August 24, 2001. At Grand Forks AFB, N.D., the Air Force implodes the last of 450 Minuteman missile silos in accordance with the terms of the first Strategic Arms Reduction Treaty.

September 8, 2001. SECAF James G. Roche, Lt. Gen. Lance W. Lord, the Assistant Vice Chief of Staff, and former first lady Nancy Reagan fly aboard the *Spirit of '76* on its final flight from Andrews AFB, Md., to San Bernardino International Airport, Calif. It will be retired and displayed at the Reagan Presidential Library in Simi Valley, Calif.

September 11, 2001. Arab terrorists hijack four U.S. passenger airliners. They crash two aircraft into the 110-story, twin-tower World Trade Center in New York and one into the Pentagon in Washington, D.C., killing more than 3,000 people. Passengers in the fourth aircraft rush the hijackers, and it crashes in a Pennsylvania field rather than another target. The FAA grounds all airliners in the U.S. for several days. President George W. Bush initiates a war on terrorism and homeland-defense efforts with combat air patrols within the U.S. under the Noble Eagle operation.

September 14, 2001. President George W. Bush authorizes the call-up of 50,000 reservists to active duty for the war on terrorism.

September 19, 2001. The USAF awards a contract for low-rate initial production of the F-22 to the Lockheed Martin Corporation. The contract called for production of 10 of the new stealthy air-superiority fighters.

October 1, 2001. The Air Force reassigns the Space and Missile Systems Center from Air Material Command to Air Force Space Command, giving the latter cradle-to-grave oversight of space systems.

October 1, 2001. Gen. Richard B. Myers, USAF, assumes command as chairman of the Joint Chiefs of Staff. A USAF general had not held that office, the highest in the military, for 19 years.

October 7, 2001. The U.S. initiated airstrikes against terrorist and Taliban targets in Afghanistan in Operation Enduring Freedom. The 509th Bomb Wing at Whiteman AFB, Mo., launches six B-2 bombers on the longest bombing missions in aviation history. The B-2s fly from Whiteman AFB over the Pacific to drop bombs on targets in Afghanistan and are recovered at Diego Garcia. The longest flight takes 44 hours. Enduring Freedom also involves B-1Bs, B-52Hs, F-15Es, KC-10s and KC-135 tankers, E-3 AWACS, EC-130s, AC-130 gunships, and MC-130 and MH-53 special operations aircraft. Naval involvement includes F-14s, F/A-18s, and AV-8 aircraft from three carriers in the Indian Ocean.

October 8, 2001. Two C-17s airdrop about 35,000 humanitarian rations over eastern and northern Afghanistan. The two airdrops are the C-17's first combat missions, first combat airdrops, and first humanitarian airdrops in Enduring Freedom. The C-17s fly from Ramstein AB more than 6,500 miles in a 22-hour round trip and were air-refueled multiple times.

October 9, 2001. In Operation Eagle Assist NATO deploys AWACS from Europe to the U.S. to help patrol American airspace after the September 11 terrorist attacks. This is the first time NATO deployed forces to the U.S. for a real contingency. By the end of the operation on May 16, 2002, seven NATO aircraft have flown 360 sorties.

October 26, 2001. The Defense Department Chooses Lockheed Martin over Boeing to build the Joint Strike Fighter. The Air Force, Navy, and Marine Corps plan to purchase thousands of the stealthy aircraft, designated the F-35.

October 31, 2001. In Operation Enduring Freedom a pair of C-17s fly over Afghanistan to deliver some 35,000 humanitarian rations. The loadmaster aboard one C-17 tosses out a single plastic-enclosed meal to symbolically represent the millionth meal airdropped. That act occurrs on the 61st C-17 humanitarian flight to help feed Afghan refugees.

November 2, 2001. During Operation Enduring Freedom, an MH-53 Pave Low helicopter crew from the 20th Special Operations Squadron earned the Mackay Trophy by rescuing another MH-53 crew whose helicopter had crashed on a rescue mission in the mountains of Afghanistan.

November 20, 2001. A 46th Test Wing F-16 from Eglin AFB, Fla., launches a joint air-to-surface standoff missile (JASSM) at the White Sands Missile Range, N. M. The successful test allows the Air Force to request initial production.

November 28, 2001. During Operation Swift Freedom C-17s land in Afghanistan on an airstrip about 80 miles south of Kandahar to deliver Navy Seabees. The operation introduces U.S. ground forces into Afghanistan.

December 3, 2001. Lockheed Martin deliver the USAF's first C-130J-30 to the 143rd Airlift Wing (Air National Guard) at Quonset State Airport, R.I. The C-130s with the "J-30" suffix have a longer fuselage than the standard J model. It can carry 128 combat troops and 8 standard 463L pallets rather than 92 troops and 6 pallets.

December 5, 2001. The USAF awards a $1.1 billion contract to Lockheed Martin to develop and demonstrate a Reliability Enhancement and Reengining Program for the C-5. The program includes hydraulic upgrades, structural improvements to ensure the life of the C-5 to 2040, and environmental control system improvements related to the C-5's Avionics Modernization Program. In the program, four C-5Bs will be equipped with the higher-thrust General Electric CF6 engines used on Boeing's 747 and 767 and the Airbus A300.

December 10, 2001. During Operation Deep Freeze, through January 20, 2002, a 50th Airlift Squadron aircrew from Little Rock AFB, Ark., is assigned to the 50th Air Expeditionary Squadron and flies the first wheeled C-130 mission from the Pegasus glacial runway in Antarctica to provide additional airlift for the National Science Foundation. The three wheeled C-130s flies 11 resupply missions between Antarctica and Christchurch International Airport, New Zealand, in that period.

December 12, 2001. A B-1B on a bombing mission to Afghanistan crashes in the Indian Ocean about 10 miles north of its island base, Diego Garcia, but the crew survives. It is the first USAF aircraft lost in the war in Afghanistan and the first B-1 lost in combat.

December 13, 2001. President George W. Bush informs Russia that the U.S. will withdraw from the Antiballistic Missile Treaty of 1972, which prevents the United States from developing and implementing a missile defense system.

December 17, 2001. USAF C-17s begin airlifting French troops from Istres Air Base, France, to Afghanistan as part of Operation Enduring Freedom.

December 26, 2001. The Pentagon approves low-rate initial production of the one-ton joint air-to-surface stand-off missile, which can penetrate enemy air defenses at ranges beyond 200 miles and destroy enemy targets with precision.

December 31, 2001. Operation Enduring Freedom/Noble Eagle Summary. From September 11 through the end of 2001, Air Mobility Command aircraft flies 1,757 airlift missions in support of America's war against terrorism outside the U.S. with C-17s and C-5s flying 45 percent and 29 percent of the missions, respectively. Tanker aircraft play a critical role by performing 953 air refueling missions—838 for KC-135s and 115 for KC-10s. For the Noble Eagle homeland defense mission in the CONUS, the Air Mobility Command completes 228 airlift missions with C-130s conducting 53 percent of the flights.

January 5, 2002. In Operation Enduring Freedom, a 437th Airlift Wing C-17 from Charleston AFB, S.C., arrives at NAS Guantánamo Bay, Cuba, with the first plane-load of materials for U.S. Navy Seabees to use in building a prison for 2,000 Taliban and Al Qaeda detainees captured in Afghanistan. In the early weeks of 2002, 21 C-17 missions from Charleston move 1.2 short tons of equipment and supplies to Guantánamo.

January 11, 2002. A 445th Airlift Wing crew in a C-141 Starlifter airlifts the first load of Taliban and Al Qaeda detainees captured in Afghanistan to the U.S. naval base at Guantánamo, Cuba.

January 11, 2002. Air Mobility Command, its Air Reserve Component, and the U.S. Transportation Command begin an 11-day exercise to test a commercial jet transport in the military aeromedical evacuation. A U.S. Airways Boeing 767 with seats and other internal fixtures removed is converted into a flying ambulance fitted with oxygen tanks, an independent electrical supply, and 87 litters. The aeromedical crew practice scenarios involving aircraft depressurization, ditching in water, and caring for a various simulated injuries. Leaving Scott AFB, Ill., the converted 767 stopps at McChord AFB, Wash., Charleston AFB, S.C., Travis AFB, Calif., Andrews AFB, Md., and Kelly AFB, Tex. The flying time between the five bases is long enough to simulate missions from overseas to the continental United States.

January 16, 2002. At the request of the Philippine government, the DOD deploys U.S. forces to train, advise, and assist Filipino troops in combating the indigenous Abu Sayyaf terrorist group. Through September 30, Air Mobility Command flies 78 missions to transport nearly 1,600 troops and more than 3,000 short tons of cargo to the Philippines.

January 23, 2002. A 305th Air Mobility Wing KC-10 from McGuire AFB, N.J., arrives at Dulles International Airport in Washington, D.C., with John Walker Lindh, a 20-year-old American accused of joining Al Qaeda and fighting in Afghanistan against U.S. forces. A U.S. Park Police helicopter takes Lindh from the Air Mobility Command flight to a detention facility in northern Virginia. On January 24 in Alexandria, Va., he is charged with conspiring to kill Americans. A C-17 flew Lindh from Kandahar to Incirlik AB, Turkey, for transfer to the KC-10.

February 4, 2002. An unmanned aerial vehicle destroys an enemy target for the first time when a remotely controlled MQ-1B Predator fires a Hellfire missile to kill a group of senior Al Qaeda members in southeastern Afghanistan.

March 1, 2002. The U.S. and its coalition partners launch Operation Enduring Freedom/Anaconda in eastern Afghanistan. Using B-52s, B-1s, AC-130s, A-10s, and F-15s for air support, the coalition ground operation tries to surround and kill Taliban and Al Qaeda forces regrouping south of Gardez.

March 1, 2002. Brig. Gen. Teresa M. Peterson, USAF, becomes the first active-duty woman to command an operational flying wing when she assumes leadership of the 305th Air Mobility Wing at McGuire AFB, N.J.

March 4, 2002. During Operation Anaconda, the Air Force experiences its first combat deaths since the Southwest Asia War of 1991. Two airmen are among seven U.S. servicemen killed in two helicopter assaults near Gardez in eastern Afghanistan.

April 18, 2002. The MC2A-X, an experimental aircraft designed for electronic communications and command and control over a combat area, makes its first flight at Hanscom AFB, Massachusetts. Since its maiden flight occurred 227 years to the day after Paul Revere's famous ride, the new aircraft is dubbed Paul Revere.

April 22, 2002. The Air Force announces a new standard wing structure that will include four groups: operations, maintenance, mission support, and medical.

May 12, 2002. Twelve C-17s from Charleston AFB, S.C., launch shortly after daybreak to fly the largest C-17 formation in history. Aircrews assigned to the 437th Airlift Wing and 315th Airlift Wing (Air Force Reserve Command) fly the multiship formation over downtown Charleston, the Ashley River, and Charleston AFB to train for their strategic brigade airdrop mission.

May 13, 2002. President George W. Bush announces an agreement between the United States and Russia that will reduce the number of nuclear weapons in their arsenals by two-thirds.

May 22, 2002. The X-45A unmanned combat air vehicle flies for the first time at Edwards AFB, Calif. It is the first unmanned aircraft designed for combat operations, such as suppression of enemy air defenses, in an extremely hostile environment.

July 10, 2002. A C-5 from the 436th Airlift Wing leaves Dover AFB, Del., for Kabul, Afghanistan, with 13,115 pounds of school supplies collected by children from 58 American schools.

July 15, 2002. The first production C-17 (P-86), fitted with the Block 13 software upgrade, arrives at McChord AFB, Wash. The Block 13 upgrade includes a state-of-the-art terrain awareness warning system (TAWS) and improved station keeping equipment (SKE). The TAWS features a terrain map to help aircrews detect and avoid terrain, while the upgraded SKE has a multifunctional display that enables 18 aircraft to fly in formation within 10 nautical miles and up to 100 aircraft to fly in formation within 100 nautical miles. The new SKEs also allows C-17 aircrews to perform formation airdrops in nearly all types of weather.

July 20, 2002. Through 2 August, five C-5s fly 23 combat missions from a forward operating location to Kandahar to redeploy more than 780 troops and 1,350 short tons of equipment from Canada's Princess Patricia's Light Infantry Regiment. The five C-5 aircraft, with three crews each from Travis AFB, Calif., and Dover AFB, Del., and a C-5 leadership package with a seventh aircrew and 44 maintenance personnel, operates from the forward operating location as the 782nd Expeditionary Airlift Squadron. A 31-person Tanker Airlift Control Element from the 615th Air Mobility Operations Group supports the C-5s at Kandahar. This redeployment demonstrates that the USAF's largest jet transport aircraft can successfully operate at a remote, austere airfield.

July 22, 2002. At McConnell AFB, Kans., the YAL-1A, destined to become the world's first directed-energy combat aircraft, completes its first flight. A modified Boeing 747, the YAL-1A is designed to carry an airborne laser to destroy enemy missiles in their boost phases.

August 15, 2002. The USAF and the Boeing Company sign a $0.8 billion multiyear contract to buy 60 more C-17s with delivery through July 2008.

August 21, 2002. The first Lockheed Martin Atlas V lifts off from Cape Canaveral, Fla., inaugurating a new type of launch missile. Part of the Air Force's Evolved Expendable Launch Vehicle program, the system uses a standard booster, adding supplemental boosters as needed for a variety of payloads.

September 11, 2002. In Operation Enduring Freedom, in the year after the terrorist attacks on American soil, Air Mobility Command aircraft fly 4,864 airlift missions to deliver 223,487 passengers and 304,801 short tons of cargo in support of this operation. C-17s and C-5s fly 48 percent and 30 percent of the missions, respectively. Moreover, KC-135s accomplish 2,760 air refueling missions and KC-10s another 176 missions in the one-year period.

October 1, 2002. U.S. Northern Command activates as a new unified command under Gen. Ralph Eberhart, USAF, who remains commander of the North American Aerospace Defense Command. Including elements of all the U.S. armed services, the new command assumes responsibility for the military protection of North America.

October 1, 2002. Gen. John Jumper, Air Force Chief of Staff, authorizes deactivation of the Peacekeeper intercontinental ballistic missile system.

October 23, 2002. The first KC-135 with the Roll-on-Beyond Line of Sight Enhancement (ROBE) relays a large quantity of tactical communications data from several F-15 fighters and an E-8C (Joint Stars) aircraft to an operations center at Hanscom AFB, Mass. The ROBE, carried on pallets aboard a KC-135 from the 319th Air Refueling Wing at Grand Forks AFB, N.D., and flying a test mission from Eglin AFB, Fla., successfully demonstrates the ROBE's ability to serve as the relay platform of a warfighting communications network.

October 24, 2002. A C-5 and aircrew from Travis AFB, Calif., fly from Andrews AFB, Md., to Tashkent International Airport, Uzbekistan, with 40 volunteer physicians from Physicians

With Heart, an affiliate of Heart-to-Heart International, and medicines, pharmaceuticals, and medical supplies worth nearly $10 million. The October 24 mission commemorates the tenth anniversary of this joint State Department and Physicians With Heart program to assist the people of Uzbekistan.

October 28, 2002. In Operation Enduring Freedom a C-141 arrives at Guantánamo Bay with the last group of Taliban and Al Qaeda detainees from Afghanistan for 2002. From January 10 to October 28, 2002, C-141s and C-17s flew 23 missions to carry 620 detainees from Kandahar to Guantánamo.

December 9, 2002. One day after Typhoon Pongsona devastates Guam with sustained winds of 150 mph, the Air Force begins airlifting relief equipment and supplies to the island. Within 10 days, 58 C-5 missions deliver 1,200 tons of humanitarian cargo to Andersen AFB.

December 21, 2002. The first C-5 Galaxy (No. 85-0004) with avionics modernization program equipment makes its maiden flight, two months ahead of schedule. Departing and recovering at Dobbins AFB, the 5.2-hour C-5 mission demonstrates the basic flying qualities of the new avionics and navigational system.

February 8, 2003. With the DOD's approval, Gen. John W. Handy, Air Mobility Command Commander, activates the Stage I Civil Reserve Air Fleet. Stage I affects 22 U.S. airline companies and 78 commercial aircraft (47 passenger and 31 wide-body planes). General Handy, however, only uses the 47-passenger aircraft to deploy American troops to the Persian Gulf. The U.S. has activated the Civil

Reserve Air Fleet only two other times in its 52-year history, Stage I on August 17, 1990 and Stage II on January 17, 1991, both for the first Persian Gulf War.

March 19, 2003. In Operation Iraqi Freedom, just after 9:30 P.M. EST, U.S. forces fire about 40 Tomahawk cruise missiles into Iraq, striking three targets around Baghdad. The attack begins the U.S.-led, multinational coalition effort to liberate the people of Iraq.

March 26, 2003. Nearly 1,000 173rd Airborne Brigade "Sky Soldiers," based at Vicenza, Italy, parachute from C-17 Globemaster IIIs into the Kurdish-controlled area of northern Iraq. It is the C-17's first insertion of paratroopers into combat. The aircraft and aircrews came from Charleston AFB, S.C., and McChord AFB, Wash.

April 19, 2003. A C-17 from McChord AFB, Wash., flown by a crew assigned to the 315th Airlift Wing (Reserve-Associate) at Charleston AFB, S.C., returns seven former U.S. Army prisoners of war to the U.S., one week after their rescue in Iraq. The C-17 flies five men and women from the 507th Maintenance Company to Fort Bliss, Tex., and two Apache helicopter crewmen from the First Cavalry Division to Fort Hood, Tex. A KC-135 and crew from the 157th Air Refueling Wing at Pease Air National Guard Base, N.H., refuel the C-17 on its trip from Ramstein AB, Germany, to the U.S.

April 21, 2003. The last of 546 Air Mobility Command KC-135 tankers complete the final Pacer CRAG modification. Eight program depot maintenance facilities ran 25 production lines to install Pacer CRAG upgrades. It took approximately 4,150 man-hours to modify each aircraft.

April 28, 2003. With the End of Operation Southern Watch, the operation to enforce a no-fly zone in southern Iraq, the Combined Air Operations Center (CAOC) at Prince Sultan AB, Saudi Arabia, with responsibility for air operations in Southwest Asia, is moved to Al Udeid AB, Qatar. On April 29, at a joint news conference in Riyadh, Secretary of Defense Donald Rumsfeld and Prince Sultan bin Abdul Aziz announce the withdrawal of all U.S. combat forces from Saudi Arabia to end 12 years of a continuous U.S. presence there. The U.S. military presence officially ends on August 26.

May 1, 2003. With the End of Major Iraqi Freedom Combat Operations, President George W. Bush declares that major combat operations in Iraq are over on the deck of the USS *Abraham Lincoln* in a speech before 5,000 crewmembers.

May 1, 2003. Operation Northern Watch ends. Between January 1, 1997 and the final Northern Watch sortie on March 17, 2003, more than 100,000 American, British, Turkish, and French airmen rotated through Incirlik AB, Turkey, to enforce UN Security Council resolutions applying to the region north of the 36th parallel in Iraq.

May 14, 2003. Air Mobility Command begins the Baghdad Express, a daily C-141 resupply mission from Ramstein AB, Germany, to Baghdad International Airport, Iraq. Two C-141s and two active-duty aircrews from the 305th Air Mobility Wing at McGuire AFB, N. J., and one aircrew from McGuire's 514th Air Mobility Wing (Reserve-Associate) are staged at Ramstein to fly the missions.

May 23, 2003. Edward C. "Pete" Aldridge, the Under Secretary of

Defense for Acquisition, Technology, and Logistics, approves an Air Force initiative to lease 100 KC-767A tankers for six years, beginning in 2006.

May 30, 2003. A 910th Airlift Wing C-130 from Youngstown Air Reserve Station, Ohio, delivers 7.5 tons of humanitarian supplies to Houari Boumedienne Airport in Algiers, the capital of Algeria. The C-130 crew, on temporary duty at Ramstein AB, Germany, delivers the first American relief supplies to the earthquake victims.

June 17, 2003. The USAF awards 34 aircrew members from Charleston AFB, S.C., the Distinguished Flying Cross (DFC) for actions in Operation Enduring Freedom. Eight recipients earn the DFC during the first night of humanitarian relief operations in Afghanistan (7 October 2001), while the others receive the DFC for inserting Marine forces at the Rhino Landing Zone near Kandahar on November 28–30, 2001. The crews flying to the landing zone performed the deepest insertion of Marines into hostile territory in Marine Corps history. They also accomplished three C-17 operational milestones: (1) the first C-17 combat landings on an unimproved dirt strip; (2) the first missions by C-17 special operations low-level aircrews in hostile conditions; and (3) the first use of night vision goggles by C-17 aircrews to make blacked-out approaches and landings in hostile territory.

June 18, 2003. The USAF releases a roadmap to retire 133 KC-135E Stratotankers and assign 100 KC-767A tankers to be leased. Under the plan, Fairchild AFB, Wash., will become the first active-duty base to receive the new KC-767As in FY 2006. By 2010, several Air Reserve Component units will also convert from E-model to R-model KC-135s to the KC-767A.

June 18, 2003. Operation Iraqi Freedom. The Air Mobility Command releases Civil Reserve Air Fleet carriers from supporting this operation. From February 8, to June 2, 2003, the 11 CRAF carriers fly 1,625 missions to airlift 254,100 troops to the Middle East and other destinations.

August 11, 2003. A C-9A Nightingale (no. 68-10959) assigned to the 375th Airlift Wing at Scott AFB, Ill., performs the last scheduled C-9 aeromedical evacuation mission. The aircraft airlifts one litter patient, a few space-available travelers, and several soldiers wounded in Iraq to their home stations in the U.S. in a 5.6-hour mission.

November 4, 2003. Travis AFB, Calif., sends the first C-5A Galaxy (No. 70-0458) to the Aerospace Maintenance and Regeneration Center (AMARC) at Davis-Monthan AFB, Ariz.

November 7, 2003. Through 11 November, four KC-135s from the 92nd Air Refueling Wing at Fairchild AFB, Wash., and one KC-135 from the 22nd Air Refueling Wing at McConnell AFB, Kans., fly India Company, 3rd Marine Battalion, from Camp Lejeune, N.C., to Djibouti to join the Combined Joint Task Force—Horn of Africa in the fight against transnational terrorists. The Marines provide security at Djibouti International Airport. The five KC-135s airlift 169 Marines and 9 short tons of cargo.

December 7, 2003. TSgt. Eric M. Olson, a 313th Airlift Squadron loadmaster, helps his aircrew safely land a C-17 struck by enemy fire while leaving Baghdad International Airport. Olson expeditiously executes emergency action checklists, tends to a wounded passenger, and calmly readies the other passengers for an emergency landing.

For his actions, he receives the Distinguished Flying Cross on January 11, 2005.

December 17, 2003. In a ceremony in Long Beach, Boeing delivers the first of eight C-17s to the 172nd Airlift Wing at Jackson, Miss. The C-17 (No. 02-1112) was also the first C-17 delivered to an Air National Guard unit. The Guard named the aircraft *The Spirit of Mississippi Minutemen* to honor those Mississippians who, historically, had remained vigilant to defend America and the Magnolia State. The delivery occurs on the 100th anniversary of the Wright brothers' historic flight at Kitty Hawk, N.C.

December 27, 2003. On 26 December, a 6.7 earthquake strikes in Iran's southeastern Kerman Province near the historic city of Bam. From December 27 to 29, the Air Mobility Command flies several humanitarian airlift missions to help the earthquake victims. A single C-5 mission flies from Travis AFB, Calif., to Kuwait City International Airport, where passengers and cargo are transferred to C-130s for airlift to Kerman AB, Iran. Nine C-130 missions and two C-17 missions eventually land at Kerman with 338,000 pounds of humanitarian cargo.

March 19, 2004. The first active-duty C-130J Hercules (No. 02-0314) arrived at the 314th Airlift Wing at Little Rock AFB, Ark.

March 25, 2004. Five C-17 airmen from the 62nd Airlift Wing at Mc-Chord AFB, Wash., win the 2003 Mackay Trophy for serving as the lead aircrew for a March 26, 2003 C-17 airdrop over northern Iraq. It is the largest combat airdrop since Operation Just Cause in December 1989. The recipients included Lt. Col. Shane Hershman and MSgt. Chris Dockery, 7th Airlift Squadron; Maj. Bob Colvin, 8th Airlift Squadron; 1st Lt. Matt Clausen, 4th Airlift Squadron; and MSgt. Shawn Brumfield, 62nd Operations Group.

March 27, 2004. NASA's X-43A unmanned aircraft becomes the first jet to break Maj. William J. Knight's X-15 record of 4,520 mph. While the X-1, Skyrocket, X-2, and X-15 were rocket-powered, the X-43 features an advanced scramjet engine, which allows it to cruise at 4,780 mph (Mach 6.83) for 11 seconds at 95,000 feet.

April 16, 2004. Near Kharbut, Iraq, in near-zero visibility, two USAF HH-60G helicopters crews search for a U.S. Army CH-47 Chinook helicopter that had crashed in a sandstorm. The brown-out conditions leave the rescue crews' infrared and night vision goggles ineffective, making navigation extremely difficult and the rescue attempt much more dangerous. In executing the rescue mission, the crews also had to navigate through multiple surface-to-air missile attacks. In the end all five Army crash survivors are located, rescued, and evacuated out of the combat zone unharmed. For that action, the crews earn the MacKay Trophy. The rescue crews, from the 41st and 38th Rescue Squadrons at Moody AFB, Ga., are Capts. Bryan Creel and Robert Wrinkle, aircraft commanders; Maj. Joseph Galletti and Capt. Greg Rockwood, copilots; TSgts. Michael Preston and SSgt. Patrick Ledbetter, flight engineers; MSgt. Paul Silver and TSgt. Thomas Ringheimer, aerial gunners; and TSgt. Matt Leigh and SSgts Vincent J. Eckert, John Griffin, and Michael Rubio and SrA Edward Ha, pararescuemen.

April 22, 2004. A 92nd Air Refueling Wing KC-135R from Fairchild AFB, Wash., assigned to the 340th Expeditionary Air Refueling Squadron, flies the first combat mission with an operational global air traffic management (GATM) system. The GATM incorporates a series of air traffic system upgrades to enable the safe operation of commercial and military aircraft in congested airspace. The system also enhances aircrew situational awareness and communications. Captain Earl Ardales commands the aircraft on the mission.

May 13, 2004. The final C-141 airdrop of paratroopers took place at the Fort Benning, Ga., drop zone. Army Pvt. Jason Stewart a native of Chandler, Ariz., becomes the last paratrooper to jump from a C-141, when he jumps from C-141C (No. 65-0229) of the 452nd Air Mobility Wing (Air Force Reserve Command) at March Air Reserve Base, Calif.

June 13, 2004. A Russian TU-154 observation aircraft lands at Travis AFB, Calif., marking the first Russian Open Skies mission over the U.S. The Open Skies Treaty, which became effective on January 1, 2002, permitts unarmed aerial observation flights over the territories of the 30 participating nations.

July 23, 2004. Congressman Jerry F. Costello (D-Ill.) announces that the 932nd Airlift Wing at Scott AFB, Ill., will receive three new C-40Cs, the military version of the Boeing 737, and three modern C-9C Nightingales. The C-9Cs will be transferred from Andrews AFB, Md., to replace the three C-9As at Scott.

August 23, 2004. The 43rd Airlift Wing at Pope AFB, N.C., and other military organizations with C-130s fly special sorties to commemorate the fiftieth anniversary of the first C-130 Hercules flight.

September 11, 2004. After the September 11, 2001, terrorist attack on America, Air Mobility Command completes 34,030 missions in support of the two combat operations Enduring Freedom and Iraqi Freedom. Air Mobility Command's organic and contracted commercial aircraft move a total of 1,771,632 troops and 1,059,498 short tons of cargo, while KC-10 and KC-135 tankers perform 11,111 air refueling missions to refuel 15,396 aircraft. The three-year airlift into Afghanistan is the third largest airlift in history, with Operation Vittles (Berlin Airlift) in 1948–1949 ranking first and Operations Desert Shield/Desert Storm (Gulf War) in 1990–1991 ranking second.

September 23, 2004. In Operation Provide Hope a C-17 from McChord AFB, Wash., delivers medicines and medical supplies worth $15 million to Kishkek International Airport in the Kyrgyz Republic. It is the largest humanitarian shipment to the republic since it became an independent state in 1991.

September 24, 2004. The first of 13 C-17s for McGuire AFB, N.J., arrive from the Boeing factory in Long Beach, Calif. The Globemaster III (No. 03-3125) is the first C-17 modified with "Block 15" upgrades, including station-keeping equipment that allows for all-weather formation flying by up to 100 aircraft.

October 3, 2004. The Lockheed Martin plant at Marietta receives the first C-5 for the reliability and enhancement and reengining program (RERP) to extend the C-5's life to 2040. The modification adds new engines and improved electrical, fuel, hydraulic, flight control, environment control systems, and structure and landing gear upgrades. The C-5B prototype aircraft (No. 86-0013) belonged to the 436th Airlift Wing at Dover AFB, Del.

October 16, 2004. NASA's X-43A breaks its 4,780 mph record of March 27, 2004, by reaching nearly Mach 10. After separating from a booster rocket at 110,000 feet, the scramjet engine accelerates the X-43 to 7,000 mph (Mach 9.8).

December 10, 2004. From Quonset State Airport, R.I., the 143rd Air Wing (Rhode Island ANG) deploys the first C-130J Super Hercules to be used in Southwest Asian combat operations.

December 17, 2004. At Edwards AFB, NASA retires its B-52B Stratofortress "mothership" air-launch aircraft after nearly fifty years of dropping advanced research vehicles. At its retirement, the aircraft is NASA's oldest aircraft and the oldest flyable B-52. At the same time, it has the lowest number of flying hours of any B-52 in operation. The aircraft first flew in June 1955, was modified to support the X-15 research aircraft program in 1959, and flew its last research mission 16 November 2004 to launch the scramjet-powered X-43A hypersonic research aircraft on its record Mach 9.6 flight over the Pacific Ocean.

December 20, 2004. The Air Force inactivates its last operational F-4 Phantom II squadron at Holloman AFB. The 20th Fighter Squadron's inactivation also ends a 33-year German-American joint fighter training program in F-4E and F-4F aircraft.

December 26, 2004 to February 23, 2005. A 9.0 earthquake in the Indian Ocean triggers an enormous tsunami with thirty-foot-high waves that slams into the coasts of thirteen nations and kills at least 300,000 people. The quake's epicenter is approximately 65 miles south of the Indonesian island of Sumatra. The death toll is greatest in Indonesia,

where at least 100,000 people perish. The tsunami kills more than 29,000 people in Sri Lanka, 10,000 in India, and 4,600 in Thailand. Another three to five million people in South Asia and Southeast Asia are left without clean water, food, shelter, and sanitation. By February 23, 2005, and the end of Unified Assistance Tsunami Relief, AMC aircraft have flown 106 missions to airlift 2,768 passengers and 3,370 short tons of cargo to Bandaranaike International Airport in Colombo, Sri Lanka, U-Tapao AB, Thailand, and Banda Aceh and Jakarta International Airports in Indonesia.

February 1–22, 2005. A 10-man crew from the 403rd Wing at Keesler AFB, Miss., crosses 33 time zones, and stops in nine locations and seven countries while circumnavigating the planet in 21 days in a C-130J Hercules. This February 1–22 flight is a first for the new airlifter.

February 4, 2005. Operation Deep Freeze. A 452nd Air Mobility Wing C-141 Starlifter from March ARB, Calif., flies the last C-141 mission to the McMurdo Research Station at the South Pole. For 39 years, C-141s carried people and equipment to Antarctica for the National Science Foundation and its research efforts there. This year alone, the C-141s carries more than 2.7 million pounds of cargo and about 3,000 passengers to McMurdo.

February 11, 2005. Langley AFB begins efforts to give the F-22 its initial operating capability.

March 1, 2005. At Langley AFB, Va., Lt. Gen. William Fraser III, the Air Combat Command Vice Commander, declares an initial operating capability for the MQ-1 Predator. The MQ-1 version is equipped with advanced sensors and armed with AGM-114 Hellfire missiles.

March 18, 2005. Brig. Gen. Jeff Reimer, the F-16 Systems Program Officer for the original F-16 contract, flies the last F-16 from the Lockheed Martin Aero Plant in Fort Worth, Tex., to Shaw AFB, S.C. It is the last of 2,231 F-16s produced for the USAF.

April 12, 2005. The C-17 makes its first airdrop to National Science Foundation scientists at the North Pole. Two C-17s complete the 12-hour, nonstop flight to airdrop about 10,000 gallons of fuel to the scientists.

April 16, 2005. Lockheed Martin returns the first U-2S upgraded through the Reconnaissance Avionics Maintainability Program (RAMP) to the 9th Reconnaissance Wing at Beale AFB, Calif. The RAMP upgrade replaces the 1960s-vintage cockpit with three 6- by 8-inch multifunction displays, an up-front control and display unit, and an independent secondary flight display system. The remaining fleet of 31 U-2S aircraft and four two-cockpit trainers will be modified by 2007.

April 18, 2005. Under Secretary of Defense for Acquisition, Technology, and Logistics, Michael W. Wynne, approves the full-rate production capability of the F/A-22.

April 19, 2005. At Edwards AFB, Calif., the 418th Flight Test Squadron successfully completes the first five-bundle low-velocity airdrop from the new C-130J Hercules. The bundles weigh 40,100 pounds and land 1,000 feet apart and only five yards off the planned impact point.

July 25, 2005. The 337th Test and Evaluation Squadron from Dyess AFB, Tex., establishes a B-1B Lancer "first" over the White Sands Missile Range,

N.M., by having two B-1Bs auto-release a joint air-to-surface standoff missile (JASSM), a single warhead self-propelled missile, and three dissimilar weapons from the same launcher. The JASSMs strike their targets after traveling more than 170 nautical miles.

August 10, 2005. At March Air Reserve Base in Riverside, Calif., the "Spirit of California" C-17 arrives. The new airlifter is the first C-17 to be stationed in California and the first to be owned by the Air Force Reserve (although the reserves have used Air Force C-17s during past missions). The Air Force plans to deliver another seven C-17s to March by the end of September.

August 18, 2005. The 7th Bomb Wing becomes the first unit to achieve initial operational capability of the JASSM cruise missile.

August 26, 2005. From Vandenberg AFB, the 576th Flight Test Squadron launches an unarmed Minuteman III as part of a developmental test demonstrating the ability to integrate a safety enhanced reentry vehicle into the existing Minuteman III weapons system.

September 14, 2005. At Creech AFB (formerly Indian Springs), Nev., the USAF starts operational testing on the latest upgrade to the MQ-1 Predator. The multiaircraft control (MAC) system allows single pilots to control two-ship and four-ship Predator sorties. Testers perform two-ship sorties September 12 and 13 and progress to four-ship sorties September 14 and 15.

September 16, 2005. Bell Helicopter/Boeing delivers the first Air Force CV-22 Osprey for electromagnetic testing to NAS Patuxent River, Md.

September 19, 2005. At Francis E. Warren AFB, Wyo., the 90th Space

Wing holds a deactivation ceremony to signal the phase-out of the Peace-keeper weapon system from the Air Force inventory.

September 20, 2005. The USAF's last remaining active-duty C-9 Nightingale (No. 876) leaves Ramstein AB, Germany, for Andrews AFB, Md. The C-9 flies to its final resting place at the Andrews Air Museum.

September 22, 2005. From Vandenberg AFB, a Minotaur rocket carries a 920-pound Defense Advanced Research Projects Agency (DARPA) military research satellite into a yearlong orbit to gather information about Earth's environment in low orbit. The Minotaur rocket was made from decommissioned first and second stages of a Minuteman II missile.

September 26, 2005. A crew of six women airmen climb aboard a C-130 Hercules together at a forward deployed location in Southwest Asia. This mission is the first time an all-female C-130 crew flew a combat mission. The crew included: Capt. Carol Mitchell, aircraft commander; 1st Lt. Siobhan Couturier, pilot; Capt. Anita T. Mack, navigator; SSgt. Josie E. Harshe, flight engineer; and loadmasters TSgt. Sigrid M. Carrero-Perez and SrA Ci Ci Alonzo. All women are assigned to the 43rd Airlift Wing at Pope AFB, N.C., and deployed to the 737th Expeditionary Airlift Squadron to fly cargo and troops in and out of Iraq, Afghanistan, and the Horn of Africa.

October 2, 2005. The Air Force delivers the first two of 40 F-15K Eagles from Boeing's F-15 production line to South Korea. Using KC-135s for refueling support, the F-15s stop at Hickam AFB, Hawaii, and Andersen AFB, Guam, on their way to South Korea.

October 8, 2005. After a devastating magnitude 7.6 earthquake devastates parts of Pakistan, airmen move to Chakala Air Base in that country to guide the delivery of tons of humanitarian relief supplies.

October 19, 2005. The USAF and the Lockheed Martin launch the last Titan IV B rocket after five decades of service, from Vandenberg AFB, Calif. The heavy-lift rocket's 200th launch carries a critical national security payload into space for the National Reconnaissance Office.

October 27, 2005 Bell Helicopter/Boeing delivers the second Air Force CV-22 to Edwards AFB, Calif. With the delivery of the third Osprey in February 2006, the Air Force Operational Test and Evaluation Center at Edwards will begin the CV-22's Operational Utility Evaluation.

November 5, 2005. The Air Force rolls out the TH-1H helicopter at Randolph AFB, Texas, home of the Air Education and Training Command's pilot instructor training. The TH-1H, the latest version of the UH-1H Huey, received extensive refurbishment with upgraded components, a new avionics suite, and a glass cockpit.

December 8, 2005. Lockheed Martin wins a $149-million contract to build a prototype unmanned near-space vehicle that will hover about 60,000 feet above Earth. The High-Altitude Airship would be about 17 times bigger than a Goodyear blimp, be untethered, be able to remain in place for one month, and be able to carry 500 pounds of sensors that could detect enemy ballistic missiles. The military has been exploring possible military uses of partly because they are far less expensive to launch and operate than satellites.

December 9, 2005. Brig. Gen. Susan Y. Desjardins becomes commandant of the Air Force Academy. One of the academy's first female graduates, she becomes with this appointment the first woman to be commandant.

December 13, 2005. An F/A-22 Raptor, flying at supersonic speed over the range at Edwards AFB, drops its first 1,000-pound guided joint direct attack munition (JDAM).

December 15, 2005. Gen. Ronald E. Keys, Commander of Air Combat Command, declares an initial operating capability for the 27th Fighter Squadron and the F-22A Raptor with the 1st Fighter Wing at Langley AFB.

December 15, 2005. At Vandenberg AFB, Calif., the 30th Space Wing activates the Western Range Operations Control Center. The WROCC will command and control all Vandenberg launches and oversee the Western Range, an area that encompasses the West Coast of the continental United States and extends more than 4,200 miles across the Pacific Ocean.

December 22, 2005. From Charleston AFB, S.C., 17 C-17s from the 437th and the 315th Airlift Wings—the largest formation of C-17 Globemaster IIIs to take off from a single base—launch into the morning skies to demonstrate the USAF's strategic airdrop capability. (The previous record was 15 C-17s from McChord AFB, Wash.) The training formation left the base to locations around the Southeast to perform required quarterly, semiannual and end-of-year training.

INDEX

Note: Page numbers in italics indicate photographs.